Active Control of Sound

P.A. Nelson

and

S.J. Elliott
Institute of Sound and Vibration Research,
The University,
Southampton, U.K.

ACADEMIC PRESS

Harcourt Brace Jovanovich, Publishers

London San Diego New York Boston
Sydney Tokyo Toronto

ACADEMIC PRESS LIMITED
24/28 Oval Road,
London NW1 7DX

United States Edition published by
ACADEMIC PRESS INC.
San Diego, CA 92101

Copyright © 1992 by
ACADEMIC PRESS LIMITED

All rights reserved
No part of this book may be reproduced in any form by photostat, microfilm, or any other means, without written permission from the publishers.

A catalogue record for this book is available from the British Library

ISBN 0–12–515425–9

Typeset by Columns Design and Production Services Limited

Transferred to digital printing 2005

Contents

Preface .... *ix*

Chapter 1 An Introduction to Acoustics

1.1 Propagation of acoustic disturbances 1
1.2 The decibel scale for the measurement of sound pressure 4
1.3 Properties of acoustic disturbances 6
1.4 The one-dimensional wave equation 9
1.5 Solutions of the one-dimensional wave equation 11
1.6 Linearity and the superposition principle 13
1.7 Specific acoustic impedance 16
1.8 Acoustic energy density and intensity 17
1.9 Standing waves 19
1.10 The three-dimensional wave equation 22
1.11 Solutions of the three-dimensional wave equation 24
1.12 Point sources of spherical radiation 26
1.13 Acoustic power output 27
1.14 Enclosed sound fields at low frequencies 29
1.15 Enclosed sound fields at high frequencies 31
1.16 The energy balance equation for an enclosure 33

Chapter 2 Frequency Analysis

2.1 Introduction 37
2.2 Fourier series 38
2.3 The Fourier integral 42
2.4 Random processes 46
2.5 The correlation coefficient and the method of least squares 52
2.6 Correlation functions for time histories 54
2.7 Spectral density functions 55
2.8 Analogue-to-digital conversion 59
2.9 The Fourier transform of sampled signals and aliasing 61
2.10 The discrete Fourier transform 65
2.11 Data truncation and windowing 67
2.12 Estimation of power spectra 69

Chapter 3 Linear Systems

3.1 Linearity and the superposition principle 72
3.2 Systems governed by ordinary differential equations 73

3.3 The Laplace transform ... 75
3.4 Properties of the transfer function ... 80
3.5 The convolution integral ... 84
3.6 Response to random inputs ... 85
3.7 The coherence function ... 89
3.8 Optimal filtering ... 89

Chapter 4 Digital Filters

4.1 Advantages of digital processing ... 92
4.2 Sampled signals ... 93
4.3 The z-transform ... 96
4.4 Digital systems ... 97
4.5 FIR and IIR filters ... 102
4.6 Frequency domain filter design ... 104
4.7 Optimal filter design ... 108
4.8 Adaptive digital filters ... 113

Chapter 5 Interference in Plane Wave Sound Fields

5.1 Introduction ... 116
5.2 The plane monopole source in an infinite duct ... 118
5.3 The cancellation of downstream radiation by using a single secondary plane monopole source ... 122
5.4 The energetics of cancellation by a single secondary plane monopole source ... 126
5.5 Absorption of sound by a plane monopole source ... 128
5.6 The minimum power output of the primary/secondary source pair ... 132
5.7 The cancellation of downstream radiation by using a pair of plane monopole sources ... 135
5.8 The mechanism of sound absorption by real sources ... 140
5.9 The plane dipole source in an infinite duct ... 143
5.10 Unidirectional radiation from a plane monopole/dipole combination ... 146
5.11 The influence of reflections from the primary source ... 148
5.12 A travelling wave model of a one-dimensional enclosed sound field ... 152
5.13 The time-averaged energy in the enclosure ... 153
5.14 The cancellation of the sound field in a one-dimensional enclosure ... 154
5.15 The absorbing termination ... 157
5.16 The active minimisation of the total acoustic energy in a one-dimensional enclosure ... 158

Chapter 6 Single Channel Feedforward Control

6.1 Equivalent block diagram ... 161
6.2 Single channel control with an independent reference signal ... 162
6.3 Control of a periodic sound at individual harmonics ... 164
6.4 Control with waveform synthesis ... 166
6.5 Single channel control of random sound ... 171
6.6 Frequency domain analysis ... 172
6.7 The effects of measurement noise ... 176
6.8 Turbulence as a source of measurement noise ... 178
6.9 Practical frequency domain design methods ... 179

6.10 An acoustical interpretation of the optimal controller 183
6.11 Other factors affecting the optimal controller 189
6.12 Time domain controller design 192
6.13 The effect of electrical delays in the controller 194
6.14 Adaptive FIR digital controllers 195
6.15 An adaptive recursive controller 199

Chapter 7 Single Channel Feedback Control
7.1 Introduction 204
7.2 Feedback control of cavity pressure fluctuations 207
7.3 A general approach to the feedback control of acoustical systems 211
7.4 The Nyquist stability criterion 213
7.5 Bode plots and relative stability 216
7.6 The design of realisable compensation filters 219
7.7 Feedback control of the sound field in an ear defender 220
7.8 Feedback control of duct-borne sound 226

Chapter 8 Point Sources and the Active Suppression of Free Field Radiation
8.1 Introduction 231
8.2 Interference between the far fields of two point monopole sources 232
8.3 The power output of two point monopole sources 235
8.4 The minimum power output of the sources 240
8.5 The minimisation of the sum of the squared pressures at a number of far field error sensors 244
8.6 The constraint of causality 247
8.7 The point dipole source 253
8.8 Point quadrupole sources 257
8.9 Multipole analysis 260
8.10 Kempton's suggestion 262
8.11 Control of periodic free field sound by using multiple sources and sensors 264
8.12 The power output of arrays of point sources 269
8.13 The minimum power output of point source arrays 271

Chapter 9 Continuous Source Distributions and the Active Absorption of Free Field Radiation
9.1 Introduction 275
9.2 The inhomogeneous wave equation 276
9.3 The Green function and the principle of reciprocity 277
9.4 The solution of the inhomogeneous wave equation 280
9.5 The solution of the inhomogeneous wave equation in an unbounded medium 281
9.6 The Kirchhoff–Helmholtz integral equation 282
9.7 Active absorption and reflection with continuous source layers 284
9.8 Active suppression of radiation from vibrating bodies 287
9.9 Active suppression of radiation from vibrating plane surfaces 288
9.10 Active suppression of scattered acoustic radiation 290
9.11 The acoustic energy balance equation 293

9.12 Sound power absorption by monopole and dipole sources 294
9.13 The superposition of maximally absorbing monopole and dipole sources 299
9.14 The discretisation of continuous absorbing surfaces 302
9.15 Sound power absorption by quadrupole sources 304
9.16 Sound power absorption by compact monopole arrays 307

Chapter 10 Global Control of Enclosed Sound Fields
10.1 Introduction 310
10.2 The eigenfunctions and boundary conditions of a rectangular enclosure 311
10.3 The sound field in a lightly damped enclosure 314
10.4 A model of the acoustic damping at the enclosure walls 316
10.5 Global active control with a single secondary source 319
10.6 Global active control in one dimension 322
10.7 Global active control with multiple secondary sources 327
10.8 Minimisation of the total acoustic potential energy in a harmonically excited two-dimensional enclosure 329
10.9 Global active control in a randomly excited enclosure 334
10.10 Minimisation of the sum of the squared pressures at a discrete number of locations 335
10.11 The influence of sensor locations 338
10.12 Mechanisms of active control at low modal densities 340
10.13 The influence of modal density and damping 343
10.14 Global control at high modal densities 347
10.15 Some examples of the application of active control in enclosed sound fields 350

Chapter 11 Local Control of Enclosed Sound Fields
11.1 Cancellation of a pure tone sound field at a point in an enclosure 356
11.2 Statistical properties of diffuse sound fields 358
11.3 The space-averaged diffuse field transfer impedance 360
11.4 The diffuse field zone of quiet 362
11.5 The statistical behaviour of the secondary source strength 366
11.6 Cancellation in the near field of a secondary source 368
11.7 Experiments on the zone of quiet in the near field of a secondary source 372
11.8 The power output of a monopole source used to cancel the pressure in its near field 374
11.9 The Olson "sound absorber" in a diffuse sound field 376

Chapter 12 Multi-channel Feedforward Control
12.1 Multi-channel control of periodic sound 379
12.2 Exact least squares solutions 381
12.3 Matrix conditioning and the shape of the error surface 384
12.4 Iterative gradient descent methods 387
12.5 Other iterative algorithms 392
12.6 Alternative cost functions 394
12.7 Multi-channel control of random sound 396

12.8 A digital filtering formulation 403
12.9 Adaptive digital filtering algorithms 407

Appendix A Little Linear Algebra
A.1 Vectors 411
A.2 Matrices 412
A.3 Determinants and the inverse matrix 414
A.4 Eigenvalue/eigenvector decomposition 415
A.5 The Hermitian quadratic form 416

References 421

Index 433

To our families

Preface

During the last two decades there has been an accelerating level of interest in the control of sound by active techniques. Many of the physical principles involved have long been established, but the technological means for the successful implementation of "active noise control" have only recently become available. The primary technological advance that has made the active control of sound a practical proposition is the development of fast digital signal processors. The availability of this new technology has acted as a considerable stimulus for research, both into the use of modern signal processing methods and into the acoustical aspects of the problem. It is one of the principle objectives of this book to bring together the results of contemporary research into these subjects and to present, in a unified fashion, work from the two disciplines of acoustics and signal processing. The emphasis of the book is very much on the fundamental scientific principles of these subjects, although examples of practical applications are presented wherever possible. Thus the reader should not expect to find a "cook book" description of the hardware and associated software necessary to implement active control. However, the algorithmic principles which form the foundation of practical systems are dealt with at some length.

The book has been written in textbook style, with the aim of catering for the widest possible range of readers who are likely to be interested in the subject. Thus the book aims to satisfy the demands of both undergraduate and postgraduate students of acoustics and signal processing, professional acoustical and electrical engineers and researchers in the field of active control. It is with this in mind that the structure of the book has been devised. The first four chapters consist of standard textbook material which can be found readily in many other books on both acoustics and signal processing. However, an effort has been made to draw together those aspects of the two subject disciplines which are prerequisites to the understanding of the more advanced topics presented in later chapters and to present them in a unified and consistent manner. Chapter 1 is intended as an introduction to acoustics for those with no previous experience or training in the field. It would thus be suitable as an introduction, for example, for an electrical engineer involved in the development of active control systems but who had no formal training in acoustics. Similarly, Chapters 2, 3 and 4 which deal with basic frequency analysis, linear systems theory and digital filters, would be suitable for students of acoustics or professional acousticians with only a little experience in these areas. The treatment of the material presented in the first four chapters is by no means exhaustive, although every effort has been made to present and describe clearly the basic principles necessary for further understanding of these subject areas. Thus, in presenting this material, appropriate references have been made to other readily available contemporary texts in order to encourage a more thorough study of these basic principles than can be afforded by the space available here.

In Chapter 5, we begin to deal specifically with the active control of sound. In this chapter, attention is restricted to an examination of the physical processes involved in the control of plane wave sound fields. It turns out that almost all the important physical aspects of active control can be described by using a series of one-dimensional examples. The capacity of actively controlled secondary sources both to reflect and to absorb sound is demonstrated, together with the underlying principles involved in the active control of enclosed sound fields. Although rigorous in its treatment, Chapter 5 has been written with the minimum of analytical sophistication and is intended to describe the physics of the active control of sound with the greatest possible clarity. Chapter 6 continues to deal with the active control of plane wave sound fields, but in this chapter we concentrate on describing the structure and performance of systems for the implementation of single channel feedforward active control. This problem has traditionally received considerable attention, not only because of the relative simplicity of the physics involved, but because of the immediate prospects for the application of the technique to practical problems such as the attenuation of low frequency sound in air-conditioning ducts. Chapter 7 continues with the topic of control system design, but in this case deals with single channel feedback control systems. This work is described within the context of classical control systems theory, since most applications of the technique to date have involved the use of analogue electronic systems and have been restricted to single channel operations. (There is thus no attempt in this book to cover the topic of modern control systems theory and deal with multiple-input multiple-output feedback control systems.) The practical problems dealt with include the application of single channel feedback control to the sound fields in ear defenders and again to the propagation of plane wave sound in ducts.

In Chapters 7 and 8 we deal with the active control of free field sound. Again the philosophy that has been adopted is to work from first principles and to delineate clearly the ultimate performance limits that are fixed by the physics of interfering sound fields. In Chapter 8 we deal with the active control of sound radiated by compact sources and use the idealisation of the point source in order to undertake a rigorous study of the possibilities of using other discrete secondary sources for free field active control. Much of this work has its origins in classical acoustics and we attempt to bring together the well-established features of the interference produced by compact arrays of monopole sources and some more recent results that have emerged from the study of this topic. Despite the fact that even present-day active control systems consist only of arrays of discrete secondary sources, much of the early research into active control was stimulated by the prospect of continuous source layers for the active absorption of sound. This is the topic with which we deal in Chapter 9. Again the subject has its foundations in classical acoustics and every effort has been made to derive the important results from first principles. In Chapter 9 we also examine the possibility of replacing continuous source layers with discrete source arrays. This also leads to an examination of the possibilities of using discrete sources for absorbing sound and we present the results of recent research into this subject.

Chapters 10 and 11 are devoted to the description of active control in enclosed sound fields. Two classical models of room acoustics are used in this description; the expression of the response of an enclosure in terms of a sum of orthogonal modes is used, and where this becomes inappropriate (at high modal densities) we adopt the notion of a diffuse sound field. Despite the obvious number of practical applications involving active control in enclosures, surprisingly little work in this area has been undertaken until fairly recently and many of the results presented have emerged from recent studies.

The final chapter of the book, Chapter 12, deals with multi-channel feedforward control systems. Such systems are applicable whenever we have some means of detecting the unwanted sound before it arrives at the position in the field to be controlled. The analysis presented is thus applicable to the control of single frequency sound at a known (or detectable) frequency (or series of harmonically related frequencies) and also to problems involving the control of non-deterministic sound fields from multiple primary sources. In describing such control systems, and indeed in any instance where multiple secondary sources are used for active control, we must necessarily resort to matrix methods of analysis. An appendix has thus been dedicated to summarising those aspects of linear algebra that are used repeatedly in the main body of the text.

Although the active control of vibration has some close associations with the active control of sound, the methods of analysis and the physical emphasis are very different and no attempt has been made to cover this topic in the current text. The general topic of active vibration control is best described within the framework of modern control systems theory with the emphasis on multi-channel feedback control (rather than the use of signal processing techniques and feedforward control described here) and is admirably treated in the recent book written by Professor L. Meirovitch (1990). A subject having even closer associations with the current material, which has also been omitted from the text, is the active control of "structure-borne sound". Again, the physical nature of wave propagation in structures is more diverse than that in fluids and this subject (which is also less mature as an area of study) could not be given adequate coverage in a book of the current length.

It only remains for the authors to thank all those who have assisted both in the production of this book and the period of research leading up to its production. The authors would firstly like to acknowledge a number of researchers in the UK whose work was highly influential in launching the authors into the study of the active control of sound. First, Professor G.B.B. Chaplin deserves much credit for pointing out, during the 1970s, the potential offered by modern electronics to the solution of a wide range of acoustical problems. This potential was also recognised by Professor J.E. Ffowcs-Williams of the University of Cambridge, who was responsible for the instigation of one of the first large scale demonstrations in 1982 of the active control of sound at a compressor station at Duxford, near Cambridge, England. This work built on the earlier research, also undertaken at the University of Cambridge, of Dr M.A. Swinbanks and Dr C.F. Ross, both of whom made important contributions to the study of the active control of sound. Dr M.J. Fisher and Professor P.E. Doak of the Institute of Sound and Vibration Research also deserve great thanks for their successful instigation, at the University of Southampton, of a programme of work on this subject. Work began during 1982 with the appointment of the authors under the Special Replacement Scheme of the Science and Engineering Research Council, the possibility of such support having been earlier suggested by Professor J.B. Large, then Director of ISVR. During the ensuing years, the authors have received tremendous help, support, advice and encouragement from their colleagues at ISVR. In this respect they would particularly like to thank Professor R.G. White (Director of the Institute during this period), Professor J.K. Hammond (Chairman of the Signal Processing and Control Group) and Dr M.J. Fisher (Chairman of the Fluid Dynamics and Acoustics Group). The authors would also like to thank their students, both undergraduate and postgraduate, from whom they have learned so much. Outstanding contributions to the ISVR research effort in the active control of both sound and vibration have been made by Chris Boucher, Andrew Bullmore, Alan Curtis, Paul Darlington, Christine Deffayet, Tim Hodges, Susan Hough, Michael Jenkins, Pamela Joplin, Phillip Joseph, Richard Silcox, Ian Stothers, Trevor Sutton, Dean Thomas and Helen Thornton. The authors are also very grateful to all those who have been kind enough to scrutinise parts of the draft manuscript of this book, point out errors and omissions and make constructive suggestions for its improvement. The authors are particularly grateful to Professor F.J. Fahy, Professor J.K. Hammond, Dr D. Guicking, Dr L. Eriksson, Dr A. Roure, Dr. S.J. Flockton, Dr. C. Carme, Professor P.D. Wheeler, Professor G.H. Koopmann, Dr. C.L. Morfey, Dr M. Tohyama, Dr R. Carter, Dr. A.J. Bullmore, Dr. A.R.D. Curtis, Mr T.J. Sutton and Mr. P.R. White. The authors are also greatly privileged to have benefited from the skill and experience, both editorial and acoustical, of Professor P.E. Doak, who was kind enough to scrutinise the entire draft manuscript and who also made many valuable suggestions for improvement. Dr. D. Guicking also deserves special acknowledgement for both checking the manuscript and for the compilation of a "reference bibliography" on the subject of active noise and vibration control. This has proved invaluable to the authors. Finally the production of this book would have been impossible if it were not for the tolerance of the authors' families, the excellent artwork of Marilyn Hicks, and the outstanding talent of Maureen Strickland whose secretarial assistance is of such enormous quality that it cannot be described adequately in words.

1 An Introduction to Acoustics

1.1 Propagation of acoustic disturbances

There are many books on acoustics that serve as an introduction to the subject. Particularly useful basic texts written primarily for undergraduate students are those by Hall (1987), Kinsler *et al.* (1982) and Dowling and Ffowcs-Williams (1983). More advanced treatments of the subject are presented, for example, by Pierce (1981), Morse and Ingard (1968) and Skudrzyk (1971). The material presented in this chapter is intended as a brief summary of the basic principles of acoustics that form the foundation for much of the work described in later chapters. In this book we restrict ourselves to the study of the active control of small amplitude sound fields in a homogeneous medium at audio frequencies. In practical terms, most implementations of active control to date have been undertaken with air as the medium through which the sound is propagating, although a number of experiments have been conducted in water. It is with this in mind that this introductory chapter has been written. The treatment given is by no means comprehensive and is relatively superficial compared to those in many of the texts referred to above. Here we will discuss briefly the physical processes involved in sound propagation and introduce some elementary ideas that will be useful in later chapters.

The physical nature of the process of propagation of acoustic waves is illustrated in Fig. 1.1. This shows a fluid enclosed in a semi-infinite tube with a sliding piston at one end. We will describe exactly what happens to the fluid when the piston begins to move. Assume the piston moves from rest with a constant velocity u. (Since a finite velocity cannot be achieved instantaneously we will assume that the piston "very rapidly" reaches velocity u.) Provided the fluid inside the tube is compressible then immediately in front of the piston the fluid becomes squeezed and its pressure is increased. The compressibility of the fluid plays a central role in the propagation of acoustic disturbances. If, for example, the fluid in the tube were incompressible and did not deform at all, then the fluid at the right hand end of the tube would move at the same instant as the piston. However, if the fluid is compressible and has inertia, it takes a finite time for the motion of the piston to be transmitted to the fluid at the right hand end of the tube. The speed at which this motion is transmitted is the *speed of sound*. In air at a temperature of 20° C, the sound speed is around $343\,\mathrm{m\,s^{-1}}$. Thus, if the tube were 100 m long the motion of the fluid due to the piston would be transmitted from one end to the other within about one-third of a second. The speed of sound in water is even faster. Although the sound speed again depends

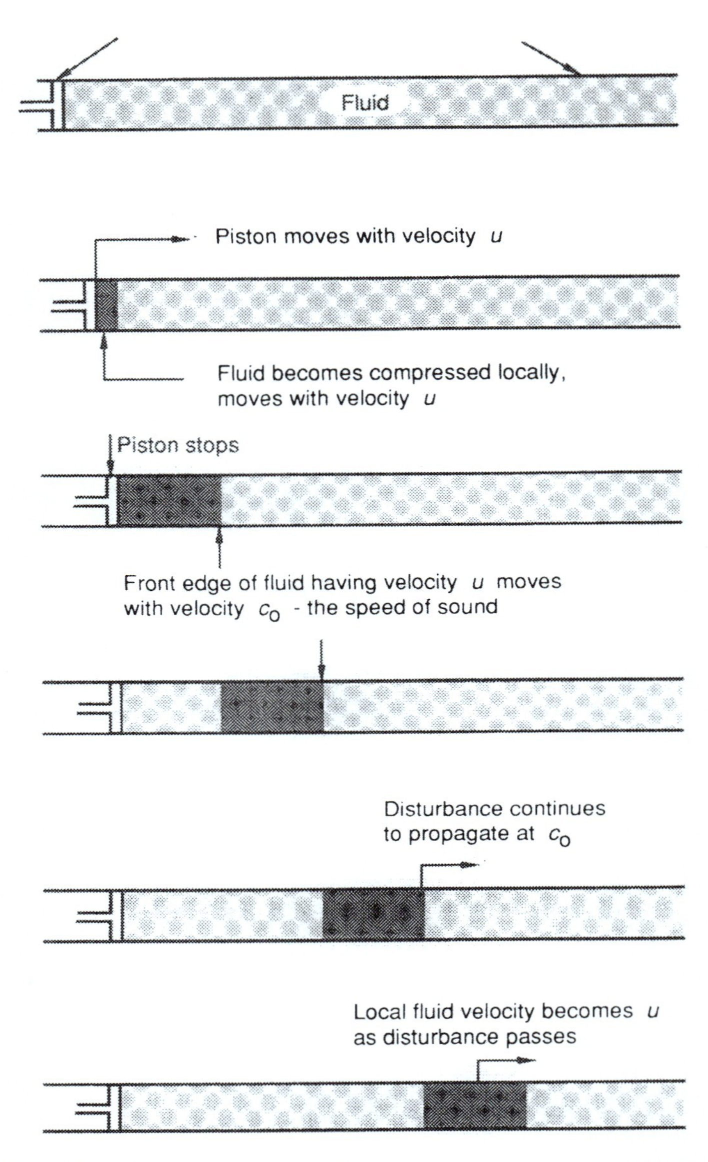

Fig. 1.1 An illustration of the propagation of a sound pulse in a semi-infinite tube. Note that the shaded area denotes a region of increased pressure in the tube.

on the ambient temperature of the fluid, and indeed the salinity in the case of sea water (see, for example, Pierce, 1981, Ch. 1), it is typically about $1500\,\mathrm{m\,s^{-1}}$.

The physical process involved is illustrated in Fig. 1.1, where the propagation of an acoustic disturbance has been "slowed down" to show what happens to the fluid in the tube after the piston is set in motion. First the fluid immediately in front of the piston becomes compressed. The increase in pressure is uniform across the cross-section of the tube. Acoustic disturbances for which the pressure is uniform in a direction normal to the direction of propagation are known as *plane waves*. The compression is then transmitted down the tube. The leading edge of the disturbance, having an increased pressure over the ambient pressure, moves at the speed c_0, the speed of sound. As the compression moves down the tube more and more of the fluid reaches the velocity u at which the piston is moving. If the piston is now suddenly brought to rest then the fluid immediately in front of the piston will also come to rest; nevertheless, the compression continues to propagate at the speed c_0, and as the compression passes down the tube it imparts a velocity u to the fluid as it passes. This illustrates the important distinction between the velocity c_0 at which the compression propagates and the velocity u which the fluid reaches as the compression passes through a given local region of fluid. The latter is known as the *particle velocity* which results from the passage of an acoustic wave. As we shall see later, the particle velocity generally has a magnitude which is very much less than the sound speed.

It is usually the motion of solid surfaces which are responsible for radiating sound, although turbulent flow can also generate sound through a process which to this day is still not fully understood. In practice, surfaces which give rise to sound radiation often move rapidly to and fro, usually vibrating about an equilibrium position. One particular form of surface vibration which is useful in the study of acoustic waves is the sinusoidal displacement of the piston depicted in Fig. 1.2. One cycle of the piston motion is illustrated which takes place during the period of time T. At time $t = 0$, the piston is at its equilibrium position but is moving with its maximum forward velocity. Since, as we shall see later, in this case the acoustic pressure increase in the fluid is directly proportional to the velocity of the piston, this forward velocity produces an equivalent pressure rise. As the piston continues to move forward then it slows down and the pressure increase produced in front of it becomes proportionately smaller. Meanwhile however, the initial increase in pressure has propagated through the fluid in the tube at the speed of sound. Thus once the piston reaches its maximum forward displacement (after a time $t = T/4$) there is a continuous distribution of pressure produced through the fluid in front of the piston. As the piston moves backwards it gathers speed and progressively *rarefies* the fluid and decreases the pressure of the fluid immediately in front of it. After a time $t = T/2$ (after half a cycle of the motion) the piston has returned to its equilibrium position and is moving with its maximum backwards velocity; it is at this time that the piston produces the maximum rarefaction of the fluid in front of it. All the time during this process, the disturbances produced are propagated away from the piston surface at the speed of sound. After one complete cycle of the motion (at time $t = T$) a continuous spatial distribution of pressure has been produced in the tube which replicates the pattern of piston velocity produced in time. The distance travelled by the sound during one cycle of motion is given by c_0T; this distance is one *wavelength* λ. Since the frequency of the motion $f = 1/T$ Hz (cycles per second), then frequency and wavelength are related by $c_0 = f\lambda$. The frequencies in which we are interested are usually those of sounds which can be detected by the human ear. This range of frequencies lies between 20 Hz and 20 kHz. This corresponds to a range of wavelengths of approximately 17 m to 17 mm in air (or 75 m to 75 mm in water). As we shall see, it is the relatively large physical dimensions of the

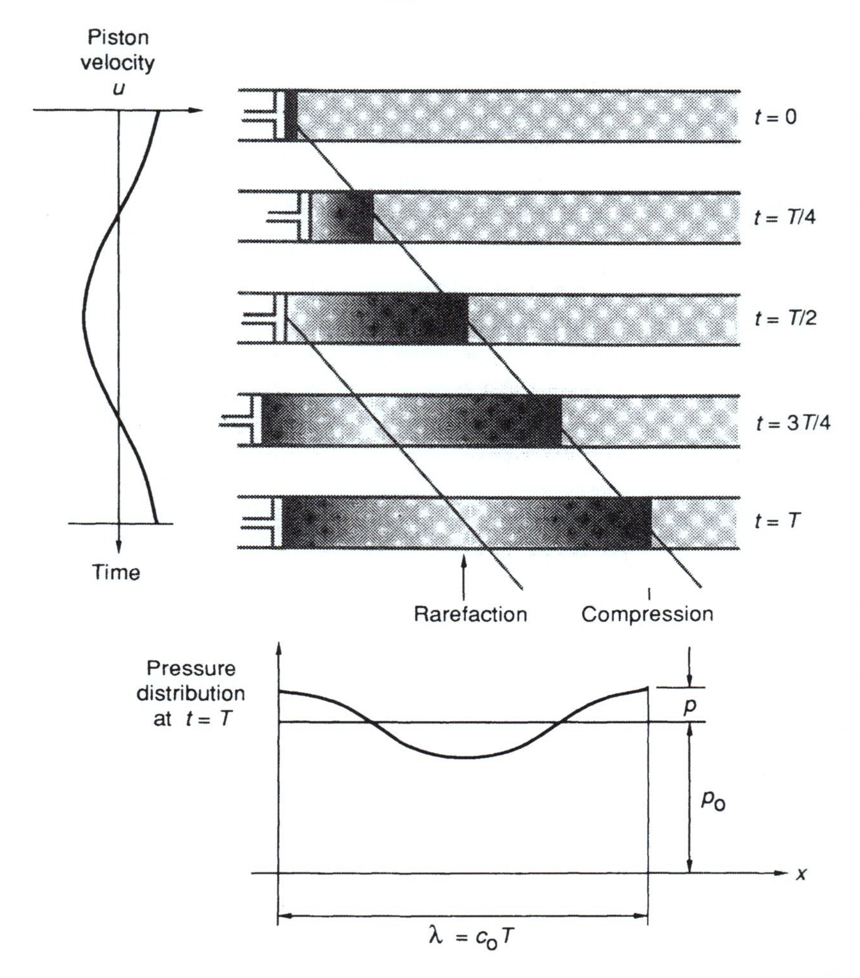

Fig. 1.2 The distribution of pressure in the tube during one cycle of harmonic piston motion. Note that at $t = T$, the spatial dependence of the acoustic pressure is of the form $\cos kx$ where $k = 2\pi/\lambda$.

wavelength at low frequencies that makes the active control of sound a practical proposition.

1.2 The decibel scale for the measurement of sound pressure

Figure 1.2 also shows the pressure distribution in the tube after one cycle of the piston motion. The pressure is shown as a variation either above or below the ambient (atmospheric) pressure in the tube. We generally denote the ambient pressure as p_0. The increase or decrease in pressure either above or below this value that is produced by the passage of an acoustic wave we shall denote as p, the acoustic pressure fluctuation. Thus, at a fixed point in the tube, as disturbances propagate past, the pressure will fluctuate with time. For the case illustrated in Fig. 1.2, this fluctuation will show a sinusoidal type of time dependence. If, however, the piston in the tube were made to vibrate in an irregular manner, with for example, a random variation in its displacement from its equilibrium position, the pressure fluctuation at a given position in the tube would also exhibit the same random variation with time but at a little time later. We shall discuss random fluctuations with time in some detail in Chapter 2 since many sources of sound of practical concern produce pressure fluctuations which can be modelled in this way. A suitable measure of the amplitude of these fluctuations is given by the mean square value of the fluctuation defined by the time average

$$\overline{p^2(t)} = \lim_{T\to\infty} \frac{1}{T} \int_{-T/2}^{T/2} p^2(t)\mathrm{d}t. \tag{1.2.1}$$

Since this measure of amplitude has dimensions of pressure squared, we often work with the "root mean square" (rms) value associated with the fluctuation, which we denote by p_{rms} and is given by the square root of the mean square pressure.

The range of rms pressure fluctuation which we encounter in practice is vast; its values vary from as small as 10^{-5} Pa to as large as 10^3 Pa and it is useful to compress this range of numbers by adopting a logarithmic measurement scale. In addition, the human perception of loudness exhibits an approximately logarithmic behaviour, with human beings judging the relative loudness of two sounds by approximately the ratio of their intensities. Thus, the quantity we use for the measurement of the sound pressure is the *sound pressure level* which is defined by

$$L_p = 10 \log_{10} \frac{\overline{p^2(t)}}{p_{\mathrm{ref}}^2} = 20 \log_{10} \frac{p_{\mathrm{rms}}}{p_{\mathrm{ref}}}, \tag{1.2.2}$$

where p_{ref} is an rms reference pressure of 20 μPa which is the reference pressure used for sound propagation in air since it defines the amplitude of a pure tone at 1 kHz which is just audible to the human ear. This "threshold of hearing" at 1 kHz thus corresponds to a sound pressure level of 0 dB. Some typical values of the appropriate sound pressure level produced by various sources are shown in Table 1.1.

Table 1.1 Typical rms pressure fluctuations and their sound pressure levels. The values shown are only rough approximations to the actual levels produced.

	p_{rms} (Pa)	L_p (dB *re* 2×10^{-5} Pa)
3 m from jet engine	200	140
Rock concert	20	120
Heavy machine shop	2	100
Vacuum cleaner	0.2	80
Conversational speech	0.02	60
Library	0.002	40
Rustling of leaves	0.0002	20
Threshold of hearing	0.00002	0

1.3 Properties of acoustic disturbances

We will now return to the situation depicted in Fig. 1.1 and describe briefly the behaviour of the fluid immediately in front of the piston as the piston is set in motion with a constant velocity u. As illustrated in Fig. 1.4, after a small time t, the piston will have moved a distance ut and the leading edge of the disturbance will be a distance c_0t from the starting position of the piston. Assume that the fluid initially has uniform pressure and density defined by p_0 and ρ_0. The compressed fluid will be increased in pressure and density by p and ρ, respectively, where ρ is the acoustic density fluctuation. For the most part, acoustic compressions take place sufficiently fast to ensure that there is no time for any heat to escape from the element of fluid that becomes compressed. Under these circumstances, where the compression can be considered isentropic (see, for example, Pierce, 1981, Ch. 1) we can assume that the density change in the fluid is determined solely by the pressure change in the fluid and is not dependent on the temperature change, even though the temperature of the fluid will be increased slightly during the course of the compression. The form of the relationship between the density and pressure will depend on the properties of the fluid. If we express the total instantaneous pressure in the fluid as $p_{tot} = p_0 + p$ and similarly the total instantaneous density as $\rho_{tot} = \rho_0 + \rho$, we can express the relationship between them as

$$p_{tot} = f(\rho_{tot}). \tag{1.3.1}$$

The second assumption that we make is that the increases p and ρ in the pressure and density are sufficiently small to be considered to be *linearly* related. That p and ρ are generally very small compared to their ambient values p_0 and ρ_0 may be demonstrated by using a simple example. The maximum acoustic pressure fluctuation p measured close to the vibrating surface of a typical automobile engine running at full speed is unlikely to exceed around 10 Pa, even though it is one of the loudest sounds we have to deal with (see Table 1.1). This compares with a value of atmospheric pressure p_0 of around 10^5 Pa. Similarly, the density fluctuation ρ will be of the order of $10^{-4}\,\mathrm{kg\,m^{-3}}$ compared to an ambient value of around $1.2\,\mathrm{kg\,m^{-3}}$. A typical graph showing the relationship between p_{tot} and ρ_{tot} is sketched in Fig. 1.3. For small changes p and ρ above and below the ambient values p_0 and ρ_0, we assume that p and ρ are directly proportional, the constant of proportionality being given by the slope of the curve $f(\rho_{tot})$ at the ambient values p_0 and ρ_0. This can be demonstrated formally by using a Taylor series expansion of the function $f(\rho_{tot})$. Thus we can write

$$f(\rho_{tot}) = f(\rho_0) + \rho\frac{\mathrm{d}f(\rho_0)}{\mathrm{d}\rho_{tot}} + \frac{\rho^2}{2!}\frac{\mathrm{d}^2 f(\rho_0)}{\mathrm{d}\rho_{tot}^2} + \ldots, \tag{1.3.2}$$

where $\mathrm{d}f(\rho_0)/\mathrm{d}\rho_{tot}$ is the gradient of the function evaluated at the density ρ_0. For small increases ρ in the density (and for a function whose rate of change of gradient $\mathrm{d}^2 f(\rho_0)/\mathrm{d}\rho_{tot}^2$ is not too large), the terms proportional to ρ^2 and higher powers of ρ may be neglected in the series expansion. Since $f(\rho_{tot}) = p_0 + p$ and $f(\rho_0) = p_0$, then with only the term proportional to ρ retained, equation (1.3.2) reduces to

$$p = \rho\,\frac{\mathrm{d}f(\rho_0)}{\mathrm{d}\rho_{tot}}. \tag{1.3.3}$$

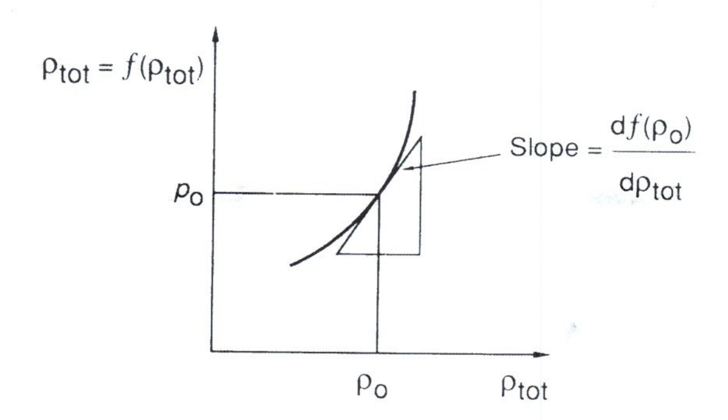

Fig. 1.3 The general form of relationship between total instantaneous pressure and density.

This expresses the linear relationship between the increases in pressure and density, the constant of proportionality being given by the rate of change of pressure with density evaluated with the fluid in the ambient (uncompressed) state.

The pressure increase p is also linearly related to the piston velocity and resulting particle velocity u. This can be demonstrated following Morse and Ingard (1968, see Ch. 6), by applying the principle of momentum conservation to the region of compressed fluid immediately in front of the piston. The net force applied to the element by the piston is given by pS where S is the cross-sectional area of the element. This can be equated to the rate of change of momentum of the fluid which shows that

$$pS = \frac{\mathrm{d}}{\mathrm{d}t}(\rho_0 c_0 t S u), \tag{1.3.4}$$

where $(\rho_0 c_0 t S)$ quantifies the total mass of the fluid in the uncompressed element (see Fig. 1.4). It follows from equation (1.3.4) that

$$p = \rho_0 c_0 u. \tag{1.3.5}$$

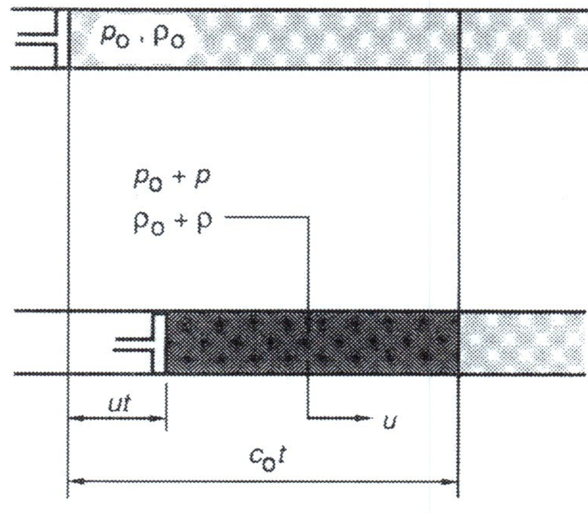

Fig. 1.4 The compression of a volume of fluid in front of the piston after the piston has been moving with a velocity u for a time t.

The constant of proportionality in this linear relationship is $\rho_0 c_0$. This quantity is known as the *characteristic acoustic impedance* of the medium. In air at atmospheric pressure and a temperature of 20° C, this has a value of $415\,\mathrm{kg\,m^{-2}\,s^{-1}}$ whilst in fresh water at 20° C, it is equal to $1.48 \times 10^6\,\mathrm{kg\,m^{-2}\,s^{-1}}$. Thus a given surface velocity will produce a much larger pressure in water than it will in air.

Finally we can establish a further linear relationship between the small density increase ρ and the particle velocity u. Again following Morse and Ingard (1968, see Ch. 6) we can apply the principle of mass conservation to the fluid element considered and equate the mass of the fluid in the element prior to its compression to its value after the time t. This gives

$$\rho_0 c_0 t S = (\rho_0 + \rho)(c_0 t - u t)S. \tag{1.3.6}$$

As we have noted above, the acoustic density increase ρ is very small compared to ρ_0. Also, in general, the particle velocity is very much smaller than c_0 (the typical velocity of the vibrating surface of the automobile engine referred to above is less than $0.1\,\mathrm{m\,s^{-1}}$). Thus in expanding the bracketed terms on the right hand side of this equation, the term proportional to ρu can be neglected, since it involves the product of two very small quantities and will be very much smaller than the other terms on the right hand side of the equation. The neglect of this term leads directly to the relationship

$$u = \rho c_0/\rho_0. \tag{1.3.7}$$

Equations (1.3.3), (1.3.5) and (1.3.7) can be combined to show that

$$p = c_0^2 \rho, \qquad c_0^2 = \frac{\mathrm{d}f(\rho_0)}{\mathrm{d}\rho_{\mathrm{tot}}}. \tag{1.3.8a,b}$$

The sound speed is therefore determined solely by the ambient properties of the fluid and is derived from the relationship between the pressure and the density defined by equation (1.2.1). Recall that a relationship of this type holds for compressions in which no heat escapes from a compressed element of fluid. In the particular case of a perfect gas (to which air is a good approximation) undergoing such an adiabatic (isentropic) compression, then the relationship between pressure and density is given by

$$p_{\mathrm{tot}} = f(\rho_{\mathrm{tot}}) = p_0 \left(\frac{\rho_{\mathrm{tot}}}{\rho_0}\right)^{\gamma}, \tag{1.3.9}$$

where γ is the ratio of the specific heats of the gas. (See, for example, Kinsler *et al.*, 1982, Appendix 9, for a detailed account of the thermodynamics of such compressions.) Differentiation of this expression and evaluation of the derivative at the density ρ_0 shows that

$$c_0^2 = \frac{\mathrm{d}f(\rho_0)}{\mathrm{d}\rho_{\mathrm{tot}}} = \frac{\gamma p_0}{\rho_0}. \tag{1.3.10}$$

The sound speed in a perfect gas is therefore very simply related to the ambient fluid pressure and density. More thorough discussions of the dependence of the sound speed on ambient fluid properties are given by, for example, Pierce (1981, Ch. 1) and Lighthill (1978, Ch. 1), who also consider the speed of sound in liquids.

1.4 The one-dimensional wave equation

We will now proceed to derive in a more formal way, the relationships between the acoustic variables p, ρ and u when they vary continuously in space and time as, for example, in the case of harmonic wave motion illustrated in Fig. 1.2. First, consider an arbitrary fixed volume of the tube through which acoustic disturbances pass. The volume has a cross section S and a length dx in the x direction. This is illustrated in Fig. 1.5. We will again denote the total instantaneous density of the fluid in the element by $\rho_{\text{tot}} = \rho_0 + \rho$ where ρ is the increase in density produced by the passage of an acoustic wave. Consider the application of the principle of conservation of mass to the element shown. The rate of mass inflow into the control volume is given by

$$\rho_{\text{tot}} uS - \left[\rho_{\text{tot}} u + \frac{\partial(\rho_{\text{tot}} u)}{\partial x} \mathrm{d}x \right] S = -\frac{\partial(\rho_{\text{tot}} u)}{\partial x} \mathrm{d}xS. \tag{1.4.1}$$

The principle of mass conservation requires that this net inflow must be balanced by an increase in mass in the element. The rate of increase of mass inside the control volume is given by $(\partial\rho_{\text{tot}}/\partial t)\mathrm{d}xS$. If we now equate the rate of mass inflow and the rate of increase of mass then it follows that

$$\frac{\partial(\rho_{\text{tot}})}{\partial t} + \frac{\partial(\rho_{\text{tot}} u)}{\partial x} = 0. \tag{1.4.2}$$

First consider the derivative $\partial(\rho_{\text{tot}} u)/\partial x$. The product $(\rho_{\text{tot}} u)$ is equal to $(\rho_0 + \rho)u$ and we again make the assumption that since the term ρu is the product of two small quantities, it is sufficiently small to be neglected. We therefore retain only the term $\rho_0 u$ and since ρ_0 is not a function of x, the derivative of this product with respect to x reduces to $\rho_0(\partial u/\partial x)$. Since ρ_0 is also not a function of t, the derivative $\partial(\rho_{\text{tot}})/\partial t$ becomes $\partial\rho/\partial t$ and equation (1.4.2) reduces to

$$\frac{\partial\rho}{\partial t} + \rho_0 \frac{\partial u}{\partial x} = 0. \tag{1.4.3}$$

This relationship is known as the *linearised* equation of mass conservation, a title which originates from the fact that the terms in the equation are only linear functions of the fluctuating quantities ρ and u.

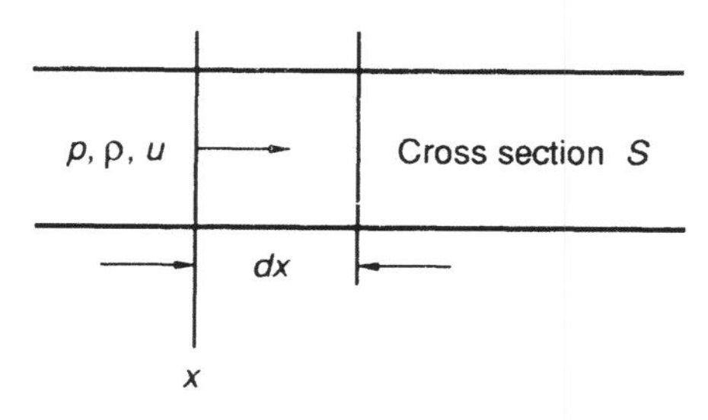

Fig. 1.5 A control volume in the tube where the acoustic variables are given by p, ρ, u at the position x.

We can also apply the principle of momentum conservation to a fluid element. In this case we will apply the principle to an element which moves with the fluid and again we will assume that the net instantaneous pressure in the fluid is given by $p_{tot} = p_0 + p$ where p is the acoustic pressure. Thus the net force on the fluid element in the positive x direction is given by

$$p_{tot}S - \left(p_{tot} + \frac{\partial p_{tot}}{\partial x}\mathrm{d}x\right)S = -\frac{\partial p_{tot}}{\partial x}\mathrm{d}xS. \tag{1.4.4}$$

The momentum conservation principle states that this net applied force must be balanced by the mass times the acceleration of the fluid element. The acceleration of the fluid element is given by $\mathrm{d}u/\mathrm{d}t = u\partial u/\partial x + \partial u/\partial t$ which accounts for the fact that the velocity of the fluid is also a function of x (see, for example, Kinsler *et al.*, 1982, Ch. 5, for a fuller discussion). However, the term $u\partial u/\mathrm{d}x$ turns out to be much smaller (by a factor of u/c_0) than $\partial u/\partial t$ and thus we can approximate the mass times the acceleration of the fluid element by $\rho_0 \mathrm{d}xS(\partial u/\partial t)$. Equating this rate of change of momentum to the net force on the fluid element therefore results in

$$\rho_0 \frac{\partial u}{\partial t} + \frac{\partial p}{\partial x} = 0. \tag{1.4.5}$$

This relationship is known as the linearised equation of momentum conservation, again because the terms in the equation are only linearly dependent on the small fluctuating quantities.

If we now differentiate equation (1.4.3) with respect to t and differentiate equation (1.4.5) with respect to x then a combination of the two resulting equations shows that

$$\frac{\partial^2 p}{\partial x^2} - \frac{\partial^2 \rho}{\partial t^2} = 0. \tag{1.4.6}$$

We again assume that the acoustic compressions are adiabatic such that the increase in pressure and increase in density are linearly related and that the square of the sound speed is the constant of proportionality such that $p = c_0^2\rho$. Substitution of this relationship into equation (1.4.6) yields the one-dimensional wave equation for the propagation of acoustic pressure fluctuations. This is given by

$$\frac{\partial^2 p}{\partial x^2} - \frac{1}{c_0^2}\frac{\partial^2 p}{\partial t^2} = 0. \tag{1.4.7}$$

This equation relates the way in which the acoustic pressure fluctuations must behave with respect to space (as a function of the co-ordinate distance x) and with respect to time t. In the derivation of this equation, the process of linearisation was a central assumption. We will now proceed to find solutions that satisfy the wave equation and demonstrate that the solutions bear out the description of the propagation of acoustic disturbances presented in Section 1.

1.5 Solutions of the one-dimensional wave equation

It is easy to show that pressure fluctuations which have a dependence on space and time described by functions f and g such that

$$p(x, t) = f(t - x/c_0) \quad \text{and} \quad p(x, t) = g(t + x/c_0), \tag{1.5.1}$$

are solutions of the one-dimensional wave equation. This can be demonstrated by differentiating the functions f and g twice with respect to x and twice with respect to t and then checking that $\partial^2 p/\partial x^2 = (1/c_0^2)\partial^2 p/\partial t^2$. These two functions also have a well-defined physical description. The first term $f(t - x/c_0)$ describes a pressure fluctuation which is travelling in the positive x direction, i.e. from left to right in Fig. 1.1. Conversely the function $g(t + x/c_0)$ describes fluctuations which travel in the negative x direction (i.e. from right to left in Fig. 1.1). For waves travelling in the positive x direction then equation (1.5.1) implies that the fluctuation $f(t)$ generated at $x = 0$ at the face of the piston in Fig. 1.1 is reproduced at a position at a distance x along the tube at a time x/c_0 later. The form of the fluctuation is preserved exactly as it travels at constant speed through the tube. This is illustrated in Fig. 1.6. This also applies to fluctuations $g(t)$ travelling in the negative x direction.

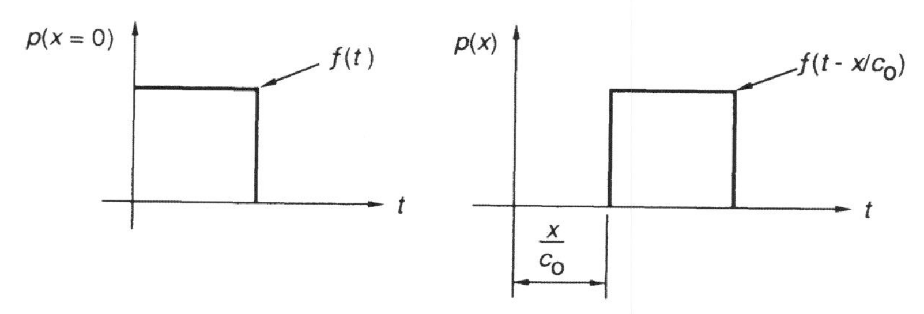

Fig. 1.6 The pressure fluctuation generated as a function of time by the propagation of the acoustic disturbance depicted in Fig. 1.1. The fluctuation generated at $x = 0$ (the piston face) is reproduced at the position a distance x along the tube at a time x/c_0 later.

A particular form of function which describes harmonic waves travelling in the positive x direction is given by the complex representation of harmonic motion described by

$$f(t - x/c_0) = \mathrm{Re}\{A\,\mathrm{e}^{j\omega(t - x/c_0)}\} = \mathrm{Re}\{A\,\mathrm{e}^{j(\omega t - kx)}\}, \tag{1.5.2}$$

where $\omega = 2\pi/T$ is the angular frequency of the harmonic fluctuation having period T and $k = \omega/c_0 = 2\pi/\lambda$ is known as the *wavenumber*. The term $\mathrm{e}^{j(\omega t - kx)}$ is a complex number which, by using the identity $\mathrm{e}^{j\theta} = \cos\theta + j\sin\theta$, can be written as

$$\mathrm{e}^{j(\omega t - kx)} = \cos(\omega t - kx) + j\sin(\omega t - kx). \tag{1.5.3}$$

Thus $\cos(\omega t - kx)$ and $\sin(\omega t - kx)$ define the real and imaginary parts of the complex number. The term A is also a complex number with real and imaginary parts $A_R = \mathrm{Re}\{A\}$ and $A_I = \mathrm{Im}\{A\}$ such that

$$A = A_R + jA_I = |A|\,\mathrm{e}^{j\phi_A}, \tag{1.5.4}$$

where $|A|$ is the modulus of the complex number A and ϕ_A is its phase. These are in turn related to A_R and A_I by

$$|A| = \sqrt{(A_R^2 + A_I^2)}, \qquad \phi_A = \tan^{-1}\frac{A_I}{A_R}. \tag{1.5.5a,b}$$

Note that A_R and A_I are both *real* numbers. In dealing with a representation of the form given by equation (1.5.2) we always recover the actual pressure fluctuation by taking the real part of the complex representation. Thus the physical pressure fluctuation is given by

$$p(x, t) = \text{Re}\{A e^{j(\omega t - kx)}\}, \tag{1.5.6}$$

and using $A = |A| e^{j\phi_A}$ together with $e^{j\theta} = \cos\theta + j\sin\theta$ shows that

$$p(x, t) = |A| \cos(\omega t - kx + \phi_A). \tag{1.5.7}$$

Thus the modulus $|A|$ of the complex number A defines the amplitude of the harmonic pressure fluctuation and ϕ_A is an arbitrary phase angle. Differentiation of this function twice with respect to both x and t will demonstrate that it is a solution to the wave equation.

Similarly, we can use complex exponential representation to describe harmonic waves travelling in the negative x direction. Thus we can write

$$g(t + x/c_0) = \text{Re}\{B e^{j\omega(t + x/c_0)}\} = \text{Re}\{B e^{j(\omega t + kx)}\}, \tag{1.5.8}$$

where B is again a complex number such that

$$B = B_R + jB_I = |B| e^{j\phi_B}. \tag{1.5.9}$$

The pressure fluctuation associated with a negative-going harmonic wave can thus be written as

$$p(x, t) = \text{Re}\{B e^{j(\omega t + kx)}\} = |B| \cos(\omega t + kx + \phi_B). \tag{1.5.10}$$

It is also useful in dealing with acoustical problems at a single frequency to define a spatially dependent *complex pressure* which in this work we will denote simply by $p(x)$. This is a complex number which is a function of position. The real pressure fluctuation can be recovered by using

$$p(x, t) = \text{Re}\{p(x) e^{j\omega t}\}. \tag{1.5.11}$$

In the particular case of positive-going harmonic plane waves discussed above, then the complex pressure $p(x) = A e^{-jkx}$, and for negative-going waves $p(x) = B e^{jkx}$. Substitution of equation (1.5.11) into the wave equation shows that

$$\left(\frac{\partial^2}{\partial x^2} - \frac{1}{c_0^2}\frac{\partial^2}{\partial t^2}\right) \text{Re}\{p(x) e^{j\omega t}\} = 0. \tag{1.5.12}$$

Since the operation of taking the real part commutes with the differential operators, we can write this equation as

$$\text{Re}\left\{\left[\frac{d^2 p(x)}{dx^2} + \frac{\omega^2}{c_0^2} p(x)\right] e^{j\omega t}\right\} = 0, \tag{1.5.13}$$

where the double differentiation with respect to time has been undertaken. This equation must be satisfied for all values of time t and in particular when $e^{j\omega t} = 1$ or $e^{j\omega t} = j$. Both these conditions can be satisfied only if the term in square brackets is zero. That is

$$\frac{d^2 p(x)}{dx^2} + k^2 p(x) = 0, \qquad (1.5.14)$$

where $k^2 = \omega^2/c_0^2$. This equation is the one-dimensional *Helmholtz equation* which must be satisfied by the complex pressure $p(x)$. It is easy to verify that $p(x) = Ae^{-jkx}$ and $p(x) = Be^{jkx}$ are solutions to this equation. A fuller discussion of the use of complex spatially dependent field variables is presented by, for example, Pierce (1981, see Ch. 1).

1.6 Linearity and the superposition principle

In our discussion of acoustic wave motion we have emphasised repeatedly that the acoustic variables pressure p, density ρ and particle velocity u are *small* perturbations. That is to say, p and ρ are very small compared to their ambient values p_0 and ρ_0 and u is very small compared to the speed of sound c_0. We have seen in Section 1.3 that, as a direct consequence of this, the variables p, ρ and u, to a very good approximation, are all *linearly* related to one another. We have also seen in Section 1.4 that we can neglect the products of small fluctuating quantities in deriving a wave equation which governs the form of acoustic pressure fluctuations in one dimension. This equation is therefore linear in the pressure fluctuation $p(x, t)$ and does not contain any non-linear terms like $p^2(x, t)$. In addition, the (partial) differential operators in the wave equation are also "linear" operators. That is to say, given a pressure fluctuation $p_1(x, t)$ and another pressure fluctuation $p_2(x, t)$, then

$$\frac{\partial^2}{\partial x^2}[p_1(x, t) + p_2(x, t)] = \frac{\partial^2}{\partial x^2}p_1(x, t) + \frac{\partial^2}{\partial x^2}p_2(x, t), \qquad (1.6.1)$$

$$\frac{\partial^2}{\partial t^2}[p_1(x, t) + p_2(x, t)] = \frac{\partial^2}{\partial t^2}p_1(x, t) + \frac{\partial^2}{\partial t^2}p_2(x, t) \qquad (1.6.2)$$

As a direct consequence of the wave equation both being linear in the fluctuation $p(x, t)$ and containing only linear differential operators, we can show that two sound pressure fluctuations, each of which satisfies the wave equation, can also be simply added together to produce another pressure fluctuation which also satisfies the wave equation. Thus, if we have

$$\left(\frac{\partial^2}{\partial x^2} - \frac{1}{c_0^2}\frac{\partial^2}{\partial t^2}\right)p_1(x, t) = 0, \qquad \left(\frac{\partial^2}{\partial x^2} - \frac{1}{c_0^2}\frac{\partial^2}{\partial t^2}\right)p_2(x, t) = 0, \qquad (1.6.3a,b)$$

then it also follows that

$$\left(\frac{\partial^2}{\partial x^2} - \frac{1}{c_0^2}\frac{\partial^2}{\partial t^2}\right)[p_1(x, t) + p_2(x, t)] = 0. \qquad (1.6.4)$$

This is known as the *principle of superposition*. That is, two sound fields which have pressure fluctuations that vary both as a function of the spatial co-ordinate x and as a function of time t, can simply be added at each position x and time t to yield the net pressure fluctuation produced. Thus, the resulting pressure fluctuation will be given by

$$p(x, t) = p_1(x, t) + p_2(x, t). \tag{1.6.5}$$

In practical terms, this means that one pressure fluctuation will not become distorted by the presence of another; having the radio on will not distort the sound of another person's voice in the same room; the two sound pressure fluctuations simply add up at each instant. There are, of course, circumstances where the neglect of the non-linear terms in the basic equations used to derive the wave equation will become significant and sound propagation is no longer governed by the wave equation derived here. This occurs in particular when dealing with pressure fluctuations of very large amplitude and with propagation over very large distances. Typically, non-linear effects will produce significant distortion of an acoustic waveform in air having a sound pressure level of 120 dB when the wave propagates over a distance of a few kilometres (see the elementary discussion presented by Hall, 1987, Ch. 9 or the more detailed accounts presented by Pierce, 1981, Ch. 11 or Morse and Ingard, 1968, Ch. 14). However, for most circumstances of practical interest, sound propagation can safely be assumed to be linear and the principle of superposition can be applied.

It is the principle of superposition that enables the active control of sound to be accomplished and we will refer repeatedly to the principle during the course of this book. In active control we are generally given an (unwanted) primary pressure fluctuation, $p_p(x, t)$ say, and, for example, we deliberately introduce another secondary fluctuation, $p_s(x, t)$, that is equal and opposite to it, i.e. such that $p_s(x, t) = -p_p(x, t)$ and therefore $p(x, t) = 0$. The extent to which this can be realised in practice is the subject matter of this work. In considering the problem, we will very often work at a single frequency and use the spatially dependent complex pressure $p(x)$ introduced above. We have seen that in one-dimensional propagation $p(x)$ must satisfy the Helmholtz equation (1.5.14). Since we are again faced with an equation which has both linear operators and is only linearly dependent on the variable $p(x)$, then the superposition principle applies and if we have two sound fields at the same frequency ω, specified by the complex pressures $p_p(x)$ and $p_s(x)$, then the net complex pressure when the two fields are added is given by

$$p(x) = p_p(x) + p_s(x). \tag{1.6.6}$$

In terms of the real and imaginary parts of these complex numbers, and without showing the explicit dependence on x, this equation can be written as

$$p_R + jp_I = (p_{pR} + jp_{pI}) + (p_{sR} + jp_{sI}), \tag{1.6.7}$$

and it therefore follows that

$$p_R = p_{pR} + p_{sR}, \qquad p_I = p_{pI} + p_{sI}. \tag{1.6.8a,b}$$

The modulus and phase of the net pressure fluctuation are then given by

$$|p(x)| = \sqrt{[(p_{p\mathrm{R}}+p_{s\mathrm{R}})^2+(p_{p\mathrm{I}}+p_{s\mathrm{I}})^2]},$$
$$\phi(x) = \tan^{-1}\left(\frac{p_{p\mathrm{I}}+p_{s\mathrm{I}}}{p_{p\mathrm{R}}+p_{s\mathrm{R}}}\right). \qquad (1.6.9\mathrm{a,b})$$

These expressions can also be written in terms of the modulus and phase of the primary and secondary pressure fluctuations by using the relationships

$$p_{p\mathrm{R}} = |p_p(x)|\cos\phi_p(x), \qquad p_{p\mathrm{I}} = |p_p(x)|\sin\phi_p(x), \qquad (1.6.10\mathrm{a,b})$$

$$p_{s\mathrm{R}} = |p_s(x)|\cos\phi_s(x), \qquad p_{s\mathrm{I}} = |p_s(x)|\sin\phi_s(x), \qquad (1.6.11\mathrm{a,b})$$

where we have assumed that $p_p(x) = |p_p(x)|\mathrm{e}^{j\phi_p(x)}$ and that $p_s(x) = |p_s(x)|\mathrm{e}^{j\phi_s(x)}$. We are usually interested in making the modulus of the net pressure fluctuation small. In particular, in order to make $|p(x)|$ equal to zero, we require not only that $p_{s\mathrm{R}} = -p_{p\mathrm{R}}$ but also that $p_{s\mathrm{I}} = -p_{p\mathrm{I}}$. This also implies by virtue of equations (1.6.10) and (1.6.11) that the secondary pressure fluctuation must have a modulus identical to that of the primary fluctuation and a phase which differs by 180°. These are the conditions for the perfect cancellation of the pressure at a point in space. Figure 1.7 shows how the level of the total pressure fluctuation depends on the relative amplitude and phase of the secondary pressure fluctuation. The figure illustrates how closely the secondary fluctuation must be matched to the primary fluctuation in its amplitude and phase in order to produce appreciable reductions in the level of the total pressure fluctuation. For example, in order to produce a 10 dB reduction in level, the amplitude of the secondary fluctuation must be within ± 2.5 dB of the primary fluctuation and its phase must differ by no more than $\pm 20°$. In general in this book we will work with the real and imaginary parts of the secondary pressure fluctuation as the variables which we control; the reasons for this will become apparent in later chapters.

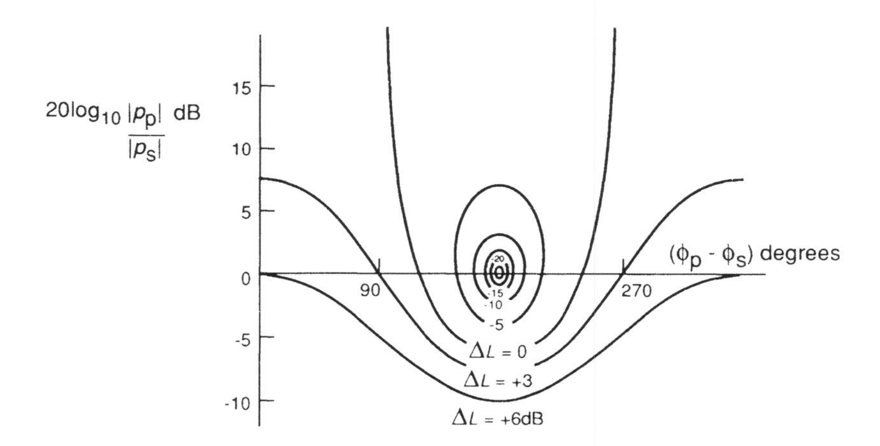

Fig. 1.7 The reduction ΔL (dB) in the amplitude of the primary pressure fluctuation as a function of the difference in amplitude and phase between the primary and secondary pressure fluctuations (after Mangiante, 1977).

1.7 Specific acoustic impedance

In addition to defining a complex acoustic pressure we can also define a spatially dependent complex acoustic particle velocity associated with the passage of a plane wave. Thus the real particle velocity fluctuation is expressed as

$$u(x, t) = \mathrm{Re}\{u(x)e^{j\omega t}\}, \tag{1.7.1}$$

where $u(x)$ is a complex number. The linearised equation of momentum conservation defined by equation (1.4.5) can therefore be written as

$$\rho_0 \frac{\partial}{\partial t}\mathrm{Re}\{u(x)e^{j\omega t}\} + \frac{\partial}{\partial x}\mathrm{Re}\{p(x)e^{j\omega t}\} = 0, \tag{1.7.2}$$

which, on using the same arguments used in Section 1.5, can also be written in the form

$$\mathrm{Re}\left\{\left(j\omega\rho_0 u(x) + \frac{\mathrm{d}p(x)}{\mathrm{d}x}\right)e^{j\omega t}\right\} = 0. \tag{1.7.3}$$

Again this equation can only be satisfied for all time provided

$$j\omega\rho_0 u(x) + \frac{\mathrm{d}p(x)}{\mathrm{d}x} = 0. \tag{1.7.4}$$

The complex particle velocity can thus be derived directly from the gradient of the complex pressure with respect to x. Thus in the case of the plane travelling wave where $p(x) = Ae^{-jkx}$,

$$u(x) = \frac{-1}{j\omega\rho_0}\frac{\mathrm{d}p(x)}{\mathrm{d}x} = \frac{1}{\rho_0 c_0}p(x), \tag{1.7.5}$$

since $k = \omega/c_0$. It also follows therefore that the actual particle velocity fluctuation is given by

$$u(x, t) = \mathrm{Re}\left\{\frac{p(x)}{\rho_0 c_0}\right\} = \frac{|A|}{\rho_0 c_0}\cos(\omega t - kx + \phi_A). \tag{1.7.6}$$

Therefore, in this particular case, the particle velocity fluctuation is of a nature identical to the acoustic pressure fluctuation and its magnitude is related to that of the pressure by a factor of $1/\rho_0 c_0$ where $\rho_0 c_0$ is the characteristic acoustic impedance of the medium. Thus in the harmonic travelling plane wave the acoustic pressure and particle velocity are always in phase. We shall see later that this is not always necessarily the case in acoustic propagation and more generally there is a complicated relationship between acoustic pressure and acoustic particle velocity in terms of the ratio of their amplitudes and their relative phases. One way in which we quantify this relationship is by defining a more general acoustic impedance which is a ratio of the complex pressure to the complex particle velocity. The *specific acoustic impedance* is then defined as

$$z(x) = \frac{p(x)}{u(x)}. \tag{1.7.7}$$

This ratio of complex numbers therefore directly quantifies the relative amplitude and phase of the pressure and particle velocity. This can be seen more clearly if we define $p(x) = |p(x)|e^{j\phi_p(x)}$ and $u(x) = |u(x)|e^{j\phi_u(x)}$. Then the expression for the impedance reduces to

$$z(x) = \frac{|p(x)|}{|u(x)|}e^{j[\phi_p(x) - \phi_u(x)]}. \tag{1.7.8}$$

Thus the modulus of the complex impedance is given by the ratio of the moduli of the complex pressure and particle velocity and its phase quantifies the difference in phase between these variables.

1.8 Acoustic energy density and intensity

To return to our basic description of acoustic wave propagation, it is evident that as compressions pass through the fluid then there is a transfer of energy from the source (the vibrating piston) to other positions in the tube. Fluid that was previously at rest gains kinetic energy, since there is a particle velocity associated with the local motion of the fluid, and also it gains potential energy as there are pressure fluctuations induced in the fluid. It is often useful to quantify these fluctuations in kinetic and potential energy and this can be done by returning to the simple acoustic compression illustrated in Fig. 1.4.

First it is straightforward to write down an expression for the kinetic energy associated with the element of fluid which is squeezed by the action of the piston. If the original volume of the uncompressed fluid is given by V_0, the instantaneous kinetic energy of the element can be written as

$$E_k(t) = \frac{1}{2}\rho_0 V_0 u^2(t), \tag{1.8.1}$$

which is simply one half times the mass of the element times its velocity squared. An expression for the potential energy is a little more difficult to derive. The potential energy produced by compressing the element of fluid from a volume V_0 to a volume V is given by

$$E_p(t) = -\int_{V_0}^{V} p(t)\mathrm{d}V, \tag{1.8.2}$$

where the minus sign is included since work is done on the element in decreasing its volume from V_0 to V. As usual it is assumed that the change in volume and density of the fluid is relatively small such that by mass conservation $(\rho_0 + \rho)V = \rho_0 V_0$ which, when products of small quantities are neglected, can be written as $\rho_0 V + \rho V_0 = \rho_0 V_0$. Thus, differentiating this expression, we can write $\mathrm{d}V/\mathrm{d}\rho = -V_0/\rho_0$ and, since the compression is adiabatic, using equation (1.3.8a) we have $\mathrm{d}V = (V_0/\rho_0 c_0^2)\mathrm{d}p$. Thus we can substitute for $\mathrm{d}V$ in equation (1.8.2) and integrate with respect to the acoustic pressure p. The integral is given by

$$E_p(t) = \frac{V_0}{\rho_0 c_0^2}\int_0^p p(t)\mathrm{d}p = \frac{V_0}{2\rho_0 c_0^2}p^2(t). \tag{1.8.3}$$

This is the expression which defines the potential energy of the element resulting from the increase in acoustic pressure p. The total instantaneous *energy density* of the fluid (i.e. the energy per unit volume) is given by the sum of $e_k(t) = E_k(t)/V_0$ and $e_p(t) = E_p(t)/V_0$ and can be written as

$$e(t) = e_k(t) + e_p(t) = \frac{1}{2}\rho_0\left[u^2(t) + \frac{p^2(t)}{(\rho_0 c_0)^2}\right]. \tag{1.8.4}$$

This expression applies at any instant in time and since both u and p are time varying this energy quantity also fluctuates as the pressure and velocity rise and fall due to the passage of acoustic disturbances. Note that for *plane wave* disturbances then $p(t)$ and $u(t)$ are simply related by $u(t) = p(t)/\rho_0 c_0$ and therefore the instantaneous energy density associated with the passage of a plane wave is given simply by $e(t) = [p^2(t)/\rho_0 c_0^2]$. It is also useful to assess the time average of these fluctuations in energy density. In the case of a plane wave whose pressure can be written as $\mathrm{Re}\{p(x)\mathrm{e}^{j\omega t}\}$, if $p(x)$ is the complex number $p_R + jp_I$ we can write

$$\begin{aligned}\mathrm{Re}\{p(x)\mathrm{e}^{j\omega t}\} &= \mathrm{Re}\{(p_R + jp_I)(\cos\omega t + j\sin\omega t)\}\\ &= p_R\cos\omega t - p_I\sin\omega t.\end{aligned} \tag{1.8.5}$$

Thus the time-averaged energy density can be written as

$$e = \frac{1}{T}\int_{-T/2}^{T/2}\frac{p^2(t)}{\rho_0 c_0^2}\mathrm{d}t = \frac{1}{\rho_0 c_0^2 T}\int_{-T/2}^{T/2}(p_R\cos\omega t - p_I\sin\omega t)^2\mathrm{d}t, \tag{1.8.6}$$

where T is one period of the harmonic motion. The product containing the term $\cos\omega t\,\sin\omega t$ integrates to zero over one period whilst the terms in $\cos^2\omega t$ and $\sin^2(\omega t)$ integrate to a value of $(T/2)$. This gives

$$e = \frac{1}{2\rho_0 c_0^2}(p_R^2 + p_I^2) = \frac{|p(x)|^2}{2\rho_0 c_0^2} \quad \text{(for plane waves)}. \tag{1.8.7}$$

The time-averaged energy density due to the passage of the plane wave is therefore related simply to the amplitude of the pressure fluctuations. It is also evident that as the waves propagate through the fluid then there is a transmission of energy from one part of the fluid to another. The time-averaged rate at which this energy is transferred is the *acoustic intensity*. This is defined as

$$I = \frac{1}{T}\int_{-T/2}^{T/2}p(t)u(t)\mathrm{d}t, \tag{1.8.8}$$

where T is again the period of one cycle of harmonic motion. Note that the acoustic intensity therefore has dimensions of force per unit area times velocity which is equivalent to the rate at which work is done per unit area by the acoustic pressure. Acoustic intensity is thus measured in $\mathrm{W\,m^{-2}}$. Also, since the instantaneous acoustic intensity $p(t)u(t)$ is proportional to the particle velocity $u(t)$ in the x direction, the acoustic intensity itself is a directed quantity and hence a vector. In the case of plane waves in the positive x direction then $u(t)$ and $p(t)$ are simply related by $u(t) = p(t)/\rho_0 c_0$ and undertaking the time average in the same way as in the case of energy density produces an expression for the intensity. This is related to the amplitude of the pressure fluctuations by

$$I = \frac{|p(x)|^2}{2\rho_0 c_0} \quad \text{(for plane waves)}. \tag{1.8.9}$$

Thus for positive going plane waves the time-averaged intensity and energy density are related by $I = c_0 e$. In general, however, the pressure and velocity are not always simply related as in the case of a plane wave. If we define a complex pressure and particle velocity at a given frequency it is useful to derive an expression for the time-averaged intensity in terms of the complex numbers $p(x)$ and $u(x)$. When undertaking time averages we again have to be careful to choose only the real part of these two quantities. We can write the real part of the pressure fluctuation as in equation (1.8.5) and similarly the real part of the velocity fluctuation is given by

$$\mathrm{Re}\{u(x)\mathrm{e}^{j\omega t}\} = u_{\mathrm{R}} \cos \omega t - u_{\mathrm{I}} \sin \omega t. \tag{1.8.10}$$

Therefore the time-averaged acoustic intensity in the positive x-direction for general harmonic fluctuations in pressure and velocity is given by

$$\frac{1}{T}\int_{-T/2}^{T/2} p(t)u(t)\,\mathrm{d}t$$

$$= \frac{1}{T}\int_{-T/2}^{T/2} (p_{\mathrm{R}} \cos \omega t - p_{\mathrm{I}} \sin \omega t)(u_{\mathrm{R}} \cos \omega t - u_{\mathrm{I}} \sin \omega t)\,\mathrm{d}t. \tag{1.8.11}$$

Note that only the products involving $\cos^2 \omega t$ and $\sin^2 \omega t$ will result in a non-zero value of the integral over the period T. The result of these integrations is therefore

$$\frac{1}{T}\int_{-T/2}^{T/2} p(t)u(t)\,\mathrm{d}t = \frac{1}{2}(p_{\mathrm{R}}u_{\mathrm{R}} + p_{\mathrm{I}}u_{\mathrm{I}}). \tag{1.8.12}$$

The right hand side of this equation is simply equal to $(1/2)\,\mathrm{Re}\{p^*(x)u(x)\} = (1/2)\,\mathrm{Re}\{p(x)u^*(x)\}$, where $p^*(x)$ and $u^*(x)$ are the complex conjugates of $p(x)$ and $u(x)$ given by $p^*(x) = p_{\mathrm{R}} - jp_{\mathrm{I}}$ and $u^*(x) = u_{\mathrm{R}} - ju_{\mathrm{I}}$. Thus in general we can use the expression

$$I = \frac{1}{2}\mathrm{Re}\{p^*(x)u(x)\} = \frac{1}{2}\mathrm{Re}\{p(x)u^*(x)\}, \tag{1.8.13}$$

to evaluate the acoustic intensity associated with harmonic fluctuations in pressure and velocity. Note that I here may be positive or negative, and its sign will indicate that energy is being transmitted in either the positive or negative x-directions.

1.9 Standing waves

Hitherto we have considered only plane propagating acoustic disturbances of the type depicted in Fig. 1.1, where there is no possibility of any reflection of the disturbances from the right hand end of the tube. Let us now deal with the case where we have a fluid in a tube which is excited by a piston at the left hand end but which is terminated by a rigid boundary at the right hand end. This is illustrated in Fig. 1.8. For harmonic excitation, one solution which will satisfy the wave equation is given by the superposition of the complex pressures associated with a positive-going and a negative-going plane wave. Thus we can write

$$p(x) = A\mathrm{e}^{-jkx} + B\mathrm{e}^{jkx}, \tag{1.9.1}$$

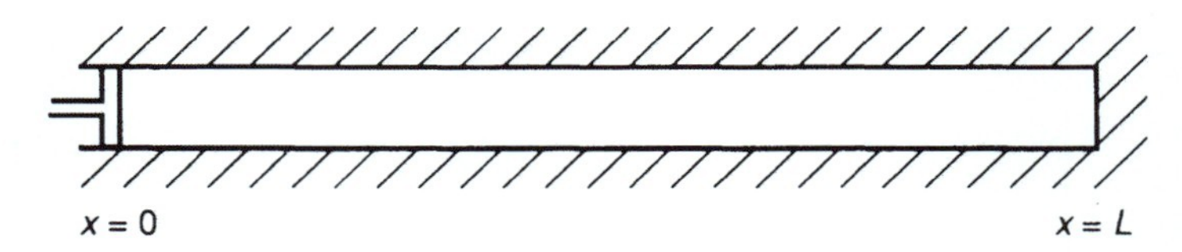

Fig. 1.8 A tube that is rigidly terminated at $x = L$ and in which the fluid is driven by a vibrating piston at $x = O$.

where A and B are arbitrary complex numbers which represent the amplitude and relative phases of waves travelling in the positive x and negative x directions respectively. Use of the linearised momentum equation (1.7.4) enables the complex particle velocities associated with each of these component travelling plane waves to be deduced. This is undertaken by evaluation of the pressure gradient associated with each of the waves as described in Section 1.7. This yields an expression for the complex particle velocity which is

$$u(x) = \frac{A\mathrm{e}^{-jkx}}{\rho_0 c_0} - \frac{B\mathrm{e}^{jkx}}{\rho_0 c_0}. \tag{1.9.2}$$

We can now determine the values of the complex constants A and B by application of the boundary conditions. Since we know that there must be zero particle velocity at the right hand end of the tube where $x = L$, then the condition $u(L) = 0$ in equation (1.9.2) results in the relationship

$$A\mathrm{e}^{-jkL} = B\mathrm{e}^{jkL}. \tag{1.9.3}$$

Similarly, we know that the particle velocity fluctuation at the left hand end of the tube is such that it must exactly match the fluctuating velocity of the piston. Thus we can use the boundary condition $u(0) = U$ where U is the complex velocity fluctuation associated with the piston. Application of this boundary condition to equation (1.9.2) results in

$$\rho_0 c_0 U = A - B. \tag{1.9.4}$$

Equations (1.9.3) and (1.9.4) can now be combined to give expressions for the complex amplitudes of the plane waves which are given by

$$A = \frac{\rho_0 c_0 U \mathrm{e}^{jkL}}{2j \sin kL}, \qquad B = \frac{\rho_0 c_0 U \mathrm{e}^{-jkL}}{2j \sin kL}. \tag{1.9.5a,b}$$

Note that in deriving these relationships use has been made of the expression $\mathrm{e}^{j\theta} - \mathrm{e}^{-j\theta} = 2j \sin \theta$. If we now substitute these expressions into equation (1.9.1) we can derive a relationship between the amplitudes of the pressure fluctuation in the tube and the amplitude of the harmonic fluctuation of the piston. Using the relationship $\mathrm{e}^{j\theta} + \mathrm{e}^{-j\theta} = 2 \cos \theta$ together with some algebra results in

$$p(x) = \frac{-j\rho_0 c_0 U \cos k(L-x)}{\sin kL}. \tag{1.9.6}$$

Similarly substitution of equations (1.9.5a) and (1.9.5b) into equation (1.9.2) yields the complex particle velocity given by

$$u(x) = \frac{U \sin k(L-x)}{\sin kL}. \qquad (1.9.7)$$

The two plane travelling waves therefore interfere to produce a *standing wave* whose pressure and particle velocity are a function of distance x along the tube.

First note that for harmonic fluctuations in the piston such that $\sin kL = 0$ both the complex pressure and particle velocity become infinite. This infinite response of the acoustic pressure fluctuations in the tube occurs at frequencies where $kL = n\pi$ where n is an integer. This corresponds to the relationship $(2\pi/\lambda)L = n\pi$ or $L = n\lambda/2$, i.e. at frequencies where an integer number of half wavelengths of sound can fit exactly into the length of the tube. Frequencies where this occurs are known as the *natural frequencies* or *resonance frequencies* of the fluid in the tube. Of course, in reality the response of the fluid in the tube will not be infinite at these frequencies since the presence of some dissipation (such as the presence of some absorbing material) will prevent the response from becoming infinite. It is also interesting to note the form of the pressure fluctuation at the natural frequencies as a function of the distance x along the tube. Thus when $kL = n\pi$ the amplitude of the pressure fluctuation as a function of x has the form $\cos n\pi(L-x)/L$ which is equivalent to $\pm\cos(n\pi x/L)$. These particular cosinusoidal functions of the pressure amplitude down the tube represent *mode shapes* corresponding to the natural frequencies of the fluid in the tube, and at each natural frequency specified by a value of the integer n a different mode shape is produced.

Also note that the specific acoustic impedance varies along the length of the tube. This is given by

$$z(x) = \frac{p(x)}{u(x)} = -j\rho_0 c_0 \cot k(L-x). \qquad (1.9.8)$$

Note that it is a purely imaginary number, i.e. the impedance has only a reactive part and no resistive part. This is in direct contrast to the case where only propagating waves in only one direction are present. A purely imaginary impedance is associated with the lack of any net flow of energy since the particle velocity is in quadrature with the pressure. Thus, if we evaluate the acoustic intensity at any frequency of excitation of the piston, at $x = 0$ (the piston face) it has a value given by

$$I = (1/2)\,\mathrm{Re}\{p^*(0)U\} = (1/2)\,|U|^2\,\mathrm{Re}\{z(0)\} = 0, \qquad (1.9.9)$$

and therefore, under steady-state conditions, at no frequency is there any net flow of energy from the piston surface. Of course, this applies only under steady-state harmonic conditions but there will be an initial flow of energy from the piston as the motion is started and this has not been described by these relationships. Again this zero intensity is in contrast to the steady flow of energy associated with the propagation of unidirectional plane waves.

1.10 The three-dimensional wave equation

In order to derive the relationship between the spatial and temporal variation of the acoustic pressure fluctuations in cases where wave propagation is not just constrained to the one-dimensional situations described so far, we again use the principles of mass and momentum conservation in order to relate the acoustic variables pressure, density and velocity in the fluid. For the three-dimensional case depicted in Fig. 1.9, any position in space can be defined by the vector $\mathbf{x}$ which has components $x_1\mathbf{i}$, $x_2\mathbf{j}$ and $x_3\mathbf{k}$, where $\mathbf{i}$, $\mathbf{j}$ and $\mathbf{k}$ are the unit vectors in the x_1, x_2 and x_3 directions, respectively. The velocity of the fluid at any given position can be denoted by the vector $\mathbf{u}$, with components $u_1\mathbf{i}$, $u_2\mathbf{j}$ and $u_3\mathbf{k}$ in the three co-ordinate directions specified by x_1, x_2 and x_3. Consideration of the balance between the net inflow of mass into a small three-dimensional element and the net increase of mass of the element leads to a similar mass conservation equation to that derived in Section 1.4. Again to the accuracy of linear approximation, the equation that results for the mass balance in a small element is

$$\frac{\partial \rho}{\partial t} + \rho_0 \left(\frac{\partial u_1}{\partial x_1} + \frac{\partial u_2}{\partial x_2} + \frac{\partial u_3}{\partial x_3} \right) = 0. \tag{1.10.1}$$

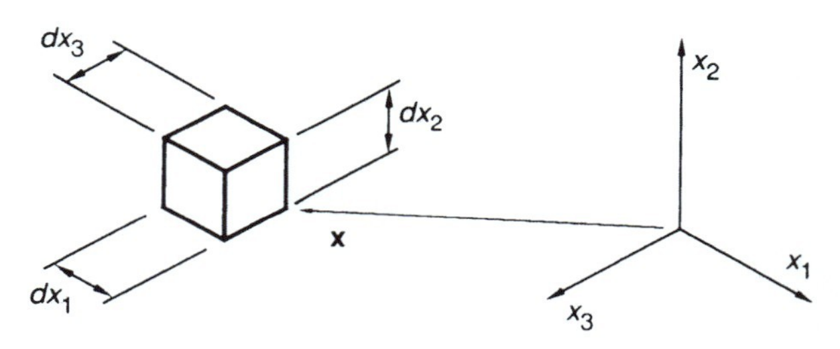

Fig. 1.9 A small control volume used to derive the three-dimensional linearised continuity equation.

This can alternatively be written in vector notation as

$$\frac{\partial \rho}{\partial t} + \rho_0 \boldsymbol{\nabla}.\mathbf{u} = 0, \tag{1.10.2}$$

where $\boldsymbol{\nabla}.$ is the divergence operator defined by $[\mathbf{i}(\partial/\partial x_1) + \mathbf{j}(\partial/\partial x_2) + \mathbf{k}(\partial/\partial x_3)]$ where $\mathbf{i}$, $\mathbf{j}$, $\mathbf{k}$ are again the unit vectors in the x_1, x_2 and x_3 directions, respectively.

This expresses essentially the same relationship between the rate of increase of density and the rate of inflow of mass that was derived for the one-dimensional case considered in more detail in Section 1.4. Similarly, one can relate the acceleration of a given element of fluid in three co-ordinate directions to the pressure gradients in those three directions by applying the principle of momentum conservation to a fluid element. In this case there are three independent momentum balances associated with each co-ordinate direction, which again to a linear approximation can be written as

$$\rho_0 \frac{\partial u_1}{\partial t} + \frac{\partial p}{\partial x_1} = 0, \quad \rho_0 \frac{\partial u_2}{\partial t} + \frac{\partial p}{\partial x_2} = 0,$$

$$\rho_0 \frac{\partial u_3}{\partial t} + \frac{\partial p}{\partial x_3} = 0. \qquad (1.10.3a,b,c)$$

These three equations can be written as one vector equation defined by

$$\rho_0 \frac{\partial \mathbf{u}}{\partial t} + \nabla p = 0, \qquad (1.10.4)$$

where ∇p has the three components $\mathbf{i}(\partial p/\partial x_1)$, $\mathbf{j}(\partial p/\partial x_2)$, $\mathbf{k}(\partial p/\partial x_3)$. This is the three-dimensional linearised equation of conservation of momentum. If one now takes the divergence of this equation (i.e. operates on the equation with the divergence operator $\nabla.$) and also if we differentiate equation (1.10.2) with respect to time we can eliminate $\nabla.\mathbf{u}$ from the two resulting equations to give

$$\nabla^2 p - \frac{\partial^2 \rho}{\partial t^2} = 0. \qquad (1.10.5)$$

Since the pressure and density are linearly related by the equation $p = c_0^2 \rho$ then we have

$$\nabla^2 p - \frac{1}{c_0^2} \frac{\partial^2 p}{\partial t^2} = 0, \qquad (1.10.6)$$

which is the wave equation which relates the spatial and temporal dependences of the pressure in a three-dimensional sound field. Note that in rectangular Cartesian co-ordinates $\nabla^2 p$ is given by the summation

$$\partial^2 p/\partial x_1^2 + \partial^2 p/\partial x_2^2 + \partial^2 p/\partial x_3^2.$$

This wave equation must be satisfied by acoustic disturbances propagated in the medium. However, in order to find solutions to this equation we must specify in more detail the form that the sound field must take. In particular, as we have demonstrated in the one-dimensional case, we have to apply boundary conditions in order to find the form of the sound field.

Finally, by using an argument that is identical to that used in Section 1.5, we can define a spatially dependent complex pressure $p(\mathbf{x})$ that, for harmonic sound, must satisfy

$$(\nabla^2 + k^2)p(\mathbf{x}) = 0, \qquad (1.10.7)$$

which is the three-dimensional Helmholtz equation. We should also note that the principle of linear superposition also applies to solutions of both the wave equation (1.10.6) and the Helmholtz equation (1.10.7). This is easily justified by following the argument presented in Section 1.6.

1.11 Solutions of the three-dimensional wave equation

The simplest kind of three-dimensional waves can be found if we assume that the propagation taking place is spherically symmetric. This is the type of wave motion that would be generated by, for example, a sphere whose surface pulsates uniformly and generates sound waves which travel radially outward from the sphere. If the radiation is assumed to be spherically symmetric then the spatial operator on the pressure in the three-dimensional wave equation can be written as

$$\nabla^2 p = \frac{1}{r^2}\frac{\partial}{\partial r}\left(r^2\frac{\partial p}{\partial r}\right), \tag{1.11.1}$$

and the wave equation for spherically symmetric wave propagation is given by

$$\frac{1}{r^2}\frac{\partial}{\partial r}\left(r^2\frac{\partial p}{\partial r}\right) - \frac{1}{c_0^2}\frac{\partial^2 p}{\partial t^2} = 0. \tag{1.11.2}$$

By using the laws of the derivatives of the product of two functions it can be shown that this equation is identical to

$$\frac{\partial^2}{\partial r^2}(rp) - \frac{1}{c_0^2}\frac{\partial^2 (rp)}{\partial t^2} = 0. \tag{1.11.3}$$

This is a very useful relationship since it has a form which is equivalent to that of the one-dimensional wave equation discussed in Section 1.4. Note however that in this case it is not just the pressure p which is the variable considered, but the product of r and p. However, we know that the solution to this equation can be written as

$$rp(r, t) = f(t - r/c_0) + g(t + r/c_0), \tag{1.11.4}$$

where again f and g are arbitrary functions. It therefore follows that the solution for the pressure can be written in the form

$$p(r, t) = \frac{f(t - r/c_0)}{r} + \frac{g(t + r/c_0)}{r}. \tag{1.11.5}$$

By analogy with the one-dimensional case, the first term in this equation represents pressure fluctuations which travel outwards from the origin of the spherical co-ordinates, whilst the second term represents pressure fluctuations which travel inwards to the origin. The second type of wave is less often encountered in practice and we are more often concerned with solutions in the three-dimensional case involving only the first term of equation (1.11.5). Again it is useful to use a harmonic function of $(t - r/c_0)$ of the type

$$p(r, t) = \mathrm{Re}\left\{\frac{A\,\mathrm{e}^{j(\omega t - kr)}}{r}\right\}, \tag{1.11.6}$$

where A is again an arbitrary complex number which specifies both the amplitude and phase of the pressure fluctuation. We can again recover the particle velocity associated with this pressure fluctuation by applying the momentum conservation equation. For the spherically symmetric case where u_r denotes the radial particle velocity component, the momentum conservation equation reduces to

$$\rho_0 \frac{\partial u_r}{\partial t} + \frac{\partial p}{\partial r} = 0. \tag{1.11.7}$$

First note that we can define a radially dependent complex pressure $p(r)$ as

$$p(r) = \frac{A\mathrm{e}^{-jkr}}{r}, \tag{1.11.8}$$

and again we recover the actual pressure fluctuation from $p(r, t) = \mathrm{Re}\{p(r)\mathrm{e}^{j\omega t}\}$. We can define similarly a complex particle velocity which has a radial component $u_r(r)$ where the actual radial particle velocity fluctuation is given by $u_r(r, t) = \mathrm{Re}\{u_r(r)\mathrm{e}^{j\omega t}\}$. In the same way that we used the one-dimensional equation of conservation of momentum in Section 1.5, it can be shown that equation (1.11.7) can be used to relate the complex pressure and particle velocity. In this case the equation reduces to

$$j\omega\rho_0 u_r(r) + \frac{\partial p(r)}{\partial r} = 0. \tag{1.11.9}$$

Differentiation of equation (1.11.8) with respect to r and substitution into equation (1.11.9) results in the expression for the complex radial particle velocity which is

$$u_r(r) = \frac{A}{j\omega\rho_0}\left(\frac{jk}{r} + \frac{1}{r^2}\right)\mathrm{e}^{-jkr}. \tag{1.11.10}$$

Note that this has both real and imaginary parts and therefore there are components of the particle velocity which are both in quadrature and in phase with the acoustic pressure. As the radial distance r becomes increasingly large then the second term in the brackets of equation (1.11.10) tends to zero more rapidly than the first and the relationship between the pressure and particle velocity reduces to that between the pressure and particle velocity in a plane one-dimensional progressive wave ($u_r(r) = p(r)/\rho_0 c_0$). This makes physical sense, since at a very long distance from the source of the waves, locally the wavefronts will appear plane. However when r becomes very small, the velocity will tend to become 90° out of phase with the pressure. This can be demonstrated more clearly by writing an expression for the specific acoustic impedance as a function of the radial distance r. This is

$$z(r) = \frac{p(r)}{u_r(r)} = \rho_0 c_0 \left(\frac{jkr}{1 + jkr}\right). \tag{1.11.11}$$

Thus as kr becomes large then the acoustic impedance tends to the value $\rho_0 c_0$ whereas when kr is very small then the acoustic impedance becomes increasingly imaginary. Note that it is the product kr which is the important parameter. This is equivalent to the ratio $2\pi r/\lambda$ and therefore it is the ratio of the radial distance to the acoustic wavelength which is important in determining the specific acoustic impedance in spherical wave propagation.

1.12 Point sources of spherical radiation

It is also useful to relate the pressure in the spherically symmetric outwardly propagating wave to the fluctuations in the velocity of the surface of the sphere which we assume to be the source of these waves. If we now apply boundary conditions such that the particle velocity described by equation (1.11.10) must exactly match the surface velocity of the pulsating sphere, and further assume that this pulsating sphere has a radius a and a complex velocity given by U_a, then the expression for the complex pressure at a can be written as

$$p(a) = \rho_0 c_0 \left(\frac{jka}{1+jka}\right) U_a = \frac{A\mathrm{e}^{-jka}}{a}. \tag{1.12.1}$$

From this we can deduce the relationship between A and U_a and it follows that the complex pressure as a function of the radial distance r from the centre of the sphere can be written as

$$p(r) = \rho_0 c_0 a \left[\frac{jka}{1+jka}\right] \frac{U_a \mathrm{e}^{-jk(r-a)}}{r}. \tag{1.12.2}$$

A very useful concept in acoustics is that of the point source. We can think of a point source by considering the sphere to be reduced steadily in radius but the surface velocity being increased such that the product of the surface velocity and the surface area of the sphere remains constant. We then define an effective complex *source strength*, such that, as $a \to 0$,

$$q = U_a 4\pi a^2 = \text{const.} \tag{1.12.3}$$

As the radius of the sphere becomes vanishingly small we can relate the pressure at a radial distance r from the origin of the waves to the source strength of this point source. Thus

$$p(r) = \rho_0 c_0 \left(\frac{jk}{1+jka}\right) \frac{q\mathrm{e}^{-jk(r-a)}}{4\pi r}, \tag{1.12.4}$$

which in the limit of ka tending to zero reduces to

$$p(r) = \frac{j\omega\rho_0 q \mathrm{e}^{-jkr}}{4\pi r}, \qquad ka \ll 1. \tag{1.12.5}$$

Note that the dimensions of complex source strength q are those of volume velocity, i.e. the product of velocity multiplied by surface area. We will sometimes find it useful in later chapters to introduce the concept of an *acoustic transfer impedance* $Z(r)$ which is the complex number relating the complex pressure to the complex strength of a point source. Thus we write

$$p(r) = qZ(r), \tag{1.12.6}$$

where it is clear from equation (1.12.5) that for free field propagation $Z(r) = j\omega\rho_0 \mathrm{e}^{-jkr}/4\pi r$. Note that this acoustic impedance has dimensions which differ from those of *specific* acoustic impedance since $Z(r)$ is a ratio of pressure to volume velocity. Using the identity $e^{j\theta} = \cos\theta + j\sin\theta$ also enables the expression for this transfer impedance to be written as

$$Z(r) = Z_0\left(\frac{\sin kr}{kr} + j\frac{\cos kr}{kr}\right), \tag{1.12.7}$$

where $Z_0 = \omega^2\rho_0/4\pi c_0$ is the real part of this transfer impedance in the limit $kr \to 0$. This form of expression will be particularly useful in later chapters and we will in general use a relationship of the form

$$p(\mathbf{x}) = q(\mathbf{y})Z(\mathbf{x}|\mathbf{y}), \tag{1.12.8}$$

where $Z(\mathbf{x}|\mathbf{y})$ is the transfer impedance relating the pressure at a position $\mathbf{x}$ to the strength of a source at position $\mathbf{y}$.

1.13 Acoustic power output

Now consider the time-averaged acoustic intensity associated with spherical wave propagation. In three-dimensional harmonic wave propagation of period T, the intensity is given by

$$\mathbf{I} = \frac{1}{T}\int_{-T/2}^{T/2} p(t)\mathbf{u}(t)\,\mathrm{d}t, \tag{1.13.1}$$

which is a vector quantity which depends on the direction of the vector $\mathbf{u}$. For the spherically symmetric propagation described above we can evaluate the intensity vector in the direction of the radial co-ordinate r. This can be expressed in terms of the complex pressure and radial particle velocity given by

$$I_r = \frac{1}{2}\mathrm{Re}\{p^*(r)u_r(r)\}. \tag{1.13.2}$$

Using the relationship between the complex pressure and radial particle velocity given by equation (1.11.11) shows that

$$I_r = \frac{|p(r)|^2}{2\rho_0 c_0}\mathrm{Re}\left\{\frac{1+jkr}{jkr}\right\} = \frac{|p(r)|^2}{2\rho_0 c_0}. \tag{1.13.3}$$

Recall that the intensity represents the average rate of energy flow per unit area of cross section normal to the direction of propagation. This therefore shows that the intensity depends only on the modulus squared of the complex pressure but, of course, this value decreases as the square of the radial distance from the source as is evident from equation (1.11.8). The total power output of a source is given by integrating the intensity travelling across a surface which totally encloses the source. The sound power output is defined by

$$W = \int_S \mathbf{I}.\mathbf{n}\,\mathrm{d}S, \tag{1.13.4}$$

where $\mathbf{n}$ is the unit normal vector pointing outwards from the closed surface S. In the case of the spherically symmetric progagation described here, then the expression of the power output can be written as

$$W = I_r(4\pi r^2), \tag{1.13.5}$$

and is therefore related to the amplitude of the pressure wave by

$$W = \frac{|A|^2}{2\rho_0 c_0 r^2} 4\pi r^2 = \frac{2\pi|A|^2}{\rho_0 c_0}. \tag{1.13.6}$$

This is clearly a constant and, as one would expect under conditions of steady harmonic motion, there is a constant rate of power output from the spherically radiating source. We can also evaluate the total power output from the source from the time-average product of the pressure at the surface of the pulsating sphere and the velocity of the surface of the sphere. The power output is given by

$$W = 4\pi a^2 \frac{1}{2}\mathrm{Re}\{p^*(a)U_a\}, \tag{1.13.7}$$

which can be written as

$$W = 4\pi a^2 \rho_0 c_0 |U_a|^2 \frac{1}{2} \mathrm{Re}\left\{\frac{-jka}{1-jka}\right\}. \tag{1.13.8}$$

Evaluating the real part of the term in curly brackets then shows that

$$W = 2\pi a^2 \rho_0 c_0 |U_a|^2 \frac{(ka)^2}{1+(ka)^2}. \tag{1.13.9}$$

This describes the efficiency with which a given surface velocity can radiate acoustic power. As the product ka becomes very large then the expression for the power output tends to the expression for a power output of a plane vibrating surface of area $4\pi a^2$ radiating plane acoustic waves. However, as the product ka becomes very small then the expression for the power output can be approximated by

$$W = 2\pi a^2 \rho_0 c_0 |U_a|^2 (ka)^2, \qquad ka \ll 1. \tag{1.13.10}$$

The product ka is simply the ratio $2\pi a/\lambda$ and therefore the radiation efficiency of the pulsating sphere depends on the ratio of the circumference of the sphere to the wavelength at the frequency of excitation. In general, for bodies which are large compared to the wavelength, sound is very effectively radiated by the vibrations of the surface but bodies which are very small compared to the wavelength are very inefficient radiators of acoustic power. In the limit of $ka \to 0$, the acoustic power output can be related to the strength of the equivalent point source and the expression for the power output reduces to

$$W = \frac{\omega^2 \rho_0}{8\pi c_0}|q|^2 = \frac{1}{2}Z_0|q|^2, \tag{1.13.11}$$

where $Z_0 = \omega^2\rho_0/4\omega c_0$ as defined in Section 1.12. Thus for a single point source in the free field the sound power output depends only on the amplitude of the source strength.

As in the case of the sound pressures encountered in practice, the range of acoustic power outputs is similarly vast. For example, the power output of a whispered voice is as small as 10^{-9} W whereas the power output of a jet engine can be as large as 10^4 W. A useful means of compressing this range is to define a logarithmic measurement scale as was the case with sound pressure. We therefore define the sound power level by

$$L_w = 10 \log_{10} \frac{W}{W_{\text{ref}}}, \tag{1.13.12}$$

where the reference sound power level W_{ref} is given by 10^{-12} W. Some typical sound power levels are shown in Table 1.2.

Table 1.2 Typical sound power outputs.

	W	dB *re* 10^{-12} W
Jet aircraft	10000	160
Chain saw	1	120
Normal voice	0.00001	70
Whisper	0.000000001	30

1.14 Enclosed sound fields at low frequencies

Now consider the solution to the three-dimensional wave equation when the sound field is in a fluid enclosed by a rigid walled rectangular box. In this case, for harmonic pressure fluctuations of a particular frequency in the enclosure, the complex pressure must satisfy the Helmholtz equation

$$\left(\frac{\partial^2}{\partial x_1^2} + \frac{\partial^2}{\partial x_2^2} + \frac{\partial^2}{\partial x_3^2}\right) p(\mathbf{x}) + k^2 p(\mathbf{x}) = 0. \tag{1.14.1}$$

In order to solve this equation we assume that the pressure can be expressed as three separate functions of x_1, x_2 and x_3, such that

$$p(\mathbf{x}) = F_1(x_1)\, F_2(x_2)\, F_3(x_3). \tag{1.14.2}$$

Substitution of equation (1.14.2) into equation (1.14.1) yields

$$\frac{1}{F_1(x_1)} \frac{\partial^2 F_1(x_1)}{\partial x_1^2} + \frac{1}{F_2(x_2)} \frac{\partial^2 F_2(x_2)}{\partial x_2^2} + \frac{1}{F_3(x_3)} \frac{\partial^2 F_3(x_3)}{\partial x_3^2} + k^2 = 0. \tag{1.14.3}$$

This sum of k^2 plus three functions which respectively depend on only one each of the three independent variables can vanish in general only if k^2 is the sum of three independent constants,

$$k^2 = k_1^2 + k_2^2 + k_3^2, \tag{1.14.4}$$

where the constant k_1, for example, is related to the function F_1 through the ordinary differential equation

$$\frac{1}{F_1(x_1)} \frac{\mathrm{d}^2 F_1(x_1)}{\mathrm{d}x_1^2} + k_1^2 = 0. \tag{1.14.5}$$

There are two other ordinary differential equations of the same form for the dependence of the pressure in the x_2 and x_3 directions. The function $F_1(x_1)$ in particular must also satisfy the rigid walled boundary conditions of zero particle velocity at $x_1 = 0$ and $x_1 = L_1$. Since the complex particle velocity is given by $u(\mathbf{x}) = (-1/j\omega\rho_0)\partial p(\mathbf{x})/\partial x_1$ then a zero particle velocity requires a zero pressure gradient $\partial p(\mathbf{x})/\partial x_1$. It is relatively easy to show that a set of solutions of equation (1.14.5) each of which satisfies the boundary conditions is $F_1(x_1) = \cos k_1 x_1$ where $k_1 = n_1\pi/L_1$ and n_1 is an integer. Similar solutions of the other two differential equations for $F_2(x_2)$ and $F_3(x_3)$ also follow. This shows that we can write a general solution of the Helmholtz equation (1.14.1) which satisfies the boundary conditions as

$$p(\mathbf{x}) = \sum_{n=0}^{\infty} A_n \cos\frac{n_1\pi x_1}{L_1} \cos\frac{n_2\pi x_2}{L_2} \cos\frac{n_3\pi x_3}{L_3}, \tag{1.14.6}$$

where the integer n denotes the trio of integers (n_1, n_2, n_3) and A_n is an arbitrary complex constant. Note the similarity of this solution to that of the problem considered in Section 1.9, where a fluid in a one-dimensional closed tube was excited by a piston. In that case we saw that at certain frequencies the pressure amplitude would become theoretically infinite and that at these frequencies an integer number of half wavelengths would define a given mode shape associated with the pressure fluctuations in the tube. In this three-dimensional case, exactly similar pressure fluctuations are generated but now for a given natural frequency there is not only a spatial dependence on the x_1 co-ordinate but also on the x_2 and x_3 co-ordinates. In fact the natural frequency associated with each of these mode shapes is found by noting that for a given combination of the integers n_1, n_2 and n_3 equation (1.14.4) can be written in the form

$$\omega_n = c_0\sqrt{\left[\left(\frac{n_1\pi}{L_1}\right)^2 + \left(\frac{n_2\pi}{L_2}\right)^2 + \left(\frac{n_3\pi}{L_3}\right)^2\right]}. \tag{1.14.7}$$

This defines the *natural frequencies* of the enclosure associated with each of the mode shapes defined by each combination of integers. In fact the solution defined by equation (1.14.6) is applicable *only* to unforced motion at these particular natural frequencies. For the perfectly rigid walled enclosure considered here with no source forcing the sound field, a sound field can exist only at the particular values of ω_n specified by equation (1.14.7). We shall see in Chapter 10 that these particular values of wave number $k_n = \omega_n/c_0$ are known as *eigenvalues* of the Helmholtz equation. Each of these eigenvalues is associated with an *eigenfunction* which defines the particular spatial pattern associated with the given value of n and which is specified by equation (1.14.6). In Chapter 10 we will show how we can use these eigenfunctions associated with rigid walled boundary conditions in order to construct a solution for the pressure field in an enclosure with walls having a high but finite impedance. The practical usefulness of the above solution is further limited by the fact that in many enclosures at the frequencies of interest there are potentially so many modal contributions. For example, Fig. 1.10 shows, for a typical enclosure, the number of modes having natural frequencies in a given bandwidth. We shall also see in Chapter 10 that when some sound-absorbing material is introduced onto the walls of the enclosure, the individual modes become damped. In addition, the modes can be excited over a range of frequencies other than those corresponding to their natural frequency. Thus at high excitation frequencies, a great many enclosure modes may be excited. Computation of the sound field by using this modal model

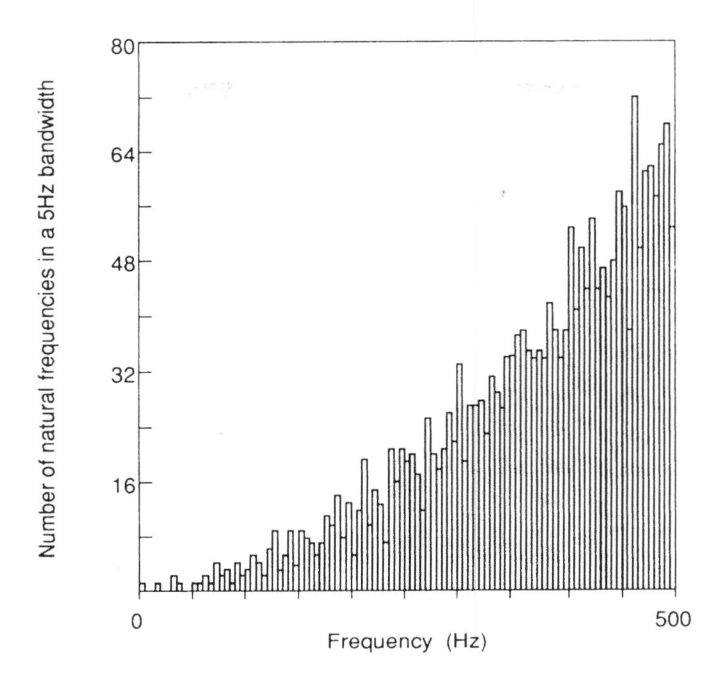

Fig. 1.10 The number of natural frequencies in each 5 Hz bandwidth up to 500 Hz for a rectangular enclosure of dimensions 10 m × 5 m × 3 m.

then becomes extremely laborious and we adopt the approach described in the next section where the sound field is assumed to consist of an infinitely large number of plane wave contributions. This is not inconsistent with the modal model presented here since each mode may be considered to consist of a number of plane wave contributions. Just as we saw in Section 1.9 that a one-dimensional standing wave can be considered to consist of two plane propagating waves, it can also be shown that the three-dimensional standing wave pattern associated with a given mode can be considered to consist of a sum of eight plane wave contributions (see, for example, Kinsler *et al.*, 1982, Ch. 9 for a full discussion).

1.15 Enclosed sound fields at high frequencies

As we discussed in the last section, in order to deal with enclosed sound fields at high frequencies (or strictly speaking, enclosures in which many modes have natural frequencies within the bandwidth of any one mode) we can adopt a model in which we assume that the sound field at a point consists of a superposition of the contributions from a number of plane waves. The plane waves are next assumed to be arriving at the point considered from *all possible propagation directions*. One can visualise this by assuming that a point in the sound field is at the centre of a sphere whose surface is divided into a very large number of segments of *equal* area. The line passing through the centre of one of these segments and the centre of the sphere defines the propagation direction associated with a particular plane wave. The field is then assumed to consist of an *infinite* number of plane waves, each associated with a *different* propagation direction. These propagation directions are defined by assuming that all the segments of equal area on the surface of the sphere become infinitesimally small. Thus the sound field at a single frequency has a complex pressure at a point (defined by the position vector $\mathbf{x}$) that is given by

$$p(\mathbf{x}) = \lim_{N \to \infty} \frac{1}{\sqrt{N}} \sum_{n=1}^{N} p_n(\mathbf{x}), \tag{1.15.1}$$

where $p_n(\mathbf{x}) = |p_n| e^{j\phi_n(\mathbf{x})}$ is the complex pressure associated with the nth plane wave and where we have arbitrarily normalised the amplitudes of the constituent plane waves by the square root of the total number N of contributing waves. Thus the constituent plane waves have moduli $|p_n|$ and phases $\phi_n(\mathbf{x})$, the latter changing as a function of position $\mathbf{x}$ as a result of the propagation of the waves through space. It is also assumed that at a given position, the phases of the individual plane waves are in a random relationship to one another. We can compute the modulus squared of the complex pressure at the position $\mathbf{x}$ by taking the product of $p(\mathbf{x})$ and its complex conjugate. Thus

$$|p(\mathbf{x})|^2 = \lim_{N \to \infty} \frac{1}{N} \left(\sum_{n=1}^{N} p_n(\mathbf{x}) \right) \left(\sum_{m=1}^{N} p_m(\mathbf{x}) \right)^*, \tag{1.15.2}$$

where the additional index m has been introduced to denote the mth plane wave. Now note that the product of terms for which $n \neq m$ takes the form

$$p_n(\mathbf{x}) p_m^*(\mathbf{x}) = |p_n||p_m| e^{j[\phi_n(\mathbf{x}) - \phi_m(\mathbf{x})]}, \tag{1.15.3}$$

which can also be written as

$$p_n(\mathbf{x}) p_m^*(\mathbf{x}) = |p_n||p_m| \{\cos[\phi_n(\mathbf{x}) - \phi_m(\mathbf{x})] + j \sin[\phi_n(\mathbf{x}) - \phi_m(\mathbf{x})]\}. \tag{1.15.4}$$

If we now move through a volume of space which has dimensions that are large compared to the acoustic wavelength at the frequency considered (but still much smaller than the dimensions of the enclosure) we will see that the terms in the curly brackets of equation (1.15.4) will have an average value of zero, since either positive or negative values of the sine and cosine terms are equally likely to occur. This is a direct result of the assumption that ϕ_n and ϕ_m are randomly related and that the two wave propagation directions are different. Thus, if we define a *local spatial average* of the modulus squared pressure, where we compute the average value of $|p(\mathbf{x})|^2$ by taking samples of the squared pressure field over a volume of space with dimensions much larger than an acoustic wavelength, then all the terms for which $n \neq m$ will average to zero and only terms for which $n = m$ contribute. The space-averaged modulus squared pressure thus becomes

$$\langle |p(\mathbf{x})|^2 \rangle = \lim_{N \to \infty} \frac{1}{N} \sum_{n=1}^{N} |p_n|^2, \tag{1.15.5}$$

where we have used $\langle\, \rangle$ to denote the operation of space averaging. We can use very similar arguments to this in deducing expressions for the space-averaged energy density and intensity in the sound field. First, we can use the expression derived in

Section 1.8 for the energy density of a three-dimensional sound field provided we interpret $u(t)$ as the instantaneous modulus of the particle velocity vector at a position in the sound field. For harmonic excitation we can use an argument identical to that presented in Section 1.8 in order to compute an expression for the time-averaged energy density. This is

$$e(\mathbf{x}) = \frac{1}{4}\rho_0\left[|u(\mathbf{x})|^2 + \frac{|p(\mathbf{x})|^2}{(\rho_0 c_0)^2}\right], \tag{1.15.6}$$

where $p(\mathbf{x})$ and $u(\mathbf{x})$ are the complex pressure and complex particle velocity modulus, respectively. If we now compute a local spatial average of this time-averaged energy density, we will again find that the contribution of terms arising from dissimilar propagation directions will average to zero. It follows that

$$<e(\mathbf{x})> = \lim_{N\to\infty}\frac{1}{N}\sum_{n=1}^{N}\frac{1}{4}\rho_0\left[|u_n|^2 + \frac{|p_n|^2}{(\rho_0 c_0)^2}\right]. \tag{1.15.7}$$

Since we know that for the nth plane wave, $|u_n| = |p_n|/\rho_0 c_0$, this shows that

$$<e(\mathbf{x})> = \lim_{N\to\infty}\frac{1}{N}\sum_{n=1}^{N}\frac{|p_n|^2}{2\rho_0 c_0^2} = \frac{<|p(\mathbf{x})|^2>}{2\rho_0 c_0^2}. \tag{1.15.8}$$

We now make a further assumption regarding the form of the model sound field. Equation (1.15.8) shows that the local space-averaged, time-average energy density can be considered to be the sum of contributions from an infinite number of plane waves. We assume that all these directions of propagation contribute *equally* to the space-averaged energy density. A sound field model comprised of plane waves of randomly related phases and coming from all possible directions which also satisfies this criterion is said to be *diffuse*. (This is the definition adopted, for example, by Pierce, 1981, see Ch. 6.) Thus, if we consider the contribution to the summation in equation (1.15.8) from a number of plane waves propagating in a range of directions (again by specifying this range of directions by dividing up the surface area of a sphere into segments of equal area) then each contribution associated with a given segment will remain the same as the areas of the segments are allowed to become infinitesimal.

1.16 The energy balance equation for an enclosure

We can use the diffuse field assumption to derive an expression for the local space average of the time-averaged intensity for the diffuse sound field model. By this we mean the intensity produced by the waves whose directions are specified by a *hemisphere* surrounding a point in space, i.e. the intensity due to the waves falling on *one side* of a hypothetical surface in the sound field. The time-averaged intensity in a direction normal to this surface can be written as

$$I(\mathbf{x}) = \frac{1}{2}\mathrm{Re}[p^*(\mathbf{x})u(\mathbf{x})], \tag{1.16.1}$$

where $u(\mathbf{x})$ is the component of the complex particle velocity in the direction normal to the surface. This velocity component can be written as

$$u(\mathbf{x}) = \lim_{N \to \infty} \frac{1}{\sqrt{N/2}} \sum_{n=1}^{N/2} \frac{|p_n|}{\rho_0 c_0} \cos \theta_n, \tag{1.16.2}$$

where θ_n defines the angle of propagation direction of the nth plane wave is the co-ordinate system specified in Fig. 1.11. If we now evaluate the local space average value of $I(\mathbf{x})$, again assuming that the terms associated with dissimilar propagation directions average to zero, the expression that results is

$$<I(\mathbf{x})> = \lim_{N \to \infty} \frac{1}{(N/2)} \sum_{n=1}^{N/2} \frac{|p_n|^2}{2\rho_0 c_0} \cos \theta_n. \tag{1.16.3}$$

We now have to make the important assumption that the contributions to the summation from all the values of $|p_n|^2$ are equal for all the directions of propagation considered. A detailed and rigorous discussion of this point is presented by Pierce (1981, Ch. 6, pp. 255–258). This amounts to assuming that, as discussed at the end of Section 1.15, the contributions to the average energy density are the same for all directions of propagation. Under these conditions, equation (1.16.3) reduces to

$$<I(\mathbf{x})> = \frac{<|p_i(\mathbf{x})|^2>}{2\rho_0 c_0} \lim_{N \to \infty} \frac{1}{(N/2)} \sum_{n=1}^{N/2} \cos \theta_n, \tag{1.16.4}$$

where we have used $p_i(\mathbf{x})$ to denote the net complex pressure due only to waves incident on one side of the hypothetical surface in the field. Again, since we assume

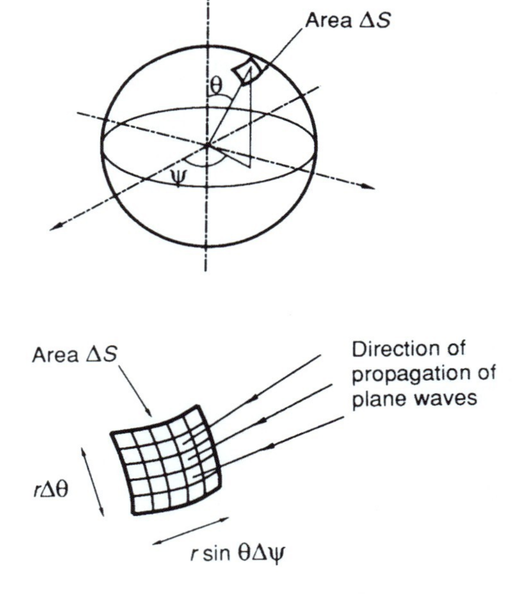

Fig. 1.11 The plane wave construction of a diffuse sound field model. The surface of the sphere of radius r surrounding a point is divided into N tiny segments of equal area. Each area element has an associated plane wave propagating in a direction defined by the normal to the surface area element. Thus the segment of area ΔS contains $(N/4\pi r^2)$ tiny elements.

that all the contributions to the total space-averaged mean square pressure are equal for all propagation directions, it follows that $<|p(x)|^2> = 2<|p_i(\mathbf{x})|^2>$ and therefore that

$$<I(\mathbf{x})> = \frac{<|p(\mathbf{x})|^2>}{4\rho_0 c_0} \lim_{N\to\infty} \frac{1}{(N/2)} \sum_{n=1}^{N/2} \cos\theta_n. \tag{1.16.5}$$

We can evaluate the summation over $\cos\theta_n$ by reducing the summation to an integral. Figure 1.11 shows an element of surface area of a sphere surrounding a point in space. The elemental area is defined by the angles $\Delta\theta$ and $\Delta\psi$. Since the number of plane waves passing through a unit area of the sphere is given by $N/4\pi r^2$, where r is the radius of the sphere, the number of plane waves passing through the element of surface area ΔS is given by

$$\frac{N}{4\pi r^2}\Delta S = \left(\frac{N}{4\pi r^2}\right) r^2 \sin\theta\Delta\theta\Delta\psi. \tag{1.16.6}$$

We therefore write the summation over all directions of propagation as a summation over all the elements of area. Thus

$$\lim_{N\to\infty} \frac{1}{(N/2)} \sum_{n=1}^{N/2} \cos\theta_n =$$

$$\lim_{\Delta\theta,\Delta\psi\to 0} \frac{1}{(N/2)} \sum_{\Delta\psi}\sum_{\Delta\theta} \frac{N}{4\pi} \cos\theta \sin\theta\Delta\theta\Delta\psi. \tag{1.16.7}$$

The summations over $\Delta\psi$ and $\Delta\theta$ can thus be reduced to integrals and

$$\lim_{N\to\infty} \frac{1}{(N/2)} \sum_{n=1}^{N/2} \cos\theta_n = \frac{1}{2\pi}\int_0^{2\pi}\int_0^{\pi/2} \cos\theta \sin\theta \mathrm{d}\theta \mathrm{d}\psi. \tag{1.16.8}$$

Performing the integration shows that the result is simply (1/2) and therefore the expression for the "one-sided" space-averaged, time-average intensity is given by

$$<I(\mathbf{x})> = \frac{<|p(\mathbf{x})|^2>}{8\rho_0 c_0} = \frac{c_0}{4}<e(\mathbf{x})>, \tag{1.16.9}$$

where we have used equation (1.15.8) in order to relate the intensity to the space-averaged, time-averaged energy density.

We can use this result to write an energy balance equation for the whole enclosure (of volume V say). Since energy must be conserved the difference in the rate of energy input (power input) to the enclosure and the rate of energy lost by the absorption of sound at the enclosure walls must be balanced by the rate of change of total room energy. The rate of energy lost as a result of the incidence of the diffuse field on a given area S_M of the wall of the enclosure can be written as $<I(\mathbf{x})>S_M\alpha_M$. The term α_M is an *absorption coefficient* which defines the ratio of the intensity absorbed by the surface to the intensity incident on the surface. Note that here we are assuming that the intensity incident on an enclosure surface is equal to the "one-sided" intensity in the diffuse field model. The total power absorbed by all the surfaces of the enclosure can thus be written as

$$<I(\mathbf{x})>[S_1\alpha_1 + S_2\alpha_2 \ldots S_M\alpha_M], \tag{1.16.10}$$

where we add all the powers absorbed by the M different surfaces comprising the walls of the enclosure. We often define an *average absorption coefficient* $\bar{\alpha}$, given by

$$\bar{\alpha} = \frac{S_1\alpha_1 + S_2\alpha_2 \; . \; . \; . \; S_M\alpha_M}{S}, \tag{1.16.11}$$

where S is the total surface area of the enclosure walls. The energy balance equation for the enclosure can then be written as

$$V\frac{\mathrm{d}}{\mathrm{d}t}<e(\mathbf{x})> = W - <I(\mathbf{x})>S\bar{\alpha}, \tag{1.16.12}$$

where W is the power input to the diffuse field and V is the volume of the enclosure. Under steady state conditions, when the power input to the diffuse field is balanced by the power lost, this leads directly to a relationship between the space-averaged squared pressure and power input to the diffuse field. This is

$$<|p(\mathbf{x})|^2> = \frac{8\rho_0 c_0}{S\bar{\alpha}}W. \tag{1.16.13}$$

The transient solution to the energy balance equation is also of significance. If the power input to the enclosure is suddenly switched off, use of equation (1.16.9) and the energy balance equation (1.16.12) shows that the space-averaged squared pressure will satisfy

$$\frac{4V}{S\bar{\alpha}c_0}\frac{\mathrm{d}}{\mathrm{d}t}<|p(\mathbf{x})|^2> = -<|p(\mathbf{x})|^2>. \tag{1.16.14}$$

The solution of this equation is an exponential decay of the form

$$<|p(\mathbf{x})|^2> = <|p(\mathbf{x})|^2>_0 \, \mathrm{e}^{-(S\bar{\alpha}c_0 t/4V)}, \tag{1.16.15}$$

where $<|p(\mathbf{x})|^2>_0$ is the initial value of the space-averaged squared pressure. If we evaluate the time taken for the squared pressure to decay to 10^{-6} times its original value (i.e. the sound pressure level decays by 60 dB) then it follows from equation (1.16.5) that this time is given by

$$T_{60} = \frac{0.161V}{S\bar{\alpha}}, \tag{1.16.16}$$

where the constant $0.161 = 4\,ln\,10/c_0$ applies for air at 20° C and c_0 in ms^{-1}. This is known as the *reverberation time* of the enclosure and equation (1.16.16) is known as the *Sabine equation* after W.C. Sabine, who first established this relationship empirically (see Sabine, 1898). This equation is useful in characterising the acoustical properties of enclosed sound fields. In fact direct measurements of the reverberation time of an enclosure can be made in order to quantify its total room absorption $S\bar{\alpha}$. This is one technique for measuring the effective (or equivalent) diffuse field absorption coefficient. Thus the reverberation time of an empty reverberant room is firstly measured and its absorption calculated. A sample of material (typically a few square metres in area) is then laid in the room and the reverberation time measured again. The difference in the calculated room absorption then enables the calculation to be made of the diffuse field absorption coefficient of the material.

2
Frequency Analysis

2.1 Introduction

A wide range of different types of dependence on time of the signals observed occur in acoustical problems. These can range from the perfectly predictable continuous sinusoidal pressure fluctuation to a random fluctuation with time whose value at any instant is in practice unpredictable from a knowledge of its previous values. These two types of "signal" represent two broad classes of fluctuation which are generally termed either "deterministic" or "random". Deterministic signals are characterised by their perfectly predictable behaviour from one time instant to the next. Representative of such signals are, for example, the periodic vibration generated by the out-of-balance forces of a constant speed rotating machine or the transient pressure fluctuation produced by the passage of the shock wave associated with a sonic boom. Random fluctuations on the other hand are those whose values cannot be predicted perfectly from the previous time history. A typical example is the noise generated by turbulent flow travelling down a pipe. Both deterministic and random signals will be considered here. It will become evident that the specification of the frequency content of periodic and transient signals is relatively straightforward but that random signals are conceptually a little more difficult to deal with.

In this latter context it may be helpful to note in passing that the "deterministic"/"random" distinction stems from fundamental assumptions in the mathematical modelling of the physical process concerned, and *not* from the observed time histories of any particular signals. In deterministic models it is a basic assumption that *all* signals are perfectly predictable in the mathematical time continuum. In random models it is a basic assumption that *no* signals are perfectly predictable in the mathematical time continuum. One can have, with equal logical validity, both a deterministic model and a random model for any physical event as qualitatively and quantitatively observed (i.e. measured) over any finite interval of time. As implied by the examples cited, for physical events which are apparently internally simple, deterministic models are usually more convenient in practice, and for physical events which are apparently too complicated internally to worry about, random models are usually more convenient in practice.

2.2 Fourier series

First we will consider the frequency content of steady periodic functions of the type described by

$$f(t) = f(t+T). \tag{2.2.1}$$

Here we have used the symbol f to refer to any physical quantity which fluctuates with time. This could be for example acoustic pressure, or density, or the displacement, velocity or acceleration of a vibrating surface. The type of time dependence described by equation (2.2.1) is sketched in Fig. 2.1. The period of the fluctuation is given by T. This period corresponds to a fundamental frequency in the fluctuation which is given by $\omega_0 = 2\pi/T$. This general periodic fluctuation can be considered to consist of a superposition of an infinite number of single frequency fluctuations, such that we can express the net dependence on time as

$$f(t) = a_0 + \sum_{n=1}^{\infty} (b_n \cos n\omega_0 t + c_n \sin n\omega_0 t). \tag{2.2.2}$$

This is the summation of a number of different frequency components all harmonically related to the *fundamental* frequency ω_0. Thus the *second harmonic* component of this series is defined by $n = 2$ and the *third harmonic* by $n = 3$ and so forth, where n takes integer values only. This is the familiar Fourier series representation of a periodic function of time (see, for example, Lynn, 1982, Ch. 2 for an introductory discussion). Note that a_0, which is the value in the series corresponding to $n = 0$, is in fact the "mean value" of the signal defined by the real number

$$a_0 = \frac{1}{T}\int_{-T/2}^{T/2} f(t)\,\mathrm{d}t. \tag{2.2.3}$$

Similarly, it is well known that the Fourier coefficients b_n and c_n are real numbers defined by

$$b_n = \frac{2}{T}\int_{-T/2}^{T/2} f(t)\cos n\omega_0 t\,\mathrm{d}t, \qquad c_n = \frac{2}{T}\int_{-T/2}^{T/2} f(t)\sin n\omega_0 t\,\mathrm{d}t. \tag{2.2.4a,b}$$

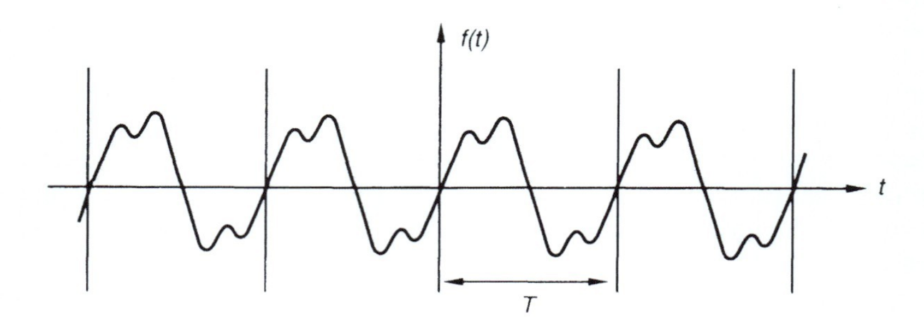

Fig. 2.1 A periodic function of time repeats exactly after every period T.

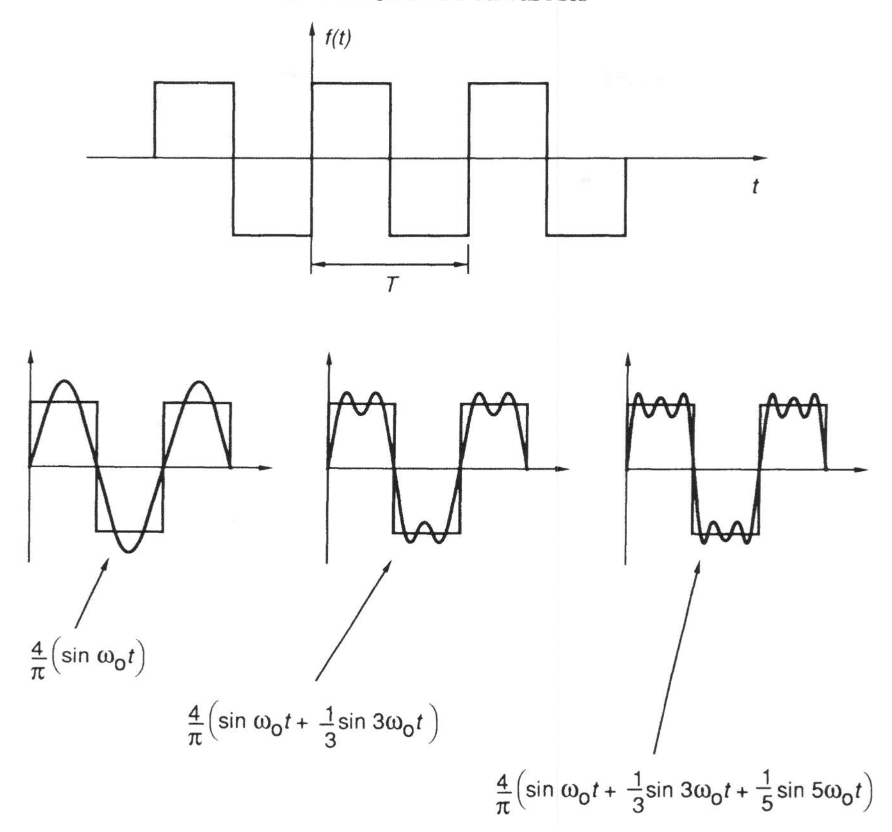

Fig. 2.2 A square wave fluctuation of unit amplitude and its approximation by successive partial sums of its Fourier series representation.

As an illustration of the manner in which a periodic function can be represented as a series of harmonically related frequencies consider the square wave fluctuation of unit amplitude depicted in Fig. 2.2. If we evaluate the coefficients a_0, b_n and c_n for this particular type of function, then a_0 and all values of b_n are zero and the series representation given by equation (2.2.2) reduces to

$$f(t) = \frac{4}{\pi}(\sin \omega_0 t + \frac{1}{3} \sin 3\omega_0 t + \frac{1}{5} \sin 5\omega_0 t \ldots). \quad (2.2.5)$$

Thus the square wave consists of a series of sinusoidal components at the odd harmonics of the fundamental frequency. Each of these components has a progressively decreasing amplitude as the frequency increases. The successive approximations to this square wave by the series of partial sums associated with these harmonics are illustrated in Fig. 2.2.

An entirely equivalent representation of equation (2.2.2) is given by using a series of complex numbers a_n such that

$$f(t) = \sum_{n=-\infty}^{\infty} a_n e^{jn\omega_0 t}, \quad (2.2.6)$$

where $a_n = (1/2)(b_n - jc_n)$ for positive values of n and $a_{-n} = (1/2)(b_n + jc_n)$ for negative values of n. For $n = 0$, a_0 remains the real number defined by equation (2.2.3). This is the complex form of the Fourier series where the integer n now takes values from minus infinity to plus infinity. The equivalence of equations (2.2.2) and (2.2.6) is readily proved by using the identities $\cos\theta = (1/2)(e^{j\theta} + e^{-j\theta})$ and $\sin\theta = (1/2j)(e^{j\theta} - e^{-j\theta})$. The net fluctuation can be considered to consist of a series of fluctuations at both "negative" and positive frequencies. The coefficients a_n are complex but each coefficient corresponding to a negative frequency $-n\omega_0$ is the complex conjugate of that corresponding to the positive frequency $n\omega_0$. The addition of these two components produces a single real fluctuation at the frequency $n\omega_0$. This can be explained by using the rotating vector representation illustrated in Fig. 2.3. Thus the component in equation (2.2.6) corresponding to $a_n e^{jn\omega_0 t}$ rotates in one direction, whilst the complex conjugate component $a_{-n}e^{-jn\omega_0 t}$ is represented by a vector rotating in the opposite direction. The imaginary parts of these two components thus add to zero always producing a real fluctuation. The notion of a "negative frequency" can thus be explained in terms of the vector which rotates in a direction opposite to that of a "positive frequency". Note that this type of representation is an alternative and entirely equivalent to the complex exponential representation of a harmonic fluctuation that was introduced in the last chapter. In Chapter 1 we used, for example, $\mathrm{Re}\{A e^{j\omega t}\}$ to represent harmonic motion, where A was a complex number. Thus in this case we could write at a given harmonic frequency $n\omega_0$

$$f_n(t) = \mathrm{Re}\{A e^{jn\omega_0 t}\} = (1/2)A_n e^{jn\omega_0 t} + (1/2)A_n^* e^{-jn\omega_0 t}, \qquad (2.2.7)$$

where we have used the fact that the real part of a complex number is given by half the sum of the number and its complex conjugate. The series representation given by equation (2.2.6) specifies that for a given harmonic frequency $n\omega_0$

$$f_n(t) = a_n e^{jn\omega_0 t} + a_{-n} e^{-jn\omega_0 t}. \qquad (2.2.8)$$

Thus $a_n = (1/2)A_n$ and $a_{-n} = (1/2)A_n^*$ and the complex coefficients are simply related to the complex number A_n. The introduction of the concept of negative frequency thus enables a real fluctuation to be expressed in terms of only complex numbers.

The relationship between the complex coefficients a_n and the fluctuation $f(t)$ is readily derived. First we multipy each side of equation (2.2.6) by the factor $e^{-jm\omega_0 t}$ where m is also an integer. This gives

$$f(t)e^{-jm\omega_0 t} = \sum_{n=-\infty}^{\infty} a_n e^{j(n-m)\omega_0 t}. \qquad (2.2.9)$$

If we now integrate both sides of this equation over one period T of the fluctuation this yields

$$\int_{-T/2}^{T/2} f(t)e^{-jm\omega_0 t}\mathrm{d}t = \sum_{n=-\infty}^{\infty} a_n \int_{-T/2}^{T/2} e^{j(n-m)\omega_0 t}\mathrm{d}t. \qquad (2.2.10)$$

Now note that $e^{j(n-m)\omega_0 t} = \cos(n-m)\omega_0 t + j\sin(n-m)\omega_0 t$. Therefore the integral over the period T on the right hand side of equation (2.2.10) will yield a non-zero

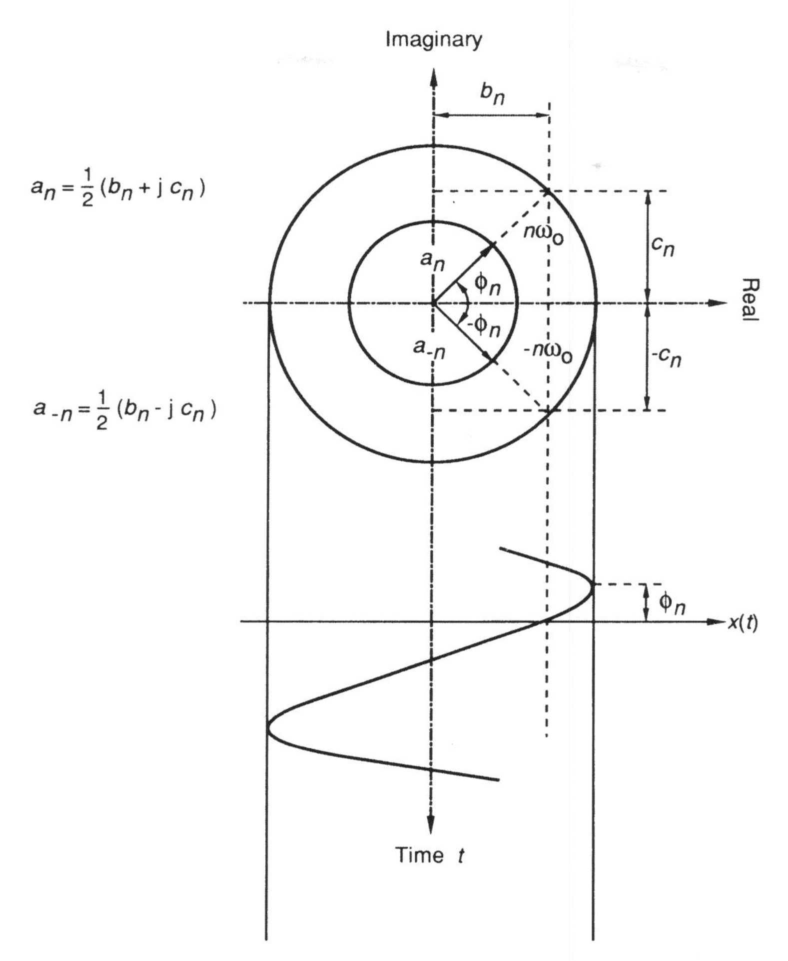

Fig. 2.3 The rotating vector representation of a component of the Fourier series at a frequency $n\omega_0$. The signal consists of the vector sum of the two components rotating in opposite directions such that their imaginary parts always add to give zero. The vector rotating clockwise can be interpreted as providing the contribution at the negative frequency $n\omega_0$.

value only under the condition that $n = m$. In this the case the value of the integral is simply T and we can write equation (2.2.10) as

$$\int_{-T/2}^{T/2} f(t)\mathrm{e}^{-jn\omega_0 t}\mathrm{d}t = a_n T. \tag{2.2.11}$$

Therefore the complex coefficients in the Fourier series are given by

$$a_n = \frac{1}{T}\int_{-T/2}^{T/2} f(t)\mathrm{e}^{-jn\omega_0 t}\mathrm{d}t. \tag{2.2.12}$$

Note that the values of the coefficients b_n and c_n defined by equation (2.2.4) also follow from this expression. It is also useful to evaluate the mean square value of the fluctuation $f(t)$ in terms of the coefficients a_n. Note that the mean square value can be written as

$$\overline{f^2(t)} = \frac{1}{T}\int_{-T/2}^{T/2} f^2(t)\mathrm{d}t = \frac{1}{T}\int_{-T/2}^{T/2} f(t)f^*(t)\mathrm{d}t, \tag{2.2.13}$$

where we are quite entitled to write $f(t)$ as $f^*(t)$ since f is a real number. If we now substitute the Fourier series representations given by equation (2.2.6) into this expression then we get

$$\overline{f^2(t)} = \frac{1}{T}\int_{-T/2}^{T/2}\left[\sum_{n=-\infty}^{\infty} a_n \mathrm{e}^{jn\omega_0 t}\right]\left[\sum_{m=-\infty}^{\infty} a_m^* \mathrm{e}^{-jm\omega_0 t}\right]. \tag{2.2.14}$$

Again we note that the integral over the period T of $\mathrm{e}^{j(n-m)\omega_0 t}$ is zero for $n \neq m$ or T for $n = m$ and therefore the expression for the mean square value of the signal becomes

$$\overline{f^2(t)} = \sum_{n=-\infty}^{\infty} |a_n|^2. \tag{2.2.15}$$

This demonstrates that the mean square value of the signal is given by the sum of the mean squared values of the frequency components defined by a_n. This is a very important principle in frequency analysis, i.e. that the total energy in the signal in the time domain can be derived from the component energies of the signal in the frequency domain. The more general expression of this principle, known as Parseval's theorem, will be dealt with in the next section.

2.3 The Fourier integral

We shall now consider functions of time which are not necessarily periodic but nevertheless are deterministic in the sense that their variation with time is perfectly predictable. We can explain the concept of the frequency content of functions of this type by extending the notion of the Fourier series discussed above. In that case we saw that for a periodic function $f(t)$ the relationship between this fluctuation and its Fourier coefficients a_n is given by

$$f(t) = \sum_{n=-\infty}^{\infty} a_n \mathrm{e}^{jn\omega_0 t}, \qquad a_n = \frac{1}{T}\int_{-T/2}^{T/2} f(t)\mathrm{e}^{-jn\omega_0 t}\mathrm{d}t. \tag{2.3.1a,b}$$

Now let us assume that the transient fluctuation with respect to time depicted in Fig. 2.4 can be considered to be just one cycle of a periodic motion where the period T tends to infinity. In this case the fundamental frequency which is given by $1/T$ can be written as $\Delta\omega/2\pi$ where the frequency $\Delta\omega$ must therefore tend to zero as T tends to infinity. The expressions for the fluctuation with respect to time and the values of the Fourier coefficients can therefore be written in the forms

$$f(t) = \sum_{n=-\infty}^{\infty}\left(\frac{a_n 2\pi}{\Delta\omega}\right)\mathrm{e}^{jn\Delta\omega t}\left(\frac{\Delta\omega}{2\pi}\right), \tag{2.3.2}$$

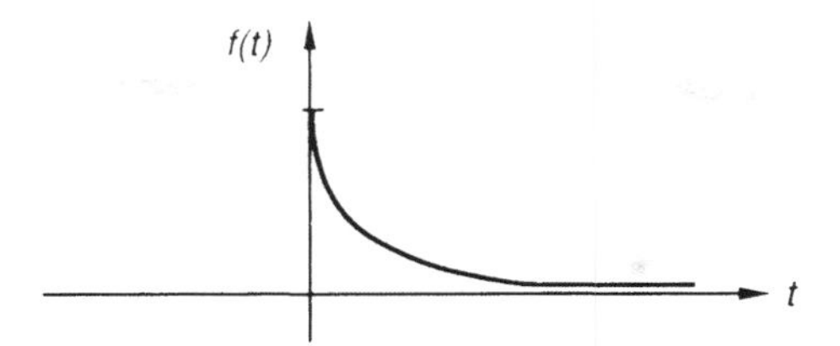

Fig. 2.4 The signal having a time dependence $f(t) = e^{-\alpha t}$, $t \geq 0$ and $f(t) = 0$, $t < 0$.

$$\frac{a_n 2\pi}{\Delta\omega} = \int_{-T/2}^{T/2} f(t) e^{-jn\Delta\omega t} dt. \tag{2.3.3}$$

If we now let $\Delta\omega$ tend to zero and write $n\Delta\omega$ as ω, then the summation in equation (2.3.2) can be written as the integral given by

$$f(t) = \frac{1}{2\pi} \int_{-\infty}^{\infty} F(\omega) e^{j\omega t} d\omega, \tag{2.3.4}$$

where $F(\omega)$ is the limiting value of $[a_n 2\pi/\Delta\omega]$ as $\Delta\omega$ tends to zero. Similarly, as $\Delta\omega$ tends to zero and T tends to infinity, then equation (2.3.3) can be written in the form

$$F(\omega) = \int_{-\infty}^{\infty} f(t) e^{-j\omega t} dt. \tag{2.3.5}$$

The quantity $F(\omega)$ is known as the *Fourier transform* of $f(t)$. Conversely the fluctuation $f(t)$ can be recovered from $F(\omega)$ by using the *inverse Fourier transform* described by equation (2.3.4). $F(\omega)$ is therefore a continuous function of frequency which specifies the frequency content of the function of time $f(t)$. Note that this Fourier transform is a complex number which can be expressed in terms of its real and imaginary parts or its modulus and phase by

$$F(\omega) = F_R + jF_I = |F(\omega)| e^{j\phi(\omega)}. \tag{2.3.6}$$

It is useful as an example to evaluate the Fourier transform of the simple function of time depicted in Fig. 2.4 and specified by

$$f(t) = \begin{Bmatrix} e^{-\alpha t}, & t \geq 0 \\ 0, & t < 0 \end{Bmatrix}. \tag{2.3.7}$$

In this case the Fourier integral can be performed easily and is given by

$$F(\omega) = \int_0^{\infty} e^{-\alpha t} e^{-j\omega t} dt = \frac{1}{\alpha + j\omega}. \tag{2.3.8}$$

Sketches showing the variation of the modulus and phase of this complex function are plotted in Fig. 2.5. Another transient signal which is useful to consider is the rectangular pulse of amplitude A and duration 2τ that is defined by

$$f(t) = \begin{Bmatrix} A, & -\tau \leq t \leq \tau \\ 0, & t < -\tau, \quad t > \tau \end{Bmatrix}. \tag{2.3.9}$$

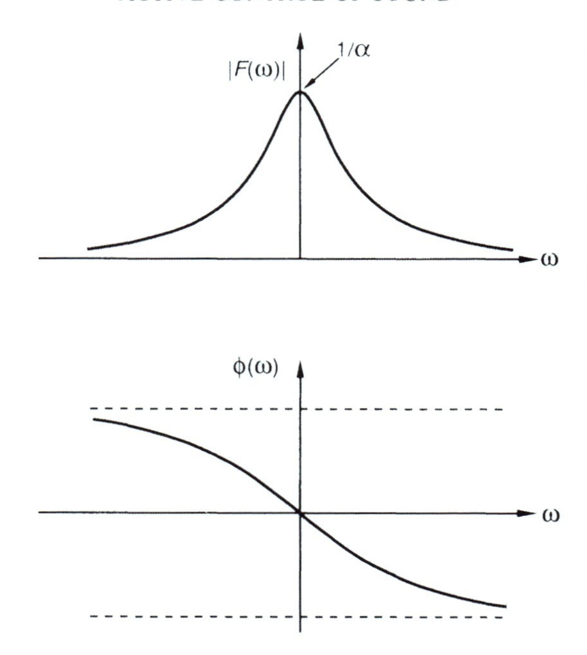

Fig. 2.5 The modulus and phase of the Fourier transform of the signal depicted in Fig. 2.4.

This pulse is also sketched in Fig. 2.6. The Fourier transform of this is again easy to evaluate and is given by

$$F(\omega) = A\int_{-\tau}^{\tau} e^{-j\omega t}dt = \frac{A}{j\omega}(e^{j\omega\tau} - e^{-j\omega\tau}). \tag{2.3.10}$$

If we now use the identity $e^{j\theta} - e^{-j\theta} = 2j \sin\theta$, the Fourier transform can be written as

$$F(\omega) = 2A\tau\left[\frac{\sin\omega\tau}{\omega\tau}\right]. \tag{2.3.11}$$

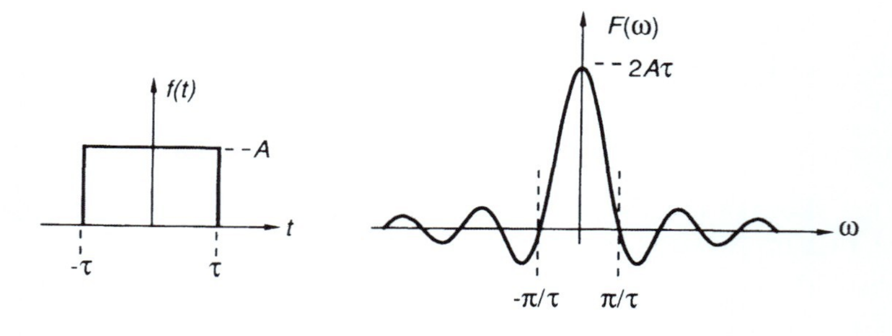

Fig. 2.6 A rectangular pulse and its Fourier transform.

Now consider the behaviour of this frequency spectrum as the pulse considered becomes taller and narrower. If we let the product $2A\tau$ remain constant whilst we progressively reduce τ, as sketched in Fig. 2.7, then in the limiting case as τ tends to zero we will have a function which is infinite at $t = 0$ and has a value of zero at all other values of time t. If we further assume that the product $A\tau$ is always maintained at unity then we have defined the *Dirac delta function* described by

$$\delta(t) = 0, \quad t \neq 0, \qquad \int_{-\infty}^{\infty} \delta(t)\mathrm{d}t = 1. \tag{2.3.12a,b}$$

If we assumed that the rectangular pulse was centred at a time t_0 then we could equivalently define the function given by

$$\delta(t-t_0) = 0, \quad t \neq t_0, \qquad \int_{-\infty}^{\infty} \delta(t-t_0)\mathrm{d}t = 1, \tag{2.3.13a,b}$$

which corresponds to a Dirac delta function at $t = t_0$. This has the useful "sifting" property given by

$$\int_{-\infty}^{\infty} f(t)\delta(t-t_0)\mathrm{d}t = f(t_0)\int_{-\infty}^{\infty} \delta(t-t_0)\mathrm{d}t = f(t_0), \tag{2.3.14}$$

where use is made of the fact that $\delta(t-t_0)$ has a zero value at all times other than $t = t_0$. We can use this property to evaluate the Fourier transform of the delta function. Thus we have

$$F(\omega) = \int_{-\infty}^{\infty} \delta(t-t_0)\mathrm{e}^{-j\omega t}\mathrm{d}t = \mathrm{e}^{-j\omega t_0}, \tag{2.3.15}$$

and therefore the Fourier transform of the delta function has a unit modulus that stretches from values of ω extending from minus infinity to plus infinity. Similarly we can show through the inverse Fourier transform that

$$\delta(t-t_0) = \frac{1}{2\pi}\int_{-\infty}^{\infty} \mathrm{e}^{j\omega(t-t_0)}\mathrm{d}\omega. \tag{2.3.16}$$

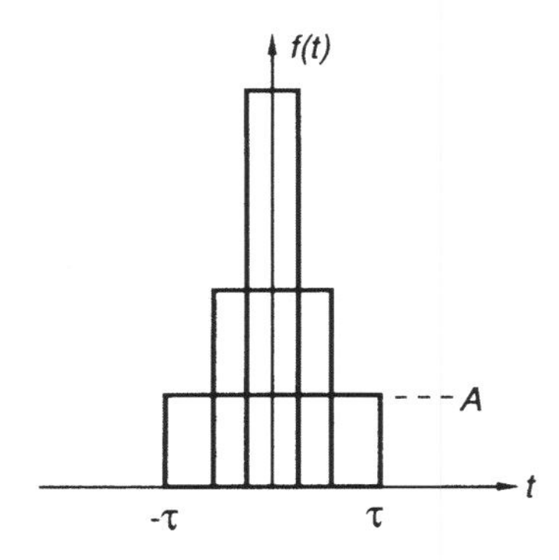

Fig. 2.7 A rectangular pulse whose duration 2τ is made smaller and whose amplitude A is increased such that the product $2A\tau$ remains constant. The limit of $\tau \rightarrow 0$ defines the Dirac delta function.

We can now use this property to express the total energy of the fluctuation $f(t)$ in terms of its Fourier spectrum $F(\omega)$. Thus again noting that we can write $f^2(t) = f(t)f^*(t)$, enables the expression for the total energy in the signal to be written as

$$\int_{-\infty}^{\infty} f^2(t)\mathrm{d}t = \int_{-\infty}^{\infty} \left[\frac{1}{2\pi}\int_{-\infty}^{\infty} F(\omega_1)\mathrm{e}^{j\omega_1 t}\mathrm{d}\omega_1\right]\left[\frac{1}{2\pi}\int_{-\infty}^{\infty} F(\omega_2)\mathrm{e}^{j\omega_2 t}\mathrm{d}\omega_2\right]^*\mathrm{d}t. \quad (2.3.17)$$

This can be rewritten, by changing the orders of integration, as

$$\int_{-\infty}^{\infty} f^2(t)\mathrm{d}t = \frac{1}{4\pi^2}\int_{-\infty}^{\infty}\int_{-\infty}^{\infty} F(\omega_1)F^*(\omega_2)\left[\int_{-\infty}^{\infty} \mathrm{e}^{j(\omega_1-\omega_2)t}\mathrm{d}t\right]\mathrm{d}\omega_1\mathrm{d}\omega_2. \quad (2.3.18)$$

Using equation (2.3.16) enables the integral in the square brackets to be replaced by $2\pi\delta(\omega_1 - \omega_2)$ and then use of the property (2.3.14) of the Dirac delta function reduces this equation to

$$\int_{-\infty}^{\infty} f^2(t)\mathrm{d}t = \frac{1}{2\pi}\int_{-\infty}^{\infty} F(\omega_1)F^*(\omega_1)\mathrm{d}\omega_1. \quad (2.3.19)$$

This can be written in the form

$$\int_{-\infty}^{\infty} f^2(t)\mathrm{d}t = \frac{1}{2\pi}\int_{-\infty}^{\infty} |F(\omega)|^2\mathrm{d}\omega \quad (2.3.20)$$

Thus the total energy in the signal is the integral over all the energies associated with the individual frequency components. This is known as Parseval's formula.

2.4 Random processes

Very often in acoustics we have to deal with fluctuations in time which are not perfectly predictable. For example, Fig. 2.8 illustrates the type of acoustic pressure fluctuation produced in the far field of a turbulent jet. The process generating the noise is associated with the random motions of the eddies in the turbulent flow. The complexity of these motions is such that the acoustic pressure fluctuations produced cannot presently be predicted in any way from the physical process since an adequately detailed knowledge of the structure of the flow at any given instant is not available. Even if such a calculation were possible in principle, the mathematics of prediction would be so complicated that it would be impossible in practice to carry it out. Such random processes thus have to be dealt with by using a statistical approach. One of the underlying concepts that is useful in dealing with random signals of this type is the notion of the ensemble of different fluctuations which *might* occur at any one time. This is represented in Fig. 2.9 where a number of time histories are depicted. These time histories can be considered as a few of the infinite number of *possible* time histories that could be generated by the turbulent jet. In practice of course only one such time history is generated but it has to be supposed that this is only one of the infinite number possible. Then, at a given time t_1 say in Fig. 2.9, we can consider the *probability* of a given value of f occurring. The notion of probability is most easily understood by using the measure known as *relative frequency*. Thus, for example, if we were tossing a coin, the relative frequency with which the value of heads was given is defined by

$$r_f = n_h/N, \tag{2.4.1}$$

where n_h defines the number of times "heads" occurs and N denotes the total number of times that the coin is tossed. Therefore as the total number of "trials" N tends to infinity it is clear that r_f converges to 1/2 as depicted in Fig. 2.10. This therefore defines the probability; it is the limiting case of the relative frequency. It is also useful to define a probability distribution function which measures the probability of f being less than or equal to a particular value f_k. Thus we define

$$P(f) = \text{Prob}[f \leq f_k], \tag{2.4.2}$$

where $\text{Prob}[f \leq f_k]$ denotes the probability of f being less than or equal to the value f_k. An example of a simple probability distribution function is depicted in Fig. 2.11 for the case of tossing a six-faced die. The probability of any one of the values from 1 to 6 occurring is 1/6, therefore the probability of the value being less than 2 is 1/6. Similarly the probability of the value being less than 3 is 2/6 and so on. In the case of the time histories depicted in Fig. 2.9 then f can take on any number of a continuum of numbers and therefore the probability distribution function may take the form depicted in Fig. 2.12 which is a continuous function of f. The probability of x lying in a certain range defined by $\Delta f = f_2 - f_1$ can be written as $\text{Prob}[f_1 < f \leq f_2]$. We can now define the *probability density function* as

$$p(f) = \lim_{\Delta f \to 0} \frac{\text{Prob}[f_1 < f \leq f_2]}{\Delta f}. \tag{2.4.3}$$

It therefore follows that in the limit as Δf tends to zero, the probability density function is defined by

$$p(f) = \frac{\mathrm{d}P(f)}{\mathrm{d}f}. \tag{2.4.4}$$

Notice that we can recover the probability of f lying in the range f_1 to f_2 by evaluating the integral of the probability density function given i.e.

$$\text{Prob}[f_1 < f \leq f_2] = \int_{f_1}^{f_2} p(f)\,\mathrm{d}f = P(f_2) - P(f_1). \tag{2.4.5}$$

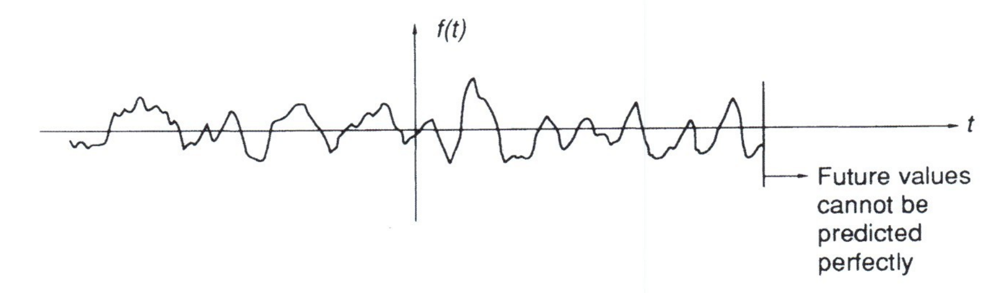

Fig. 2.8 A random fluctuation with time.

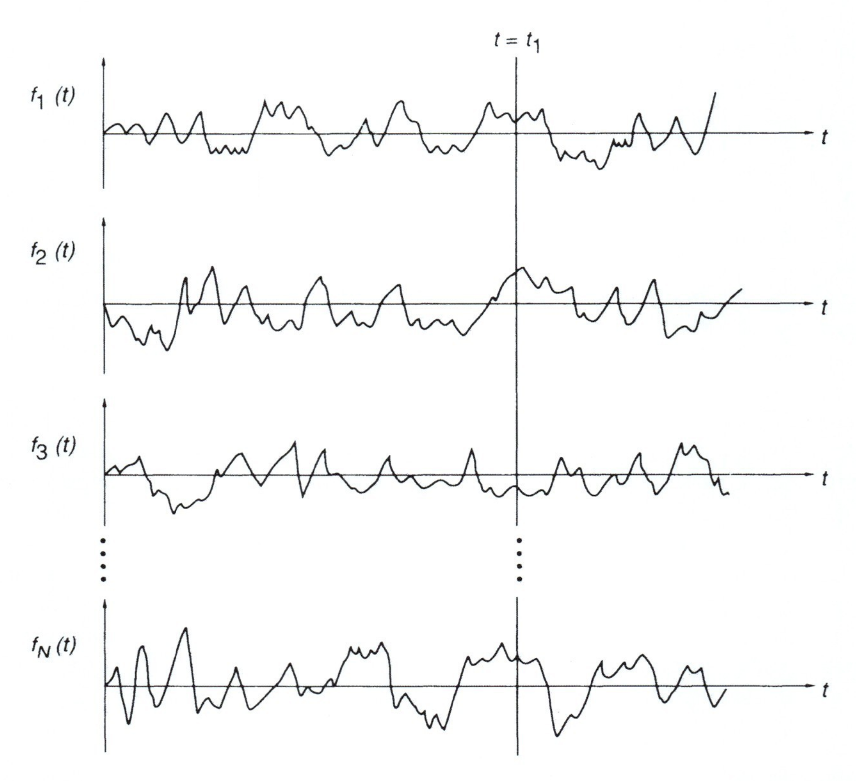

Fig. 2.9 The ensemble of time history records which are associated with a random function of time. The values of $f(t)$ occurring at $t = t_1$ are shown and can be interpreted as contributions to the infinite ensemble of values that $f(t)$ can take at $t = t_1$.

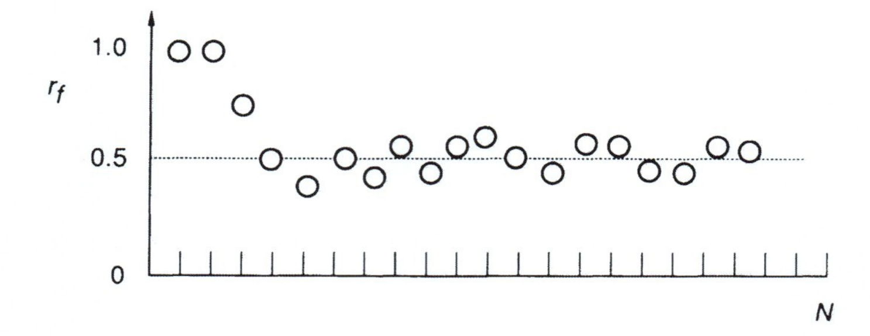

Fig. 2.10 The relative frequency with which "heads" occurs during one experiment in which a coin is tossed a number of times N. As $N \rightarrow \infty$, then $r_f \rightarrow 0.5$ and this defines the probability of "heads" occurring.

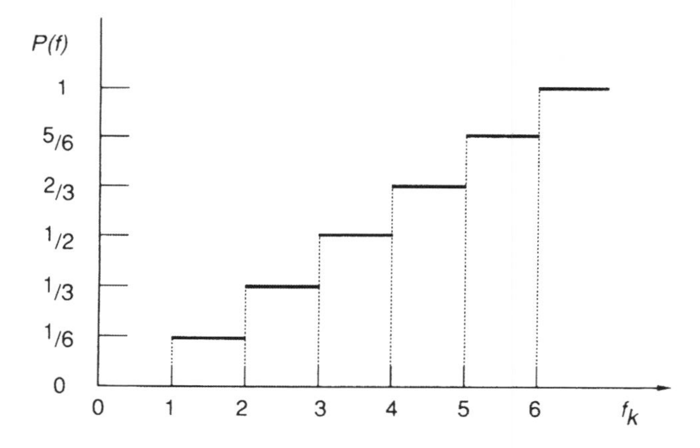

Fig. 2.11 The probability distribution function associated with tossing a six-faced die. $P(f)$ shows the probability of occurrence of a value f which is less than or equal to a specified value f_k.

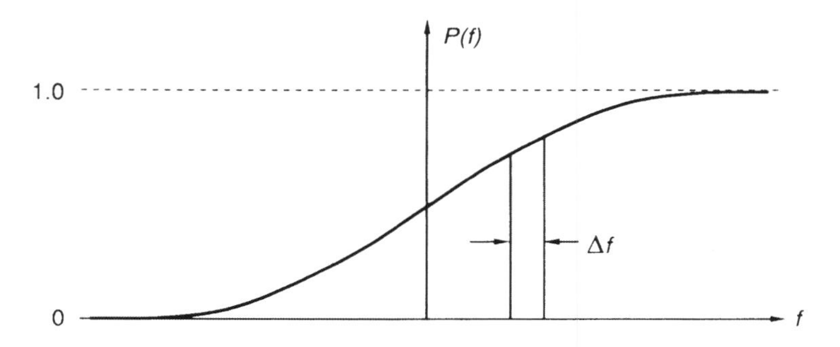

Fig. 2.12 The probability distribution function associated with a continuous range of values of f.

Note that we are also certain that $P(-\infty) = 0$ and $P(\infty) = 1$, and therefore it follows that

$$\int_{-\infty}^{\infty} p(f)\,\mathrm{d}f = 1 \tag{2.4.6}$$

Another important statistical measure is the *mean* or *expected value* of the process. Assume again that f can only take a certain number of discrete values denoted by f_k. The mean of the number of "outcomes" f_k resulting from a series of N trials can be written as

$$\mu_f = \frac{1}{N}\sum_{k=1}^{M} f_k n_k, \tag{2.4.7}$$

where n_k is the number of times that a given f_k occurs and M is the total number of possible values f_k. We can recognise the ratio n_k/N as the relative frequency r_f and as the total number of trials N tends to infinity we can replace r_f by $p(f_k)\Delta f_k$. Therefore we can write the expression for the mean value as

$$\mu_f = \sum_{k=1}^{M} f_k p(f_k)\Delta f_k. \tag{2.4.8}$$

If we now assume that the discrete variable f_k is replaced by the continuous variable f then the summation over the values of f_k in equation (2.4.8) can be replaced by an integral such that the mean or expected value for the process can be denoted by

$$\mu_f = E[f] = \int_{-\infty}^{\infty} fp(f)\,\mathrm{d}f. \tag{2.4.9}$$

Here we have used the notation $E[f]$ to denote the expected value of f. Similarly we can define the expected value of, for example, f^2 by

$$E[f^2] = \int_{-\infty}^{\infty} f^2 p(f)\,\mathrm{d}f. \tag{2.4.10}$$

This defines the *mean square* value of f. We define the *variance* of f by

$$\sigma_f^2 = E[(f-\mu_f)^2] = \int_{-\infty}^{\infty} (f-\mu_f)^2 p(f)\,\mathrm{d}f. \tag{2.4.11}$$

The variance is a measure of the average squared deviation of f from its mean value μ_f. Note that the mean square value and the variance of a signal are equal for a signal that has zero mean.

One particular probability density function of significance is the Gaussian probability density function or the *normal distribution* which is defined entirely by its mean value μ_f and its *standard deviation* σ_f. This is given by

$$p(f) = \frac{1}{\sqrt{2\pi}\sigma_f}\mathrm{e}^{-(f-\mu_f)^2/2\sigma_f^2}. \tag{2.4.12}$$

The form of this probability density function is plotted in Fig. 2.13. This bell-shaped curve of the probability density function is characteristic of many naturally occurring random processes. In fact, the central limit theorem (see Mood *et al.*, 1976, Ch. 6) roughly speaking, states that any random process which results from the addition of a number of independent random processes (each having different probability density functions), will tend to have a Gaussian probability density function.

Note that all these properties of the random variable f have been deduced by assuming that f can take on any number of values at a given time t and therefore in order to evaluate these statistical properties we would strictly speaking have to somehow gain access to and record the infinite number of possible time histories which could occur. This would enable the evaluation of the extent to which the statistical measures such as mean and variance themselves varied with time t. Processes whose statistical properties do change with time are called *non-stationary*. Thus, for example, a non-stationary process could result in a time history whose mean and variance changed as a function of time in the manner sketched in Fig. 2.14. Conversely a random process is called *stationary* (in the strict sense) if all of its

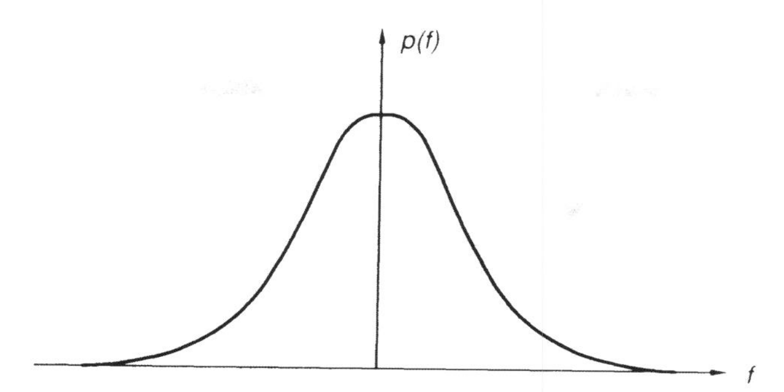

Fig. 2.13 A probability density function associated with a continuous range of values of f. This is related to the probability distribution function by $p(f) = \mathrm{d}P(f)/\mathrm{d}f$.

statistical properties are invariant to a shift of the time origin. This is a very strict restriction on the stationarity of the signal and very often in practice we deal with fluctuations with time which are only *weakly stationary*. Such processes are those in which only some of the statistical properties (such as mean and variance) are invariant with time (see, for example, the discussion presented by Newland, 1984, Ch. 2).

Another property of fluctuations with time which is very important from the practical point of view is that of *ergodicity*. Clearly when undertaking real experiments, then one does not have access to the infinite number of simultaneous time histories that "might have occurred". More usually we have only a single fluctuation with respect to time. A process is called ergodic if the average values associated with a single sample time history are the same as those found from averaging over the ensemble of time histories. Thus, for example, we can compute the mean or expected value of $f(t)$ from the relationship

$$\mu_f = E[f(t)] = \lim_{T \to \infty} \frac{1}{T} \int_{-T/2}^{T/2} f(t)\,\mathrm{d}t. \tag{2.4.13}$$

Similarly we can compute the mean square value of $f(t)$ from the expression

$$\overline{f^2(t)} = E[f^2(t)] = \lim_{T \to \infty} \frac{1}{T} \int_{-T/2}^{T/2} f^2(t)\,\mathrm{d}t. \tag{2.4.14}$$

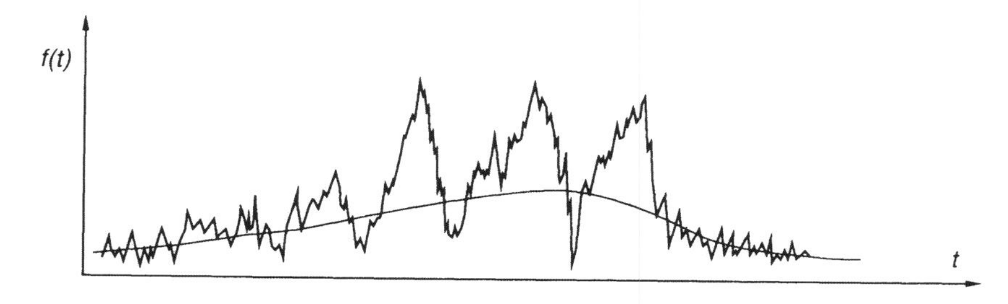

Fig. 2.14 A non-stationary random signal whose mean value and mean square value are both time dependent.

Thus for a stationary ergodic process we can evaluate these properties of the signal simply by undertaking the time average operation.

2.5 The correlation coefficient and the method of least squares

The notion of the degree of correlation between two random variables is of great importance in the analysis of random acoustic pressure fluctuations. The idea of correlation can be best explained by assuming that we have two simultaneously occurring random functions of time defined by $f(t)$ and $g(t)$. Again, the statistical properties of these functions at a given time t_1 are characterised by the total ensemble of functions of $f(t)$ and $g(t)$ that "could have occurred". Let us assume for the moment that both these ensemble averages have zero mean values, that is $\mu_f = E[f] = 0$ and $\mu_g = E[g] = 0$. Also let us assume that we can plot graphs of g against f for a particular time t where a number of values of f and g are taken from the ensemble of total values. It could be that the resulting graph (or "scatter diagram") is of the form depicted in Fig. 2.15(a). In this instance there would appear to be some relationship between f and g. On the other hand, it could be that the graph has the form depicted in Fig. 2.15(b) where it appears that there is very little relationship between f and g. In the first case then g and f would appear as a first approximation, to be linearly related. The correlation is a direct measure of exactly how closely these two functions are linearly related. Thus for example we could perform a least squares fit to this relationship by assuming a linear relationship between g and f of the form

$$g = Kf, \tag{2.5.1}$$

where K is some constant to be determined. If we denote the error, or deviation of a

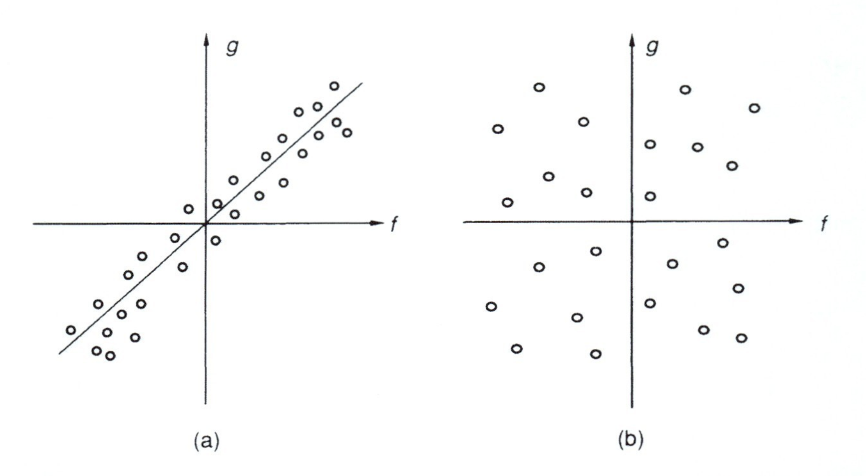

Fig. 2.15 Values of the two variables f and g are plotted against each other for cases when the variables are (a) well correlated and (b) uncorrelated.

given value of g from this assumed linear relationship by δ, then this must be given by

$$\delta = g - Kf. \tag{2.5.2}$$

The average value of the square of this deviation from the linear relationship can be written as the function

$$J = E[\delta^2] = E[(g - Kf)^2], \tag{2.5.3}$$

which can be expanded to give

$$J = E[g^2] + K^2 E[f^2] - 2KE[fg]. \tag{2.5.4}$$

The best fit to the linear relationship is given by a slope of line K which minimises this sum of squared deviations. This function has a *quadratic* dependence on K and can be written as

$$J = AK^2 + 2bK + c, \tag{2.5.5}$$

where $A = E[f^2]$, $b = -E[fg]$ and $c = E[g^2]$ where all three terms are real numbers. Thus, if we plot a graph of J against K, the curve will be parabolic in shape. The minimum of this function can be found by equating to zero the derivative of the function with respect to K. Thus

$$\frac{\mathrm{d}J}{\mathrm{d}K} = 2AK + 2b = 0, \tag{2.5.6}$$

and therefore the value of K that minimises J is given by $K_0 = -b/A$. That this stationary point is a minimum (and not a maximum or point of inflection) of the function can be checked by ensuring that $\mathrm{d}^2J/\mathrm{d}K^2$ is positive. Thus we require A to be positive if the minimum is to be defined. We are sure that a minimum exists because $A = E[f^2]$ must be positive, being proportional to the square of the values of f. The value of J at the minimum defined by $K_0 = -b/A$ is given by substitution of this value back into equation (2.5.5). This gives the minimum value $J_0 = c - b^2/A$. Thus the value of the slope K_0 that minimises the sum of the deviations and the corresponding minimum value J_0 are given respectively by

$$K_0 = -b/A = E[fg]/E[f^2], \tag{2.5.7}$$

$$J_0 = c - b^2/A = E[g^2] - \frac{(E[fg])^2}{E[f^2]}. \tag{2.5.8}$$

Writing $E[f^2] = \sigma_f^2$ and $E[g^2] = \sigma_g^2$ (for these zero mean processes) enables us to non-dimensionalise this minimum value by σ_g^2 such that we have

$$\frac{J_0}{\sigma_g^2} = 1 - \left[\frac{E[fg]}{\sigma_f \sigma_g}\right]^2. \tag{2.5.9}$$

The quantity $E[fg]/\sigma_f\sigma_g = \rho_{fg}$ is known as the *normalised correlation coefficient* between f and g. Thus if f and g are perfectly correlated then $\rho_{fg} = 1$ and the minimum value of the sum of the squared deviations from the straight line is zero. This of course corresponds to the case when f and g fall perfectly on the straight line relationship $g = Kf$. Conversely if there is no correlation between g and f then $\rho_{fg} = 0$ and the scatter in the variance $E[\delta^2]$ is simply equal to the variance σ_g^2. Naturally in many cases the correlation coefficient ρ_{fg} will fall somewhere between these two extremes. The plot shown in Fig. 2.15(a) corresponds to a correlation coefficient which is relatively high but is clearly not unity since the values do not fall perfectly on to a straight line.

2.6 Correlation functions for time histories

In the case considered above we assumed that the values of f and g were chosen from the ensembles making up the possible occurrences of f and g at a given time t_1 say. It is also possible to define a correlation coefficient between the two variables f and g when f and g are chosen from the ensembles representing the two processes at two different times t_1 and t_2. In this case we can write the normalised correlation coefficient as

$$\rho_{fg}(t_1, t_2) = \frac{E[f(t_1)g(t_2)]}{\sigma_f \sigma_g}. \tag{2.6.1}$$

The relationship between f and g at two different times is very often expressed as the *cross-correlation function* which is simply given by the numerator of equation (2.6.1) and written as

$$R_{fg}(t_1, t_2) = E[f(t_1)g(t_2)]. \tag{2.6.2}$$

If the processes f and g involved are time stationary, then the value of this cross-correlation function will depend only on the separation between the times t_1 and t_2 and not on their absolute values. Thus in such cases we can write $t_2 = t_1 + \tau$, where τ is simply the time lag or separation between the two sample times, and the expression for the cross-correlation function can be written as

$$R_{fg}(\tau) = E[f(t)g(t+\tau)]. \tag{2.6.3}$$

It is also useful to note that it is not only possible to evaluate the correlation of $f(t)$ with that of $g(t+\tau)$ but also the value of $f(t)$ with itself at a time τ later, i.e. we can conceive of an *autocorrelation function* defined by

$$R_{ff}(\tau) = E[f(t)f(t+\tau)]. \tag{2.6.4}$$

This is a very important function in the analysis of time series, since as we shall see later it enables us to quantify the frequency content of random signals. One important property of the autocorrelation function that we should note is that its value at the lag $\tau = 0$ is given by the mean square value of the signal for stationary processes. Thus we have

$$R_{ff}(0) = E[f^2(t)]. \tag{2.6.5}$$

This in turn is equal to the variance of a zero-mean process. Another useful property of the autocorrelation function is that it is symmetric with respect to the value of the time lag τ. This can be demonstrated by writing, again for a stationary process,

$$R_{ff}(\tau) = E[f(t)f(t+\tau)] = E[f(t-\tau)f(t)] = R_{ff}(-\tau). \qquad (2.6.6)$$

Similarly we can deduce some useful properties of the cross-correlation function. First note that

$$R_{fg}(\tau) = E[f(t)g(t+\tau)] = E[f(t-\tau)g(t)], \qquad (2.6.7)$$

and similarly for the function $R_{gf}(\tau)$ we have

$$R_{gf}(t) = E[g(t)f(t+\tau)] = E[g(t-\tau)f(t)]. \qquad (2.6.8)$$

These relationships show that

$$R_{fg}(\tau) = R_{gf}(-\tau), \qquad R_{gf}(\tau) = R_{fg}(-\tau). \qquad (2.6.9a,b)$$

Note that for a stationary ergodic process we can again evaluate the cross-correlation (or autocorrelation) function by undertaking an average with respect to time. Since in general we will not have the whole ensemble of records available, we evaluate the cross-correlation function from

$$R_{fg}(\tau) = \lim_{T\to\infty} \frac{1}{T}\int_{-T/2}^{T/2} f(t)g(t+\tau)\mathrm{d}t, \qquad (2.6.10)$$

and similarly for the autocorrelation function. We shall now see in a more detailed way how the autocorrelation function quantifies the frequency content of a random signal.

2.7 Spectral density functions

The autocorrelation function of a random process is directly related to its frequency content by the Fourier integral relationships given by

$$S_{ff}(\omega) = \int_{-\infty}^{\infty} R_{ff}(\tau)\mathrm{e}^{-j\omega\tau}\mathrm{d}\tau, \qquad (2.7.1)$$

$$R_{ff}(\tau) = \frac{1}{2\pi}\int_{-\infty}^{\infty} S_{ff}(\omega)\mathrm{e}^{j\omega\tau}\mathrm{d}\omega. \qquad (2.7.2)$$

In these equations $S_{ff}(\omega)$ is known as the *power spectral density* of the signal $f(t)$ and it is the Fourier transform of the autocorrelation function $R_{ff}(\tau)$. The inverse relationship also holds, that is the autocorrelation function is recovered from the power spectral density by the inverse Fourier transform relationship described by equation (2.3.4). As we will demonstrate below the power spectral density function is a measure of the distribution of the energy in the signal through the frequency range. This can perhaps most easily be understood by observing that with a value of $\tau = 0$, equation (2.7.2) can be written as

$$R_{ff}(0) = \frac{1}{2\pi}\int_{-\infty}^{\infty} S_{ff}(\omega)\mathrm{d}\omega = E[f^2(t)], \qquad (2.7.3)$$

which therefore shows that the integral over all frequencies of the power spectral density gives the total power in the signal, i.e. the mean square value of the signal.

One can think of the power spectral density as the mean square value associated with a given small bandwidth $\Delta\omega$ and it therefore has units of mean square value per unit frequency.

The relationship between the time history of a random signal, its autocorrelation function and its power spectral density is illustrated in Fig. 2.16(a) and (b). The time history of a *narrow band* random process is sketched in Fig. 2.16(a). This type of process contains only frequencies in a relatively narrow range. The signal can be thought of as consisting of many different sinusoidal signals whose amplitudes and phases are randomly related, but because all these frequencies are very similar, there is a relatively high degree of correlation between one value in the time history and the next. The autocorrelation function therefore takes the oscillatory form depicted which decays only slowly as a function of the lag τ. Fourier transformation of this signal from the τ domain to the frequency domain then identifies the narrow band of frequencies contained in the signal such that the power spectral density is only large in this narrow frequency range. Note that for a signal containing only one sinusoidal component, the autocorrelation function would be perfectly cosinusoidal and the power spectral density would consist of a delta function at the frequency of the single component. The case of a "broad band" random signal containing components over a wide frequency range is illustrated in Fig. 2.16(b). In this case the time history shows a small degree of correlation between one instant and the next and its autocorrelation function decays rapidly. Fourier transformation then yields the broad band power spectral density in the frequency domain.

Now note that equation (2.7.1) can be written in the form

$$S_{ff}(\omega) = \int_{-\infty}^{\infty} R_{ff}(\tau)[\cos\omega\tau - j\sin\omega\tau]\mathrm{d}\tau. \tag{2.7.4}$$

Since we have shown above (see equation (2.6.6)) that $R_{ff}(\tau)$ is an even function of τ then the imaginary part of this expression always integrates to zero leaving

$$S_{ff}(\omega) = \int_{-\infty}^{\infty} R_{ff}(\tau)\cos\omega\tau\,\mathrm{d}\tau = 2\int_{0}^{\infty} R_{ff}(\tau)\cos\omega\tau\,\mathrm{d}\tau. \tag{2.7.5}$$

Thus $S_{ff}(\omega)$ is a purely real quantity, as one would expect from a quantity related to the energy of the signal, and since $R_{ff}(\tau)$ is an even function it follows from the properties of Fourier transforms that $S_{ff}(\omega)$ is also an even function such that

$$S_{ff}(\omega) = S_{ff}(-\omega). \tag{2.7.6}$$

In view of this relationship it is also often useful to define a single-sided spectral density which is valid only for positive frequencies and this is given by

$$G_{ff}(\omega) = \begin{Bmatrix} 2S_{ff}(\omega), & \omega \geq 0 \\ 0, & \omega < 0 \end{Bmatrix}. \tag{2.7.7}$$

The notion of a single-sided spectral density is illustrated in Fig. 2.17.

One particular form of fluctuation which is of interest is that which contains energy uniformly distributed over all frequencies from minus infinity to plus infinity. Such a signal is known as *white noise* (by analogy with white light) and its spectral density has a constant value over frequency, given by, for example, S_w. This can be

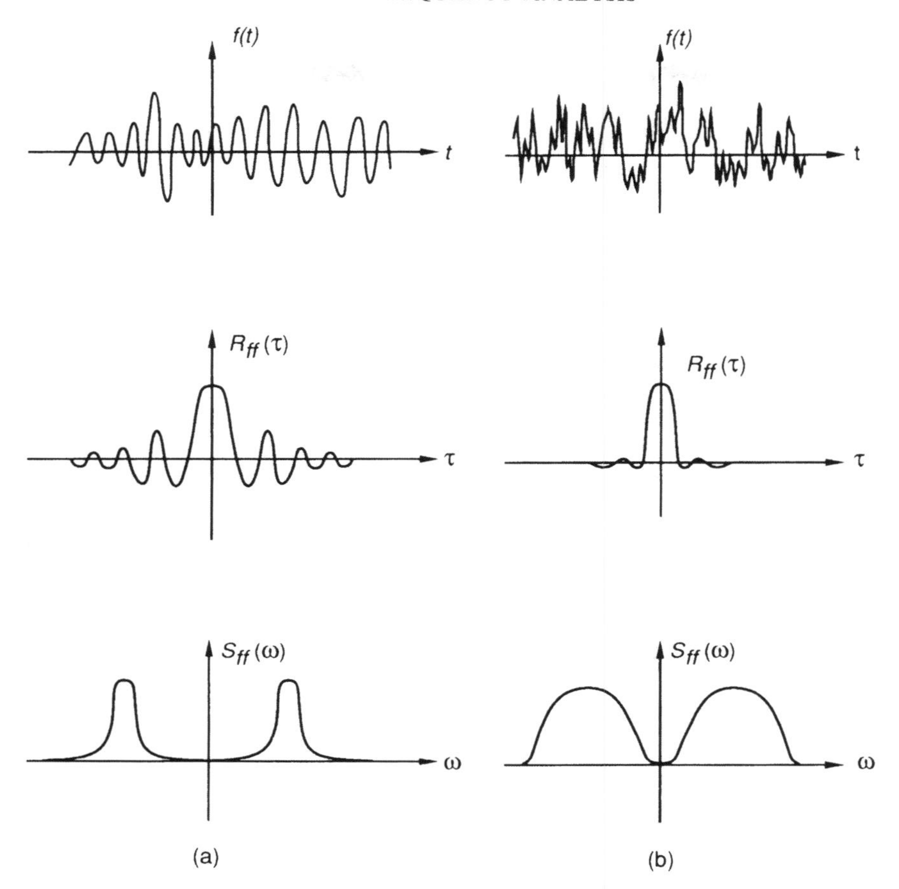

Fig. 2.16 An illustration of the time history, autocorrelation function and power spectral density associated with (a) a narrow band random process and (b) a broad band random process.

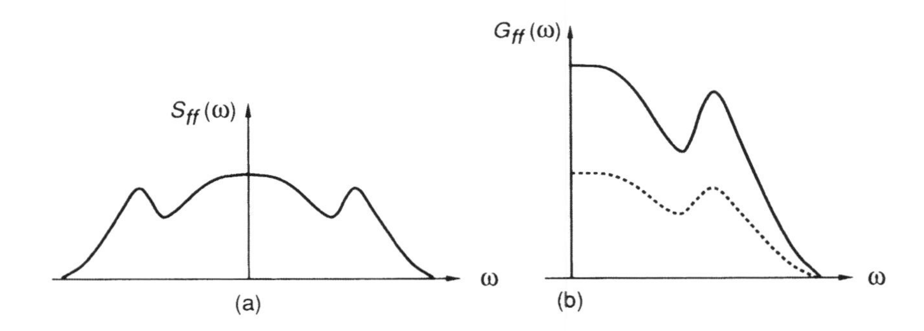

Fig. 2.17 (a) A power spectral density function and (b) its "single-sided" representation.

thought of as the limiting case of the broad band random signal discussed above. Substitution of this value S_w into equation (2.7.2) shows that the autocorrelation function of white noise is given by

$$R_{ff}(\tau) = \frac{S_w}{2\pi} \int_{-\infty}^{\infty} e^{j\omega\tau} d\omega = S_w \delta(\tau). \tag{2.7.8}$$

White noise therefore has an autocorrelation function consisting only of a delta function at $\tau = 0$ and the signal has the property that its value at any particular time is totally unrelated to its value at any previous instant.

We can also define an equivalent *cross-spectral density* function associated with two signals as the Fourier transform of the cross-correlation function. Again we can recover the cross-correlation function by inverse Fourier transformation of the cross-spectral density function. These relationships can be written as

$$S_{fg}(\omega) = \int_{-\infty}^{\infty} R_{fg}(\tau) e^{-j\omega\tau} d\tau, \tag{2.7.9}$$

$$R_{fg}(\tau) = \frac{1}{2\pi} \int_{-\infty}^{\infty} S_{fg}(\omega) e^{j\omega\tau} d\omega. \tag{2.7.10}$$

It can also be shown by using the relationship $R_{fg}(\tau) = R_{fg}(-\tau)$ that $S_{gf}(\omega) = S^*_{gf}(\omega)$, and $S_{fg}(\omega) = S^*_{fg}(\omega)$.

There is another very important, and perhaps more direct, relationship between these spectral density functions and the frequency content of random signals. Let us assume that we have two random signals $f(t)$ and $g(t)$ and that we take time samples of a finite duration T from the ensemble of records of f and g where $f_k(t)$ and $g_k(t)$ are samples from each of the ensemble of records. It is possible to take these finite time histories and define finite Fourier transforms of these records which are given by the expressions

$$F_k(\omega) = \int_0^T f_k(t) e^{-j\omega t} dt, \qquad G_k(\omega) = \int_0^T g_k(t) e^{-j\omega t} dt. \tag{2.7.11}$$

It can be shown via a straightforward, although somewhat lengthy, proof (see, for example, Bendat and Piersol, 1986, Ch. 5) that the cross-spectral density function is given by

$$S_{fg}(\omega) = \lim_{T\to\infty} E\left[\frac{1}{T} F^*_k(\omega) G_k(\omega)\right], \tag{2.7.12}$$

where the expectation operator refers to the average over the ensemble of records. The principle above also applies to the case where we wish to estimate the auto power spectral density from a number of finite Fourier transforms of each set of data $F_k(t)$. In this case the expression for the power spectral density can be written as

$$S_{ff}(\omega) = \lim_{T\to\infty} E\left[\frac{1}{T} |F_k(\omega)|^2\right]. \tag{2.7.13}$$

This very important relationship shows that only if we have data lengths T which are of infinite duration and if we average over a whole ensemble of such records we can recover the power spectral density from Fourier transformation of these data lengths. Nevertheless, as we shall see later, then we can derive an *estimate* $\hat{S}_{ff}(\omega)$ of the power spectral density associated with a random signal by taking a series of data lengths of finite duration T and averaging their results. In later chapters in this work it will be convenient to abbreviate the expression (2.7.12) by writing simply

$$S_{fg}(\omega) = E[F^*(\omega) G(\omega)], \tag{2.7.14}$$

where the expectation operator implies both the ensemble averaging operation and the use of individual time histories of infinite length.

2.8 Analogue-to-digital conversion

During the last decade analogue techniques for spectral analysis have been largely superseded by digital methods. Thus the analogue signal, for example the voltage from a microphone, is converted into a digital representation which can then be subject to much more detailed and extensive analysis of the signal through the use of digital computers. There are now many systems available in which a computer is used to process and analyse analogue data when converted into digital form. In recent years there has also been an extensive implementation of special purpose digital hardware to undertake specific processing tasks. The use of such hardware in spectral analysis was made particularly attractive by the advent of the *fast Fourier transform* (FFT) (Cooley and Tukey, 1965). This is a very rapid numerical technique for computing the Fourier spectrum of a sampled time history of finite length.

The process of analogue-to-digital conversion is illustrated in Fig. 2.18. This shows a typical arrangement of equipment for undertaking the process. First of all the analogue voltage is passed through an analogue low pass filter known as an *anti-aliasing filter*. This removes frequencies in the signal which, as we shall see later, are too high to be observed by the process of sampling the signal at discrete points in time. The filtered signal, with the high frequencies having been removed, is then passed into a sample-hold device, which selects the value of the analogue voltage at discrete points in time and holds this value whilst the process of analogue to digital conversion takes place. The conversion process involves quantising the *level* of the signal at each discrete point in time selected by the sample-hold device. Thus each level of quantisation is associated with a binary representation in a digital word whose length will be determined by the number of quantisation levels chosen. Having a discrete number of quantisation levels results in an inherent inaccuracy in the representation of the signal. One way of interpreting this inaccuracy is to consider it to result in *quantisation noise*. This is illustrated in Fig. 2.19. The level of quantisation noise will depend on the number of "bits" chosen to represent the total range of the analogue signal. A more detailed quantification of this process can be deduced by considering the quantisation noise as having idealised statistical properties. First we assume that the quantisation signal $e(n)$, where n denotes a sample number at a particular time, is such that

$$E[e(n)e(n+m)] = 0, \qquad m \neq 0. \tag{2.8.1}$$

That is the quantisation noise at a sample at a time index $n+m$ is totally uncorrelated with the sample at the value n. Strictly speaking, this is true only when

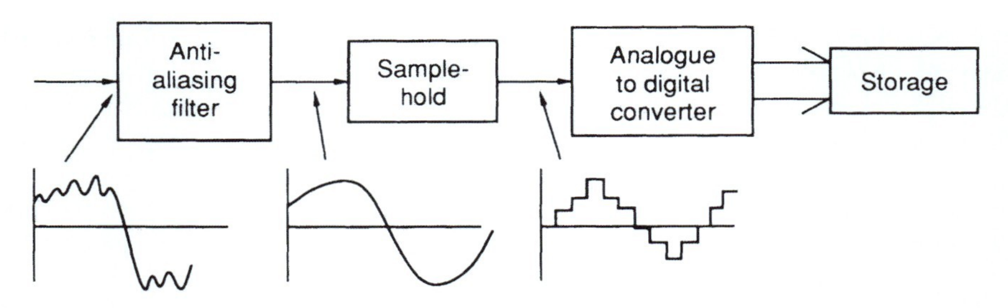

Fig. 2.18 The process of analogue-to-digital conversion. An anti-aliasing filter removes the high frequencies in the signal that cannot be detected by the sampling process. The sample-hold device samples the analogue signal at discrete points in time and holds the value while the analogue-to-digital conversion is undertaken.

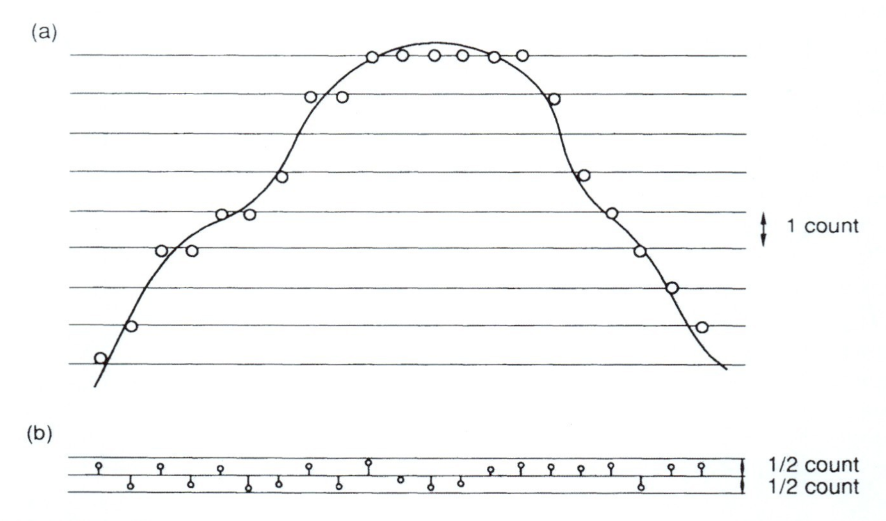

Fig. 2.19 (a) The effective quantisation in level produced by an analogue-to-digital converter with (b) the resulting "quantisation noise" signal.

the change in the time history from one sample to another is much greater than a quantisation interval, although it is approximately true for the waveform sketched in Fig. 2.19. Further we assume that the probability density function associated with the quantisation noise is given by

$$p(e) = 1, \qquad (-1/2) \leq e < (1/2). \tag{2.8.2}$$

Thus the quantisation noise signal has an equal probability of lying anywhere between the limits specified by the quantisation interval. In addition we assume the mean value of the quantisation signal is zero. That is

$$\mu_e = E[e] = \int_{-\infty}^{\infty} ep(e)\mathrm{d}e = 0. \tag{2.8.3}$$

The variance of the signal is calculated relatively easily if we make these assumptions. This is given by

$$\sigma_e^2 = E[(e-\mu_e)^2] = \int_{-\infty}^{\infty} e^2 p(e)\mathrm{d}e = \int_{-1/2}^{1/2} e^2 \mathrm{d}e, \tag{2.8.4}$$

for the assumptions chosen. Evaluating this integral between the limits given shows that

$$\sigma_e^2 = \left[\frac{e^3}{3}\right]_{-1/2}^{1/2} = \frac{1}{12}. \tag{2.8.5}$$

This variance is in units corresponding to the quantisation interval. If we assume that the total range of the analogue-to-digital converter is specified by V (volts) for example, then if we use an N bit representation of this signal level, one quantisation interval will correspond to $V/(2^N-1)$ volts since the number of quantisation intervals will be given by 2^N. Thus the variance associated with the quantisation noise is equal to $(1/12)(V/(2^N-1))^2$ volts squared. It is useful to compare this mean square value of quantisation noise with the mean square value associated with the sine wave whose amplitude is such that the entire range V of the converter is encompassed. In such a case the amplitude of the sine wave is $V/2$ and therefore its mean square value is $V^2/8$. We can therefore define an equivalent signal to noise ratio (for such a perfectly scaled sinusoidal signal) that is given by

$$10\log_{10}\frac{V^2}{8}\frac{(12)(2^N-1)^2}{V^2} \approx 20N\log_{10}2 + 1.76\,\mathrm{dB}, \tag{2.8.6}$$

where the approximation is good provided $N \geqslant 1$. Thus the effective signal to noise ratio associated with an N bit converter can be very approximately calculated from the expression $6N$dB. For example, an 8 bit converter has a signal to noise ratio of 48 dB whilst a 12 bit converter has a signal to noise ratio of 72 dB. Practical analogue-to-digital converters can now operate with at least 14 bit and very often 16 bit resolution thus giving an acceptably high signal to noise ratio for even the most demanding practical applications.

2.9 The Fourier transform of sampled signals and aliasing

We will now consider in more detail the implication of sampling a continuous signal at discrete points in time. The process of sampling can be represented as shown in Fig. 2.20, which illustrates how the process can be viewed as multiplication of the analogue signal by a series of impulses. Thus we can express the sampled signal in the form

$$f_s(t) = f(t)\sum_{n=-\infty}^{\infty}\delta(t-nT), \tag{2.9.1}$$

where T is the sampling period, or time interval between samples. We will now calculate the form of the Fourier spectrum of this sampled signal. This can be accomplished by evaluating the Fourier transform of the sampled signal given by

$$F(e^{j\omega T}) = \int_{-\infty}^{\infty} [f(t) \sum_{n=-\infty}^{\infty} \delta(t-nT)] e^{-j\omega t} dt, \tag{2.9.2}$$

where the notation $F(e^{j\omega T})$ is conventionally used to represent the Fourier transform of the signal $f(t)$ sampled at the interval T. The reasons for this will become apparent when we deal with digital systems in Chapter 4. Using the properties of the delta function we can write this expression in the form

$$F(e^{j\omega T}) = \sum_{n=-\infty}^{\infty} f(nT) e^{-j\omega nT}, \tag{2.9.3}$$

where $f(nT)$ is now the discrete variable defined by

$$f(nT) = \int_{-\infty}^{\infty} f(t)\delta(t-nT)dt. \tag{2.9.4}$$

Equation (2.9.3) expresses the usual form of the Fourier transform of a sampled signal. We will discuss the numerical calculation of this quantity in the next section. However, return to the representation of the Fourier transform given by equation (2.9.2). We can consider the series of impulses to be a periodic signal which has a fundamental frequency given by $\omega_s = 2\pi/T$ where ω_s is the frequency at which samples of the analogue signal are taken. We can therefore write this series of impulses in the form of its Fourier series which can be expressed as (see Section 2.2)

$$\sum_{m=-\infty}^{\infty} a_m e^{jm\omega_s t}.$$

In this case the complex Fourier coefficients a_m are given by (see equation (2.2.12))

$$a_m = \frac{1}{T}\int_{-T/2}^{T/2} \delta(t) e^{-jm\omega_s t} dt = \frac{1}{T}. \tag{2.9.5}$$

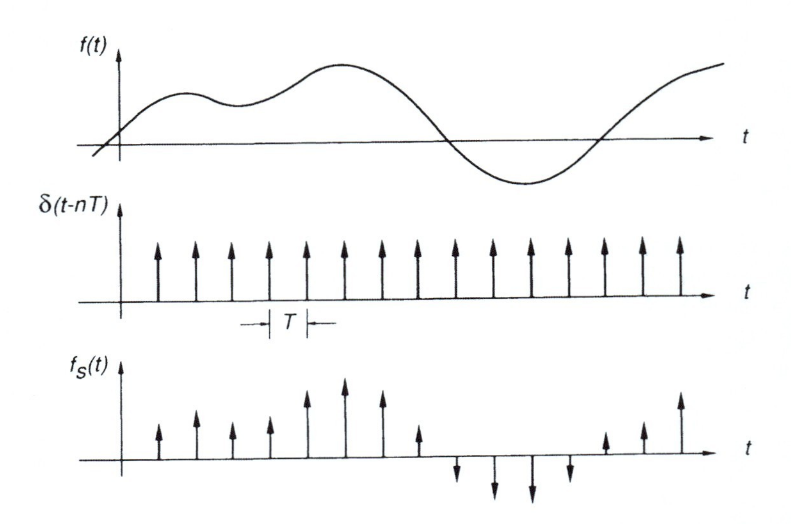

Fig. 2.20 Ideal sampling represented as the multiplication of the original analogue signal by an impulse train with a sampling period of T seconds.

Substitution of this value of the Fourier transform coefficients into the Fourier series representation of the signal shows that we can represent the impulse train by

$$\sum_{n=0}^{\infty} \delta(t-nT) = \sum_{m=-\infty}^{\infty} \frac{1}{T} e^{jm\omega_s t}. \tag{2.9.6}$$

Substitution of this into equation (2.9.2) then shows that

$$F(e^{j\omega T}) = \int_{-\infty}^{\infty} f(t) \left[\sum_{m=-\infty}^{\infty} \frac{1}{T} e^{jm\omega_s t} \right] e^{-j\omega t} dt. \tag{2.9.7}$$

We can now rearrange the order of integration such that the Fourier spectrum of the sampled signal can be written in the form

$$F(e^{j\omega T}) = \frac{1}{T} \sum_{m=-\infty}^{\infty} \left[\int_{-\infty}^{\infty} f(t) e^{-j(\omega - m\omega_s)t} dt \right]. \tag{2.9.8}$$

This is exactly equivalent to expressing the Fourier spectrum in the form of a sum of Fourier spectra which repeats at integer numbers of the sampling frequency and can be written as

$$F(e^{j\omega T}) = \frac{1}{T} \sum_{m=-\infty}^{\infty} F(\omega - m\omega_s). \tag{2.9.9}$$

This periodic structure of the Fourier spectrum is illustrated in Fig. 2.21. From this representation of the net spectrum of the sampled signal it becomes evident that a problem arises if the basic analogue signal from which the samples are taken contains frequencies which are higher than *half* the sampling frequency. In this case if one observes the spectrum of the sampled signal between the frequency range from zero to $\omega_s/2$ then not only will the original spectrum be observed but also a contribution will be seen from the other spectra in the periodic sequence of spectra. This phenomenon is known as *aliasing* and the distortion in the resulting spectrum is due to the fact that high frequencies in the data are indistinguishable from lower

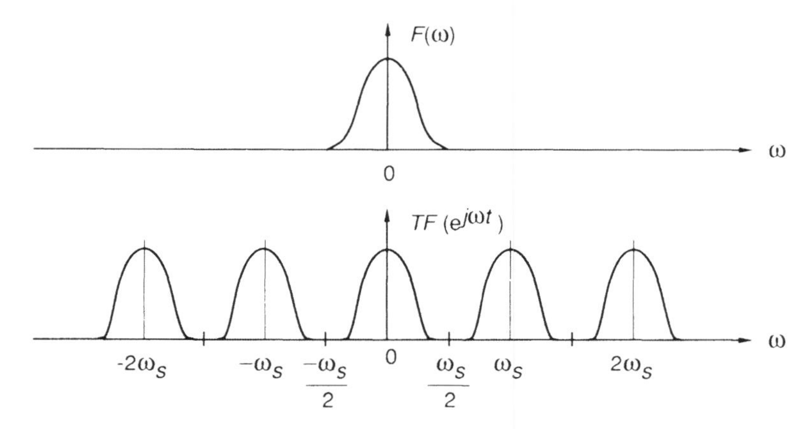

Fig. 2.21 The Fourier spectrum of the sampled signal showing its periodically repeated form at intervals of the sampling frequency ω_s.

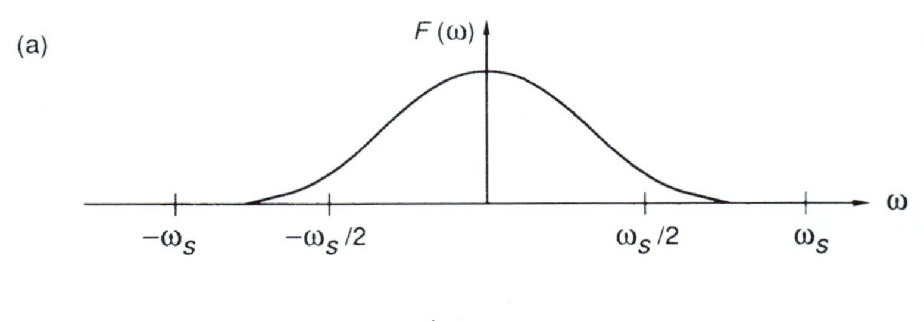

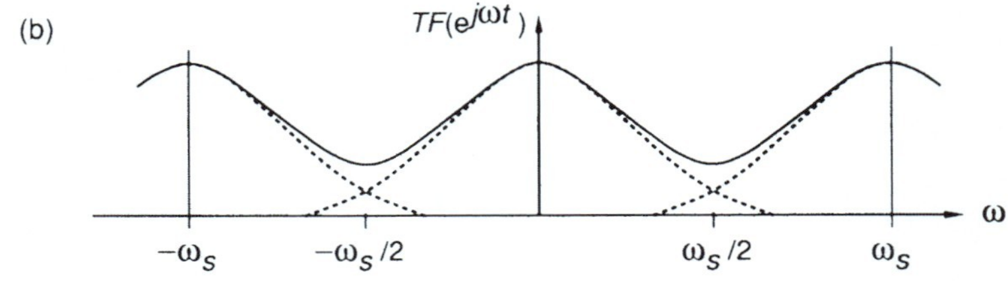

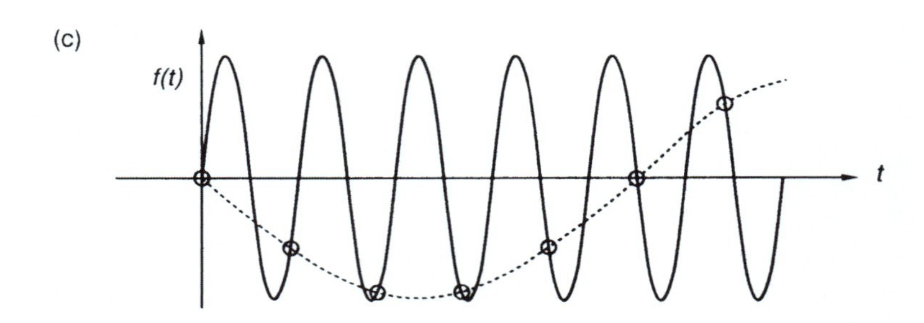

Fig. 2.22 Aliasing, showing (a) the spectrum of the original signal containing frequencies greater than $\omega_s/2$, (b) the resulting spectrum of the sampled signal which contains contributions from the aliased spectra and (c) a time interpretation showing how higher frequencies are indistinguishable from lower frequencies if the sample rate is not fast enough.

frequencies if the sampling rate is not fast enough. This is also illustrated in Fig. 2.22. The problem of aliasing can be avoided provided the data is sampled at a frequency which is more than twice that of the highest frequency component present in the continuous time signal. Therefore it is usually the practice to firstly filter out these high frequencies using an analogue filter prior to their conversion into digital form. As we have seen already these filters are known as anti-aliasing filters. Finally note that the highest frequency component present is sometimes referred to as the *Nyquist frequency*. Twice this frequency is known as the *Nyquist sampling rate*.

2.10 The discrete Fourier transform

Once we have sampled the analogue data and ensured that the frequencies that are too high for the sampling to account for have been removed, then as we saw in Section 2.7, we can compute an estimate of the power spectral density of a random signal by Fourier transforming a finite length of data. The discrete Fourier transform provides a convenient numerial technique for undertaking this Fourier transformation. Equation (2.9.3) enables the calculation of the Fourier spectrum resulting from sampling the continuous signal $f(t)$ so as to produce an infinite number of values of the discrete variable $f(nT)$, where n stretches from minus infinity to plus infinity. Of course we cannot calculate this spectrum in practice since we obviously cannot deal with an infinite number of data points. Nevertheless, we can take a sample of our series of data points which stretches over a finite number of points such that n varies from zero to $N-1$ say, where N is the total number of data points chosen. Then we can *estimate* the Fourier spectrum defined by equation (2.9.3) by truncating the summation in the equation and expressing it in the form

$$\hat{F}(\mathrm{e}^{j\omega T}) = \sum_{n=0}^{N-1} f(nT)\mathrm{e}^{-j\omega nT}, \tag{2.10.1}$$

where we have used $\hat{F}(\mathrm{e}^{j\omega T})$ to denote the estimate of $F(\mathrm{e}^{j\omega t})$. We could now in principle evaluate the entire spectrum. However, the (periodically repeating) spectrum consists of a continuous range of frequencies ω and it is possible in practice only to choose a finite number of values of ω for which we can compute $\hat{F}(\mathrm{e}^{j\omega T})$. Recall that the sampling frequency is given by $\omega_s = 2\pi/T$. In choosing the specific values of frequency for which we evaluate $\hat{F}(\mathrm{e}^{j\omega T})$, we split the frequency range between zero and ω_s into the same number N of which we have time samples. Thus the frequency range is split into N values spaced at a frequency increment of $2\pi/NT$. We evaluate the Fourier spectrum at each of $k(2\pi/NT)$ values of frequency where k is the index associated with each discrete frequency chosen. Thus we evaluate numerically

$$\hat{F}(\mathrm{e}^{jk(2\pi/NT)T}) = \sum_{n=0}^{N-1} f(nT)\mathrm{e}^{-jk(2\pi/NT)nT}. \tag{2.10.2}$$

Since the number of frequencies for which we evaluate this quantity is determined by the number of samples in the data length chosen, then it is more usual to write this expression in the compact form

$$F(k) = \sum_{n=0}^{N-1} f(n)\mathrm{e}^{-j[(2\pi nk)/N]}. \tag{2.10.3}$$

The justification for this choice of discrete frequencies for which we evaluate the transform is that there is an inverse transform such that we can exactly recover the values $f(n)$ from the values of $F(k)$. This inverse transform is given by

$$f(n) = \frac{1}{N}\sum_{k=0}^{N-1} F(k)\mathrm{e}^{j(2\pi nk)/N}, \qquad 0 \le n \le N-1. \tag{2.10.4}$$

The proof of the inverse transform relationship given by equation (2.10.4) is relatively straightforward but will not be given here. The net form of the sequence

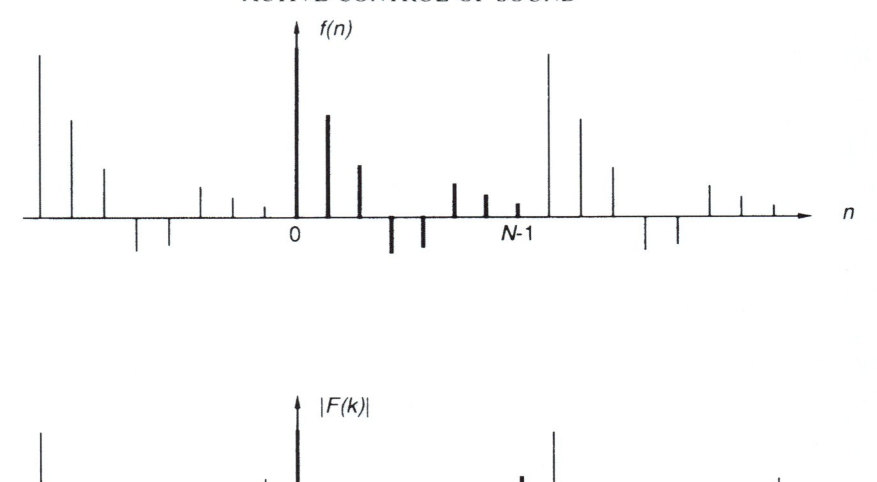

Fig. 2.23 An illustration of the discrete Fourier transform. Note that the N original data points are assumed to be associated with a periodically repeating data sequence. The discrete Fourier transform is evaluated at N frequencies in the range between zero frequency and the sampling frequency. This spectrum repeats periodically in the same way as that illustrated in Fig. 2.21.

$f(n)$ is illustrated in Fig. 2.23. Outside the range of n given by $(0 \leq n \leq N-1)$, it can be shown that the sequence can be considered to be periodic with a period which corresponds to the length of the finite sequence of data chosen. Similarly, the resulting Fourier spectrum evaluated at the same number of discrete frequencies also has a periodic structure as illustrated in Fig. 2.23.

Now consider the numerical process which has to be undertaken in order to evaluate the discrete Fourier transform given by equation (2.10.3). The multiplication of the sample value $f(n)$ by the exponential term has to be undertaken for both N samples and N frequencies and thus there are N^2 multiplications involved in undertaking this process. The fast Fourier transform (FFT) is a computer algorithm for calculating this extremely efficiently. The details of the algorithm will not be given here but essentially if a number of samples N is such that $N = 2^m$ where m is an integer then it can be shown that the number of multiplications required by using the FFT is reduced to $N.\log_2 N$ (see, for example, the discussion presented by Newland, 1984, Ch. 12).

The simplest form of FFT algorithm does in fact involve the use of a number of samples N which is 2^m. Thus, for example, typically $N = 2^{10} = 1024$ samples. In this case then the number of multiplications in using the fast Fourier transform is reduced by a factor of roughly 100 from that which would be accomplished simply by using equation (2.10.5) as it stands. It is largely due to this vastly increased efficiency of computation that the process of digital evaluation of spectral densities has become so prevalent. Prior to the advent of the fast Fourier transform, the power spectrum was very often evaluated by using direct Fourier transformation of the autocorrelation function. The accuracies and errors involved in using such techniques are well documented (see, for example, Bendat and Piersol, 1986).

2.11 Data truncation and windowing

In evaluating the discrete Fourier transform we have seen that only a finite number of data points is used. This is equivalent to looking at the original signal through a time "window", as illustrated in Fig. 2.24. This process can be described by the expression

$$f_w(t) = f(t)w(t). \tag{2.11.1}$$

Figure 2.24 shows the rectangular window function or "box car" function. It is a function $w(t)$ which has a zero value outside of a certain range of time of duration T_w and has a unit value within that time. It is important to note that simply applying a window of this sort automatically changes the spectral content of the data which we are examining. This can be described in a more formal way through considering the form of the Fourier transforms of the original data $F(\omega)$ and the Fourier transform of the window given by $W(\omega)$. We can write equation (2.11.1) in the form

$$f_w(t) = \left[\frac{1}{2\pi}\int_{-\infty}^{\infty} F(\omega_1)e^{j\omega_1 t}d\omega_1\right]\left[\frac{1}{2\pi}\int_{-\infty}^{\infty} W(\omega_2)e^{j\omega_2 t}d\omega_2\right]. \tag{2.11.2}$$

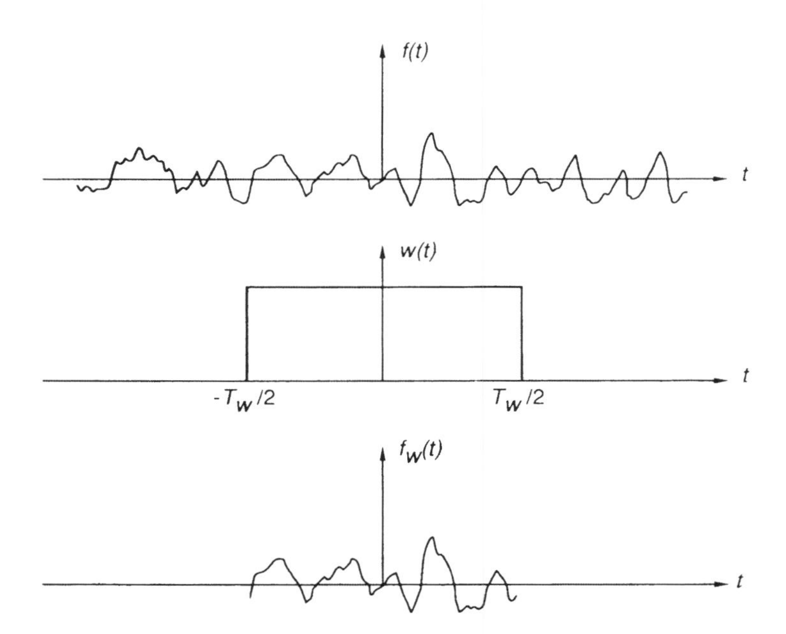

Fig. 2.24 The process of data truncation through multiplication of the original signal $f(t)$ by the rectangular window function $w(t)$.

The Fourier transform of $f_w(t)$ can therefore be written as

$$F_w(\omega) = \frac{1}{(2\pi)^2} \int_{-\infty}^{\infty} \int_{-\infty}^{\infty} \int_{-\infty}^{\infty} F(\omega_1) W(\omega_2) e^{j(\omega_1 + \omega_2)t} e^{-j\omega t} d\omega_1 d\omega_2 dt. \quad (2.11.3)$$

Evaluating the integral with respect to time t gives

$$F_w(\omega) = \frac{1}{2\pi} \int_{-\infty}^{\infty} \int_{-\infty}^{\infty} F(\omega_1) W(\omega_2) \delta(\omega - \omega_1 - \omega_2) d\omega_1 d\omega_2. \quad (2.11.4)$$

Using the sifting property of the delta function in undertaking the integration with respect to ω_2 then shows that

$$F_w(\omega) = \frac{1}{2\pi} \int_{-\infty}^{\infty} F(\omega_1) W(\omega - \omega_1) d\omega_1. \quad (2.11.5)$$

This relationship formally expresses the way in which the spectrum of the original data $F(\omega)$ becomes modified by a "spectral window" $W(\omega)$ which is associated with the time windowing function. Let us illustrate this with a specific example. Assume firstly that the raw signal $f(t)$ is simply a cosine wave which has a Fourier transform given by

$$F(\omega) = \pi\delta(\omega - \omega_0) + \pi\delta(\omega + \omega_0). \quad (2.11.6)$$

It can also be shown that the Fourier transform of the box car function described above is given by

$$W(\omega) = T_w \frac{\sin(\omega T_w/2)}{\omega T_w/2}. \quad (2.11.7)$$

Substitution of these relationships into equation (2.11.5) yields

$$F_w(\omega) = \frac{T_w}{2} \int_{-\infty}^{\infty} [\delta(\omega_1 - \omega_0) + \delta(\omega_1 + \omega_0)] \frac{\sin(\omega - \omega_1) T_w/2}{(\omega - \omega_1) T_w/2} d\omega_1. \quad (2.11.8)$$

Evaluation of the integral by using the properties of the delta function shows that the Fourier spectrum of the windowed cosine function now has the form

$$F_w(\omega) = \frac{T_w}{2} \left[\frac{\sin(\omega - \omega_0) T_w/2}{(\omega - \omega_0) T_w/2} + \frac{\sin(\omega + \omega_0) T_w/2}{(\omega + \omega_0) T_w/2} \right]. \quad (2.11.9)$$

The form of this function is sketched in Fig. 2.25. It is evident that no longer does the spectrum of the signal exhibit a simple delta function at the frequency ω_0. The mere act of windowing the data has caused this spectral energy to "smear" into other frequency components which are represented by the continuous function of frequency shown. It is also clear that as the window length T_w becomes very large compared to the number of cycles in the cosine wave then the less serious is this "leakage" of spectral energy from the frequency of the signal. This relationship between the length of the data window and the frequency of the data being observed dictates the "spectral resolution" with which we can observe the data. One can, however, prevent too much leakage from the main frequency being observed into the "side lobes" evident in the spectrum of $F_w(\omega)$ by having a smoother form of data window. Many such windows have been proposed; one typical example which is

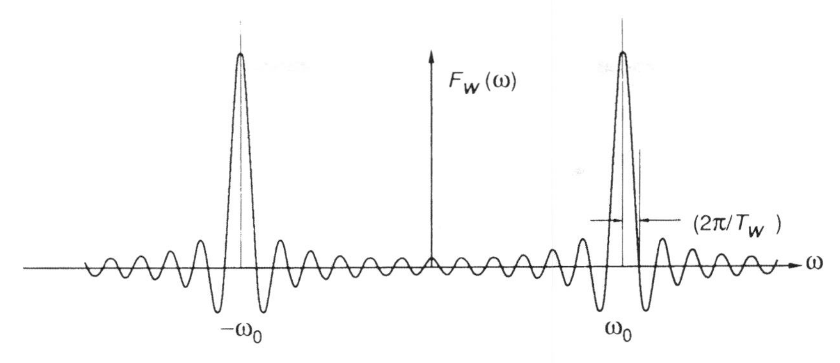

Fig. 2.25 The Fourier spectrum of a cosinusoidal signal truncated with a rectangular window of duration T_w. Note that as T_w is increased the "leakage" of the frequency ω_0 into other frequency components becomes progressively decreased.

commonly used is the Hanning window which is specified in Fig. 2.26. It is also true to say, however, that in choosing a data window such as this, one often sacrifices resolution in the frequency domain for the sake of suppressing leakage in the spectrum. This subject will be returned to in Chapter 4.

2.12 Estimation of power spectra

We saw in Section 2.7 that the power spectral density is related to the Fourier spectrum of an ensemble of finite data lengths through the relationship

$$S_{ff}(\omega) = \lim_{T\to\infty} \frac{1}{T} E\,[|F_k(\omega)|^2]. \tag{2.12.1}$$

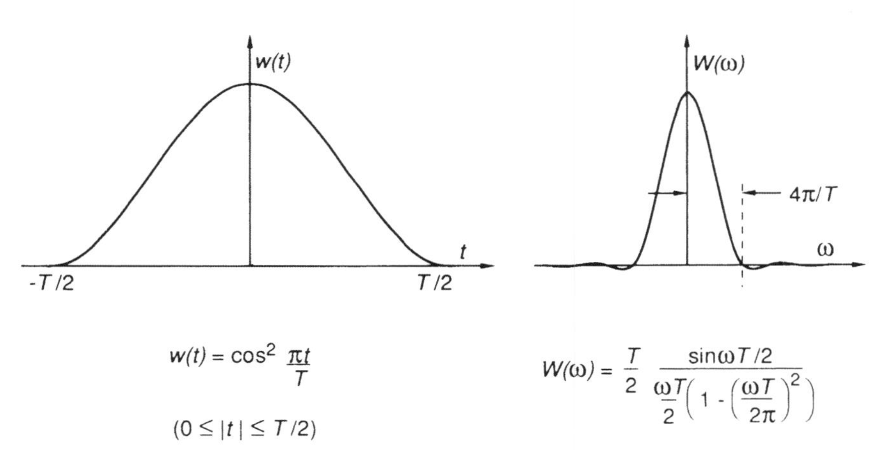

Fig. 2.26 Illustration of the Hanning window $w(t)$ for the truncation of data and its resulting spectral window $W(\omega)$.

Thus in order to obtain a true value of the spectrum we have to use data lengths which are of infinite duration and we have to average over the infinite ensemble of records. This is clearly impracticable and most modern techniques for estimating the power spectrum use a finite length of data T which is further split into M segments of data each of duration T_w. The approximation to (2.12.1) which is the estimate of power spectrum is given by

$$\hat{S}_{ff}(\omega) = \frac{1}{M} \sum_{m=1}^{M} \frac{1}{T_w} |F_m(\omega)|^2. \tag{2.12.2}$$

Thus we simply take M sets of data and evaluate the "periodogram" associated with each data set by performing a discrete Fourier transform on each of these lengths of data. The spectral estimate is then given by averaging all these periodograms. It is useful to describe the statistical aspects of the accuracy to which we can evaluate this estimate. First, the contribution to the total estimate from each periodogram can be written in the form

$$|F_m(\omega)|^2 = F_{m\mathrm{R}}^2 + F_{m\mathrm{I}}^2 \tag{2.12.3}$$

The summation given by equation (2.12.2) can thus be written as

$$\hat{S}_{ff}(\omega) = \frac{1}{MT_w}(F_{1\mathrm{R}}^2 + F_{1\mathrm{I}}^2 + F_{2\mathrm{R}}^2 + F_{2\mathrm{I}}^2 + \ldots F_{M\mathrm{R}}^2 + F_{M\mathrm{I}}^2). \tag{2.12.4}$$

It is now assumed that all the real and imaginary parts of the Fourier spectrum $F_{m\mathrm{R}}$ and $F_{m\mathrm{I}}$ are random variables having a Gaussian probability distribution with zero mean. Furthermore, it is also assumed that all the terms $F_{m\mathrm{R}}$ and $F_{m\mathrm{I}}$ are statistically independent from one another. Taking the expectation of equation (2.12.4) shows that

$$E[\hat{S}_{ff}(\omega)] = \frac{1}{MT_w}(E[F_{1\mathrm{R}}^2] + E[F_{1\mathrm{I}}^2] + \ldots E[F_{M\mathrm{R}}^2] + E[F_{M\mathrm{I}}^2]). \tag{2.12.5}$$

Since all the variables $F_{m\mathrm{R}}$ and $F_{m\mathrm{I}}$ are assumed to have zero mean, then $E[F_{m\mathrm{R}}^2]$ and $E[F_{m\mathrm{I}}^2]$ are simply the variances of $F_{m\mathrm{R}}$ and $F_{m\mathrm{I}}$. It is now assumed that these are all equal to one another, and therefore assuming $E[\hat{S}_{ff}(\omega)] = S_{ff}(\omega)$, it follows from equation (2.12.5) that each of the terms $E[F_{m\mathrm{R}}^2]/T_w$ and $E[F_{m\mathrm{I}}^2]/T_w$ will be given by $S_{ff}(\omega)/2$. We can therefore write

$$\frac{2M\hat{S}_{ff}(\omega)}{S_{ff}(\omega)} = \chi_k^2 = x_1^2 + x_2^2 + x_3^2 \ldots x_k^2, \tag{2.12.6}$$

where the variable $2M\hat{S}_{ff}(\omega)/S_{ff}(\omega)$ now has a *chi-squared* statistical distribution having k *degrees of freedom* since it consists of the sum of the squares of k independent Gaussian random variables x each having unit variance and zero mean. It is well known (see, for example, Newland, 1984, Ch. 9) that such a statistical distribution has a variance that is given by $2k$. Thus we can write

$$\frac{4M^2}{S_{ff}^2(\omega)} E[(\hat{S}_{ff}(\omega) - S_{ff}(\omega))^2] = 2k. \tag{2.12.7}$$

Since in this case the number of degrees of freedom $k = 2M$, then it follows that

$$\frac{E[(\hat{S}_{ff}(\omega) - S_{ff}(\omega))^2]}{S_{ff}^2(\omega)} = \frac{1}{M}. \tag{2.12.8}$$

The value of M is given by $M = T/T_w$ and if we define the effective resolution bandwidth of our spectral estimate to be related to the window length T_w by $B = 1/T_w$, we see that the accuracy of the power spectral estimate can be expressed in the form

$$\frac{E[(\hat{S}_{ff}(\omega) - S_{ff}(\omega))^2]}{S_{ff}^2(\omega)} = \frac{1}{BT}. \tag{2.12.9}$$

Thus although the resolution is fixed by the window length T_w the statistical accuracy is improved by continuing to take greater and greater total lengths of data defined by the data length T. The statistical properties of the χ^2 distribution can be used further in order to establish the confidence limits associated with estimating the power spectrum when using this type of technique (see, for example, Newland, 1984, Ch. 9). It should also be emphasised that this technique is not the only method of evaluating the power spectrum but it does represent one of the most commonly used methods.

3
Linear Systems

3.1 Linearity and the superposition principle

In this chapter we will review briefly some of the important results of linear systems theory. More comprehensive discussions can be found, for example, in the books by Kuo (1966), Lynn (1982), Papoulis (1981) and Richards (1979). We will postpone a discussion of "digital" systems, i.e. those which operate on *sequences* of numbers, until the following chapter, and concentrate here on those which operate on *continuous* time signals: "analogue" systems. The first task is to define linearity. Consider a system with a single input waveform, $f(t)$, producing a single output waveform, $g(t)$, which is defined by some arbitrary transformation, H, of the input signal:

$$g(t) = H[f(t)]. \tag{3.1.1}$$

The arbitrary transformation H encompasses a host of possible operations including, for example, simple multiplication by a constant, squaring, differentiation, integration, or even "clipping" of the input signal $f(t)$. However, the important distinction that we must make is that the system is said to be *linear* with respect to its input (and the operation H is said to be linear) if the system obeys the *principle of superposition*. That is if we have two inputs to the system given by $f_1(t)$ and $f_2(t)$ then

$$H[f_1(t) + f_2(t)] = H[f_1(t)] + H[f_2(t)]. \tag{3.1.2}$$

Thus the output signal from a linear system, when two input signals are applied simultaneously, is the sum of the output signals which would be caused by applying each of the two input signals on their own. Thus, of the operations referred to above, multiplication by a constant, differentiation and integration are all linear operations whilst squaring and clipping are not. A consequence of the superposition principle is that if the amplitude of the input signal to a linear system is doubled, then the amplitude of the output signal is also doubled: i.e., the output signal amplitude increases linearly with the input signal amplitude. In general then, for a linear system

$$H[af_1(t) + bf_2(t)] = aH[f_1(t)] + bH[f_2(t)], \tag{3.1.3}$$

where a and b are constants.

The principle of superposition is of central importance in the active control of sound and we have already discussed this in some detail in Chapter 1 (see Section

1.6). Although, as we have seen in Chapter 1, acoustic disturbances are strictly only approximately linear, the non-linear components are, in practice, negligible for normal sound pressure levels. A more important source of non-linearity in active sound control systems is the electroacoustic transducers. Moving coil loudspeakers, in particular, will generate distortion levels of about 0.1–1% at low frequencies, even if good quality units are used, and care is taken to prevent rattles and leaks in the enclosure. In the discussions of practical control systems in Chapters 6 and 12, we will assume that such non-linearities can be ignored and the principle of superposition (equation (3.1.3)) can be applied. In practice, when using good quality transducers working within their dynamic range, this assumption is a reasonable one, provided attenuations of greater than 20–40 dB are not being contemplated.

3.2 Systems governed by ordinary differential equations

Much of the theory for linear systems has been derived within an electrical engineering framework, with voltages and currents as the signal variables. It will be helpful here to use the same formulation in order to interpret the response of acoustic systems, with acoustic pressure and volume velocity as signal variables. The "analogies" between acoustical and electrical systems were exhaustively explored in the 1940s (see, for example, Firestone, 1938, 1954; Olson, 1943). We will not, however, attempt here to interpret the physical behaviour of acoustic systems purely in terms of an equivalent electrical circuit. The physical behaviour of acoustic systems in active control applications is too subtle for this to be worthwhile. However, the obvious parallels between electrical and some acoustical systems will be pointed out, in order to demonstrate how the tools of linear systems theory can be applied to the problem in hand.

Acoustical systems are generally defined by the partial differential equations derived in Chapter 1, which lead to the acoustic wave equation. Wave propagation implies that the acoustical variables are delayed in their propagation from one part of an acoustical system to another. When the dimensions of the acoustical system are sufficiently small, this time delay is sufficiently small to be neglected compared to the period of the signal at the highest frequency of interest. Under these conditions the partial differential equations governing an acoustical system can be reasonably well approximated by a manageable number of ordinary differential equations, and the resulting model of the system is said to have *lumped elements*. These ordinary differential equations will be linear if the system is linear. We make the additional assumption that the coefficients of the differential equation are not functions of time, so that the system is *time invariant*.

As an example, consider the compression of the air in the small cavity illustrated in Fig. 3.1. If the frequencies of interest are sufficiently low, so that no wave motion can exist in the cavity, its response can be modelled accurately by using an ordinary differential equation. In terms of the modal response of the cavity, we will see that this is equivalent to ignoring all but the mode corresponding to the modal integers $n_1 = n_2 = n_3 = 0$ in the analysis presented in Section 1.14 of Chapter 1. Under these conditions and for small adiabatic compressions of the air volume, the pressure increase in the cavity $p(t)$ (over the original atmospheric pressure p_0) due to a

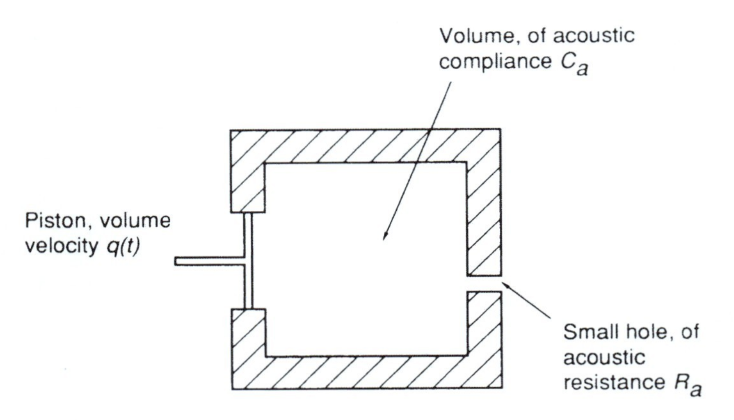

Fig. 3.1 The compression of the air in a small leaky cavity, driven by a piston.

decrease in volume of the cavity $V(t)$ (compared to the original volume V_0) is given approximately by the linearised equation

$$p(t) = \left(\frac{\gamma p_0}{V_0}\right) V(t) = \left(\frac{\rho_0 c_0^2}{V_0}\right) V(t), \tag{3.2.1}$$

where γ is the ratio of principal specific heats of the gas in the cavity, and γp_0 is the adiabatic bulk modulus which can also be written as $\rho_0 c_0^2$ (see Section 1.3).

The rate of change of volume is equal to the total volume velocity, $q_T(t)$, supplied to the cavity, so $\mathrm{d}V(t)/\mathrm{d}t = q_T(t)$, and

$$\frac{\mathrm{d}p(t)}{\mathrm{d}t} = \frac{\rho_0 c_0^2}{V_0} q_T(t) = \frac{q_T(t)}{C_\mathrm{a}}, \tag{3.2.2}$$

where $C_\mathrm{a} = V_0/\rho_0 c_0^2$ is known as the *acoustic compliance* of the cavity. The total volume velocity flowing into the cavity of Fig. 3.1 is made up of the volume velocity of the driving piston, $q(t)$ say, and the volume velocity lost through the small hole, which acts as the "leak" in the cavity through which flow can escape. Provided the volume velocity through the small hole is not too great (so that the system acts linearly) and the frequency is not too high (so that the inertance of the air in the hole can be ignored), the hole will act as a frequency-independent acoustic resistance R_a (see the discussion presented by Olson, 1957). Provided the alternating pressure outside the cavity is negligible, the volume velocity driven through the small hole will be equal to $p(t)/R_\mathrm{a}$. So the total volume velocity flowing into the cavity becomes

$$q_T(t) = q(t) - p(t)/R_\mathrm{a}, \tag{3.2.3}$$

and the ordinary differential equation which describes this simple acoustical system can be written using equation (3.2.2) as

$$C_\mathrm{a}\frac{\mathrm{d}p(t)}{\mathrm{d}t} + \frac{1}{R_\mathrm{a}} p(t) = q(t). \tag{3.2.4}$$

To take a purely electrical example for comparison, the relationship between the voltage and the current in the electrical circuit shown in Fig. 3.2 is determined by the differential equations governing the electrical capacitance and resistance. Thus

$$i_{\mathrm{C}}(t) = C_{\mathrm{e}}\frac{\mathrm{d}v(t)}{\mathrm{d}t} \quad \text{and} \quad i_{\mathrm{R}}(t) = \frac{1}{R_{\mathrm{e}}}v(t), \tag{3.2.5a,b}$$

where $v(t)$ denotes the fluctuating electrical voltage, and $i_{\mathrm{C}}(t)$ and $i_{\mathrm{R}}(t)$ the fluctuating currents in the capacitance (C_{e}) and resistance (R_{e}). Combining these with the continuity equation for the electrical current (Kirchoff's law): $i(t) = i_{\mathrm{C}}(t) + i_{\mathrm{R}}(t)$, the differential equation describing the whole electrical system becomes

$$C_{\mathrm{e}}\frac{\mathrm{d}v(t)}{\mathrm{d}t} + \frac{1}{R_{\mathrm{e}}}v(t) = i(t). \tag{3.2.6}$$

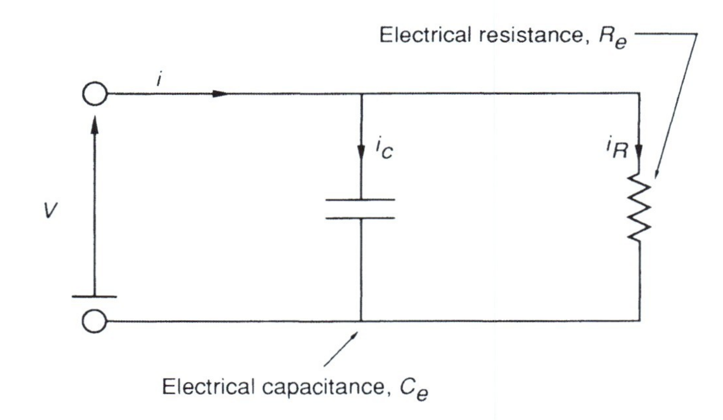

Fig. 3.2 Electrical circuit whose governing ordinary differential equation is the same as that of the acoustic system of Fig. 3.1.

The parallels between the acoustical network in Fig. 3.1 and the electrical network in Fig. 3.2 are clear from a comparison of the differential equations (3.2.4) and (3.2.6) defining their behaviour. The algebraic tools developed in order to gain insight into the behaviour of the electrical circuits of Fig. 3.2 can now be used directly for analysing the acoustic system depicted in Fig. 3.1.

3.3 The Laplace transform

The Laplace transform is a generalisation of the Fourier transform, which often can more easily be made convergent for classes of signals of interest than the Fourier transform in its usual simple form adopted here. (Note that when the sophisticated form of the Fourier transform is used, with the "frequency" ω in it regarded from the outset as a complex variable, the Fourier transform can be regarded as more, or at least equally, general. See, for example, the discussion presented by Morse and

Feshbach, 1953, Section 4.8.) We will use the Laplace transform here to transform ordinary differential equations, which are the most basic representation of the response of analogue systems, into algebraic equations. These algebraic equations are easier to manipulate than the original differential equations, and can be used to deduce the response of the system to sinusoidal or impulsive excitation.

The (one-sided) Laplace transform of a function of time $f(t)$ is defined to be (see, for example, Kuo, 1966, Ch. 6; Lynn, 1982, Ch. 3)

$$F(s) = \int_0^\infty f(t)\mathrm{e}^{-st}\mathrm{d}t, \tag{3.3.1}$$

where s is a complex variable whose real and imaginary parts are conventionally denoted by σ and ω, so that

$$s = \sigma + j\omega. \tag{3.3.2}$$

Note that for a causal function (i.e. a function $f(t)$ for which $f(t) = 0$ for $t<0$), the Laplace transform reduces formally to the Fourier transform if the substitution $s = j\omega$ is made, i.e.

$$F(j\omega) = F(s)|_{s = j\omega}. \tag{3.3.3}$$

To take a simple example, the Laplace transform of the waveform defined by the equation

$$f(t) = \begin{Bmatrix} \mathrm{e}^{-at}, & t \geq 0 \\ 0, & t<0 \end{Bmatrix}, \tag{3.3.4}$$

where the variable "a" is real (plotted in Fig. 3.3 for positive a) is

$$F(s) = \int_0^\infty \mathrm{e}^{-at}\,\mathrm{e}^{-st}\mathrm{d}t = \left[\frac{\mathrm{e}^{-(s+a)t}}{-(s+a)}\right]_0^\infty. \tag{3.3.5}$$

Notice that $\mathrm{e}^{-(s+a)t} = \mathrm{e}^{-(\sigma+a)t}\,\mathrm{e}^{-j\omega t}$, so that the upper limit value of the result given in equation (3.3.5) will only be finite if $(\sigma + a) > 0$, i.e.

$$\sigma > -a. \tag{3.3.6}$$

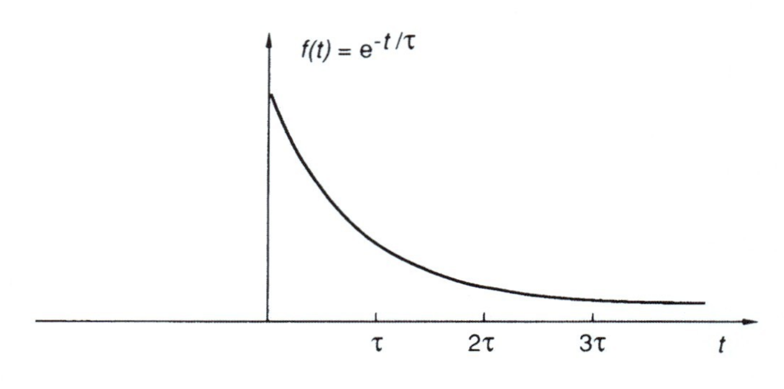

Fig. 3.3 A decaying exponential waveform.

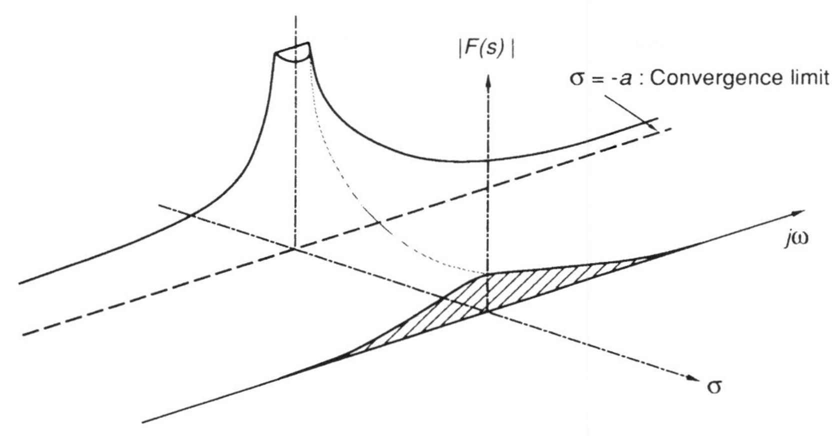

Fig. 3.4 The modulus of the Laplace transform of the signal in Fig. 3.3, plotted against the real and imaginary parts of the Laplace variable $s = \sigma + j\omega$. The Laplace transform converges for $s > -a$, i.e. forward of the dashed line. The modulus of the Fourier transform is a slice through the surface on the $j\omega$ axis and is denoted by the hatched section.

This provides the convergence condition for the Laplace transform of this signal, and assuming this condition is fulfilled, the integral in equation (3.3.5) converges and in this case the Laplace transform reduces to

$$F(s) = \frac{1}{s+a}. \tag{3.3.7}$$

The Laplace transform has a complex value for all σ and ω, and so is difficult to plot in general. The modulus of $F(s)$ is sketched in Fig. 3.4 as a surface in the σ, $j\omega$ plane for values of σ within the region of convergence $\sigma > -a$. The Fourier transform of this signal also converges for $a > 0$, and a geometrical interpretation of the modulus of the Fourier transform can be obtained from Fig. 3.4 by taking a "slice" through the surface on the $j\omega$ axis, i.e. for $\sigma = 0$.

The real power of the Laplace transform is demonstrated when signals which do not have a convergent Fourier transform are considered. If, for example, the waveform defined by equation (3.3.4) were a diverging exponential ($a < 0$), a region of convergence of the Laplace transform would still exist, but on the other side of the $j\omega$ axis. The ability of the Laplace transform to describe such signals (which could, for example, be generated as the outputs of unstable systems) in a mathematically correct way makes it a powerful tool in the analysis of the stability of linear systems.

Some important properties of Laplace transforms (following Kuo, 1966, Ch. 6) are listed below, where $\mathscr{L}[\ ]$ is used to denote the Laplace transform operator.

Linearity

$$\mathscr{L}[af_1(t) + bf_2(t)] = a\mathscr{L}[f_1(t)] + b\mathscr{L}[f_2(t)], \tag{3.3.8}$$

where a and b are constants.

Differentiation

$$\mathscr{L}\left[\frac{\mathrm{d}f(t)}{\mathrm{d}t}\right] = s\mathscr{L}[f(t)] - f(0). \tag{3.3.9}$$

Integration

$$\mathscr{L}[\int_0^t f(t)\mathrm{d}t] = \frac{1}{s}\mathscr{L}[f(t)]. \tag{3.3.10}$$

Convolution

$$\mathscr{L}[\int_{-\infty}^{\infty} f_1(\tau)f_2(t-\tau)\mathrm{d}\tau] = \mathscr{L}[f_1(t)]\ \mathscr{L}[f_2(t)]. \tag{3.3.11}$$

These properties can be used to calculate the Laplace transform of the linear differential equations defining a lumped parameter system. For example, the acoustical cavity described in Section 3.2 is governed by the differential equation

$$C_\mathrm{a}\frac{\mathrm{d}p(t)}{\mathrm{d}t} + \frac{1}{R_\mathrm{a}}p(t) = q(t), \tag{3.3.12}$$

which can be transformed term by term to give

$$sC_\mathrm{a}P(s) + \frac{1}{R_\mathrm{a}}P(s) = Q(s), \tag{3.3.13}$$

where it has been assumed that the initial condition, $p(0)$, is zero.

The s-domain *transfer function* of the system is defined to be the ratio of the Laplace transform of the output of the system, to the Laplace transform of its input. For the example above, the transfer function, $H(s)$, of the cavity is

$$H(s) = \frac{P(s)}{Q(s)} = \frac{R_\mathrm{a}}{1 + sC_\mathrm{a}R_\mathrm{a}} = \frac{1/C_\mathrm{a}}{s + 1/R_\mathrm{a}C_\mathrm{a}}. \tag{3.3.14}$$

Apart from the constant $1/C_\mathrm{a}$, this function is exactly the same as the Laplace transform of the exponential signal considered above (equation (3.3.7)) with $a = 1/R_\mathrm{a}C_\mathrm{a}$. Since $R_\mathrm{a}C_\mathrm{a} > 0$, the *frequency response* of the system is well defined, and is obtained by setting $s = j\omega$ in the transfer function equation (3.3.14) to give

$$H(j\omega) = \frac{R_\mathrm{a}}{1 + j\omega R_\mathrm{a}C_\mathrm{a}}. \tag{3.3.15}$$

The modulus and phase of this function are plotted in Fig. 3.5.

The response of a linear system to an impulsive excitation plays an important role in the theory of linear systems. The impulse function (or formally the Dirac delta function) $\delta(t)$ was introduced in Section 2.3 of Chapter 2 as the limiting case of a large narrow pulse, and can be defined by the equations

$$\delta(t) = 0 \quad \text{if} \quad t \neq 0, \quad \text{and} \quad \int_{-\infty}^{\infty} \delta(t)\mathrm{d}t = 1. \tag{3.3.16}$$

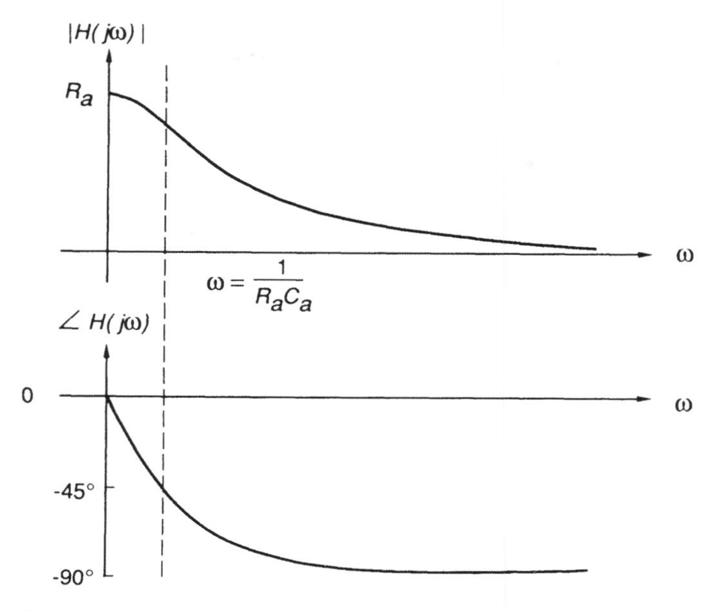

Fig. 3.5 The frequency response of the acoustic cavity.

The Laplace transform of such a function is unity. The Laplace transform of the output of a system excited by an impulse is thus the same as the s-domain transfer function. The time domain *impulse response* $h(t)$ of a system can be obtained from the inverse Laplace transform of its transfer function, and is defined to be

$$h(t) = \frac{1}{2\pi j}\int_{\sigma - j\infty}^{\sigma + j\infty} H(s)e^{st}\mathrm{d}s. \tag{3.3.17}$$

This equation can be evaluated analytically by using contour integration and the calculus of residues (the reader is referred to the discussion presented by Papoulis, 1981, Ch. 7). In practice, however, the inverse Laplace transform for model problems is most often deduced from standard tables of Laplace transforms. Such tables are readily available and presented, for example, by Kuo (1966) and Lynn (1982).

For the acoustic cavity considered here, a Laplace transform of identical form to its transfer function was obtained as the forward transform of the exponential signal (equation (3.3.4)). So this waveform must also be the inverse transform of the transfer function of the cavity, and thus be equal to the impulse response of the cavity. The output variable in this example is the pressure and the input variable is the volume velocity of the driving piston. An impulse of volume velocity has a convenient physical interpretation here as an instantaneous movement of the piston, i.e. a step function in its displacement. So, if the time history of the total volume $V_{\mathrm{tot}}(t)$ of the cavity is

$$V_{\mathrm{tot}}(t) = \begin{Bmatrix} V_0, & t<0 \\ V_0 + V, & t\geq 0 \end{Bmatrix}, \tag{3.3.18}$$

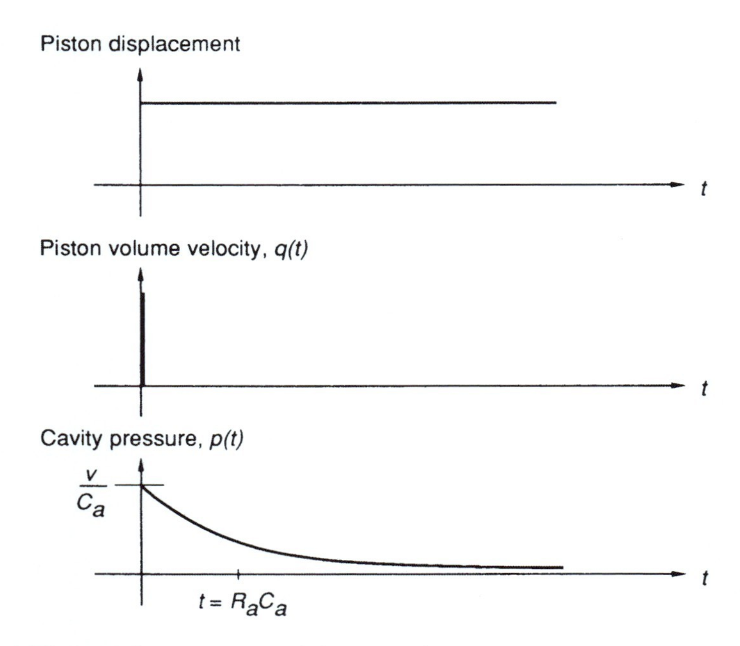

Fig. 3.6 The transient response of the acoustic cavity.

then the volume velocity is given by

$$q(t) = V\delta(t), \tag{3.3.19}$$

and therefore the solution for the resulting pressure can be written as

$$p(t) = \frac{V}{C_a} e^{-t/R_a C_a}. \tag{3.3.20}$$

If the time constant $R_a C_a$ is very large, the pressure rise in the cavity will persist for a long time. This rise in pressure is given by V/C_a, which is exactly the same as that deduced from equation (3.2.1) above. The waveforms for the piston displacement, volume velocity and cavity pressure are shown in Fig. 3.6.

3.4 Properties of the transfer function

The system transfer function, deduced from the Laplace transform of the differential equation defining the fundamental physical behaviour of the system, has been shown to be a useful intermediary in calculating the frequency response and impulse response of a system. We have assumed, however, in making the substitution $s = j\omega$ in order to evaluate the frequency response, that the Laplace transform of the system's impulse response converges in this region, i.e. the system, does have a defined frequency response. Before discussing the conditions under which this assumption is valid we introduce a very convenient way of summarising the nature of a system's response.

If the transfer function of a system defined by a linear ordinary differential equation of finite order is evaluated (and the differential equation has constant coefficients so that the system is time invariant), it can be expressed as the ratio of two polynomials in s, given by $N(s)$ and $D(s)$, where

$$H(s) = \frac{N(s)}{D(s)}. \tag{3.4.1}$$

The roots of the polynomial $D(s)$ are known as the *poles* of the transfer function $H(s)$, and the roots of the polynomial $N(s)$ are known as the *zeros* of $H(s)$. If $N(s)$ and $D(s)$ are factorised into first order terms, the transfer function may be written as

$$H(s) = \frac{K(s-z_1)(s-z_2)\ \ldots}{(s-p_1)(s-p_2)\ \ldots}, \tag{3.4.2}$$

where K is a constant, z_1, z_2, etc. are the zeros and p_1, p_2, etc. are the poles of $H(s)$, both of which are either real, or appear as complex conjugate pairs. (A more detailed discussion is presented by, for example, Papoulis, 1981, Ch. 4.) A pole-zero diagram is a representation of the s-plane with the poles marked as crosses and the zeros marked as circles. For example, the transfer function considered previously

$$H(s) = \frac{1}{s+a}, \tag{3.4.3}$$

has no zeros and one pole at $s = -a$. We have seen that this system has an impulse response given by

$$h(t) = \begin{Bmatrix} e^{-at}, & t \geq 0 \\ 0, & t < 0 \end{Bmatrix}. \tag{3.4.4}$$

The pole-zero diagram for this system with $a > 0$ is shown in Fig. 3.7(a). Comparing this to the contour plot of the transfer function for the same system, in Fig. 3.4, reveals that the pole-zero diagram could be considered as a plan view of the contour plot: the singular values of s where $H(s) \rightarrow \infty$ are marked with crosses and the points at which $H(s) = 0$, if they exist, are marked with circles.

A system is *stable* with respect to its output if its impulse response is absolutely integrable (see the discussion presented by, for example, Kuo, 1966, Ch.10). Thus a stable system requires that its impulse response $h(t)$ satisfies

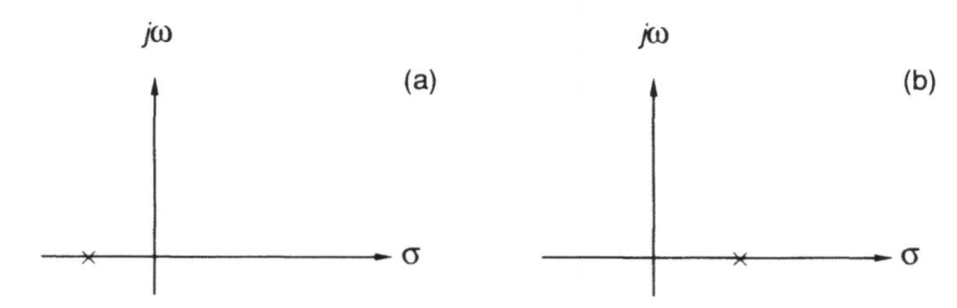

Fig. 3.7 The pole positions for a stable (a) and unstable (b) first order system.

$$\int_0^\infty |h(t)|\,dt < \infty. \tag{3.4.5}$$

In the example above, the system is clearly stable according to this definition, and also according to common sense, provided $a > 0$. If $a \leq 0$, the system is *unstable* according to this definition. The pole-zero diagram corresponding to the unstable version of this first order system is shown in Fig. 3.7(b). Although illustrated by example here, it can be demonstrated that *any* system whose transfer function has all its poles on the left hand side of the s-plane (i.e. for $\sigma < 0$) must be stable, and any system which has any pole in the right hand half plane is unstable. For more detailed discussions, see, for example, Kuo (1966, Ch. 10) and Papoulis (1981, Ch. 4). The stability of a system can thus be assessed by inspection of its pole-zero diagram. Only if all the poles are in the left hand half plane will $H(s)$ converge on the $j\omega$ axis, so a system must be stable if it is to have a defined frequency response.

Another property of the system which can be inferred directly from the pole-zero diagram is whether the system is *minimum phase*. This means that of all the possible linear systems with a given magnitude of frequency response, there exists one whose phase response is the smallest possible. This is known as the minimum phase system. It can be shown that systems with complex conjugate pairs of zeros placed on either one side of the $j\omega$ axis or the other, have identical magnitudes to their frequency responses (see, for example, Papoulis, 1981, Ch. 4). It is, however, only the system which has all its zeros in the left hand half plane which is minimum phase. This is illustrated for a simple example in Fig. 3.8, in which the magnitude and phase responses of the two systems are also shown. One of the practically important consequences of a system being minimum phase is that, provided it has no zeros on the $j\omega$ axis, its inverse, i.e. the system with transfer function

$$\frac{1}{H(s)} = \frac{D(s)}{N(s)}, \tag{3.4.6}$$

is stable. This follows since the numerator $N(s)$ of $H(s)$ is now the denominator of $1/H(s)$, so the zeros of a minimum phase $H(s)$, which are all in the left hand half plane, become the poles of $1/H(s)$, which are similarly positioned so that $1/H(s)$ is stable.

The influence of the poles and zeros on the frequency response can be directly deduced from the pole-zero diagram by using the graphical construction described, for example, by Kuo (1966, Ch. 8). The frequency response of the system at a frequency of ω_0 is given by

$$H(j\omega_0) = \frac{K(j\omega_0 - z_1)(j\omega_0 - z_2)\ldots}{(j\omega_0 - p_1)(j\omega_0 - p_2)\ldots}. \tag{3.4.7}$$

Each of the factors $(j\omega_0 - z_i)$ or $(j\omega_0 - p_k)$ can be represented as vectors in the pole zero diagram from the zero z_i or the pole p_k to a point $j\omega_0$ on the imaginary axis, as illustrated in Fig. 3.9. We can express these vectors in polar form as

$$j\omega_0 - z_i = N_i e^{j\psi_i} \quad \text{and} \quad j\omega_0 - p_k = M_k e^{j\theta_k}, \tag{3.4.8a,b}$$

and write the frequency response as

$$H(j\omega_0) = \frac{K(N_1 N_2 \ldots)}{(M_1 M_2 \ldots)} e^{j(\psi_1 + \psi_2 + \ldots - \theta_1 - \theta_2 - \ldots)}, \tag{3.4.9}$$

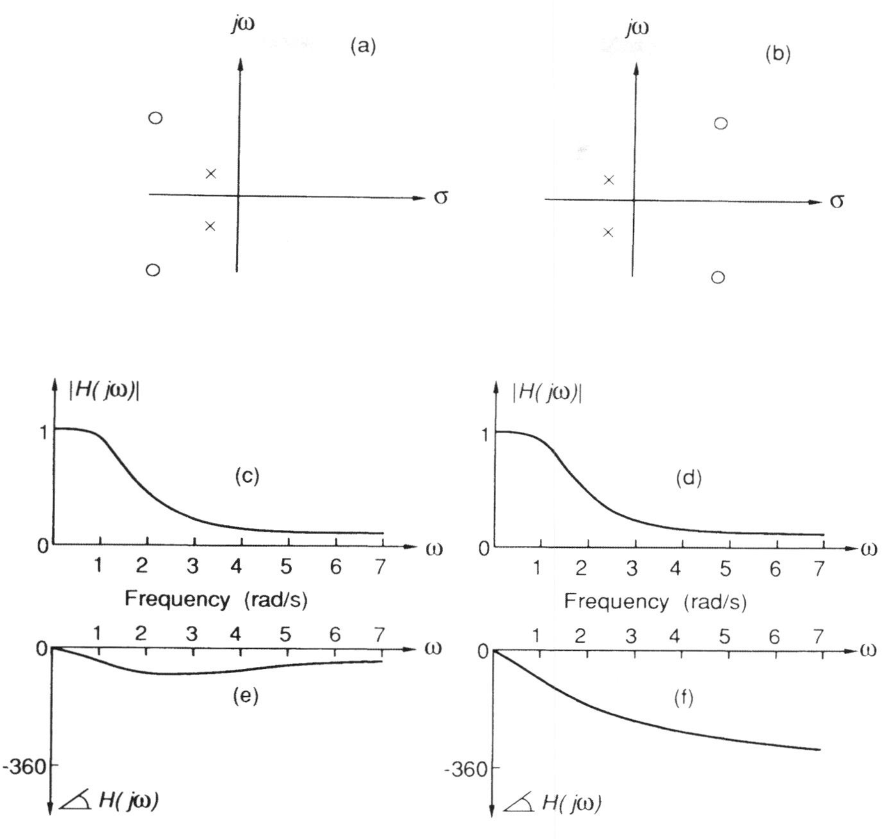

Fig. 3.8 The pole-zero diagrams of two systems which are minimum phase (a) and non-minimum phase (b), together with the moduli (c) and (d), and phases of their frequency responses (e) and (f).

which, as we have said, is valid only if all poles lie in the left hand half plane, i.e. $90° > \theta_k > -90°$ for all poles. Thus the magnitude of the frequency response will be dictated by the *distances* of the poles and zeros from the point on the $j\omega$ axis corresponding to the frequency of interest. The closer are the poles and the further are the zeros from this point on the $j\omega$ axis, the greater will be the response of the system. The phase of the frequency response is dictated by the sum of the phase leads ψ_i introduced by the zeros and phase lags in θ_n introduced by the poles. In the example shown in Figs 3.8(a) and 3.9, where there are as many zeros as poles, the phase contributions become equal and opposite as $\omega \rightarrow \infty$ and the phase of the frequency response tends to zero at high frequencies.

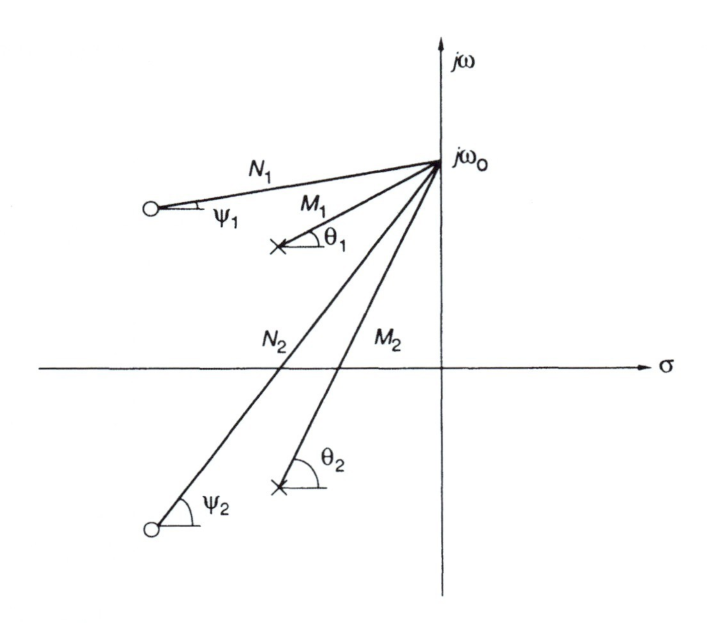

Fig. 3.9 The influence of the poles and zeros on the frequency response at $\omega = \omega_0$.

3.5 The convolution integral

Suppose we approximate some continuous waveform $f(t)$ by a series of pulses

$$f(t) \approx \sum_{n=-\infty}^{\infty} f_n p(t - n\Delta t), \tag{3.5.1}$$

where the pulse waveform is defined to be

$$p(t) = \begin{Bmatrix} 1, & 0<t<\Delta t \\ 0, & \text{otherwise} \end{Bmatrix}, \tag{3.5.2}$$

and where Δt is the time duration of the pulse and f_n is a constant weighting factor for the nth pulse equal to the value of $f(t)$ at $t = n\Delta t$. Let the response of some linear, time invariant system to an isolated pulse waveform be $r(t)$, then using the principle of superposition and the fact that the system is time invariant, the response of such a system to a sequence of delayed and weighted pulses will be

$$g(t) \approx \sum_{n=-\infty}^{\infty} f_n r(t - n\Delta t). \tag{3.5.3}$$

As the width of the pulse is decreased to zero, we can replace the pulse function in equation (3.5.1) by the Dirac delta function and the summation by an integral. Thus the input to the system can be written as

$$f(t) = \int_{-\infty}^{\infty} f(\tau)\delta(t-\tau)\mathrm{d}\tau. \tag{3.5.4}$$

If the response of the linear system to an impulse is $h(t)$, then we can now express the output of the system given by equation (3.5.3) as

$$g(t) = \int_{-\infty}^{\infty} f(\tau)h(t-\tau)\mathrm{d}\tau, \tag{3.5.5}$$

or alternatively, as

$$g(t) = \int_{-\infty}^{\infty} h(\tau)f(t-\tau)\mathrm{d}\tau. \tag{3.5.6}$$

If the system is *causal*, i.e. its impulse response is zero for all t less than zero, such that

$$h(t) = 0 \quad \text{for} \quad t<0, \tag{3.5.7}$$

then the output of the system can be written as

$$g(t) = \int_{0}^{\infty} h(\tau)f(t-\tau)\mathrm{d}\tau. \tag{3.5.8}$$

Each of equations (3.5.5), (3.5.6) and (3.5.8) is a form of the *convolution integral*, which is sometimes denoted by

$$g(t) = f(t) * h(t) = h(t) * f(t). \tag{3.5.9}$$

The response of a linear system to deterministic signals has now been derived in the time, Fourier and Laplace domains. The important results are summarised in Fig. 3.10. Figure 3.11 also shows the connections between the properties of the system which have been discussed in the preceding sections.

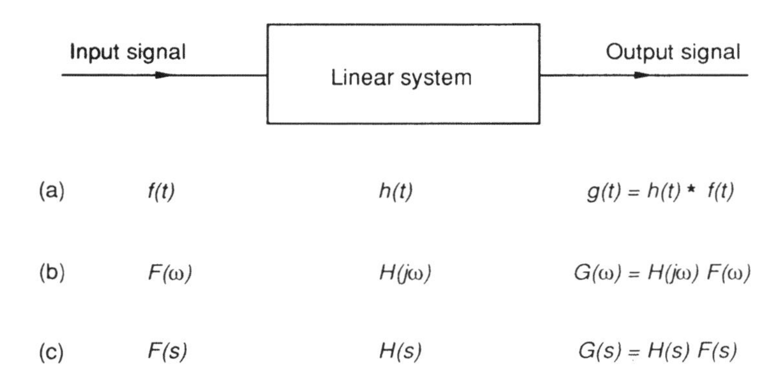

Fig. 3.10 The response of a linear system expressed in three domains: (a) time (via the convolution integral and impulse response), (b) Fourier (via multiplication by the complex frequency response) and (c) Laplace (via multiplication by the s-domain transfer function).

3.6 Response to random inputs

If the input signal to a linear system is random in nature, the output waveform, or its instantaneous spectrum, are of little use in describing the statistical properties of the

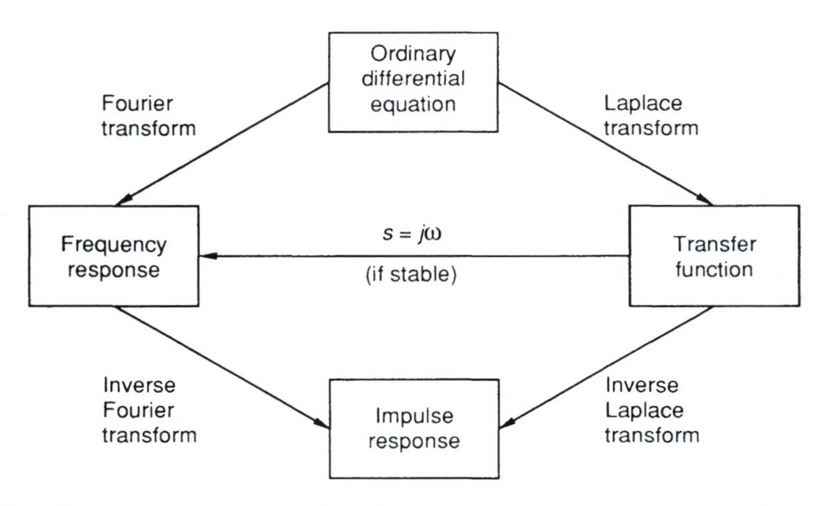

Fig. 3.11 A diagrammatic representation of the connections between the various properties of a linear system.

resulting signal. The important statistical parameters in the time domain are the auto- and cross-correlations, and in the frequency domain they are the auto- and cross-spectra introduced in Chapter 2. In this section, we derive the values of these functions between the input and output signals of a linear time invariant system driven by a random process. A more comprehensive discussion is presented, for example, by Newland (1984, Ch. 7).

The autocorrelation of the output waveform is defined as

$$R_{gg}(\tau) = E[g(t)g(t+\tau)], \tag{3.6.1}$$

which, by using equation (3.5.6), can be written in terms of the input waveform and the impulse response of the linear system to give

$$R_{gg}(\tau) = E\left[\int_{-\infty}^{\infty} h(\tau_1)f(t-\tau_1)\mathrm{d}\tau_1 \int_{-\infty}^{\infty} h(\tau_2)f(t+\tau-\tau_2)\mathrm{d}\tau_2\right]. \tag{3.6.2}$$

Since the impulse response is not a function of time, the expectation operator can be used to write this expression as

$$R_{gg}(\tau) = \int_{-\infty}^{\infty}\int_{-\infty}^{\infty} h(\tau_1)h(\tau_2)E[f(t-\tau_1)f(t+\tau-\tau_2)]\mathrm{d}\tau_1\mathrm{d}\tau_2. \tag{3.6.3}$$

The expectation $E[f(t-\tau_1)f(t+\tau-\tau_2)]$ can be recognised as the autocorrelation function of the input signal at a lag of $(\tau+\tau_1-\tau_2)$ so that by rearranging the order of integration

$$R_{gg}(\tau) = \int_{-\infty}^{\infty} h(\tau_1)\left[\int_{-\infty}^{\infty} h(\tau_2)R_{ff}(\tau+\tau_1-\tau_2)\mathrm{d}\tau_2\right]\mathrm{d}\tau_1. \tag{3.6.4}$$

By substituting $\tau_1 = -u$, this can be identified as the convolution of $h(-u)$ with the function in square brackets, which is itself the convolution of $h(\tau)$ with $R_{ff}(\tau)$. The autocorrelation of the output signal can thus be written as

$$R_{gg}(\tau) = h(\tau) * h(-\tau) * R_{ff}(\tau). \tag{3.6.5}$$

The power spectral density of the output signal can be obtained by taking the Fourier transform of equation (3.6.6) and using the fact that convolution in the time domain is equivalent to multiplication in the frequency domain. This results in

$$S_{gg}(\omega) = H(j\omega)H^*(j\omega)S_{ff}(\omega) = |H(j\omega)|^2 S_{ff}(\omega). \tag{3.6.6}$$

We have also used the fact that the Fourier transform of a time-reversed signal is the complex conjugate of the Fourier transform of the signal itself. (Note that the symbol * when used as a superscript refers to complex conjugation.)

The cross-correlation between the output and input is given by

$$R_{fg}(\tau) = E[f(t)g(t+\tau)]. \tag{3.6.7}$$

Again, expressing the output in terms of the impulse response of the linear system enables this expression for the cross-correlation function to be written as

$$R_{fg}(\tau) = E\left[f(t)\int_{-\infty}^{\infty} h(\tau_1)f(t+\tau-\tau_1)\mathrm{d}\tau_1\right]$$

$$\int_{-\infty}^{\infty} h(\tau_1)E[f(t)f(t+\tau-\tau_1)]\mathrm{d}\tau_1. \tag{3.6.8}$$

This expression can in turn be reduced to

$$R_{fg}(\tau) = \int_{-\infty}^{\infty} h(\tau_1)R_{ff}(\tau-\tau_1)\mathrm{d}\tau_1 = h(\tau) * R_{ff}(\tau). \tag{3.6.9}$$

The cross-correlation function is thus the convolution of the autocorrelation function of the input signal with the system's impulse response. By taking the Fourier transform of the equation for the cross-correlation in equation (3.6.9), the expression for the cross-spectral density between the input and output of the system is obtained. This is given by

$$S_{fg}(\omega) = H(j\omega)S_{ff}(\omega). \tag{3.6.10}$$

A summary of the various measures of the response of a linear system to random excitation is provided in Fig. 3.12.

The observed output of the physical system may be corrupted by uncorrelated measurement noise, as illustrated in Fig. 3.13. We can express the measured output as

$$\hat{g}(t) = g(t) + n(t), \tag{3.6.11}$$

where we assume that the noise signal $n(t)$ is uncorrelated with the input signal $f(t)$, and therefore we can write

$$E[n(t)f(t+\tau)] = 0, \quad \text{and} \quad E[N^*(\omega)F(\omega)] = 0, \tag{3.6.12}$$

where the expectation operator used in the second of these expressions implies the operation defined by equation (2.7.12) of Chapter 2. The cross-spectral density between the input and observed output signals can thus be written as

$$S_{f\hat{g}}(\omega) = E[F^*(\omega)(G(\omega)+N(\omega))] = E[F^*(\omega)G(\omega)], \tag{3.6.13}$$

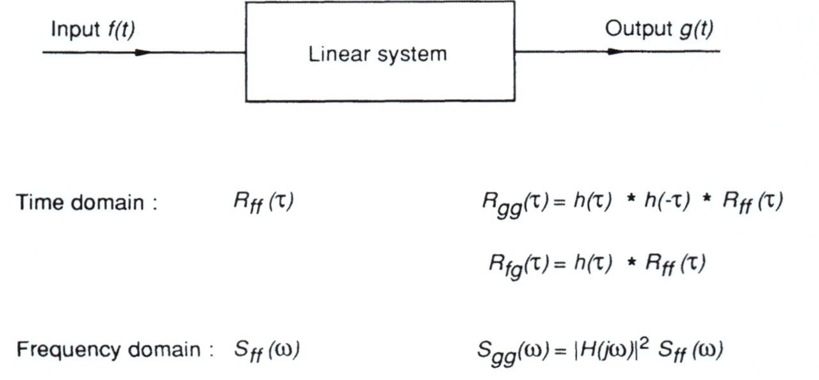

Fig. 3.12 The statistical behaviour of a linear system driven by a random input signal in the time domain (in terms of correlation functions) and the frequency domain (in terms of spectral densities).

so, even with corrupting output noise, we have the same relationship for the cross-spectral density between the input and the output as that given by equation (3.6.10). Thus

$$S_{f\hat{g}}(\omega) = S_{fg}(\omega) = H(j\omega)S_{ff}(\omega). \tag{3.6.14}$$

The frequency response of an unknown system with output noise can thus be estimated without any bias error (i.e. the expectation of the estimate gives the true value) from the cross-spectral density between the input and observed output, and the power spectral density of the input. This estimate can be expressed as

$$H_1(j\omega) = \frac{S_{f\hat{g}}(\omega)}{S_{ff}(\omega)}. \tag{3.6.15}$$

Randall (1987) for example has also discussed the case in which uncorrelated measurement noise is present on the observed *input* signal $\hat{f}(t)$ and has demonstrated that the transfer function estimate

$$H_2(j\omega) = \frac{S_{gg}(\omega)}{S_{g\hat{f}}(\omega)}, \tag{3.6.16}$$

is unbiased under these conditions.

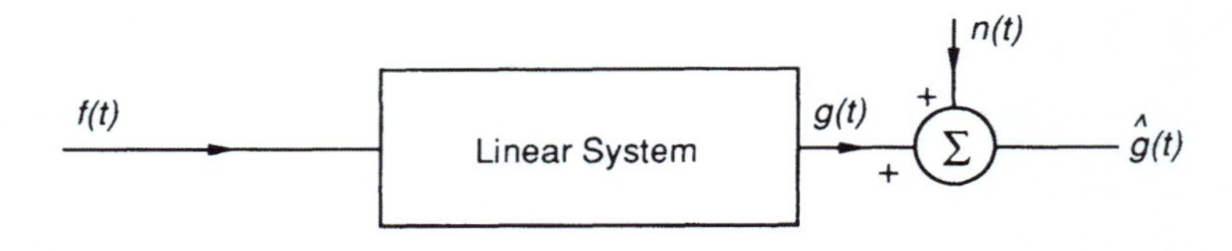

Fig. 3.13 Block diagram of a linear system whose output is corrupted by uncorrelated measurement noise.

3.7 The coherence function

When the output of a linear system is corrupted with uncorrelated measurement noise, as illustrated in Fig. 3.13, the power spectral density of the output signal is

$$S_{\hat{g}\hat{g}}(\omega) = S_{gg}(\omega) + S_{nn}(\omega). \tag{3.7.1}$$

The term $S_{\hat{g}\hat{g}}(\omega)$ is proportional to the total power output of the system at frequency ω and $S_{gg}(\omega)$ is proportional to the power output of the system due to the original input signal, $f(t)$, at the same frequency. Using equations (3.6.6) and (3.6.14), note that we can write

$$S_{gg}(\omega) = |H(j\omega)|^2 S_{ff}(\omega) = \frac{|S_{f\hat{g}}(\omega)|^2}{S_{ff}(\omega)}. \tag{3.7.2}$$

The ratio of the power output of the system due to the input $f(t)$, to the total power output, at frequency ω, is thus given by

$$\frac{S_{gg}(\omega)}{S_{\hat{g}\hat{g}}(\omega)} = \frac{|S_{f\hat{g}}(\omega)|^2}{S_{ff}(\omega)S_{\hat{g}\hat{g}}(\omega)} = \gamma^2_{f\hat{g}}(\omega). \tag{3.7.3}$$

This quantity is known as the *ordinary coherence function* (or just "coherence") relating f and $\hat{g}$. The noise signal, $n(t)$, may be more widely interpreted than just measurement noise. It could, for example, be caused by non-linearities in the response of the system. The coherence can then be interpreted as the ratio of the output power *linearly* derived from $f(t)$ to the total output power. $S_{gg}(\omega)$ can thus be interpreted as the *coherent output power*, that is the output power spectrum caused by signals which are coherent with $f(t)$. This is given by

$$S_{gg}(\omega) = \gamma^2_{f\hat{g}}(\omega)S_{\hat{g}\hat{g}}(\omega). \tag{3.7.4}$$

If no source of corrupting noise is present, so that $\hat{g}(t) = g(t)$ and the system is linear, then $\gamma^2_{f\hat{g}}(\omega)$ will be unity. The ratio of the output power due to the corrupting noise to the total output power at frequency ω, can also now be expressed as

$$\frac{S_{nn}(\omega)}{S_{\hat{g}\hat{g}}(\omega)} = \frac{S_{\hat{g}\hat{g}}(\omega) - S_{gg}(\omega)}{S_{\hat{g}\hat{g}}(\omega)} = 1 - \gamma^2_{f\hat{g}}(\omega), \tag{3.7.5}$$

and the "signal to noise ratio" is

$$\frac{S_{gg}(\omega)}{S_{nn}(\omega)} = \frac{\gamma^2_{f\hat{g}}(\omega)}{1 - \gamma^2_{f\hat{g}}(\omega)}. \tag{3.7.6}$$

A further discussion of the above relationships has been presented by, for example, Randall (1987).

3.8 Optimal filtering

In this section we review briefly the consequences of adjusting the impulse response of an analogue filter so that its output is as close as possible to some desired signal. More detailed treatments have been given by Van Trees (1968, Ch. 3, Ch. 6) and

Orfanides (1985, Ch. 4). Consider the block diagram of Fig. 3.14. The object is to design a system with a *causal* impulse response so that the difference between the desired signal, $d(t)$, and the filter output

$$\hat{d}(t) = \int_0^\infty h(\tau_1) f(t-\tau_1) \mathrm{d}\tau_1, \tag{3.8.1}$$

is in some way minimised. Papoulis (1984, Ch. 13), for example, has shown that adjusting the impulse response to minimise the expectation of the squared error signal given by

$$E[e^2(t)] = E[(d(t) - \hat{d}(t))^2], \tag{3.8.2}$$

is equivalent to arranging for the filter to remove all components of the error signal, $e(t)$, which are causally correlated with the input signal, $f(t)$. Alternative treatments are presented by, for example, Van Trees (1968, Ch. 6) and Orfanides (1985, Ch. 4). Once the filter has been optimally adjusted, the cross-correlation between the current error signal and all previous values of the input signals will thus be zero and under these conditions we have

$$E[e(t) f(t-\tau_2)] = 0 \quad \text{for} \quad 0 < \tau_2 < \infty. \tag{3.8.3}$$

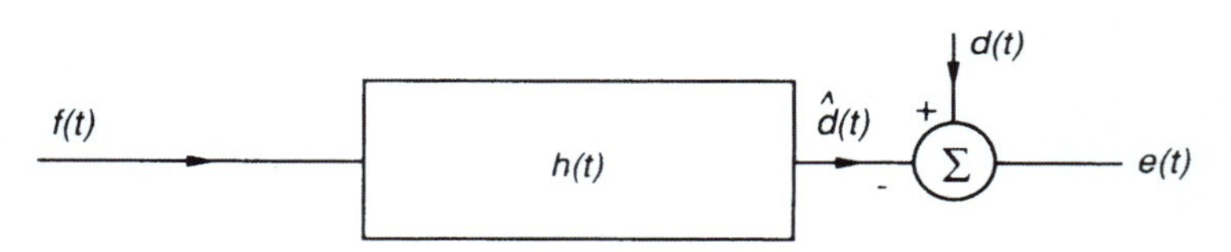

Fig. 3.14 A causal linear system operation on an input reference signal, $f(t)$, to produce an estimate, $\hat{d}(t)$, of desired signal, $d(t)$.

This is known as the *orthogonality principle* (Papoulis, 1984, Ch. 13). Substituting the expressions for $e(t)$ and $\hat{d}(t)$ into this equation gives

$$E\left[(d(t) - \int_0^\infty h(\tau_1) f(t-\tau_1) \mathrm{d}\tau_1) f(t-\tau_2)\right] = 0. \tag{3.8.4}$$

Since the impulse response does not depend on time, the expectations of the products of the time varying signals may be taken, and written respectively as the cross-correlation between the desired and input signals, and the autocorrelation of the input signal. This enables equation (3.8.4) to be written as

$$R_{fd}(\tau_2) - \int_0^\infty h(\tau_1) R_{ff}(\tau_2 - \tau_1) \mathrm{d}\tau_1 = 0, \qquad 0 < \tau_2 < \infty, \tag{3.8.5}$$

where the property $R_{df}(-\tau_2) = R_{fd}(\tau_2)$ has been used. Equation (3.8.5) is a *Wiener–Hopf integral equation*, and the necessary and sufficient condition for $E[e^2(t)]$ to be minimised is that the impulse response $h(\tau_1)$ satisfies this equation.

One interesting way of looking at the action of this optimal filter is that it splits the desired signal into two components given by

$$d(t) = e(t) + \hat{d}(t). \tag{3.8.6}$$

One component, $\hat{d}(t)$, is perfectly correlated with the input signal; it has been derived from $f(t)$ by linear filtering. The other component is perfectly *uncorrelated* with the input signal according to equation (3.8.3). The two components of $d(t)$ are thus themselves uncorrelated and are sometimes said to be "orthogonal". This decomposition of a signal into two components, one of which is perfectly correlated and one of which is perfectly uncorrelated with a reference signal, will be used in a rather different context in Chapter 11.

If the desired signal were the input waveform at some time T in the *future* ($d(t) = f(t+T)$) the problem above reduces to that of *predicting* the future value of the input signal from the waveform $f(t)$ and all previous values of this waveform, as implied by the convolution in equation (3.8.1). Under these circumstances, the Wiener–Hopf integral equation (3.8.5) reduces to

$$R_{ff}(\tau_2 + T) + \int_0^\infty h(\tau_1) R_{ff}(\tau_2 - \tau_1)\mathrm{d}\tau_1 = 0, \qquad 0 < \tau_2 < \infty. \tag{3.8.7}$$

We shall deal with the solution of this equation in Section 8.6 within the context of the active cancellation of far field radiation of random sound. We could also calculate the optimal prediction of $f(t+T)$ obtainable from the single, instantaneous measurement $f(t)$ only, in which case equation (3.8.1) becomes

$$\hat{d}(t) = hf(t), \tag{3.8.8}$$

where h is a scalar constant, and equation (3.8.4) becomes

$$E[(f(t+T) - hf(t))f(t)] = 0.$$

This in turn implies that

$$R_{ff}(T) = hR_{ff}(0), \tag{3.8.9}$$

and the optimal prediction of $f(t+T)$ from the single measurement $f(t)$ is thus given by multiplying $f(t)$ by a constant given by $h = R_{ff}(T)/R_{ff}(0)$. A similar result for the optimal prediction from a point measurement of a *spatially* varying random variable will be used in Section 11.4.

4
Digital Filters

4.1 Advantages of digital processing

In this chapter we aim to provide a brief introduction to the properties of digital filters and the conventional methods used to design them. Only an outline of the various topics will be provided in order to provide a clear framework for later development and specific application of the material to active sound control problems. More details may be found in numerous textbooks, for example, Oppenheim and Schafer (1975), Rabiner and Gold (1975), Bozic (1979), Lynn (1982), and Bellanger (1984).

In the electronic controllers for active control systems increasing use is being made of digital filters. The reasons for this are similar to the reasons for the replacement of analogue signal processing by digital signal processing (DSP) in so many other areas of application, and these are briefly discussed below.

Flexibility

Most implementations of digital filters are now programs running on a programmable digital processor. It is comparatively easy to modify this software to change the coefficients of the digital filter and so adjust the response of the system. If the hardware and software have been well designed, it should also be possible to change the order, or even the structure, of the digital filter by reasonably straightforward program changes.

Adaptability

As well as implementing a fixed coefficient digital filter on the programmable signal processor, algorithms for the adaptation of the filter coefficients can be programmed to run on the same hardware, and operate alongside the digital filter program. The ability to implement self-adaptive controllers on a single device is increasingly making practical active control systems available at a reasonable cost.

Accuracy

The arithmetic precision which is commonly available in modern digital processors (typically 16 bits) can be used to specify very accurate filter coefficients, giving the fine control over the modulus and phase response of the controller at each frequency which is important for active control.

Stability

In fully adaptive control systems, the controller can (in principle) correct for variations in the filter response due to temperature variations, component ageing, etc. If the controller is not fully adaptive the very stable characteristics of digital filters, whose response depends only on fixed coefficient values and a crystal controlled clock, can be of great advantage. In order to prevent aliasing, however, digital controllers are often equipped with anti-aliasing and reconstruction filters, which are usually analogue devices. The analogue components in these filters can be prone to drift problems with temperature and age, which may well compromise the stability of the digital part of the system (see the discussion presented by Swinbanks, 1985).

Cost

There are now a large number of microprocessors available specifically designed for digital signal processing applications. Because of the size of the market in telecommunications and other mass production areas, the cost of these devices is moderate and has decreased considerably over the past few years in many cases. Technological advances in semiconductor device design and fabrication have meant that the performance of these digital signal processing devices has also been increasing rapidly. The processing performance available for a fixed real price has at least doubled every 2 years for the past decade (Texas Instruments, 1987).

4.2 Sampled signals

Before discussing the properties and design of digital filters, it is worthwhile to review briefly the important features of the sampled signals upon which they operate. In Chapter 2, sampled signals were introduced as analogue signals which had been passed through an analogue-to-digital converter. Here we will discuss some sampled signals, or sequences as they should properly be called, which are generated from simple mathematical formulae. These simple sequences help to orientate oneself in the properties of the digital world; their properties also are widely used in later chapters.

A general sequence is denoted as $f(n)$, where n can only take integer values, and so $f(n)$ is only defined for $n = 0, \pm 1, \pm 2$, etc. We begin with the unit step function defined by

$$f(n) = u(n) = \begin{Bmatrix} 1, & n \geq 0 \\ 0, & n < 0 \end{Bmatrix}, \tag{4.2.1}$$

and which is plotted in Fig. 4.1. The Fourier transform of any sequence is defined (from Chapter 2) as

$$F(e^{j\omega T}) = \sum_{n=-\infty}^{\infty} f(n) e^{-j\omega T n}, \tag{4.2.2}$$

where ω is the frequency and T is the sample time, although it is the product ωT called the *normalised frequency* which is the important variable. If $f(n)$ is the unit step sequence, this summation does not converge and so the Fourier transform of the unit step sequence does not exist.

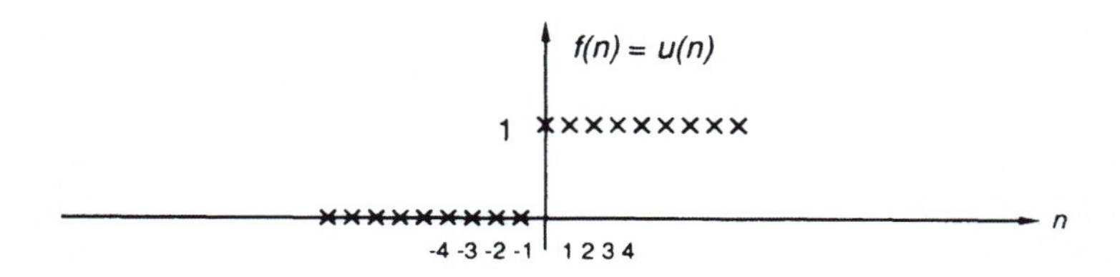

Fig. 4.1 The unit step sequence.

Two unit step sequences together can be used to describe a *digital pulse* such that

$$f(n) = u(n+M) - u(n-M). \tag{4.2.3}$$

The total number of non-zero samples in such a symmetric digital pulse is $2M+1 = N$, say. Such a sequence is illustrated in Fig. 4.2 with $N = 7$. The Fourier transform for this sequence does converge (see, for example, Oppenheim and Schafer, 1975, Ch. 1) and is given by

$$F(e^{j\omega T}) = \frac{\sin(\omega TN/2)}{\sin(\omega T/2)}, \tag{4.2.4}$$

which is entirely real in this case, since $f(n)$ is symmetrical about $n = 0$, and is illustrated for $N = 7$ in Fig. 4.3. If $\omega T/2 \ll 1$, this function is very much like the $(\sin x)/x$ or "sinc" function, that was obtained by Fourier transformation of an analogue pulse in Chapter 2. For $\omega T/2 \gtrsim 1$, it becomes significantly larger than its analogue equivalent until the function begins to repeat itself after $\omega T = \pi$. It is periodic every $\omega T = 2\pi$ as is the Fourier transform of any sequence. The "main lobe" of this Fourier transform lies between $-2\pi/N < \omega T < 2\pi/N$.

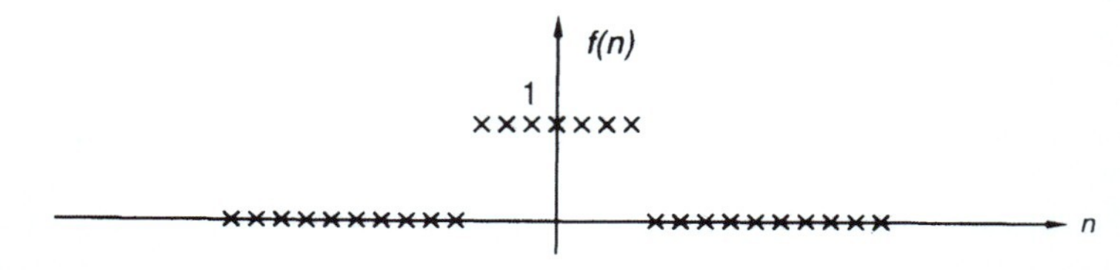

Fig. 4.2 Symmetrical digital pulse of seven samples duration.

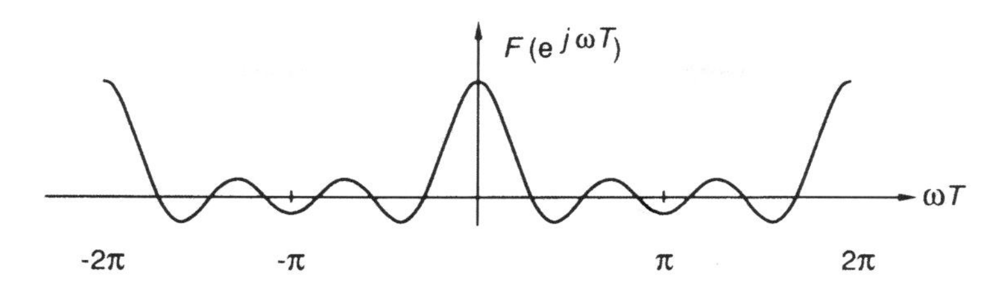

Fig. 4.3 The Fourier transform of the digital pulse.

A particularly important special case of such a digital pulse is when $N = 1$, in which case it becomes the unit sample sequence

$$f(n) = \delta(n) = \begin{Bmatrix} 1, & n = 0 \\ 0, & n \neq 0 \end{Bmatrix}, \tag{4.2.5}$$

sometimes called the *digital impulse*, and shown in Fig. 4.4. The simple analytic form of this impulse function (the *Kronecker delta function*) is in contrast to the mathematical complexity involved in defining the analogue impulse function (the Dirac delta function), and this leads to a mathematical description of digital systems which is, in many ways, easier than the equivalent analogue system. The Fourier transform of the unit sample sequence, obtained from equation (4.2.4) with $N = 1$, is $F(e^{j\omega T}) = 1$, i.e. it has an exactly uniform spectrum.

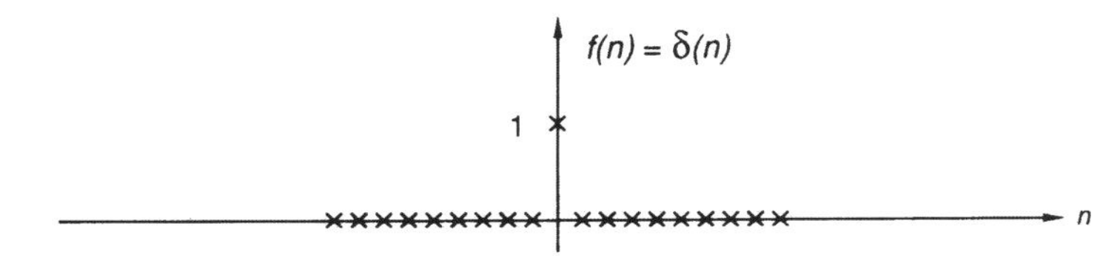

Fig. 4.4 The unit sample sequence or digital impulse.

The last sequence we consider is of the form

$$f(n) = \begin{Bmatrix} a^n, & n \geq 0 \\ 0, & n < 0 \end{Bmatrix}, \tag{4.2.6}$$

which can also be written as $f(n) = u(n)a^n$ for all n and which is illustrated in Fig. 4.5 for $a = 0.6$ and $a = 1.4$. It is called the *digital exponential* because the sampled version of an analogue exponential waveform can be written as

$$f(n) = e^{nT/\tau} = a^n, \tag{4.2.7}$$

where T is the sample time and τ is the time constant, which may be positive (diverging) or negative (converging). The term a is greater than 1 for a diverging sequence (e.g. $a = 1.4$ in Fig. 4.5) and less than 1 for a converging sequence (e.g. $a = 0.6$ in Fig. 4.5). The Fourier transform of this sequence converges only for $|a| < 1$, in which case

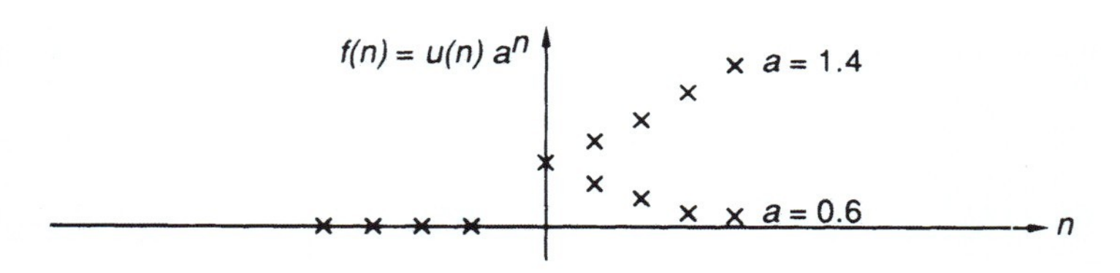

Fig. 4.5 The converging and diverging exponential sequence.

$$F(e^{j\omega T}) = [1 - ae^{-j\omega T}]^{-1}. \tag{4.2.8}$$

The modulus of this function is plotted for $a = 0.6$ in Fig. 4.6, and displays a general "low pass" characteristic. If $-1 < a < 0$ the sequence a^n converges but with each sample being of alternate sign and it cannot then be interpreted as a sampled exponential as defined in equation (4.2.7). In fact, its Fourier transform under these conditions then displays a "high pass" characteristic.

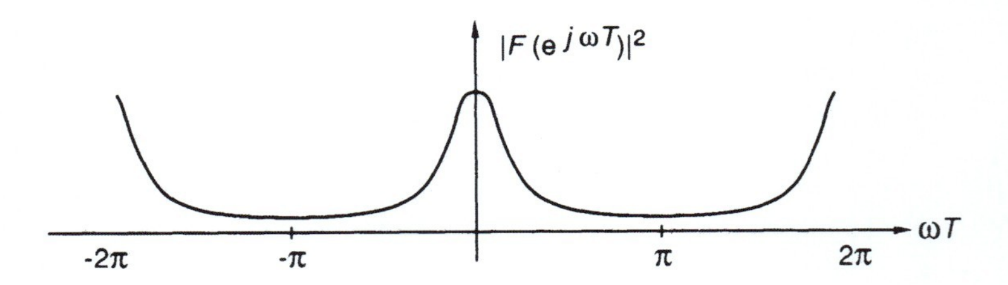

Fig. 4.6 The modulus of the Fourier transform of a converging exponential sequence with $a = 0.6$.

4.3 The *z*-transform

The need for a transform domain description of a step sequence, or diverging exponential sequence, motivates a generalisation of the Fourier transform of a sequence which is analogous to the Laplace transform for analogue systems. This is known as the *z-transform*, and the z-transform of a sequence $f(n)$ is defined to be

$$F(z) = \sum_{n=-\infty}^{\infty} f(n)z^{-n}, \tag{4.3.1}$$

where z is a complex variable. Strictly, this is the two-sided z-transform (see Oppenheim and Schafer, 1975, Ch. 2). Its region of convergence for the sequence defined by equation (4.2.6) is for all $|z| > a$, for example. For all sequences whose Fourier transforms are defined, the z-transform must converge for $|z| \geq 1$. Indeed, the Fourier transform can be considered as a special case of the z-transform with z set equal to $e^{j\omega T}$ such that

$$F(e^{j\omega T}) = F(z)|_{z = e^{j\omega T}}. \tag{4.3.2}$$

This relationship is the motivation for the notation $F(e^{j\omega T})$ introduced in Section 2.9 for the Fourier transform of a sequence.

The inverse z-transform is evaluated by contour integration or (more usually) by using a table of standard z-transforms as listed, for example, by Oppenheim and Schafer (1975). An important property of the z-transform is its form for *shifted* sequences. For example, the z-transform of the sequence $f(n)$ shifted, or delayed, by one sample, $f(n-1)$, is

$$\sum_{n=-\infty}^{\infty} f(n-1)z^{-n} = \sum_{m=-\infty}^{\infty} f(m)z^{-(m+1)} = z^{-1}F(z), \tag{4.3.3}$$

where $m = n-1$ and $F(z)$ is the z-transform of the unshifted sequence $f(n)$. The term z^{-1} is called the delay operator and block diagrams of digital systems with delays often contain elements labelled z^{-1}, despite the fact that sequences, rather than z-transforms, are being operated on. This usage, illustrated in Fig. 4.7, is so common we will retain it here.

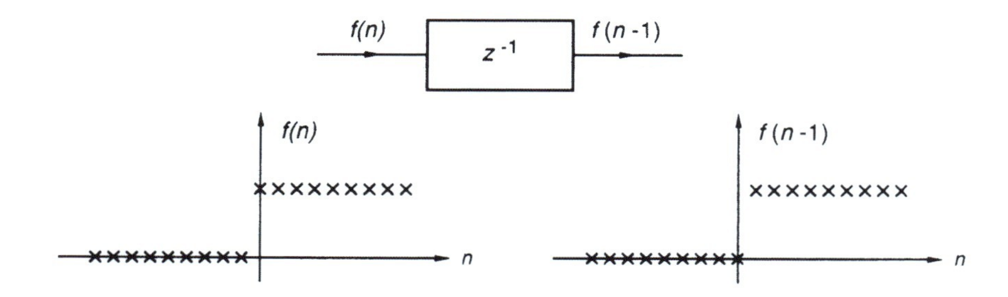

Fig. 4.7 The block diagram used to denote a single sample delay.

4.4 Digital systems

A digital system operates on a sequence of numbers (the digital input signal) to produce another sequence of numbers (the digital output signal). Figure 4.8 illustrates such a system in general, where the output sequence $g(n)$ is given by the input sequence, $f(n)$, transformed by some operator H. Thus we write

$$g(n) = H[f(n)]. \tag{4.4.1}$$

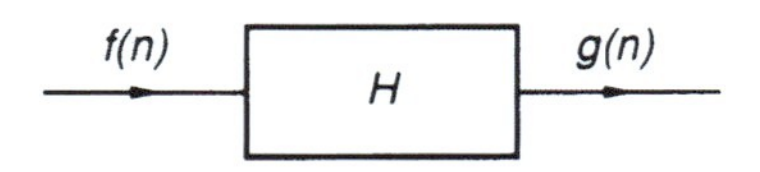

Fig. 4.8 A digital system producing an output sequence $g(n)$ by operating on an input sequence $f(n)$.

The unit delay operator is a simple example of such an operator, in which case $g(n) = f(n-1)$. The unit delay operator also displays two important properties. First, it is linear and therefore

$$H[af_1(n)+bf_2(n)] = aH[f_1(n)]+bH[f_2(n)], \qquad (4.4.2)$$

where a and b are constants, and second, it is time (or shift) invariant, and thus

$$\text{if} \quad g(n) = H[f(n)], \qquad \text{then} \quad H[f(n-m)] = g(n-m). \qquad (4.4.3a,b)$$

We now define the response of a linear, time invariant digital system to the digital impulse, $\delta(n)$, to be

$$g(n) = H[\delta(n)] = h(n), \qquad (4.4.4)$$

where H is a linear time invariant operator and where $h(n)$ is called the *impulse response* of the system. Using the properties of the Kronecker delta function we can write the input sequence in the form

$$f(n) = \sum_{i=-\infty}^{\infty} f(i)\delta(n-i). \qquad (4.4.5)$$

By using the defining characteristics of a linear, time invariant system, equations (4.4.2) and (4.4.3), the output of such a system to the input sequence $f(n)$ can be written as

$$g(n) = \sum_{i=-\infty}^{\infty} f(i)h(n-i), \qquad (4.4.6)$$

which is called the *discrete convolution*. By a simple change of variable it may also be expressed as

$$g(n) = \sum_{i=-\infty}^{\infty} h(i)f(n-i). \qquad (4.4.7)$$

These expressions have clear parallels with the convolution process dealt with for continuous time signals in Section 3.5 of the last chapter. We can also define similarly the causality and stability of a digital system. A digital system is *causal* if $h(i) = 0$ for $i<0$. For such a system we can write equation (4.4.7) as

$$g(n) = \sum_{i=0}^{\infty} h(i)f(n-i). \qquad (4.4.8)$$

Also, following the discussion of analogue systems presented in Section 3.4 of Chapter 3, a digital system is said to be stable if

$$\sum_{n=-\infty}^{\infty} |h(n)|<\infty. \qquad (4.4.9)$$

If the input to such a system is of the form $f(n) = \mathrm{e}^{j\omega_0 Tn}$, i.e. a complex sinusoid at frequency ω_0, the output that follows from equation (4.4.7) is given by

$$g(n) = \mathrm{e}^{j\omega_0 Tn} \sum_{i=-\infty}^{\infty} h(i)\mathrm{e}^{-j\omega_0 Ti}. \qquad (4.4.10)$$

The complex sinusoidal input sequence thus retains its general form in passing through the system but is modified in magnitude and phase by the complex term

$$H(e^{j\omega_0 T}) = \sum_{i=-\infty}^{\infty} h(i)e^{-j\omega_0 Ti}, \tag{4.4.11}$$

which is called the *frequency response* of the digital system, and is seen to be equal to the Fourier transform of its impulse response. Digital systems specified in terms of their frequency response are often called *digital filters*.

A general class of digital system which obeys the linear and time invariant properties is that defined by the causal *difference equation*

$$g(n) = \sum_{i=0}^{I-1} a_i f(n-i) + \sum_{k=1}^{K} b_k g(n-k), \tag{4.4.12}$$

in which the output is the weighted sum of the current and $I-1$ previous input samples and the K previous output samples. The system is said to be *recursive* since it employs a degree of feedback of previous outputs to the current input. The $I+K$ constants a_i and b_k are known as the *coefficients* of the digital system. The frequency response and impulse response of such a system can be obtained via the z-transform of equation (4.4.12). Using the delay property of the z-transform (equation (4.3.3)) and assuming zero initial conditions, we have

$$G(z) = F(z)\sum_{i=0}^{I-1} a_i z^{-i} + G(z)\sum_{k=1}^{K} b_k z^{-k}, \tag{4.4.13}$$

where $F(z)$ and $G(z)$ are the z-transforms of the input and output sequences. The z-domain transfer function of the digital system can now be written as

$$\frac{G(z)}{F(z)} = H(z) = \frac{\sum_{i=0}^{I-1} a_i z^{-i}}{1 - \sum_{k=1}^{K} b_k z^{-k}}. \tag{4.4.14}$$

If the input signal is a digital impulse, which has a z-transform of unity, the output would be the impulse response and can therefore be obtained from the inverse z-transform of this transfer function. The frequency response can also be obtained simply by substituting $z = e^{j\omega T}$ in this expression. Another way in which the properties of such systems are often illustrated is by plotting, in the complex z-plane, the roots of the denominator of $H(z)$ (which are known as the *poles* of $H(z)$) and the roots of the numerator of $H(z)$ (which are known as the *zeros* of $H(z)$) as a *pole-zero diagram*.

The relationship among the properties of linear digital systems are summarised in Fig. 4.9, which may be compared with the equivalent diagram for analogue systems shown in Fig. 3.11. The simplest first order recursive system is governed by a recursive difference equation of the form

$$g(n) = f(n) + b_1 g(n-1). \tag{4.4.15}$$

Substitution of $f(n) = \delta(n)$ and evaluation of the difference equation for a few terms soon demonstrates that the impulse response of such a system is

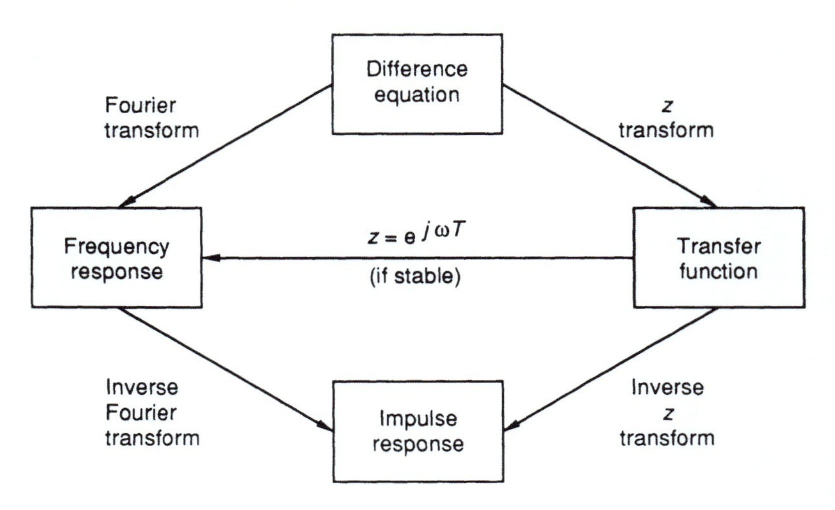

Fig. 4.9 The relationship between the various properties of a linear digital system defined by a difference equation.

$$g(n) = \begin{Bmatrix} b_1^n, & n \geq 0 \\ 0, & n < 0 \end{Bmatrix}. \tag{4.4.16}$$

The form of this digital exponential has been plotted in Fig. 4.5, together with its Fourier transform in Fig. 4.6, which in this case is the frequency response of this first order recursive system. We have seen that the response has a low-pass characteristic if $1 > b_1 > 0$ and a high-pass characteristic if $0 > b_1 > -1$. If $|b_1| > 1$ the impulse response diverges and the system is unstable. The transfer function of this first order recursive system follows from equation (4.4.14) and is given by

$$H(z) = \frac{1}{1 - b_1 z^{-1}} = \frac{z}{z - b_1}. \tag{4.4.17}$$

The system has a zero at $z = 0$ and a pole at $z = b_1$. The pole-zero diagram for this system is shown in Fig. 4.10.

The next example of a digital system that it is worthwhile to consider briefly is the second order recursive system governed by the difference equation

$$g(n) = f(n) + b_1 g(n-1) + b_2 g(n-2). \tag{4.4.18}$$

If $b_1^2 + 4b_2 \geq 0$, the two poles of the system are real (see, for example, Bellanger,

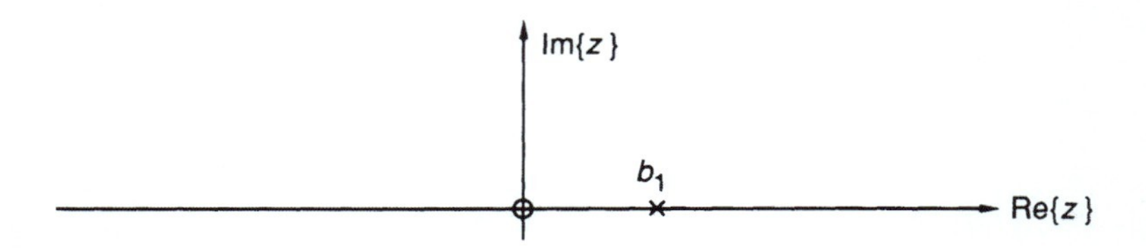

Fig. 4.10 Pole-zero diagram of a first order recursive digital system.

1984), and the impulse response is the sum of two real exponential sequences. The system under these circumstances could be implemented as the sum of two first order systems with real coefficients, and the response deduced from the discussion above. The system is then said to be "overdamped" by analogy with the response of an analogue second order system.

If $b_1^2 + 4b_2 < 0$ (so that b_2 must be negative), the two poles of the system are complex, and form a conjugate pair. The system now starts to display more interesting behaviour and is said to be underdamped. If we change the variables in the difference equation so that

$$g(n) = f(n) + 2r\cos(\omega_0 T)g(n-1) - r^2 g(n-2), \tag{4.4.19}$$

where $r = \sqrt{-b_2}$ and $\cos(\omega_0 T) = b_1/2\sqrt{-b_2}$, the transfer function can be written as

$$H(z) = \frac{1}{1 - 2r\cos(\omega_0 T)z^{-1} + r^2 z^{-2}}. \tag{4.4.20}$$

There are thus two zeros of the system at the origin of the z-plane, and by finding the roots of the denominator, it can be shown that the poles lie at $z = re^{\pm j\omega_0 T}$. This is illustrated in Fig. 4.11. (More details of the algebra involved in the above manipulations are given by, for example, Rabiner and Gold, 1975, Ch. 2.) The frequency response can be deduced from $H(z)$ by making the substitution $z = e^{j\omega T}$ and is plotted, for $r = 0.95$ and $\omega_0 T = \pi/4$, in Fig. 4.12. The impulse response of the system is given by (see again, for example, Rabiner and Gold, 1975):

$$h(n) = \begin{Bmatrix} cr^n \sin(\omega_0 T(n+1)), & n \geq 0 \\ 0, & n < 0 \end{Bmatrix}, \tag{4.4.21}$$

where $c = (\sin\omega_0 T)^{-1}$, and is plotted, for the values of r and $\omega_0 T$ used above, in Fig. 4.13. This system is clearly unstable for $|r| > 0$, in which case the poles lie outside a circle in the z-plane defined by the expression $|z| = 1$, which is known as the *unit circle*. Similarly, the stability of the first order system defined by equation (4.4.17) is assured only if $|b_1| < 1$: i.e., its pole lies within the unit circle.

Any transfer function, of arbitrary complexity, can be factorised into first and second order terms with real coefficients. The poles and zeros of each of these terms will be the same as those of the original transfer function. First and second order systems are stable provided their poles lie within the unit circle, and this criterion must thus define the stability of any transfer function.

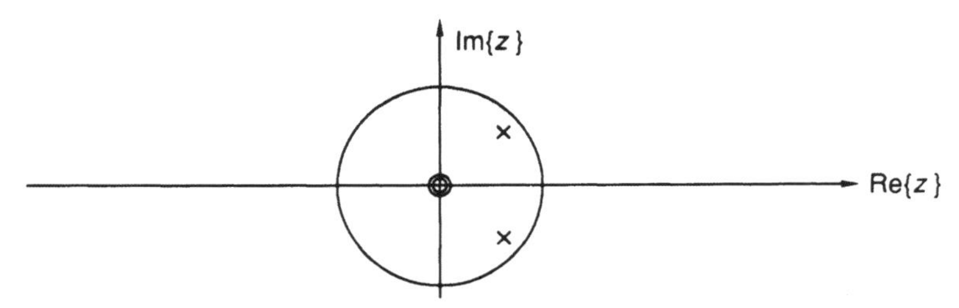

Fig. 4.11 Pole-zero diagram of a stable but underdamped second order digital system, also showing the unit circle.

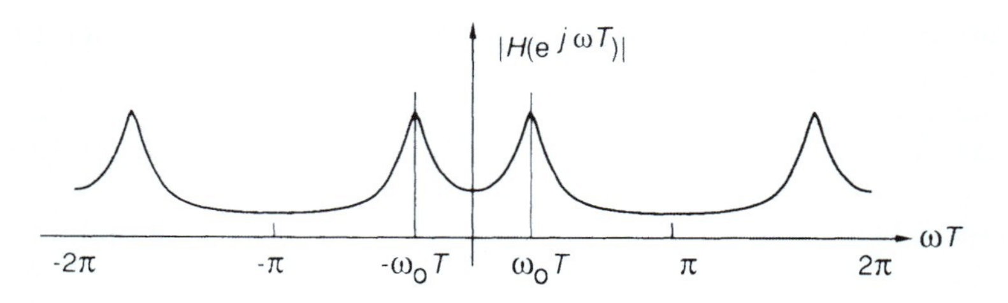

Fig. 4.12 The modulus of the frequency response of the underdamped second order system.

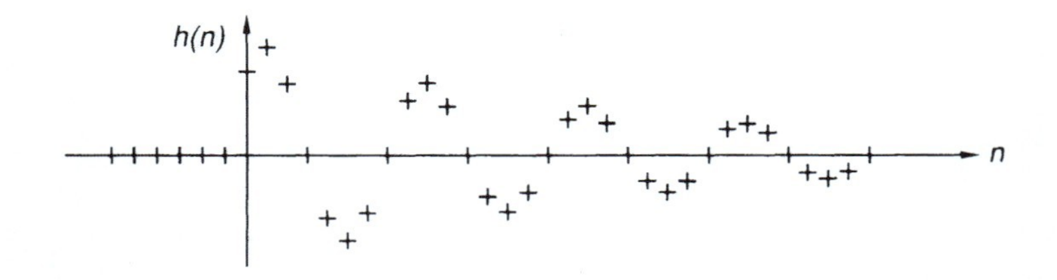

Fig. 4.13 The impulse response of the underdamped second order system.

4.5 FIR and IIR filters

A digital filter whose impulse response is zero after some finite number of samples is said to have a *finite impulse response*, and be an FIR filter. The most common implementation of such a filter is to use a digital system whose output is the weighted sum of a finite number of previous inputs, so that

$$g(n) = \sum_{i=0}^{I-1} a_i f(n-i), \tag{4.5.1}$$

where the filter has I coefficients. The impulse response of such a filter is

$$h(n) = \begin{Bmatrix} a_n, & 0 \leqslant n \leqslant I-1 \\ 0, & \text{otherwise} \end{Bmatrix}, \tag{4.5.2}$$

and so its output can be written as a finite form of the convolution in equation (4.4.8)

$$g(n) = \sum_{i=0}^{I-1} h(i) f(n-i). \tag{4.5.3}$$

This is sometimes expressed in the form of a vector inner product such that

$$g(n) = \mathbf{h}^{\mathrm{T}}\mathbf{f}(n) = \mathbf{f}^{\mathrm{T}}(n)\mathbf{h}, \tag{4.5.4}$$

where T denotes the vector transpose and where the vectors $\mathbf{h}$ and $\mathbf{f}(n)$ are defined by

$$\mathbf{h}^{\mathrm{T}} = [h(0), h(1), \ldots, h(I-1)],$$
$$\mathbf{f}^{\mathrm{T}}(n) = [f(n), f(n-1), \ldots, f(n-I+1)]. \tag{4.5.5a,b}$$

A further discussion on the use of vectors and the associated notation is presented in the Appendix.

FIR filters implemented by using equation (4.5.1) are also called “all zero”, “moving average” (MA), “non-recursive”, or “transversal”. A block diagram showing an implementation of the difference equation defined above is shown in Fig. 4.14. Such filters were originally implemented by using analogue delay lines, with outputs, taken after various delays, being weighted and summed together. The coefficients of FIR filters are still sometimes called “tap weights”. Although FIR filters can be implemented by charge coupled devices and surface wave acoustic devices, the most common physical implementation, at the frequencies of interest here, is currently digital. Either dedicated shift registers and digital multipliers are used or (more usually) a general purpose (DSP) processor, with the coefficients and past input values stored in memory locations, and the difference equation above implemented as a program. Some important properties of FIR digital filters are as follows:

(1) they are always stable, for bounded coefficients;
(2) if the impulse response is symmetric about half the filter length, the phase response of the filter corresponds to a pure delay and the filter is said to have *linear phase*;
(3) the changes in the frequency response of the filter caused by small changes in any of its coefficients are small and easily predictable.

A digital filter whose impulse response never decays to exactly zero is said to have an *infinite impulse response* and be an IIR filter. The most common implementation of such filters is to use the full difference equation:

$$g(n) = \sum_{i=0}^{I-1} a_i f(n-i) + \sum_{k=1}^{K} b_k g(n-k). \tag{4.5.6}$$

IIR filters implemented by using this equation are also called “pole-zero” or “autoregressive moving average” (ARMA) or “recursive”. One practical implementation of the difference equation above is shown on Fig. 4.15. A number of others are discussed in the textbooks by, for example, Rabiner and Gold (1975) and Oppenheim and Schafer (1975). The important properties of IIR filters are as follows:

(1) they are not always stable; they may have poles outside the unit circle;
(2) they cannot be linear phase;

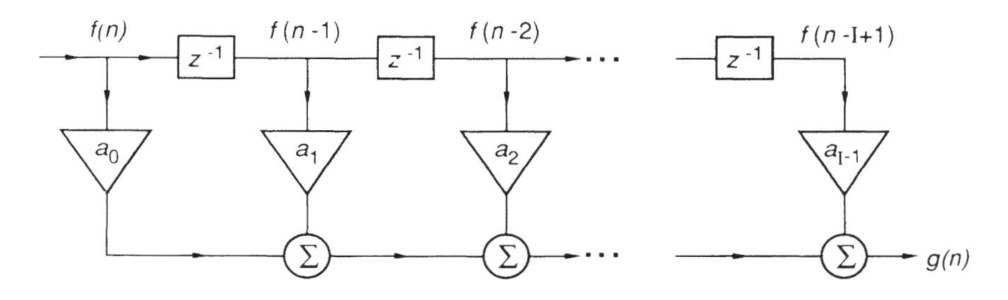

Fig. 4.14 Implementation of an FIR filter as the weighted sum of past input values.

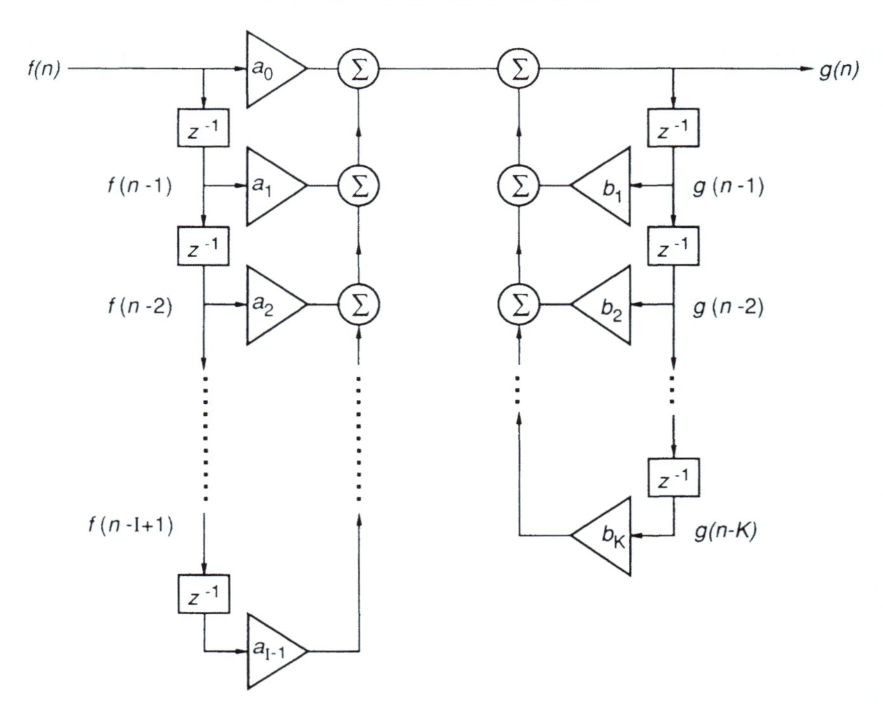

Fig. 4.15 Direct implementation of an IIR filter.

(3) large changes in the frequency response can be caused by small changes in any of the coefficients b_k, if the poles are close to the unit circle.

Resonant behaviour and sharp transitions between pass bands and stop bands can generally be obtained by using IIR filters with a smaller total number of filter coefficients than is the case with FIR filters.

4.6 Frequency domain filter design

Conventional methods of digital filter design begin with a specification in the frequency domain, generally in terms of dividing the frequency response into stop bands and pass bands and specifying the allowable pass band ripple and stop band attenuation (see, for example, Oppenheim and Schafer, 1975, Ch. 5 and Rabiner and Gold, 1975, Ch. 3, Ch. 4). Generally speaking, a stop band is the range of frequencies over which the filter response is required to be as close to zero as possible, and in a pass band the filter response is required to be as close to unity as possible. The design process is one of selecting the form (FIR or IIR) and order (number of coefficients) of the required filter, and using some mathematical design procedure to calculate the coefficients of the filter which meet the required frequency domain specification.

A specification in these terms is rarely available for digital filters which are to operate as controllers in active control systems. A brief review of one such design method (that uses windowing techniques) is worthwhile, however, since it illustrates some of the basic trade-offs inherent in digital filter design. This method has also been used, in modified form, in the design of digital controllers for the active control of sound in ducts (Roure, 1985). The design of digital FIR filters by windowing involves the initial specification of the "desired" frequency response which we can write, for example, as $H_d(e^{j\omega T})$. The impulse response of the system with this frequency response is then calculated by using the inverse Fourier transform to obtain

$$h_d(n) = \frac{T}{2\pi}\int_{-\pi/T}^{\pi/T} H_d(e^{j\omega T})\, e^{j\omega Tn} d\omega. \tag{4.6.1}$$

For simple forms of $H_d(e^{j\omega T})$, e.g. ideal low pass, high pass or band pass filters, this integral can be evaluated analytically. Several problems in implementing a practical FIR filter with the desired frequency response become apparent when this ideal impulse response is examined. First the impulse response will be symmetrical about $n = 0$ if no phase response has been specified in $H_d(e^{j\omega T})$. This can be overcome relatively easily by adding a linear phase shift to the desired frequency response, which will add delay to $h_d(n)$. Even after this operation, however, the desired impulse response will probably have finite (if small) values for large positive and negative values of n. In order to truncate these so that the filter can be implemented in FIR form, with (say) I coefficients, the ideal impulse response is multiplied by a "window" function, $w(n)$, of length I samples to give the impulse response of the FIR filter which is to be implemented. Thus we write

$$h(n) = h_d(n)w(n). \tag{4.6.2}$$

The filter coefficients are thus calculated directly by using

$$a_i = h_d(i)w(i), \qquad 0 \leqslant i \leqslant I-1. \tag{4.6.3}$$

Multiplication in one Fourier domain corresponds to convolution in the other, and the frequency response of the filter that is actually implemented will be given by

$$H(e^{j\omega T}) = \frac{1}{2\pi}\int_{-\pi}^{\pi} H_d(e^{j\theta})W(e^{j(\omega T-\theta)})d\theta, \tag{4.6.4}$$

where $W(e^{j\omega T})$ is the Fourier transform of the window function, $w(n)$. Figure 4.16 illustrates the effects of convolving the frequency response of an ideal low pass filter with the Fourier transform of a "rectangular" window (i.e. $w(n) = 1$ for $0 \leqslant n \leqslant I-1$ and $w(n) = 0$ otherwise) which we have seen corresponds to the "digital sinc" function. The truncation of the ideal impulse response in the time domain gives rise to two effects in the frequency domain. First, the cut-off of the implemented filter is no longer perfectly sharp, but is controlled by the width of the main lobe of $W(e^{j\omega T})$. Second, the values of $H(e^{j\omega T})$ in the pass and stop bands are no longer constant (at 1 and 0, respectively) but are controlled by the side lobe structure of $W(e^{j\omega T})$.

For a rectangular window the main lobe of $W(e^{j\omega T})$ has a width of $\omega T = 4\pi/I$, and the ripples added to the stop and pass bands have a peak linear value of about 0.09.

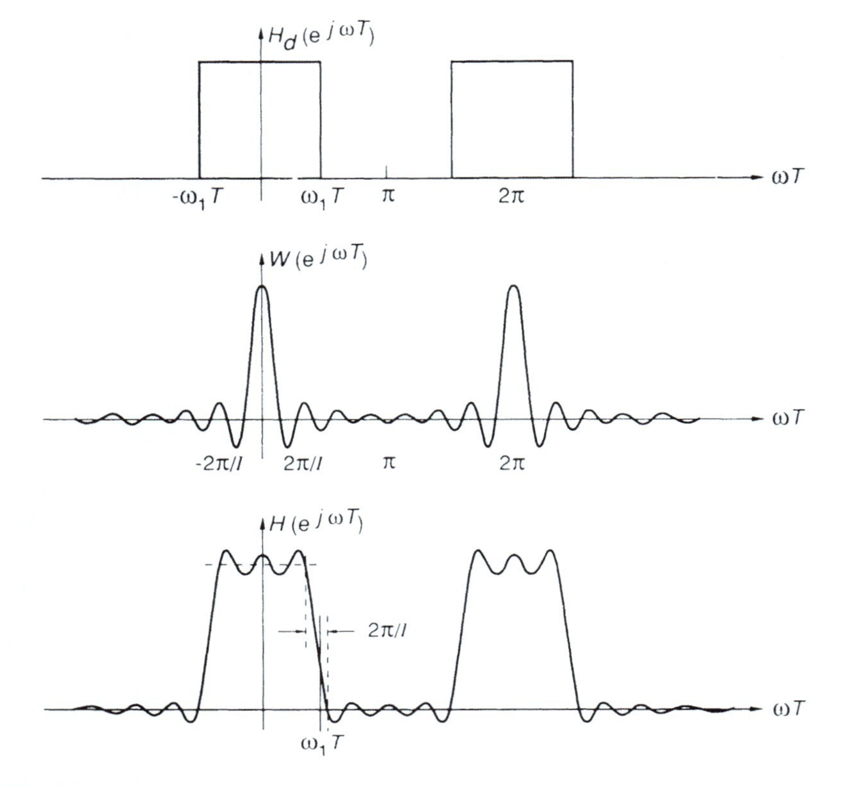

Fig. 4.16 Frequency response of desired filter, window function and convolution of the two, which is equal to the frequency response of the implemented filter.

This causes ripples in the pass band whose maximum value is about 0.74 dB, and a minimum stop band attenuation of about -21 dB. These ripples are an inherent effect of the truncation of the impulse response, and are similar to the Gibbs phenomenon found in the time histories of periodic signals synthesised from a finite number of harmonic components.

A more gentle truncation of the impulse response can reduce the effects of these ripples, but only at the expense of increasing the width of the main lobe (for a given filter length, I). Considerable effort has been expended in manipulating the window function so that the resulting frequency response is "best" for particular applications. Durrani and Nightingale (1972) and Harris (1978), for example, list the properties of dozens of windows. The basic trade-off between the stop band attenuation and the width of the transition from pass band and stop band always remains. Table 4.1 lists three of the most famous window types and their associated main lobe widths, and peak pass band ripple and peak stop band attenuation (more details are presented by, for example, Oppenheim and Schafer, 1975, Ch. 5).

Although ripples in the pass band and stop band response are inevitable with FIR digital filters, the window design method always gives ripples of equal (linear) magnitude in the two bands. This means that windows which achieve high stop band

Table 4.1 The frequency domain characteristics of some common windows, of length I samples, for the design of FIR filters.

Window type	Transition width of main lobe	Maximum pass band ripple (dB)	Minimum stop band attenuation (dB)
Rectangular	$4\pi/I$	0.74	−21
Hanning	$8\pi/I$	0.05	−44
Blackman	$12\pi/I$	0.002	−74

attenuations inevitably give extremely low pass band ripple, as shown in Table 4.1. More sophisticated design methods, such as the Remez exchange algorithm (see the description presented by Rabiner and Gold, 1975, Ch. 3), allow the ratio of the linear ripple in the two bands to be specified as other than unity, and can also give optimal performance for a given filter length by generating equi-ripple frequency responses. Such complicated design methods are rarely useful in active control applications since the frequency domain specification of the controller is usually itself subject to measurement errors and may change with time.

The design of digital IIR filters from a frequency domain specification is usually performed via an equivalent analogue filter design. This is because a considerable wealth of analytical and practical experience exists in designing analogue filters which give a good trade-off between the parameters which are important in filter design, such as transition width, pass band ripple, stop band attenuation, group delay, etc. Widely used prototype analogue filters are of the Butterworth, Chebyshev and elliptic type. Having performed the analogue filter design from the frequency domain specification, the question then is how "best" to transform the analogue filter into digital form. This problem is essentially one of mapping the poles, and sometimes the zeros, of the analogue filter from the complex s-plane to the complex z-plane. Mappings which have been suggested (see, for example, Oppenheim and Schafer, 1975, Ch. 5) include the impulse invariant mapping, the differential mapping and the bilinear transform. The latter has many useful properties and is widely used in practice. The bilinear transformation from the Laplace variable (s) to the z-transform variable (z) is defined by the equation

$$z = \frac{1 + (T/2)s}{1 - (T/2)s}, \tag{4.6.5}$$

where T is the sample time. Some important properties of this transformation, as illustrated in Fig. 4.17, are as follows:

(1) The left hand side of the s-plane is mapped into the interior of the unit circle on the z-plane, so that the stability of an analogue system is preserved.

(2) The $j\omega$ axis from $\omega = 0$ to ∞ in the s-plane is mapped once onto the upper half of the unit circle of the z-plane from $z = 1$ to $z = -1$. Although the frequency axis is compressed or "warped" in the digital filter, there is a one-to-one correspondence between the frequency response of the analogue filter and its digital equivalent. (See, for example, Rabiner and Gold, 1975, Ch. 4 for a fuller description of this point.)

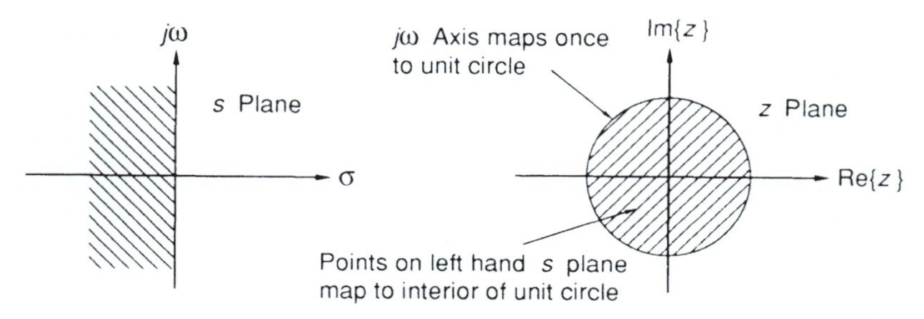

Fig. 4.17 The mapping from the complex s-plane to the complex z-plane which corresponds to the bilinear transform, used for converting analogue filter designs into digital filter designs.

4.7 Optimal filter design

Frequency domain design methods are dependent upon having a reliable specification for the frequency response of the filter in terms of pass bands (where the desired response is unity) and stop bands (where the desired response is zero). If the digital filter is being designed in order to separate sequences corresponding to useful signals and corrupting noise, and the spectra of these two sequences do *not* overlap, then formulation of the frequency domain specification is fairly straightforward. If, however, the spectra of the "signal" and "noise" *do* overlap, as illustrated in Fig. 4.18 for example, no simple frequency domain specification can be deduced for the filter required. In the example shown the filter should obviously have some form of low pass characteristic but the cut-off frequency is a trade-off between passing too much of the noise sequence and attenuating too much of the signal sequence. The first thing which must be done in designing such a filter is to quantify this trade-off.

Consider the general time domain estimation problem illustrated in Fig. 4.19 where the *reference* sequence $f(n)$ is fed to an FIR digital filter, with I coefficients h_i, whose output $g(n)$ is subtracted from a *desired* sequence, $d(n)$, to form an error sequence, $e(n)$ given by

$$e(n) = d(n) - \sum_{i=0}^{I-1} h_i f(n-i). \tag{4.7.1}$$

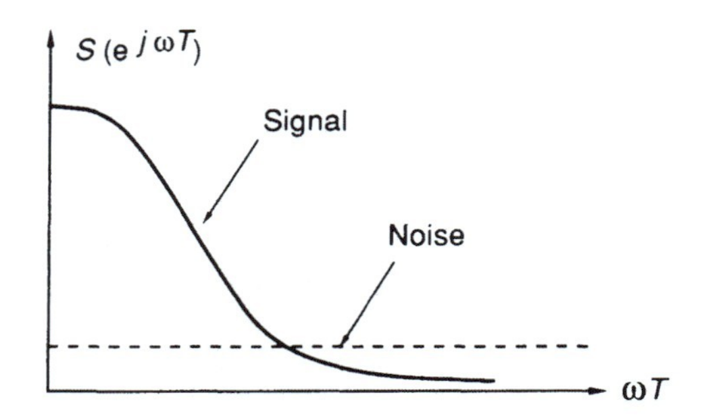

Fig. 4.18 An example where the spectrum of the signal overlaps with the spectrum of the noise.

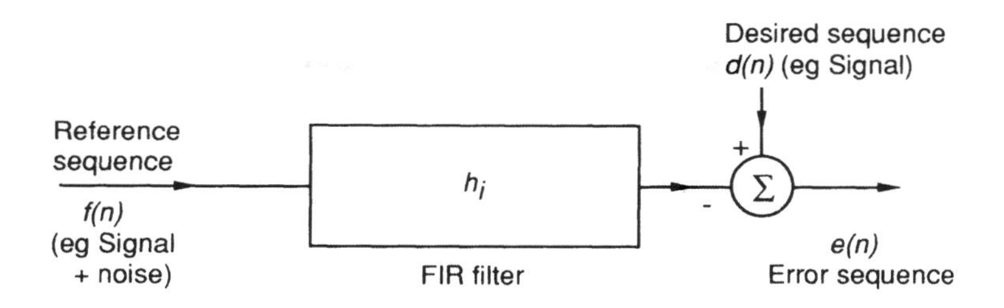

Fig. 4.19 The block diagram of a general estimation problem.

Using the inner product representation for the discrete convolution (introduced in equations (4.5.4)) enables the error sequence to be written as

$$e(n) = d(n) - \mathbf{h}^{\mathrm{T}}\mathbf{f}(n) = d(n) - \mathbf{f}^{\mathrm{T}}(n)\mathbf{h}. \tag{4.7.2}$$

This general formulation can be used to solve the particular problem of best separating signal from noise, if the reference sequence is taken to be the sum of the signal and noise sequences whose filtered output is made as close as possible to the desired sequence (which is taken to be the signal sequence alone) by reducing the error sequence, $e(n)$. The filter design problem is now to adjust each h_i to best reduce the error sequence, $e(n)$, according to some criterion. A very common and useful criterion is that each h_i is adjusted in order to minimise the mean square error sequence given by $E[e^2(n)]$ where E denotes an average over ensembles of the random error sequence $e(n)$. This of course assumes that $f(n)$ and $d(n)$ are themselves random sequences. Using equation (4.7.2) this mean square error can be written as

$$E[e^2(n)] = E[(d(n) - \mathbf{h}^{\mathrm{T}}\mathbf{f}(n))^2], \tag{4.7.3}$$

and expanding out the square using the two forms of equation (4.7.2) gives

$$E[e^2(n)] = E[d^2(n) - 2\mathbf{h}^{\mathrm{T}}\mathbf{f}(n)d(n) + \mathbf{h}^{\mathrm{T}}\mathbf{f}(n)\mathbf{f}^{\mathrm{T}}(n)\mathbf{h}]. \tag{4.7.4}$$

Since the filter coefficients in $\mathbf{h}$ are assumed time invariant it follows that we can write this expression as

$$E[e^2(n)] = E[d^2(n)] - 2\mathbf{h}^{\mathrm{T}}E[\mathbf{f}(n)d(n)] + \mathbf{h}^{\mathrm{T}}E[\mathbf{f}(n)\mathbf{f}^{\mathrm{T}}(n)]\mathbf{h}. \tag{4.7.5}$$

This equation shows that the mean square error is a *quadratic* function of each of the filter coefficients, h_i. In the appendix it is shown that this equation has a unique global minimum associated with it, provided the scalar quantity $\mathbf{h}^{\mathrm{T}}E[\mathbf{f}(n)\mathbf{f}^{\mathrm{T}}(n)]\mathbf{h}$ is positive for all possible values of the vector $\mathbf{h}$, which implies that the matrix $E[\mathbf{f}(n)\mathbf{f}^{\mathrm{T}}(n)]$ is *positive definite* (see, for example, Noble, 1969 and the discussion presented in the appendix). If $d(n) = 0$ then the mean square error becomes

$$E[e^2(n)] = \mathbf{h}^{\mathrm{T}}E[\mathbf{f}(n)\mathbf{f}^{\mathrm{T}}(n)]\mathbf{h}, \tag{4.7.6}$$

which is non-zero for all non-zero $\mathbf{h}$ provided there are at least half as many frequency components in $f(n)$ as there are coefficients in $\mathbf{h}$, i.e. the input is *persistently exciting* or *spectrally rich* (see, for example, the discussions presented by

Goodwin and Sin, 1984, Honig and Messerschmitt, 1984 and Treichler *et al.*, 1987). The mean square error plotted against any two of the filter coefficients will thus give a "bowl"-shaped surface (the *error surface*) as shown in Fig. 4.20, and the error surface can, mathematically, be thought of as extending into I dimensions for the general minimisation problem, although it cannot be visualised if I is greater than 2. One important consequence of this shape of error surface is that its only stationary point is a minimum, and so the optimal set of coefficients, which reduces the mean square error to its minimum value, can be found by setting the differential of $E[e^2(n)]$ with respect to each h_i to zero simultaneously. The approach is thus a generalisation to multiple dimensions of the method of least squares described in Chapter 2. The derivative of the mean square error with respect to one particular filter coefficient, h_k, can be deduced by differentiating equation (4.7.1) so that it follows that

$$\frac{\partial E[e^2(n)]}{\partial h_k} = 2E\left[e(n)\frac{\partial e(n)}{\partial h_k}\right] = -2E[e(n)f(n-k)]. \tag{4.7.7}$$

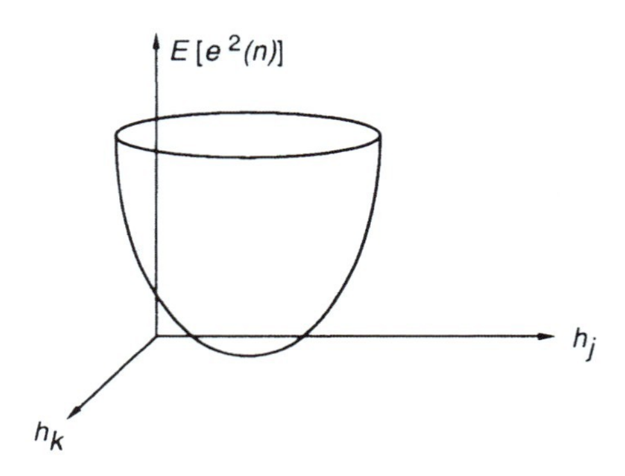

Fig. 4.20 The quadratic surface formed by plotting mean squared error against any two filter coefficients.

Again using equation (4.7.1) in order to expand $e(n)$ and taking the expectations of the products of the sequences gives

$$\frac{\partial E[e^2(n)]}{\partial h_k} = -2\left[R_{fd}(k) - \sum_{i=0}^{I} h_i R_{ff}(k-i)\right], \tag{4.7.8}$$

where $R_{fd}(m)$ is the digital cross-correlation between the reference and desired sequences, at a lag of m samples, and $R_{ff}(m)$ is the digital autocorrelation of the reference signal, at a lag of m samples, both defined by analogy with the cross- and autocorrelations for continuous signals in Chapter 2. Thus $R_{fd}(m) = E[f(n)d(n+m)]$ and $R_{ff}(\mathrm{m}) = E[f(n)f(n+m)]$.

In order to derive the optimal set of filter coefficients, equation (4.7.8) must be set to zero for all values of k from 0 to $I-1$, so

$$R_{fd}(k) - \sum_{i=0}^{I} h_{i(\mathrm{opt})} R_{ff}(k-i) = 0, \qquad 0 \leqslant k \leqslant I-1. \tag{4.7.9}$$

This set of equations is often called the *normal equations* and is the discrete time form of the Wiener–Hopf equation introduced in Section 3.8 of Chapter 3. In contrast to the continuous time Wiener–Hopf equation, however, equation (4.7.9) can be solved in closed form. Each of the equations in (4.7.9), which must be simultaneously satisfied, can be explicitly written down in the form

$$\begin{aligned} R_{fd}(0) &= h_{0(\mathrm{opt})}R_{ff}(0) + h_{1(\mathrm{opt})}R_{ff}(1) + \ldots + h_{I-1(\mathrm{opt})}R_{ff}(I-1) \\ R_{fd}(1) &= h_{0(\mathrm{opt})}R_{ff}(1) + h_{1(\mathrm{opt})}R_{ff}(0) + \ldots + h_{I-1(\mathrm{opt})}R_{ff}(I-2) \\ &\ \ \vdots \\ R_{fd}(I-1) &= h_{0(\mathrm{opt})}R_{ff}(I-1) + h_{1(\mathrm{opt})}R_{ff}(I-2) + \ldots h_{I-1(\mathrm{opt})}R_{ff}(0) \end{aligned} \quad (4.7.10)$$

where the symmetric nature of the autocorrelation function has been used. This set of simultaneous equations can be put into matrix form

$$\begin{bmatrix} R_{fd}(0) \\ R_{fd}(1) \\ \cdot \\ \cdot \\ R_{fd}(I-1) \end{bmatrix} = \begin{bmatrix} (R_{ff}(0) & R_{ff}(1) & \ldots & R_{ff}(I-1) \\ R_{ff}(1) & R_{ff}(0) & \ldots & R_{ff}(I-2) \\ \cdot & & & \\ \cdot & & & \\ R_{ff}(I-1) & R_{ff}(I-2) & \ldots & R_{ff}(0) \end{bmatrix} \begin{bmatrix} h_{0(\mathrm{opt})} \\ h_{1(\mathrm{opt})} \\ \cdot \\ \cdot \\ h_{I-1(\mathrm{opt})} \end{bmatrix} \quad (4.7.11)$$

which can be expressed simply as the matrix equation

$$\mathbf{r} = \mathbf{R}\mathbf{h}_{\mathrm{opt}}, \quad (4.7.12)$$

where the definitions of the vectors **r** and **h** and the matrix **R** can be readily deduced. On the assumption for the moment that **R** is not singular, the matrix equation can be solved for $\mathbf{h}_{\mathrm{opt}}$ to give

$$\mathbf{h}_{\mathrm{opt}} = \mathbf{R}^{-1}\mathbf{r}. \quad (4.7.13)$$

The filter that has these optimal coefficients is often called the "Wiener filter" after the pioneering work of Wiener in the 1940s (Wiener, 1949). The matrix **R** has a very regular structure and is said to be of *Toeplitz* form. Very efficient algorithms have been developed in order to invert such matrices (see, for example, Markel and Gray, 1976).

Although the various steps in this derivation have been spelt out for clarity, it should be noted that in terms of the matrix error equation (4.7.5) above, the vector **r** and the matrix **R** are defined to be

$$\mathbf{r} = E\,[\mathbf{f}(n)d(n)] \quad \text{and} \quad \mathbf{R} = E\,[\mathbf{f}(n)\mathbf{f}^{\mathrm{T}}(n)], \quad (4.7.14\text{a,b})$$

so that **R** is guaranteed to be positive definite (provided $f(n)$ is spectrally rich as discussed above), and hence it must be invertible. The mean square error can now be written as the quadratic function

$$E[e^2(n)] = E[d^2(n)] - 2\mathbf{h}^{\mathrm{T}}\mathbf{r} + \mathbf{h}^{\mathrm{T}}\mathbf{R}\mathbf{h}, \qquad (4.7.15)$$

whose minimum value is shown in the appendix to be given when $\mathbf{h} = \mathbf{h}_{\text{opt}}$, as in equation (4.7.13) above. This more direct path to the solution illustrates the advantages of working in terms of a matrix formulation of a standard quadratic form. A vector of cross-correlations between the error sequence and input sequence, $E[f(n-k)\ e(n)]$ for $k = 0$ to $I-1$, can be defined as

$$E[\mathbf{f}(n)e(n)] = E[\mathbf{f}(n)(d(n) - \mathbf{f}^{\mathrm{T}}(n)\mathbf{h})] = \mathbf{r} - \mathbf{R}\mathbf{h}. \qquad (4.7.16)$$

If the filter is optimally adjusted in order to minimise the mean square error, so that $\mathbf{h} = \mathbf{h}_{\text{opt}}$ (equation (4.7.12)), this vector of cross-correlations is zero. This lack of correlation between the error and input signals for an optimally adjusted filter is another example of the orthogonality principle which has already been used, in Chapter 3, in the equivalent continuous time problem. It should be noted that the optimal filter coefficients are defined using the equation above under both the constraints of causality and finite filter length. Although the autocorrelation and cross-correlation functions of the sequences $f(n)$ and $d(n)$ are required in the equation as a measure of their statistical behaviour, they may be obtained, in practice, from Fourier transformation of their power and cross-spectral densities, exactly as for the continuous time case in Chapter 2.

If we were to relax the constraints of causality and finite filter length, the set of equations which must be satisfied in order to minimise the mean square error would become

$$R_{fd}(k) = \sum_{i=-\infty}^{\infty} h_{i(\text{opt})} R_{ff}(k-i), \qquad -\infty < k < \infty. \qquad (4.7.17)$$

Taking the Fourier transform of both sides of this equation shows that

$$\sum_{k=-\infty}^{\infty} R_{fd}(k)\mathrm{e}^{-j\omega Tk} = \sum_{k=-\infty}^{\infty} \sum_{i=-\infty}^{\infty} h_{i(\text{opt})} R_{ff}(k-i)\mathrm{e}^{-j\omega Tk}. \qquad (4.7.18)$$

The order of the summations can be rearranged to show that

$$\sum_{k=-\infty}^{\infty} R_{fd}(k)\mathrm{e}^{-j\omega Tk} = \sum_{i=-\infty}^{\infty} h_{i(\text{opt})}\mathrm{e}^{-j\omega Ti} \sum_{l=-\infty}^{\infty} R_{ff}(l)\mathrm{e}^{-j\omega Tl}, \qquad (4.7.19)$$

where $l = k - i$. The Fourier transforms of the auto- and cross-correlation functions can be recognised as the auto- and cross-spectral densities of the sequences and the remaining term is the Fourier transform of the impulse response of the *unconstrained* optimal filter which is therefore equal to its frequency response. Thus in the frequency domain we have

$$S_{fd}(\mathrm{e}^{j\omega T}) = H_{\text{opt}}(\mathrm{e}^{j\omega T}) S_{ff}(\mathrm{e}^{j\omega T}). \qquad (4.7.20)$$

can now return to our original problem (Fig. 4.19) in which the desired sequence, $d(n)$, is equal to the signal only, and the reference sequence $f(n)$ is equal to the sum of the signal and an uncorrelated noise component $v(n)$ so that

$$f(n) = d(n) + v(n). \qquad (4.7.21)$$

Under these circumstances the expression for the power spectral density of the sequence $f(n)$ is given by

$$S_{ff}(\mathrm{e}^{j\omega T}) = S_{dd}(\mathrm{e}^{j\omega T}) + S_{vv}(\mathrm{e}^{j\omega T}), \tag{4.7.22}$$

and the cross-spectral density of the sequences $f(n)$ and $d(n)$ can be written as

$$S_{fd}(\mathrm{e}^{j\omega T}) = S_{dd}(\mathrm{e}^{j\omega T}). \tag{4.7.23}$$

The frequency response of the unconstrained filter which minimises the mean square difference between the actual filter output and that desired is thus

$$H_{\mathrm{opt}}(\mathrm{e}^{j\omega T}) = \frac{S_{dd}(\mathrm{e}^{j\omega T})}{S_{dd}(\mathrm{e}^{j\omega T}) + S_{vv}(\mathrm{e}^{j\omega T})}. \tag{4.7.24}$$

For the signal and noise spectra illustrated in Fig. 4.18, the frequency response of the optimal unconstrained filter is clearly similar to the power spectrum of the signal sequence, with a general low pass characteristic but having a response somewhat less than unity at low frequencies and somewhat greater than zero at high frequencies. Although it does not have a simple spectral interpretation, the optimal causal filter of finite length, whose coefficients can be calculated by using equation (4.7.13), will have a similar frequency response to this.

4.8 Adaptive digital filters

As an alternative to computing the auto- and cross-correlation functions of the reference and desired sequences, and performing the matrix inversion necessary to compute the optimal filter coefficients in one step (equation 4.7.13) a number of iterative or adaptive methods have been used to change the filter coefficients gradually in such a way that the mean square error is reduced. Such adaptive digital filters have been widely used in many applications including telecommunications, array beamforming and control. A number of excellent books have been published which describe their properties and applications (see, for example, Goodwin and Sin, 1984; Widrow and Stearns, 1985; Cowan and Grant, 1985; Haykin, 1986). A brief introduction to one important and widely used algorithm for the adaptation of an FIR digital filter will be presented here as an extension of the formulation of the previous section. The modifications to the basic algorithm which are necessary when it is used in active sound control applications will be discussed in Chapter 6.

An important property of the error surface associated with an FIR digital filter, when used in the type of estimation problem discussed above, is that it is quadratic, with a unique global minimum. Simple gradient descent algorithms are then guaranteed to converge towards the global optimum, provided they are stable. One of the simplest such algorithms is based on the method of steepest descent, which for the ith coefficient of an FIR filter can be written as

$$h_i(\mathrm{new}) = h_i(\mathrm{old}) - \mu \left.\frac{\partial E[e^2(n)]}{\partial h_i}\right|_{h_i = h_i(\mathrm{old})}, \tag{4.8.1}$$

where μ is a convergence coefficient. The gradient term can be written as in equation (4.7.7) as

$$\left.\frac{\partial E\,[e^2(n)]}{\partial h_i}\right|_{h_i\,=\,h_i(\text{old})} = -2E\,[e(n)f(n-i)]\Big|_{h_i\,=\,h_i(\text{old})}, \qquad (4.8.2)$$

where $e(n)$ has been expanded out, the expectation evaluated and the gradient set to zero to deduce the optimal, Wiener, filter above. The expressions above could also be used as the basis for an iterative algorithm for adjusting the filter coefficients, in which the expectation in the expression for the true gradient is evaluated by using a time average (assuming that the signals are stationary). This procedure would avoid the need to invert a matrix but would still involve a lengthy averaging process at each iteration. A crucial step taken by Widrow and Hoff in 1960 was to recognise that an estimate of this true gradient given by differentiating the *instantaneous* squared error with respect to the filter coefficients, can be calculated for every new sample, and (on average) will adjust the coefficients in such a way that the mean square error is reduced. The algorithm using this gradient estimate to update all of the filter coefficients at every sample time has become known as the *LMS algorithm* and may be written as

$$h_i(n+1) = h_i(n) + \alpha e(n)f(n-i), \qquad \text{for all } i, \qquad (4.8.3)$$

where α is a convergence coefficient and $h_i(n)$ is the value of the ith coefficient in the FIR filter at the nth sample time. This algorithm is also an example of the descriptively named *stochastic gradient algorithm* (see, for example, Cowan and Grant, 1985, Ch. 3). If the output of the filter is expressed in vector terms by

$$y(n) = \mathbf{h}^{\mathrm{T}}(n)\mathbf{f}(n), \qquad (4.8.4)$$

where

$$\mathbf{h}^{\mathrm{T}}(n) = [h_0(n), h_1(n), \ldots, h_{I-1}(n)], \qquad (4.8.5)$$

is the "impulse response" vector of the *time varying* digital filter, then the error sequence becomes

$$e(n) = d(n) - \mathbf{h}^{\mathrm{T}}(n)\mathbf{f}(n), \qquad (4.8.6)$$

and the LMS algorithm can be expressed in vector terms as

$$\mathbf{h}(n+1) = \mathbf{h}(n) + \alpha e(n)\mathbf{f}(n). \qquad (4.8.7)$$

These equations demonstrate that the number of arithmetic operations necessary to implement the LMS adaptive filter is only about twice the number needed to calculate $e(n)$ using a fixed FIR filter with an equal number of coefficients. The conventional block diagram of such an adaptive digital filter is illustrated in Fig. 4.21 where the feedback from the error signal to the filter coefficients is explicitly shown.

The convergence properties of the LMS algorithm have been widely discussed. It turns out that the conditions for the filter coefficients to converge, in the mean, towards the optimal (Wiener) solution are generally less stringent than the conditions for the variance of the filter coefficients, and hence the mean square error to converge, in the mean (Haykin, 1986). The actual convergence condition derived by Haykin is rather complicated and assumes that the reference signal is uncorrelated

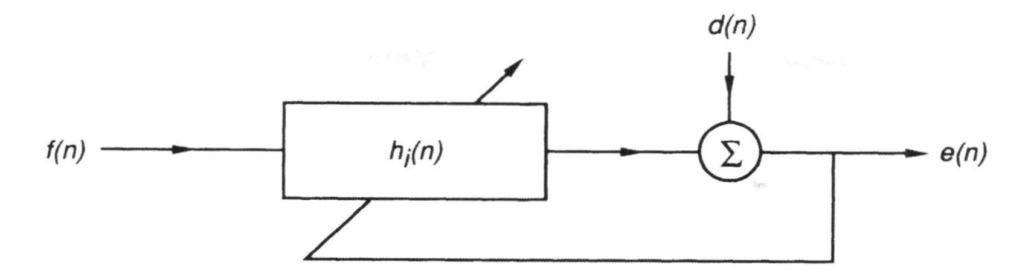

Fig. 4.21 Diagrammatic representation of an adaptive digital filter.

with the variations in $h(n)$. A simplification which provides a useful "rule of thumb" for the convergence of the LMS algorithm is that the algorithm will usually converge towards the Wiener solution if the convergence coefficient α satisfies

$$0 < \alpha \lesssim 1/IE\,[f^2(n)], \tag{4.8.8}$$

where $E\,[f^2(n)]$ is the mean square value of the reference sequence, which is equal to $R_{ff}(0)$, and I is the number of adaptive filter coefficients. Many properties of the convergence of the LMS algorithm are common to any steepest descent method, and these will be discussed in more detail in Chapter 11. Not only does this adaptive algorithm converge towards the optimal solution for stationary signals, but it is capable of "tracking" changes in the statistics of the signals under non-stationary conditions (see, for example, Widrow *et al.*, 1976). More complicated algorithms using the equivalent of Newton's method for their adaptation have been developed and are known as *recursive least squares* (*RLS*) algorithms. Although they overcome some of the problems of a steepest descent algorithm such as the LMS, in a stationary environment, recent studies have shown that they are potentially worse than the LMS at tracking non-stationary signals (a discussion of this has been presented by Bershad and Macchi, 1989).

5
Interference in Plane Wave Sound Fields

5.1 Introduction

Plane wave sound propagating in a rigid walled waveguide was one of the first forms of acoustic field to be identified as a candidate for active control. Paul Lueg filed for a US patent in 1934 which was later granted (Lueg, 1936) and in which he described a technique for controlling sound by the introduction of additional sound. The illustration page of Lueg's patent is reproduced in Fig. 5.1. Lueg's "Fig. 1" illustrates his scheme for controlling plane wave sound in a duct or tube. The components of this system are comprised of a microphone for detecting the waves propagating down the tube and a loudspeaker for emitting the cancelling sound. The electronic system connecting the two ("*V*" in Fig. 1) is responsible for producing the necessary delay and amplification of the detected signal in order to ensure that the loudspeaker emits an inverted replica of the detected wave at the instant the wave arrives at the loudspeaker. When a compression arrives at the loudspeaker, it produces a rarefaction of equal magnitude, thus cancelling the net pressure fluctuation.

It is interesting to observe that Lueg's system contains all the components necessary to accomplish active control of plane wave sound in a duct and if the element "*V*" is broadly interpreted as having the transfer function of the necessary controller, then the diagram accurately describes modern systems for the control of noise in ventilation ducts which, at the time of writing, have only just come onto the market. This delay of over 50 years between the recognition of a potentially useful idea and its implementation on a commercial basis is largely due to the technological difficulty in implementing the electronic system "*V*", a difficulty which has only recently been overcome with the advent of relatively inexpensive fast digital signal processors. A full discussion of the techniques for designing the appropriate system "*V*" will be given in the next chapter.

In this chapter we will restrict ourselves to an examination of the acoustical consequences of implementing active control of plane wave fields. Our starting point will be the problem of controlling plane wave sound propagation in a long duct, a problem which has received considerable attention in the years since Lueg's patent. The degree of attention devoted to this problem is probably because it appears to be relatively easy to understand acoustically and it has always had an obvious practical application in the control of low frequency duct-borne sound, noise which is in practice particularly difficult to control by passive means. First we will investigate in more detail the acoustical processes involved in introducing cancelling sound in the

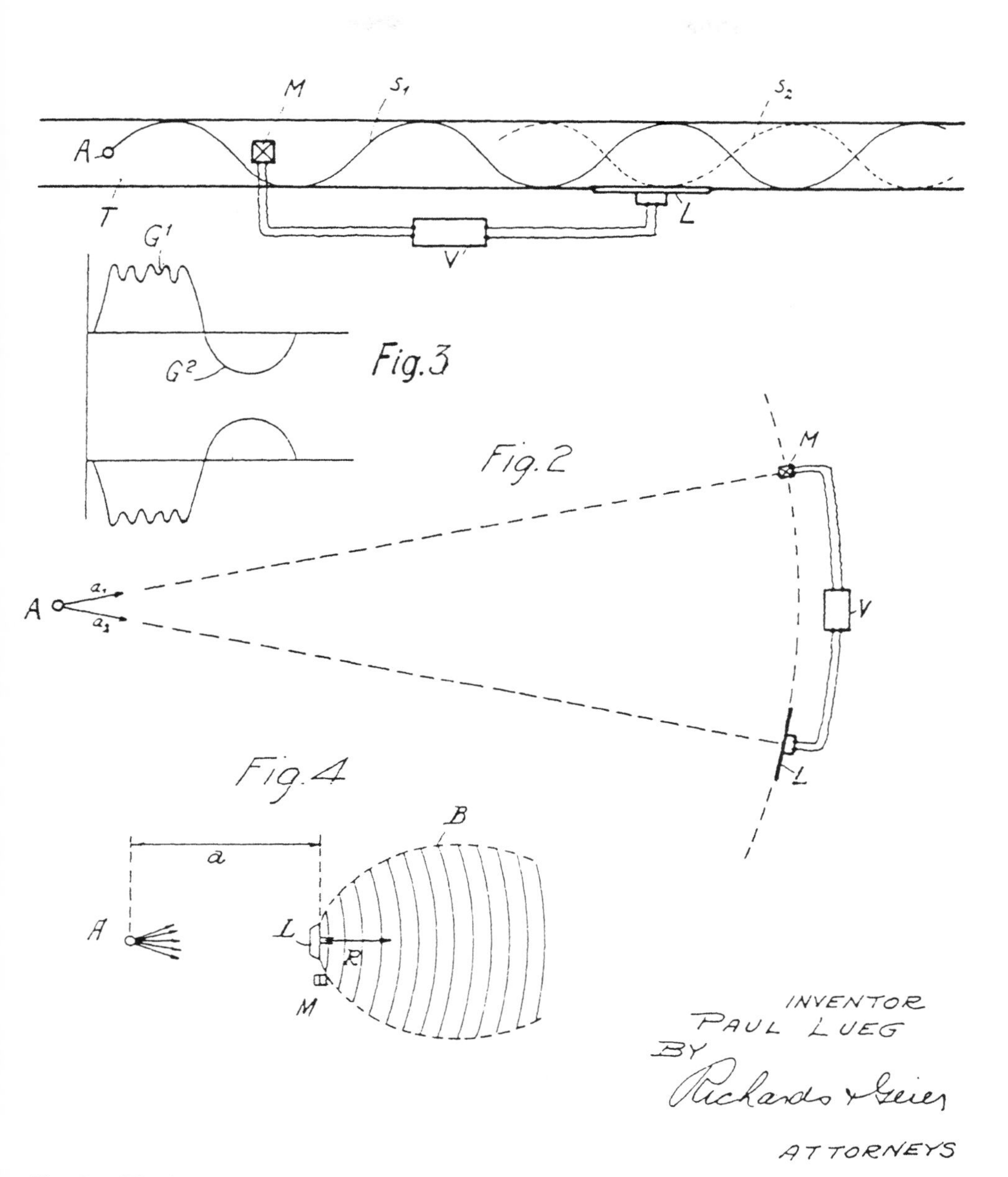

Fig. 5.1 The illustration page from Lueg's 1936 patent.

manner described by Lueg. As we shall see, this technique results in cancellation of the field downstream of the secondary source but has the net effect of reflecting the incident sound back upstream. This reflected sound will then be detected by the microphone and interfere with its task of detecting downstream propagating sound.

Two approaches were later suggested in order to avoid this potential difficulty. The first was suggested by Jessel and Mangiante (1972). This was an application of the general theory developed by both Jessel (1968) and Malyuzhinets (1969) for controlling three-dimensional sound fields through the use of a continuous layer of monopole and dipole sources to provide perfect absorption of incident sound. This general theory will be described in more detail in Chapter 9 although the principles of the approach can be explained relatively easily in the one-dimensional case, and this is another of the objectives of this chapter. However, the practical implementation of this approach has its difficulties, and a method of overcoming these was suggested by Swinbanks (1973). In this arrangement, two secondary sources are used to generate sound radiation in only the downstream direction. Downstream sound can thus be cancelled whilst no upstream radiation is produced although, as we shall see, such an arrangement has a limited operational bandwidth.

Both these approaches to the problem of controlling duct-borne sound will be described in this chapter. Some effort will be made to explain clearly their mechanisms of operation through analyses conducted in both the frequency domain and the time domain. The time domain approach usually involves describing the impulse response of the system considered, an approach which was shown by Curtis *et al.* (1985, 1987) to be a very effective means of illustrating the mechanism of active control. Particular attention will be given to describing the energetics of the various acoustical configurations since the problem of "where the energy goes" has historically caused concern amongst the lay population (if not the scientific community!). Here we hope to make clear the energy flows involved in the operation of active control and in particular to describe the mechanism of absorption of sound by real loudspeakers.

Finally, we will introduce some basic concepts associated with the active control of enclosed sound fields, a problem which in spite of its clear practical applicability has not been investigated in detail until relatively recently (see Nelson *et al.*, 1987b). Much of the physics of active control can be described using a relatively simple plane wave model of a one-dimensional enclosed sound field and again we follow closely the work of Curtis *et al.* (1987) in describing the processes involved. This work serves as an introduction to the material presented in Chapter 10, where we will introduce a more general modal model of low frequency enclosed sound fields.

5.2 The plane monopole source in an infinite duct

In describing the interference of sound fields in one dimension we will use a plane monopole source as our basic source element (Nelson and Elliott, 1989a). Such a source is illustrated in Fig. 5.2. In reality, the active control of sound in ducts is generally accomplished by using a loudspeaker as a secondary source of sound and this is typically mounted on one wall of a rectangular duct. Most work in this field

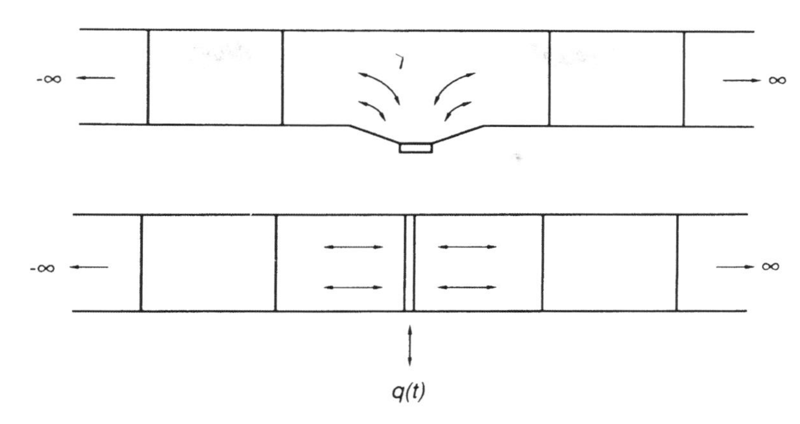

Fig. 5.2 A loudspeaker on a duct wall idealised as a plane monopole source. The plane source simulates the action of the loudspeaker at low frequencies in driving plane waves in the duct in both the upstream and the downstream directions. The plane source can be visualised as a pair of massless pistons forced to oscillate apart by the introduction of a fluctuating volume flow $q(t)$ introduced between them.

has been restricted to plane wave propagation where the frequency of the sound is such that its wavelength is much longer than the largest cross-sectional dimension of the duct. If the loudspeaker is driven at frequencies higher than that corresponding to an acoustic wavelength which is less than twice this largest cross-sectional dimension, then higher order modes of propagation will be produced (see Kinsler *et al.*, 1982, for an introductory description of higher order mode propagation). Work on the active control of higher order mode propagation has to date been mostly theoretical (see, for example, Fedoryuk, 1975; Boiko and Ivanov, 1976; Mazanikov *et al.*, 1977; Urusovskii, 1977a, b, 1980; Arzamasov *et al.*, 1982; Nelson and Elliott, 1991) although some limited results on the control of two propagating modes were presented by Mazanikov and Tyutekin (1976) and more comprehensive work has recently been presented by Eriksson *et al.* (1989) and Silcox (1990).

If we restrict our attention to frequencies less than the "cut-on" frequency of the first transverse mode, then a loudspeaker on a duct wall will generate plane waves propagating both upstream and downstream and only in the immediate region of the loudspeaker will higher order modes of the duct be excited. These modes constitute the near field of the loudspeaker which decays exponentially with distance upstream and downstream as one moves away from the position of the source. (Detailed investigations of the loudspeaker near field have been undertaken by Trinder and Nelson, 1983, Tichy *et al.*, 1984, Galland and Sunyach, 1986 and Kazakia, 1986.) If we ignore the details of the oscillatory flow produced by the loudspeaker cone in its immediate vicinity, then the net effect on the sound field can be represented by a plane source which effectively generates the same plane wave field as the loudspeaker in the upstream and downstream directions. The plane source can be visualised as two massless pistons separated by an infinitesimal distance which are forced to oscillate apart by the introduction of a fluctuating volume flow. This is illustrated in Fig. 5.2. The rate of volume flow which forces the pistons apart is the same net rate of volume flow produced by the loudspeaker in the low frequency limit when the acoustic wavelength becomes infinitely long compared to the loudspeaker dimensions (Berengier and Roure, 1980). Thus, as flow is introduced, the two massless pistons are pushed apart and as flow is extracted, they are pulled together. Assume that the pistons comprising the plane source are located at a position y as illustrated in Fig. 5.3, where y is used to denote a co-ordinate position along the x-axis (and not as a co-ordinate axis that is orthogonal to the x-axis). We denote $U(y_+)$ as the complex velocity in the positive x-direction of the right hand piston in Fig. 5.3, and we know from Chapter 1 that the complex pressure and particle velocity fluctuations produced in an infinite duct can be written as

$$\left.\begin{aligned} p(x) &= \rho_0 c_0 U(y_+) e^{-jk(x-y)} \\ u(x) &= U(y_+) e^{-jk(x-y)} \end{aligned}\right\}, \qquad x>y. \tag{5.2.1}$$

We shall continue to refer to this region as that which is "downstream" from the source. For the region $x<y$ where waves propagate "upstream" the complex pressure and particle velocity can be written as

$$\left.\begin{aligned} p(x) &= -\rho_0 c_0 U(y_-) e^{jk(x-y)} \\ u(x) &= U(y_-) e^{jk(x-y)} \end{aligned}\right\}, \qquad x<y. \tag{5.2.2}$$

where $U(y_-)$ is the complex velocity (again in the positive x-direction) of the left hand piston in Fig. 5.3. Note that a positive value of $U(y_-)$ is associated with a negative value of pressure, i.e. positive-going motions of the left hand piston generate upstream travelling rarefactions. A sketch of the instantaneous pressure and particle velocity distributions at a given point in one cycle of harmonic motion of the pistons is shown in Fig. 5.4. At any given instant the pressure is continuous across the source but the particle velocity "jumps" from the velocity of the left hand piston to that of the velocity of the right hand piston. The magnitude of this jump is determined by the volume flow injected between the two pistons. We should also note that the plane monopole source is completely transparent to any incident sound radiation; since the hypothetical pistons comprising the source have no mass they will move exactly with the particle velocity fluctuations associated with any other

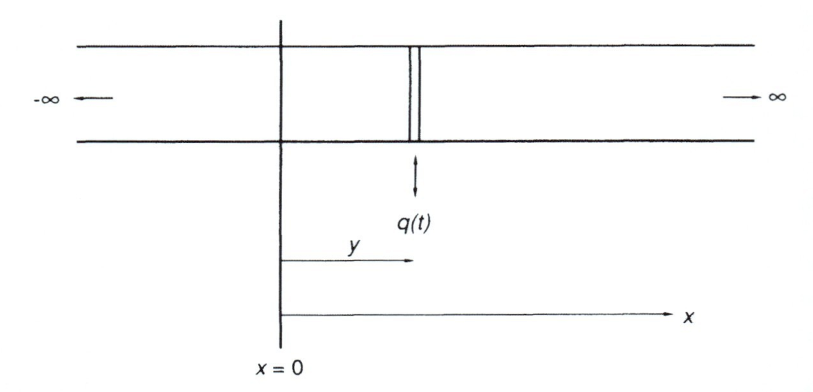

Fig. 5.3 The co-ordinate system used in describing the pressure field generated by a plane monopole source.

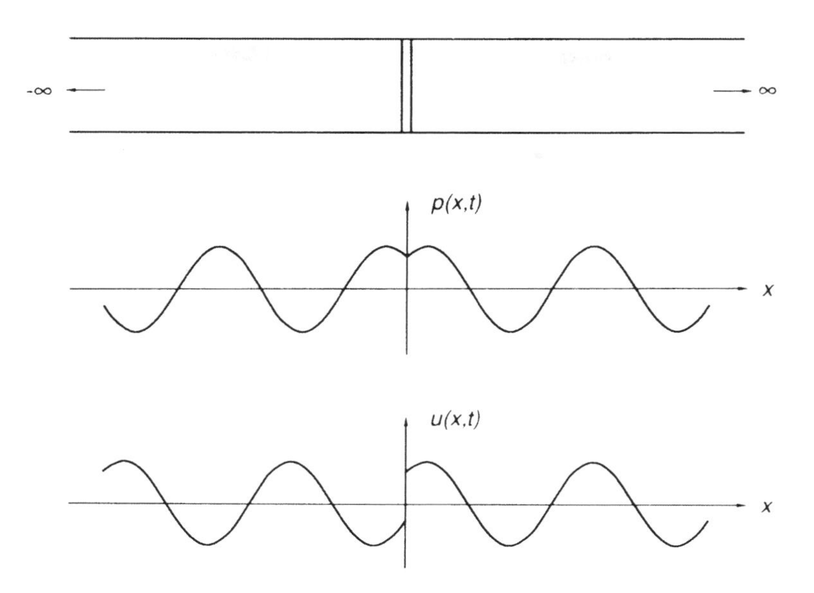

Fig. 5.4 The pressure and particle velocity distributions produced by a plane monopole source at a particular instant during a cycle of harmonic motion. Note that the pressure is continuous across the source but that there is a discontinuity in particle velocity.

wave propagating down the duct. In addition, and most importantly, we will make the assumption that the volume flow injected between the two pistons is *independent* of the pressure produced on the pistons. This amounts to assuming that the loudspeaker being modelled by the plane source has an infinite internal acoustic impedance.

A compact means of writing the above expressions for the complex pressure generated by the source is given by recognising that at $x = y$, since the pressure is continuous across the source, then it follows from equations (5.2.1) and (5.2.2) that $p(y) = \rho_0 c_0 U(y_+) = -\rho_0 c_0 U(y_-)$. Thus it follows that $U(y_+) = -U(y_-)$, and if we define $q(y)$ as the volume velocity introduced into the duct by the source, we can write $U(y_+) = -U(y_-) = q(y)/2S$, where S is the cross-sectional area of the duct. The total pressure field can thus be written in the form

$$p(x) = q(y)\frac{\rho_0 c_0 e^{-jk|x-y|}}{2S}. \tag{5.2.3}$$

This conveniently relates the complex pressure produced at a given position x to the volume velocity $q(y)$ of a source at a given position y. Since the magnitude of the pressure fluctuations clearly relate to the magnitude of the volume velocity fluctuations, we refer to $q(y)$ as the complex *source strength* of the plane monopole. The variables $q(y)$ and $p(x)$ can be considered the input and output respectively of a linear time invariant system. The frequency response function of this system can therefore be written as $H(j\omega) = \rho_0 c_0 e^{-jk|x-y|}/2S$. The inverse Fourier transform of this frequency response function yields the impulse response of the system. It is well known from the theory of Fourier transforms (see Section 2.3) that a frequency dependent function of the form $ae^{j\omega b}$ has an inverse Fourier transform given by $a\delta(t+b)$ where a and b are arbitrary real constants and $\delta(\ )$ denotes the Dirac delta function. Since $k = \omega/c_0$, it follows that the impulse response of this system can be expressed as

$$h(t) = \frac{\rho_0 c_0}{2S}\,\delta\left(t - \frac{|x-y|}{c_0}\right). \tag{5.2.4}$$

This impulse response is thus the pressure produced at x due to an impulse in volume velocity of a source at y. This particular transformation from the frequency domain to the time domain will be used repeatedly in this chapter. It also follows (see Section 3.5) that any particular input due to source strength fluctuations $q(y,t)$ is given by the convolution integral

$$p(x,t) = \int_{-\infty}^{\infty} h(t-\tau)q(y,\tau)\mathrm{d}\tau, \tag{5.2.5}$$

where τ is the usual additional time variable. Thus the pressure fluctuation produced is related to the source strength fluctuation by

$$p(x,t) = \frac{\rho_0 c_0}{2S}\int_{-\infty}^{\infty} \delta\left(t - \frac{|x-y|}{c_0} - \tau\right) q(y,\tau)\mathrm{d}\tau, \tag{5.2.6}$$

and using the properties of the delta function then shows that

$$p(x,t) = \frac{\rho_0 c_0}{2S}\, q\left(y, t - \frac{|x-y|}{c_0}\right). \tag{5.2.7}$$

This demonstrates that any fluctuation in volume flow which is introduced in order to squeeze the two pistons apart is reproduced exactly in the pressure fluctuation generated in the duct at a time $|x-y|/c_0$ later. This time lag is, of course, the time taken for the sound pulse to travel either up or down the duct from y to x.

Finally it should be noted that this simple relationship between volume velocity fluctuations and pressure fluctuations does not hold for the propagation of acoustic waves in three dimensions. A similar relationship can be derived, however, for the pressure fluctuations produced by a point monopole source in an unbounded medium. Using a similar argument to that presented above, it can be shown that in this case the pressure fluctuation produced is directly related to the *time derivative* of the volume velocity fluctuation. This follows from equation (1.12.5) of Chapter 1.

5.3 The cancellation of downstream radiation by using a single secondary plane monopole source

The simplest of all active sound control problems can now be discussed by using our idealisation of a plane monopole source. As depicted in Fig. 5.5, let us assume that we have a primary monopole source of complex strength q_p at $x = 0$ and a secondary plane monopole source of strength q_s placed downstream at $x = L$, both sources being located in an infinite duct. The complex pressure produced by the two sources acting independently can be written as

$$p_p(x) = \frac{\rho_0 c_0}{2S} q_p \mathrm{e}^{-jk|x|}, \qquad (5.3.1)$$

$$p_s(x) = \frac{\rho_0 c_0}{2S} q_s \mathrm{e}^{-jk|x-L|}. \qquad (5.3.2)$$

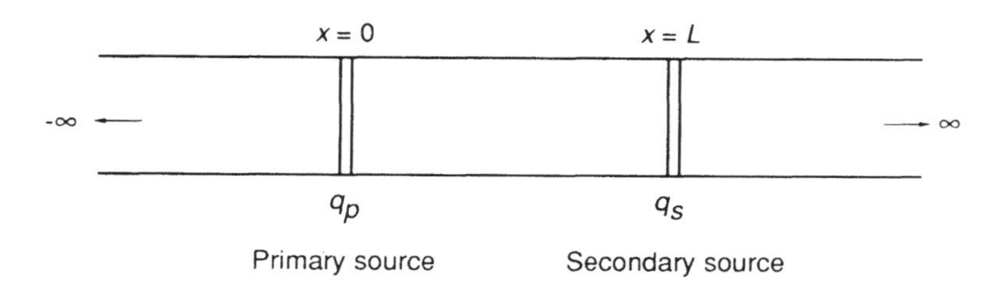

Fig. 5.5 Primary and secondary plane monopole sources in an infinite duct.

We will now use the principle of superposition to calculate the net sound field produced by the two sources. This is the central assumption which is necessary for the implementation of active control and has been discussed in Section 1.6. Thus, we superpose the pressure fields due to the primary and secondary sources such that the total complex pressure can be written as

$$p(x) = p_p(x) + p_s(x). \qquad (5.3.3)$$

Now we can choose a secondary source strength which will ensure that the pressure field downstream of the secondary source will become zero, i.e. that

$$p(x) = 0, \qquad x \geq L. \qquad (5.3.4)$$

This requirement, together with equation (5.3.3) and equations (5.3.1) and (5.3.2) can be written as

$$\frac{\rho_0 c_0}{2S} q_p \mathrm{e}^{-jkx} + \frac{\rho_0 c_0}{2S} q_s \mathrm{e}^{-jk(x-L)} = 0, \qquad x \geq L. \qquad (5.3.5)$$

It therefore follows that the necessary relationship between the complex strengths of the secondary and primary sources is given by

$$q_s = -q_p \mathrm{e}^{-jkL}. \qquad (5.3.6)$$

In exactly the same way that we deduced the relationship in the time domain between the pressure produced by a given source and its complex strength in the last section, we can use the Fourier transform pair consisting of $a\mathrm{e}^{j\omega b}$ and $a\delta(t-b)$ to deduce the time domain relationship between the two sources. It follows that for an arbitrary fluctuation in the primary source strength $q_p(t)$ then the requisite secondary source strength to provide perfect cancellation of downstream radiation is given by

$$q_s(t) = -q_p(t - L/c_0). \qquad (5.3.7)$$

Thus, in order to cancel radiation downstream of the secondary source, the secondary source strength fluctuation must be identical to the primary source strength fluctuation but of opposite sign and delayed by a time L/c_0. This is the time taken for sound to travel from the primary source to the secondary source. A physical interpretation of this process is very easy to give. Figure 5.6 illustrates, by using this time domain analysis, the process by which the sound field is controlled. First we assume that the primary source emits a short duration pulse of sound which will travel both upstream and downstream as the component massless pistons associated with the source are forced apart. The upstream propagating pulse will continue propagating unimpeded towards infinity. Once the downstream propagating pulse reaches the secondary source, then the secondary source will emit precisely the opposite form of pulse to the primary source, which again propagates both upstream and downstream. In the downstream direction the secondary source pulse exactly cancels that due to the primary source, whilst the upstream propagating pulse will travel upstream unimpeded and back past the primary source. The net effect of the secondary source is to reflect sound perfectly although a compression will be

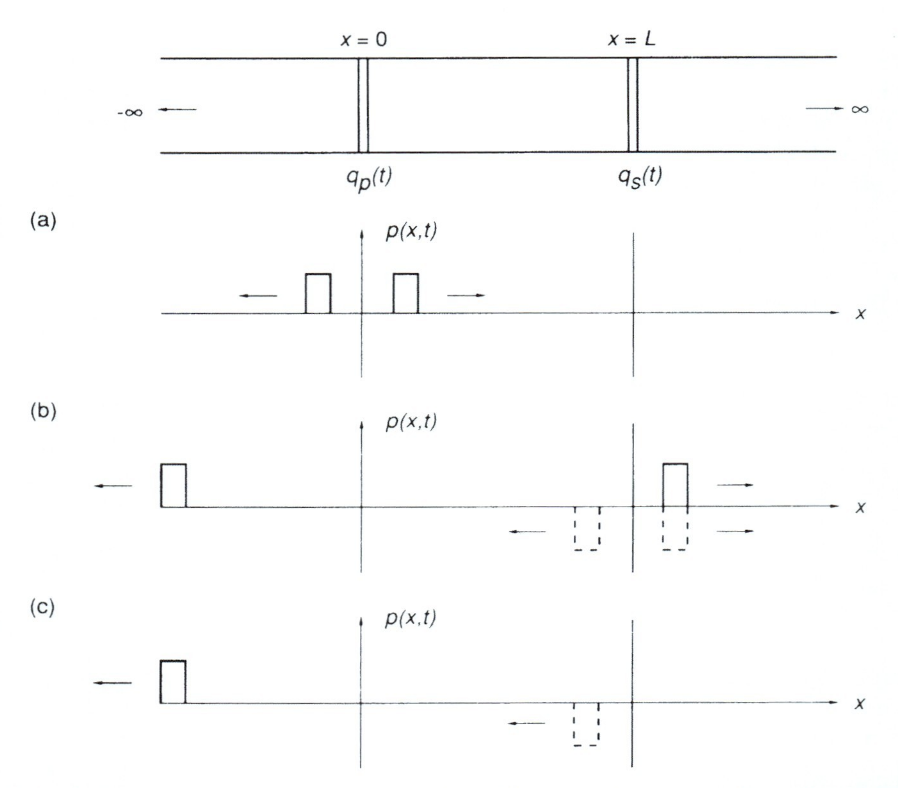

Fig. 5.6 A time domain description of the action of the secondary source in producing perfect cancellation of the sound pressure fluctuations in the downstream region $x>L$. The instantaneous pressure distribution at a time t just after the primary source has radiated a rectangular pulse of sound is shown in (a) and at a time $t+L/c_0$ is shown in (b). The net result is shown in (c).

reflected as a rarefaction. The secondary source provides a "pressure release" boundary condition, since the pressure at $x = L$ is maintained at zero by the action of the secondary source.

Further insight into the process of downstream cancellation can be given by reverting to a frequency domain description. By using the relationship between the complex strengths of the secondary and primary sources given by equation (5.3.6) and substituting this back into equation (5.3.3) it can be shown that for the region between the primary and secondary sources the net complex pressure takes the form

$$p(x) = \frac{\rho_0 c_0}{2S} q_p e^{-jkL}[e^{jk(L-x)} - e^{-jk(L-x)}], \qquad 0 \leq x L. \tag{5.3.8}$$

This shows that between the primary and secondary sources the pressure field consists of two travelling waves of equal amplitudes and opposite directions of propagation which combine to form a standing wave (see Chapter 1, Section 1.9). The expression for the complex pressure upstream of the primary source is given by

$$p(x) = \frac{\rho_0 c_0}{2S} q_p[1 - e^{-j2kL}]e^{jkx}, \qquad x \leq 0. \tag{5.3.9}$$

This shows that upstream of the primary source at a given frequency then there is only an upstream propagating plane wave. The amplitude of this plane wave varies depending upon the value of the parameter kL. In fact, the amplitude falls to zero

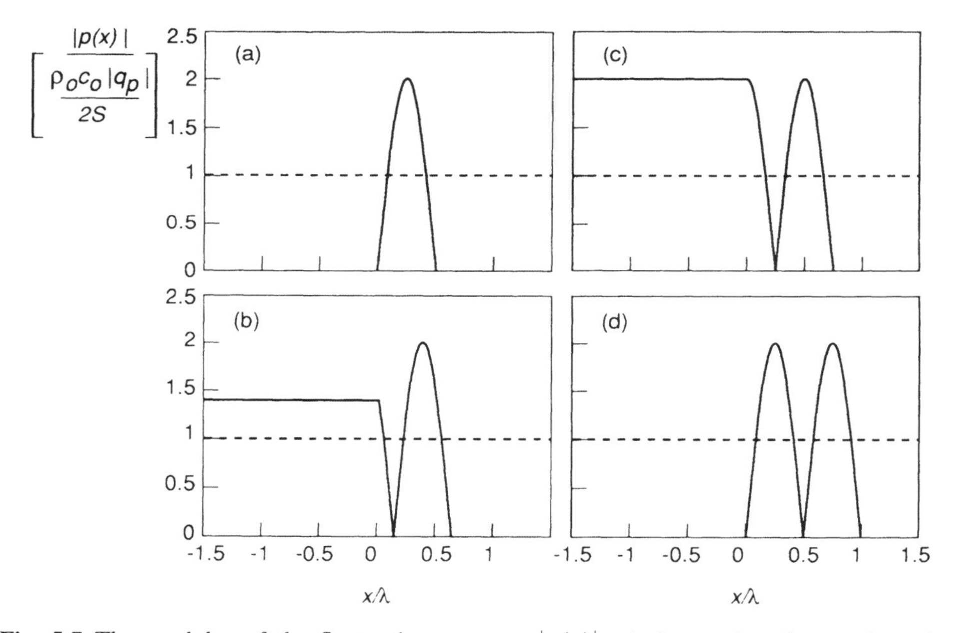

Fig. 5.7 The modulus of the fluctuating pressure $|p(x)|$ relative to that due to the primary source (– – – – –) when a single secondary plane monopole source at $x = L$ is used to cancel the pressure fluctuations in the region $x > L$. The primary source is at $x = 0$ and results are shown for various positions of the secondary source: (a) $L = \lambda/2$, (b) $L = 5\lambda/8$, (c) $L = 3\lambda/4$, (d) $L = \lambda$.

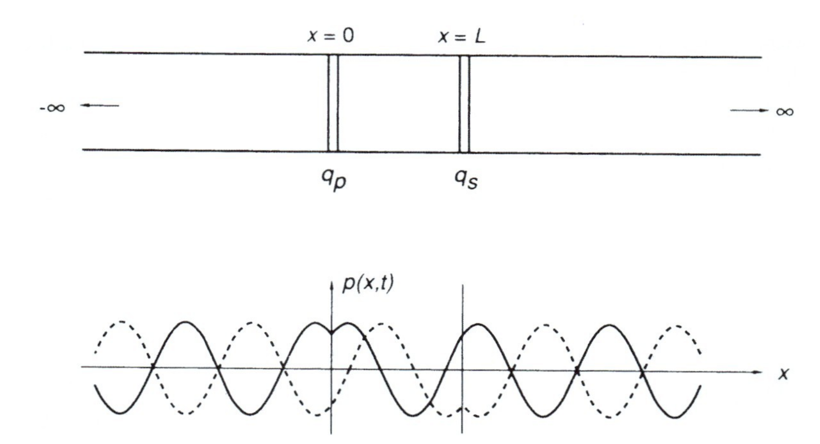

Fig. 5.8 The instantaneous pressures due to the primary (————) and secondary (– – – – –) sources at one point during a cycle of harmonic motion. The sources are separated by exactly one wavelength and the diagram illustrates that perfect cancellation can be achieved in both the upstream ($x<0$) and downstream ($x>L$) regions of the duct. This can be achieved whenever the sources are separated by a distance equal to an integer number of half-wavelengths.

for values of kL such that $kL = n\pi$. (This can be shown by recognising that $e^{-j2kL} = \cos 2kL - j\sin 2kL$.) Therefore the net sound field at a given frequency produced by the interference of the primary and secondary sound fields has a strong dependence on the separation between the primary and secondary sources relative to the wavelength. Figure 5.7 shows the modulus of the complex pressure relative to the modulus of the complex pressure produced by the primary field only in the absence of control for various separation distances between the two sources. It can be seen that when the primary and secondary sources are separated by a distance which is exactly an integer number of half wavelengths (i.e. when $kL = n\pi$) then the sound field will be cancelled both upstream and downstream, leaving a standing wave field between the two sources. This cancellation process can be explained by observing the instantaneous pressure fluctuation at a particular time during one period of the motion, as illustrated in Fig. 5.8. This shows that in the particular case of a source separation of exactly one wavelength, then cancellation of the sound field both upstream and downstream is possible.

5.4 The energetics of cancellation by a single secondary plane monopole source

Now we will consider in a little more detail the time-averaged power outputs of the primary and secondary sources when downstream cancellation of the sound field is produced. It is relatively easy to calculate the power output of a plane monopole source from a knowledge of the net pressure at the position of the source and its source strength. This can be demonstrated by using the following argument. First, note that for a source at a position y the net acoustic intensity at a position y_+, which is slightly downstream from y, is given by

$$I(y_+) = \frac{1}{2}\text{Re}\{p^*(y_+)u(y_+)\}, \tag{5.4.1}$$

where p and u are the complex pressure and particle velocity, respectively. Similarly, the net intensity just upstream from y can be written as

$$I(y_-) = \frac{1}{2}\text{Re}\{p^*(y_-)u(y_-)\}. \tag{5.4.2}$$

Since we are dealing with a lossless system where no energy dissipation can occur, then the net power output of a given source must be equal to

$$W = S[I(y_+) - I(y_-)]. \tag{5.4.3}$$

This is where our idealisation of a plane monopole source again is of great assistance. As the distance between y_+ and y_- vanishes, then any contribution to the particle velocity at the position of the plane source due to waves impinging on the source will be the same on both sides of the source. The only difference in particle velocity from one side of the source to another must be related directly to the strength of the source. Thus, we can write

$$S[u(y_+) - u(y_-)] = q(y), \tag{5.4.4}$$

and it therefore follows that the net power output of the source can be written simply as

$$W = \frac{1}{2}\text{Re}\{p^*(y)q(y)\}. \tag{5.4.5}$$

Now let us calculate the power outputs of the primary and secondary sources when downstream cancellation is produced. We have shown that under conditions of downstream cancellation the pressure at the position of the secondary source is exactly zero and therefore *the secondary source neither radiates nor absorbs any net power*. The power output of the primary source can be calculated by using equation (5.3.9) to deduce the pressure at the position of the primary source. Therefore the power output can be written as

$$W_p = \frac{1}{2}\text{Re}\left\{\frac{\rho_0 c_0}{2S} q_p^*(1 - e^{j2kL})\, q_p\right\}. \tag{5.4.6}$$

Since we know that $\text{Re}\{(1 - e^{j2kL})\} = (1 - \cos 2kL) = 2\sin^2 kL$, it therefore follows that

$$W_p = \frac{\rho_0 c_0}{2S} |q_p|^2 \sin^2 kL. \tag{5.4.7}$$

It is also easy to show that the power output of the primary source in the absence of any control can be written as $W_{pp} = \rho_0 c_0 |q_p|^2/4S$. Thus, the ratio of the primary source power output with the control operating to the primary source power output in the absence of control can be written as

$$\frac{W_p}{W_{pp}} = 2\sin^2 kL. \tag{5.4.8}$$

It is therefore evident that the primary source power output is doubled at some frequencies, these frequencies corresponding to separation distances such that $kL = (2n+1)\pi/2$ where n is an integer, i.e. at separation distances of $\lambda/4$, $3\lambda/4$, $5\lambda/4$, and so forth. The primary source power output is also *zero* at frequencies such that the separation distance is an integer number of half wavelengths. This occurs at $kL = n\pi$ which is the condition for zero upstream radiation discussed in Section 5.3. It is clear that the action of the secondary source at a given frequency is simply to modify the pressure fluctuations produced at the position of the primary source and thereby modify its net power output. In simple terms, if the pressure fluctuations at the primary source are driven to zero, the primary source will no longer have anything to "push against" (i.e. it will become "unloaded") and it will therefore do no work on the surrounding medium and emit no sound power.

5.5 Absorption of sound by a plane monopole source

It will now be shown that in the primary/secondary source arrangement discussed in the previous section, the secondary source can be made to *absorb* sound, although the level of absorption produced is such that this strategy is not attractive as a means of preventing downstream radiation. In order to calculate the power output of the secondary source when the primary and secondary sources are arranged as shown in Fig. 5.5, we first note that the total pressure at the position of the secondary plane monopole source is given by the sum of the contributions from the primary source ($\rho_0 c_0 q_p e^{-jkL}/2S$) and that due to the secondary source itself ($\rho_0 c_0 q_s/2S$). Using equation (5.4.5) then shows that the power output of the secondary source can be written as

$$W_s = \frac{1}{2}\,\mathrm{Re}\left\{\frac{\rho_0 c_0}{2S}(q_s + q_p e^{-jkL})^* q_s\right\}. \tag{5.5.1}$$

This equation can be put into a more useful form if we recall that the real part of a complex number is equal to half the sum of the number itself and its complex conjugate. Thus we can write

$$W_s = \frac{\rho_0 c_0}{8S}[(q_s + q_p e^{-jkL})^* q_s + (q_s + q_p e^{-jkL}) q_s^*)]. \tag{5.5.2}$$

Expansion of this expression then shows that

$$W_s = \frac{\rho_0 c_0}{8S}[2|q_s|^2 + (q_p e^{-jkL})^* q_s + (q_p e^{-jkL}) q_s^*)]. \tag{5.5.3}$$

This equation can be written in the general form

$$W_s = A|q_s|^2 + b^* q_s + q_s^* b, \tag{5.5.4}$$

where A is a positive real number equal to $\rho_0 c_0/4S$ and b is the complex number ($\rho_0 c_0 q_p e^{-jkL}/8S$).

This expression can be better understood if we write it in terms of the real and imaginary parts of the complex variable q_s. Thus if we put $q_s = q_{sR} + jq_{sI}$ and also $b = b_R + jb_I$, the equation reduces to

$$W = A(q_{sR}^2 + q_{sI}^2) + 2b_R q_{sR} + 2b_I q_{sI}. \tag{5.5.5}$$

A plot is shown in Fig. 5.9 of the power output W as a function of the real and imaginary parts (q_{sR} and q_{sI}) of the secondary source complex strength. This shows that the power output of the secondary source is a *quadratic* function of the complex secondary source strength. This form of quadratic function is central to many of the analyses presented in this book. A full discussion of quadratic functions of a complex variable is given in the appendix. Perhaps the most important property of this type of function is that, provided the value of A in equation (5.5.4) is positive, then the function has a *unique minimum* value. This is found by differentiating the function with respect to both the real and imaginary parts of the complex variable (q_{sR} and q_{sI} in this case) and setting the resulting derivatives to zero (see the appendix). The value of the complex variable which defines the minimum is then given simply by $(-b/A)$. Thus using the equations for A and b above, in this case the value of q_s which minimises W is given by

$$q_{s0} = -\frac{1}{2} q_p e^{-jkL}. \tag{5.5.6}$$

Substitution of this optimal value of the complex variable back into a function of the form given by equation (5.5.4) shows that the minimum value is equal to $-|b|^2/A$. In this case then the minimum power output is given by

$$W_{s0} = -\frac{\rho_0 c_0}{16S} |q_p|^2. \tag{5.5.7}$$

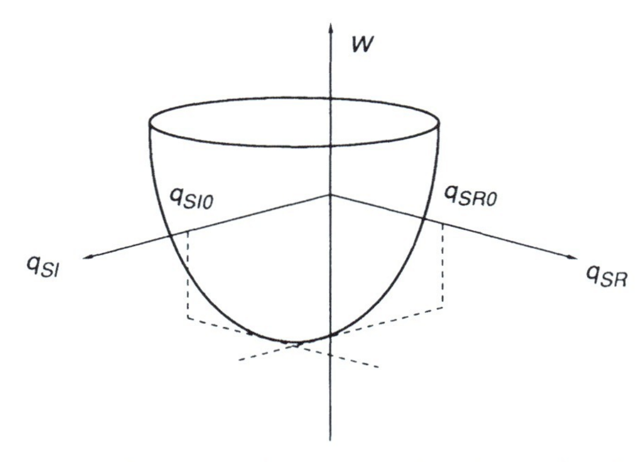

Fig. 5.9 The power output of the secondary source plotted as a function of the real and imaginary parts (q_{sR} and q_{sI}) of the complex secondary source strength. The minimum in the function has a negative value corresponding to the maximum power that can be absorbed by the secondary source.

First note that this has a *negative* value. Also, since $\rho_0 c_0 |q_p|^2/4S$ is the power output W_{pp} due to the primary source in the absence of the secondary source, then it also follows that

$$W_{s0} = -\frac{1}{4} W_{pp}. \tag{5.5.8}$$

Thus, under optimal conditions, the secondary source power output is exactly *minus* one-quarter of the primary source power output, the negative value of the secondary source power output corresponding to an *absorption* of acoustic power.

A useful means of describing this behaviour can again be found by observing the action of the secondary source in the time domain. Apart from the factor of one-half, equation (5.5.6) describes a secondary source strength which is related to the primary source strength in a manner which is identical to that necessary for downstream cancellation (equation (5.3.6)). Thus we can use the same transformation into the time domain to show that in this case

$$q_{s0}(t) = -\frac{1}{2} q_p(t - L/c_0). \tag{5.5.9}$$

Thus, in order to maximise its absorption the secondary source must produce an output which is exactly half that of the primary source but inverted and delayed by a time L/c_0. A description of this process is shown in Fig. 5.10. If the primary source emits a pulse of unit amplitude travelling downstream (and one travelling upstream) then the secondary source will emit a negative pulse of amplitude one-half at the instant the pulse from the primary source arrives at the position of the secondary source. This leaves a positive pulse of amplitude one-half travelling downstream and a negative pulse of amplitude one-half travelling upstream. Since the energies of these pulses are proportional to the *square* of their amplitudes (see Section 1.8), then the two pulses contribute energies of $(1/2)^2 + (-1/2)^2 = (1/2)$ where the energy associated with the original downstream travelling pulse is unity. Thus, half of the energy of the downstream travelling pulse is absorbed; this corresponds to one-quarter of the energy associated with the two pulses (travelling upstream and downstream) emitted by the primary source. Thus, although a plane monopole source (and thus a loudspeaker) can indeed absorb incident sound, the maximum it can absorb from a plane wave in an infinite duct is half the energy of the incident sound. Under these conditions it allows one-quarter of the incident energy to be transmitted and one-quarter to be reflected. Identical conclusions have been reached by Guicking *et al*. (1989) in calculating the maximum energy that can be absorbed by a point force applied to an infinite thin beam transmitting propagating flexural waves.

Finally, it is worth calculating the power output of the primary source when the secondary source strength is adjusted in order maximally to absorb sound. The primary source power output under these conditions can be written as

$$W_p = \frac{1}{2} \mathrm{Re}\left\{ \frac{\rho_0 c_0}{2S} (q_p + q_{s0} \mathrm{e}^{-jkL})^* q_p \right\}, \tag{5.5.10}$$

which on substitution of equation (5.5.6) can be reduced to

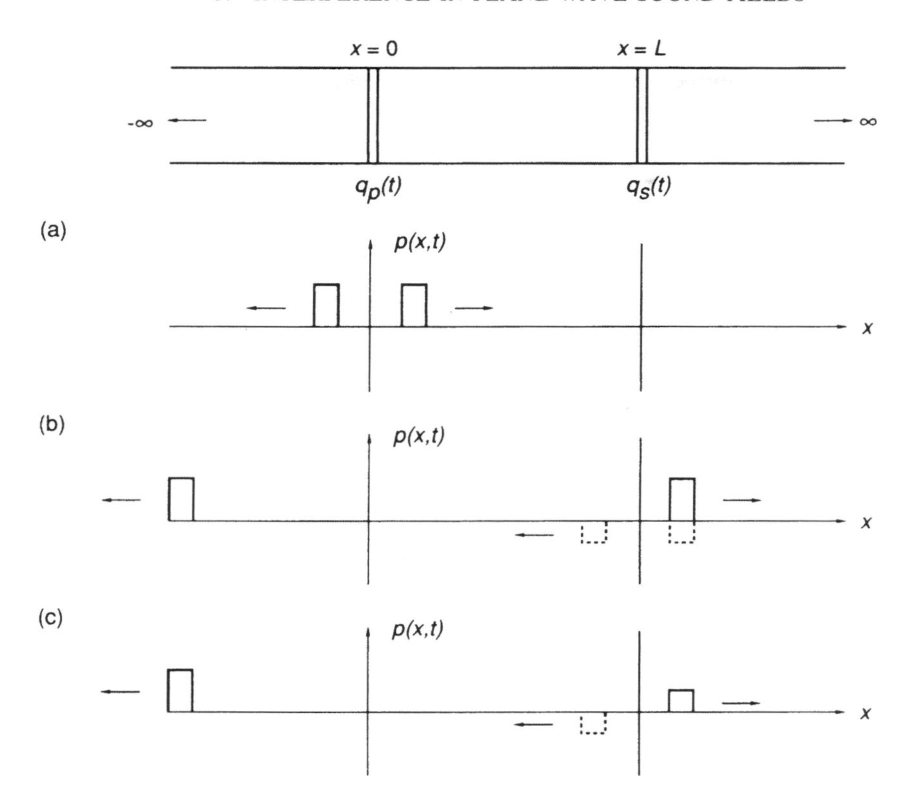

Fig. 5.10 A time domain description of the action of a plane monopole source whose strength is optimised to maximise its sound power absorption. The instantaneous pressure at a time t just after the primary source has radiated a pulse of sound is shown in (a) and in (b) at a time $t + (L/c_0)$. The net effect of the secondary source action is shown in (c).

$$W_p = W_{pp}\left(1 - \frac{1}{2}\cos 2kL\right), \tag{5.5.11}$$

where $W_{pp} = \rho_0 c_0 |q_p|^2/4S$ is again the primary source power output in the absence of the secondary source. Thus the effect on the primary source is to modify its power output between a minimum of one-half of its original value to a maximum of one and a half times its original value, depending on the separation of the two sources relative to the acoustic wavelength. So for a spacing of $L = n\lambda/4$, the effect of the absorbing secondary source is to "suck" *more* sound power from the primary source than it would otherwise radiate into an infinite duct. When averaged over frequency however, the primary source power output remains unchanged, an observation which is consistent with the time domain description of the process depicted in Fig. 5.10.

5.6 The minimum power output of the primary/secondary source pair

Before turning to a secondary source arrangement which *is* effective in absorbing substantial incident energy, we will again consider the primary and secondary source arrangement discussed above and show that there is a solution to the problem of producing the least possible power output from *both* sources. This amounts to adjusting the secondary source strength to ensure that the smallest possible radiation escapes in both the downstream and the upstream directions. Again we solve the problem of minimising a quadratic function but in this case the function considered is the *total* power output of the source combination. Thus we add together the secondary and primary source power outputs such that the total power output can be written as

$$W = \frac{1}{2}\,\mathrm{Re}\left\{\frac{\rho_0 c_0}{2S}(q_s + q_p e^{-jkL})^* q_s\right\} + \frac{1}{2}\,\mathrm{Re}\left\{\frac{\rho_0 c_0}{2S}(q_p + q_s e^{-jkL})^* q_p\right\}, \quad (5.6.1)$$

where we have again used the property of the plane monopole source that its power output can be calculated from the time-averaged product of its strength and the pressure acting upon it. The first and second terms in the curly brackets represent the power outputs of the secondary and primary sources respectively. Using the identities $\mathrm{Re}\{a\} = \frac{1}{2}\{a + a^*\}$ and $(e^{-jkL} + e^{jkL}) = 2\cos kL$ enables this expression to be reduced to

$$W = \frac{\rho_0 c_0}{4S}[|q_s|^2 + (q_p \cos kL)^* q_s + q_s^*(q_p \cos kL) + |q_p|^2]. \quad (5.6.2)$$

This is again a quadratic function of the complex secondary source strength which in this case can be written in the general form

$$W = A|q_s|^2 + b^* q_s + q_s^* b + c, \quad (5.6.3)$$

where, in this case, the positive real number A is again $\rho_0 c_0/4S$, the complex number b is $\rho_0 c_0 q_p \cos kL/4S$ and c is a positive real number given by $\rho_0 c_0|q_p|^2/4S$. The latter is equal to the power output of the primary source in the absence of the secondary source. Since A is positive, this quadratic function again has a unique minimum associated with the optimal value of q_s given by $-b/A$. Thus, in this case

$$q_{s0} = -q_p \cos kL. \quad (5.6.4)$$

The corresponding minimum power output (see the appendix) is given by $c - |b|^2/A$ and is thus equal to

$$W_0 = \frac{\rho_0 c_0|q_p|^2}{4S}(1 - \cos^2 kL), \quad (5.6.5)$$

or, since $\rho_0 c_0|q_p|^2/4S = W_{pp}$, we can write

$$\frac{W_0}{W_{pp}} = (1 - \cos^2 kL). \quad (5.6.6)$$

Thus, employing the optimal secondary source strength specified by equation (5.6.4) ensures that we never produce a net power output greater than W_{pp}. The worst case (where no reduction in power output can be produced) occurs when $\cos kL = 0$ or $kL = (2n+1)\pi/2$ where n is an integer, i.e. at source separation distances of $\lambda/4$, $3\lambda/4$, $5\lambda/4$ and so forth. In this case the optimal secondary source strength is *zero*, i.e. any non-zero value of secondary source strength can only increase the total power radiated into the medium. The best condition occurs when $\cos kL = 1$ or when the sources are separated by an integer number of half wavelengths, which as we have already seen, corresponds to the condition when perfect cancellation occurs in both directions in the duct. Note, however, that on average, if we integrate $\cos^2 kL$ over a frequency range such that kL varies from zero to π, then the minimum value of the power output is one-half that of the primary source alone. This can be explained by again observing the source behaviour in the time domain. First note that the optimal secondary source strength $(-q_p \cos kL)$ can be written as

$$q_{s0} = -\frac{1}{2} q_p (\mathrm{e}^{+jkL} + \mathrm{e}^{-jkL}). \tag{5.6.7}$$

Transformation into the time domain then shows that

$$q_{s0}(t) = -\frac{1}{2} q_p(t + L/c_0) - \frac{1}{2} q_p(t - L/c_0). \tag{5.6.8}$$

This action is illustrated in Fig. 5.11. To achieve this optimal solution, the primary source must firstly emit a negative pulse having an amplitude exactly half that of the primary pulse, but at a time L/c_0 *before* the primary pulse is emitted. This negative pulse arrives at the primary source exactly at the time the primary source pulse is emitted, thus reducing the amplitude of the upstream propagating pulse by one-half. When the downstream propagating pulse emitted by the primary source arrives at the secondary source, the secondary source again emits a negative pulse of half the amplitude of the primary pulse leaving pulses of amplitude of one-half propagating both downstream and upstream. The net energy released into the duct through this action is thus that associated with two pulses of amplitude (1/2) and two pulses of amplitude (− 1/2), i.e. the pulses having energy proportional to $[(1/2)^2 + (1/2)^2 + (-1/2)^2 + (-1/2)^2] = 1$. This is *half* that associated with two pulses of unit amplitude emitted by the primary source alone.

Note, however, that this average reduction in energy radiated is achieved by having a secondary source which acts *non-causally* with respect to the primary source, i.e. the secondary source has to emit sound prior to its emission by the primary source. This is very often the consequence of conducting an analysis in the frequency domain; optimal solutions are not necessarily constrained to causal actions of the secondary source. Nevertheless, this example does illustrate the two mechanisms by which a secondary source can reduce energy input into the medium. First the energy radiated by the primary source is reduced by the arrival of a pulse emitted some time before by the secondary source; the effective "loading" on the primary source is modified by the introduction of the secondary pulse such that the primary source has less pressure to work against. The second means of reducing the energy input to the medium is through absorption by the secondary source which, in this example, occurs in exactly the same way as that described in the last section.

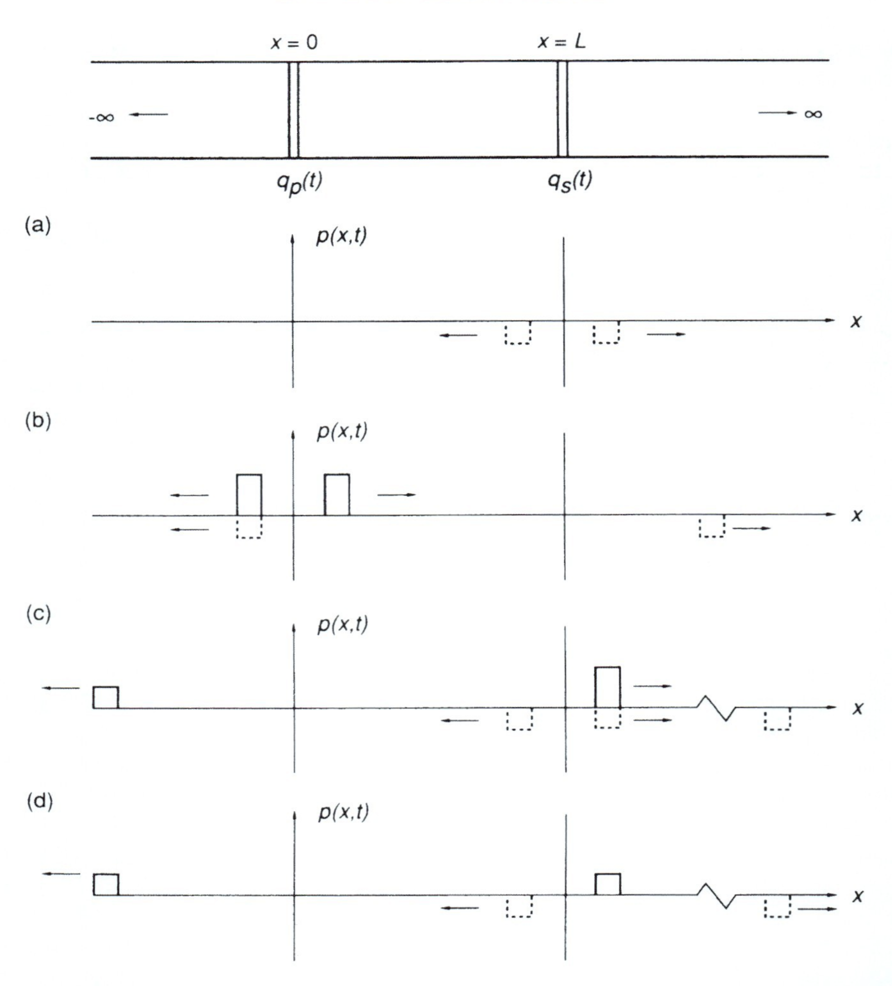

Fig. 5.11 A time domain description of the action of the secondary source in producing the minimum possible power output of the source combination. The pressure produced by the secondary source which acts at a time (L/c_0) *before* the output of the primary source is shown in (a). The pressure distribution just after the action of the primary source is shown in (b) and at a time (L/c_0) after this is shown in (c). The net effect of the process is shown in (d).

Finally it is worth noting that if the individual power outputs of the two sources are calculated from the two terms in curly brackets in equation (5.6.1), it is found that when the secondary source strength has its optimal value q_{s0} given by equation (5.6.4), then the power output of the secondary source is exactly *zero*. Thus, under these optimal conditions the power input to the medium is equal to that radiated by the primary source.

5.7 The cancellation of downstream radiation by using a pair of plane monopole sources

It is evident from the analysis presented in Section 5.3 that using a single plane monopole secondary source to cancel downstream radiation produces a reflection of incident primary sound. It was considered desirable by many workers to avoid such reflections since they would potentially contaminate the detection of any downstream propagating sound from a primary source and lead to difficulties in implementing control of the sound field. This will be discussed in much greater detail in Chapter 6. However, it was also recognised (Swinbanks, 1973) that two secondary sources could be used to produce only downstream radiation. It is obvious from the analysis of Section 5.3 that this is possible. Thus, for example, if we have a secondary source of strength q_{s1} located at $x = L$, and a further secondary source of strength q_{s2} located at $x = L + d$ (see Fig. 5.12) then we can always ensure zero radiation in the *upstream* direction by employing a strength of the first secondary source which is related to the second secondary source by

$$q_{s1} = -q_{s2}e^{-jkd}, \tag{5.7.1}$$

or, equivalently, in the time domain by

$$q_{s1} = -q_{s2}(t - d/c_0). \tag{5.7.2}$$

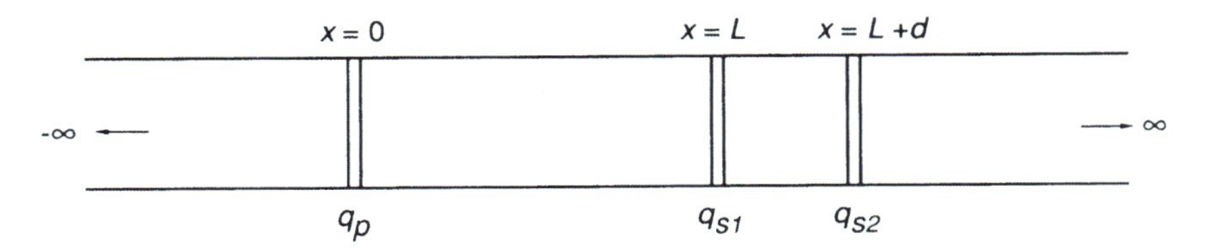

Fig. 5.12 Two plane monopole secondary sources of strengths q_{s1} and q_{s2} arranged to cancel downstream radiation from a primary source of strength q_p.

Thus, the first secondary source must always have a strength which is an inverted version of the second secondary source strength but delayed by a time d/c_0. This is the time taken for any sound to travel from the second secondary source upstream to the first secondary source. Using this relationship between the two secondary sources always ensures that there will be zero upstream radiation from the source pair. Since the net complex pressure downstream of the second secondary source is given by

$$p(x) = \frac{\rho_0 c_0}{2S} q_{s1} e^{-jk(x-L)} + \frac{\rho_0 c_0}{2S} q_{s2} e^{-jk(x-(L+d))}, \qquad x \geq L + d, \tag{5.7.3}$$

then it follows that if we use equation (5.7.1) the downstream radiation produced whilst ensuring zero upstream radiation will be given by

$$p(x) = \frac{\rho_0 c_0}{2S} q_{s2} e^{-jk(x-L)}(2j \sin kd), \qquad x \geq L + d, \tag{5.7.4}$$

Notice that there are some frequencies where this downstream radiation will also be zero. These frequencies correspond to values of kd when $\sin kd = 0$, which correspond to the condition when the two sources are separated by an integer number of half wavelengths. Therefore, although two secondary sources can be used to produce finite downstream radiation but zero upstream radiation, this is achievable only over a limited bandwidth. Figure 5.13 shows a plot as a function of frequency of the output of the source combination in the downstream direction. Swinbanks (1973) clearly recognised this bandwidth limitation and showed that the effective operational bandwidth could be extended by using an array of *three* secondary sources. Such extended arrays of sources have also been considered by other workers (see, for example, Eghtesadi and Leventhall, 1983, and Kazakia, 1986) but will not be examined in detail here.

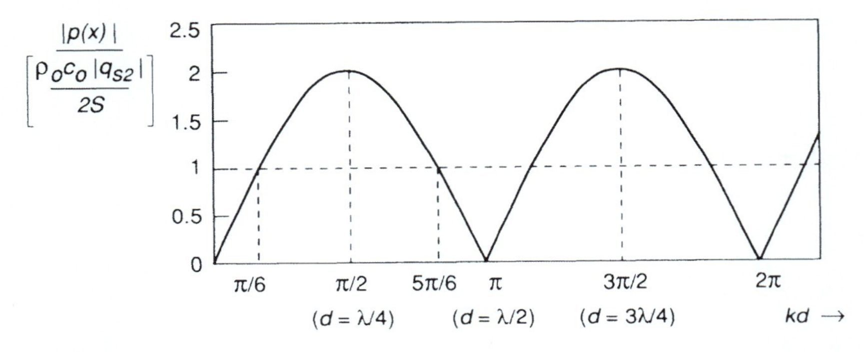

Fig. 5.13 The modulus of the downstream pressure fluctuations produced by a pair of plane monopole sources when their strengths are adjusted to produce zero upstream radiation (after Swinbanks, 1973).

Now let us calculate the secondary source strengths required to ensure that zero pressure is produced downstream when we have an upstream primary source in operation. Thus, downstream of the second secondary source the expression for the total complex pressure can, by using the superposition principle, be written as

$$p(x) = \frac{\rho_0 c_0}{2S} q_p e^{-jkx} + \frac{\rho_0 c_0}{2S} q_{s2} e^{-jk(x-L)}(2j \sin kd), \qquad x \geq L + d. \quad (5.7.5)$$

If we now apply the condition $p(x) = 0$ for $x > L + d$, then it follows that the necessary value of the second secondary source strength is given by

$$q_{s2} = \frac{-q_p e^{-jkL}}{2j \sin kd}. \quad (5.7.6)$$

Notice that this source strength becomes infinite at frequencies where $\sin kd = 0$, i.e. where the two secondary sources are separated by an integer number of half wavelengths. Using this relationship and the relationship given by equation (5.7.1), we can substitute back to find the net pressure field in the duct when the two secondary sources are adjusted to cancel downstream radiation. After some algebra it can be shown that

$$p(x) = \frac{\rho_0 c_0}{2S} q_p \mathrm{e}^{-jk|x|}, \qquad x \leq L, \qquad (5.7.7)$$

$$p(x) = \frac{\rho_0 c_0 q_p \mathrm{e}^{-jkL}}{j4S \sin kd} [\mathrm{e}^{-jk(x-(L+d))} - \mathrm{e}^{jk(x-(L+d))}], \quad L \leq x \leq (L+d), \qquad (5.7.8)$$

$$p(x) = 0, \qquad x \geq L + d. \qquad (5.7.9)$$

Thus, for all values of $x \leq L$ the complex pressure is exactly that which is produced by the primary source in the absence of the operation of any secondary sources. Between the two secondary sources, however, we can see from equation (5.7.8) that two travelling waves are produced which add together to give a standing wave. This net pressure field is illustrated in Fig. 5.14 for various values of the secondary source separation distance. Clearly, for frequencies such that the wavelength is close to an integer number of half wavelengths, a standing wave of large amplitude is produced between the two secondary sources. This amplitude, which corresponds to the requirement for large secondary source strengths at these frequencies, becomes infinite when $d = n\lambda/2$.

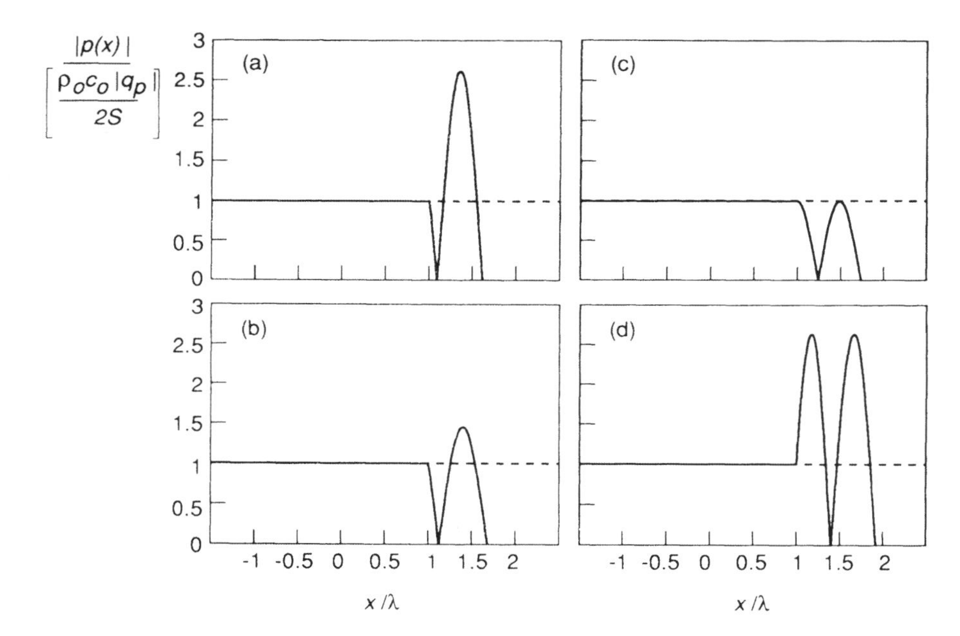

Fig. 5.14 The modulus of the fluctuating pressure $|p(x)|$ when a pair of plane monopole sources at $x = L$ and $x = (L+d)$ are used to provide perfect cancellation of the pressure fluctuations in the region $x > (L+d)$. The primary source is at $x = 0$ and the first secondary source is placed such that $L = \lambda$. Results are shown relative to the pressure fluctuation produced by the primary source alone (– – – – –) for various positions of the secondary source: (a) $d = 9\lambda/16$, (b) $d = 5\lambda/8$, (c) $d = 3\lambda/4$, (d) $d = 15\lambda/16$.

It is also interesting to consider the power output of the sources in these circumstances. Again recall that we need know only the net complex pressure and net source strength at a particular source to evaluate its power output. First it is obvious that the second secondary source of strength q_{s2} has zero power output since the pressure there is always zero. Also, since the pressure at the primary source is identical to that which is produced in the absence of control, the primary source power output is completely unaffected by the action of the secondary sources. One therefore must conclude that it is the first secondary source of strength q_{s1} which is responsible for absorbing exactly the power output radiated in the downstream direction by the primary source. This absorption process can be illustrated by using an example in the time domain. The relationship given by equation (5.7.6) which expresses the complex strength of the second secondary source relative to the complex strength of the primary source can equally be interpreted as defining the frequency response function relating these two sources. The same function relates the Fourier transforms of the fluctuations associated with both sources. Thus, we can write

$$Q_{s2}(\omega) = \frac{-Q_p(\omega)\mathrm{e}^{-jkL}}{2j\ \sin kd}, \tag{5.7.10}$$

where $Q_{s2}(\omega)$ and $Q_p(\omega)$ are the Fourier transforms of some arbitrary transient source strength fluctuations $q_{s2}(t)$ and $q_p(t)$ respectively. In order to enable an interpretation to be made of the action of the secondary sources in the time domain let us choose a primary source strength output whose Fourier transform is some function $F(\omega)$ multiplied by $2j\ \sin kd$. Thus, assume a primary source strength output whose Fourier transform is given by

$$Q_p(\omega) = 2j\ \sin kdF(\omega) = (\mathrm{e}^{jkd} - \mathrm{e}^{-jkd})F(\omega). \tag{5.7.11}$$

We know from the theory of Fourier transforms that if $F(\omega)$ is the Fourier transform of the function $f(t)$ in the time domain, then $F(\omega)\mathrm{e}^{j\omega b}$ is the Fourier transform of the function $f(t+b)$. Thus we are assuming that our primary source output is a function of time having the form

$$q_p(t) = f(t + d/c_0) - f(t - d/c_0). \tag{5.7.12}$$

It also follows from substituting equation (5.7.11) into equation (5.7.10) that for this choice of primary source output the Fourier transform of the second secondary source output is given by

$$Q_{s2}(\omega) = -F(\omega)\mathrm{e}^{-jkL}. \tag{5.7.13}$$

We can again use the same transformation into the time domain and it therefore follows that

$$q_{s2}(t) = -f(\mathrm{t} - L/c_0). \tag{5.7.14}$$

Now we can see the action of the secondary source in the time domain. This is illustrated in Fig. 5.15. First the primary source emits two pulses, one at $t = -d/c_0$ and one at $t = +d/c_0$. These pulses both travel downstream towards the secondary sources. The first pulse passes through the first secondary source unimpeded. When it reaches the second secondary source it is exactly cancelled by the pulse which is emitted at a time $t = L/c_0$. Note that the pulse has travelled for a time $(L+d)/c_0$ since its initiation at the primary source at time $t = -d/c_0$. The net effect of the emission of the cancelling pulse by the second secondary source is to produce zero pressure downstream, but a pulse is also emitted in the upstream direction which continues to travel towards the first secondary source. It arrives at the first secondary source at exactly the same instant as the second pulse emitted by the primary source. At that time the first secondary source emits a pulse which cancels both the upstream propagating pulse and downstream propagating pulse, leaving zero

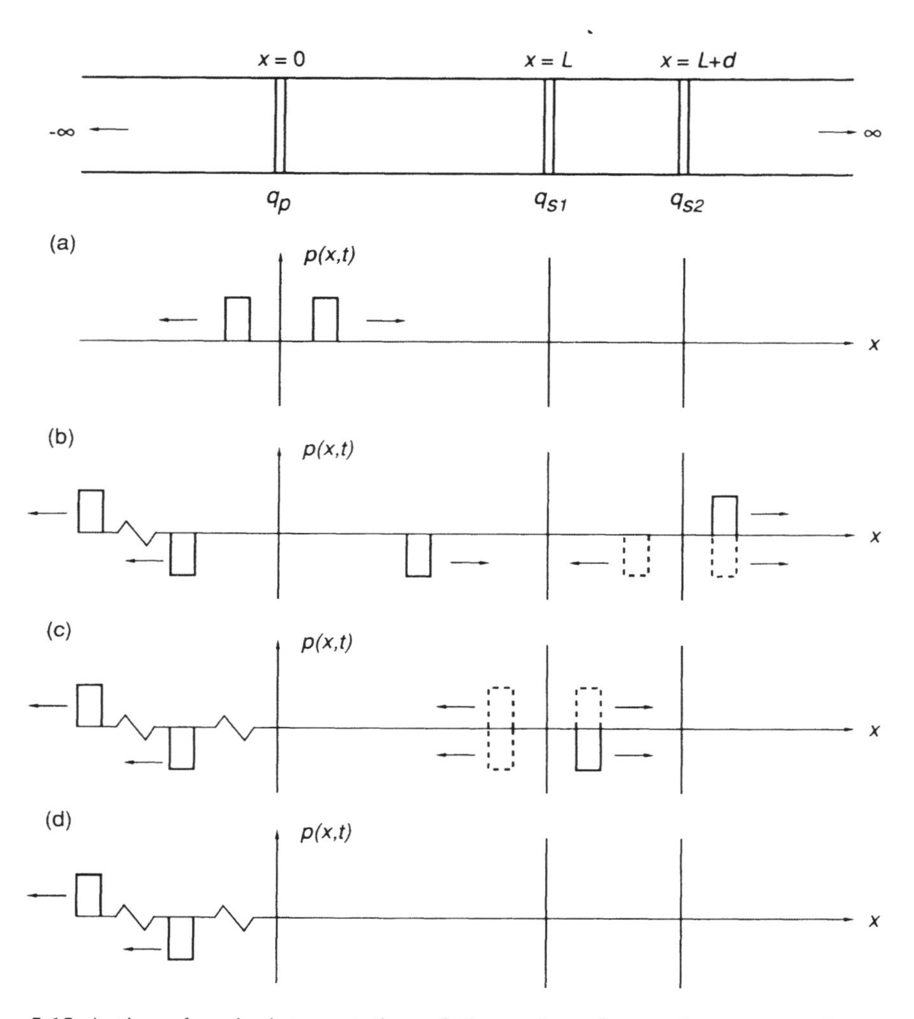

Fig. 5.15 A time domain interpretation of the action of two plane monopole secondary sources in perfectly absorbing incident sound. A pulse of sound is emitted by the primary source as shown in (a) and a cancelling pulse is emitted by the second secondary source (q_{s2}) at a time $(L+d/c_0)$ later as shown in (b). By this time the primary source has emitted a second pulse which converges on the first secondary source (q_{s1}) at the same instant as the pulse reflected from q_{s2}. Both pulses are then effectively absorbed by q_{s1} as shown in (c).

pressure in both directions. Through this action the first secondary source simultaneously absorbs both the pulse reflected from the second secondary source and the second pulse emitted from the primary source.

5.8 The mechanism of sound absorption by real sources

As we have seen, sound can be absorbed by the action of a secondary source. It is appropriate at this stage to make clear the action by which this can occur in real acoustic sources. Let us therefore go deeper into our source model and "take apart" the two massless pistons comprising our ideal plane monopole and look at the mechanism which drives the fluctuating flow which forces the two surfaces to oscillate to and fro. The action of a real loudspeaker on a duct wall, for example, can be represented as shown in Fig. 5.16. The analysis presented here follows closely that presented by Shepherd *et al.* (1986). The cone of the loudspeaker can be represented as a piston having mass M_m which is supported on a suspension which has a net stiffness K_m and a damping R_m. The cone is forced to and fro by passing a current through a coil which is attached to the cone and held in the magnetic field of a permanent magnet. The net effect of passing the current through the coil is to produce an oscillatory force on the cone. This force has to overcome the inertia, stiffness and damping force associated with the cone suspension. In addition, it will encounter another force due to any acoustic pressure fluctuations which load the surface of the cone. The net equation of motion of the loudspeaker cone can therefore be written as

$$j\omega M_m u + R_m u + \frac{K_m u}{j\omega} = Bli - pA, \tag{5.8.1}$$

where harmonic motion has been assumed and u is the complex cone velocity, i is the complex current flowing in the coil and p is the complex acoustic pressure which acts over the area A of the cone. The term Bl is the product of the magnetic flux density and the length of the coil such that Bli quantifies the net force transmitted to the piston by the action of the coil in the magnetic field. If we define $Z_m = j\omega M_m + R_m + K_m/j\omega$ as the mechanical impedance and $Z_a = pA/u$ as the mechanical impedance due to the acoustic conditions "seen" by the piston, then this equation can be written more compactly as

$$(Z_m + Z_a)u = Bli. \tag{5.8.2}$$

Similarly, we can write an effective equation of motion for the current flowing in the coil as a function of the electrical properties of the coil. Thus if the coil has an electrical resistance R and inductance L, we can write an equation in terms of the complex voltage v applied to the coil such that

$$j\omega Li + Ri = v - Bliu. \tag{5.8.3}$$

The term $Bliu$ quantifies the back e.m.f. produced in the coil by its motion in the magnetic field. If we define $Z_e = j\omega L + R$ as the blocked electrical impedance of the

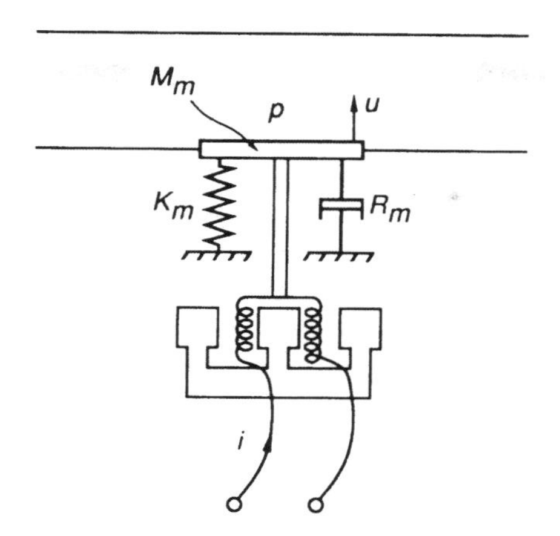

Fig. 5.16 A model of a real loudspeaker mounted on a duct wall. The cone of the loudspeaker is assumed to be a rigid piston of mass M_m and area A mounted on a suspension of stiffness K_m and damping R_m. The piston is forced to oscillate by the force produced when a current i is passed through the coil suspended in the field of a permanent magnet.

coil, we can write this equation more compactly as

$$Z_e i = v - Bliu. \tag{5.8.4}$$

If we now combine equations (5.8.2) and (5.8.4) it follows that

$$v = Z_e i + \frac{(Bl)^2 i}{(Z_m + Z_a)}, \tag{5.8.5}$$

where $(Bl)^2/(Z_m + Z_a)$ is known as the *motional impedance* of the loudspeaker. The electrical power supplied to the coil is given by (1/2) $\mathrm{Re}\{i^* v\}$, so taking the product of i^* with equation (5.8.5) and taking half the real part shows that the electrical power input can be written as

$$W_e = \frac{1}{2} \mathrm{Re}\left\{ |i|^2 Z_e + \frac{(Bl)^2 |i|^2}{Z_m + Z_a} \right\}. \tag{5.8.6}$$

A more useful form of this equation can be found by using equation (5.8.2) to show that $|i|^2 = |u|^2 |Z_m + Z_a|^2/(Bl)^2$. Substitution of this value of $|i|^2$ then shows that

$$W_e = \frac{1}{2} \mathrm{Re}\{|i|^2 Z_e + |u|^2 Z_m + |u|^2 Z_a\}, \tag{5.8.7}$$

where use has been made of the relationship $\mathrm{Re}\{|a|^2/a\} = \mathrm{Re}\{a\}$. This equation can also be written as

$$W_e = \frac{1}{2} |i|^2 R + \frac{1}{2} |u|^2 R_m + \frac{1}{2} \mathrm{Re}\{pq^*\}, \tag{5.8.8}$$

where $q = uA$ is the total volume velocity produced by the piston velocity fluctuations. The terms on the right hand side of this equation now have a clear physical interpretation in terms of the electrical and mechanical power dissipated and the acoustical power radiated respectively. It is therefore evident that the electrical power input to the loudspeaker is dissipated through three principal mechanisms: electrical, mechanical and acoustical. Note that, as we have seen, it is entirely possible for the third of these, given by $(1/2)\,\mathrm{Re}\{pq^*\}$, to take a *negative* value. Under these circumstances it is clear that less electrical power will be required to sustain a given piston velocity amplitude $|u|$. Of course in real loudspeakers it is generally the electrical and mechanical losses which consume the bulk of the electrical power.

A thorough analysis of this problem has been conducted by Shepherd *et al.* (1986) who used the above simple model to estimate the electrical power dissipated by two loudspeakers configured to act as a pair of secondary sources in order to absorb sound. The results of their experiments (see Fig. 5.17) are compared with their predictions based on the above model and measurements of the electrical and mechanical properties of the loudspeakers. The electrical power consumed is seen to vary from between a minimum of about fifteen times the acoustic power that would be radiated by a source of the same strength as the loudspeaker in an infinite duct to up to ten thousand times this value, with a strong dependence on frequency. Since the acoustic power output is such a small fraction of the electrical power consumed, the results do not show any systematic difference in the electrical power consumption between the two loudspeakers. The minimum in the electrical power consumption occurred at the natural frequency of the cone suspension system, when relatively little electrical energy is required to sustain the cone motion. This property was used by Ford (1983), who succeeded in measuring the small reduction in electrical power consumption when the upstream secondary source functions as a sound absorber in the two secondary sources configuration.

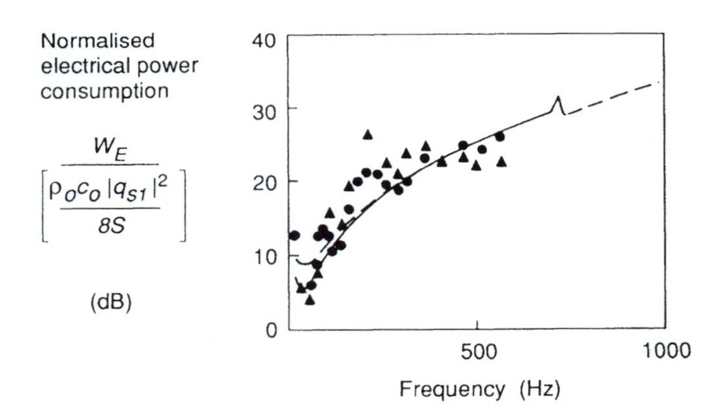

Fig. 5.17 Experimental results and theoretical predictions presented by Shepherd *et al.* (1986) of the electrical power consumption of the loudspeakers used as secondary sources to cancel downstream radiation. Results show the measured power consumption of the first secondary source (q_{s1}, ●) and the second (downstream) secondary source (q_{s2}, ▲) and the theoretical value for both sources (————) computed using equation (5.8.8) (and measured values of the loudspeaker properties). Also shown is the theoretical power consumption (– – – –) for a single loudspeaker operating alone in an infinite duct.

5.9 The plane dipole source in an infinite duct

We will now describe an alternative means of generating sound. Rather than using a plane monopole source where two massless pistons are forced apart in an oscillatory fashion, it is also possible to conceive of a single massless piston which is simply oscillated to and fro in the duct. This produces both upstream and downstream propagating waves, but the downstream propagating waves are out of phase with the upstream propagation. Let us suppose that we have a massless piston which is situated at a position y in the duct and which produces a downstream propagating pressure field and associated particle velocity field which are given by

$$\left.\begin{aligned} p(x) &= A\mathrm{e}^{-jk|x-y|} && (5.9.1) \\ u(x) &= \frac{A}{\rho_0 c_0}\mathrm{e}^{-jk|x-y|} && (5.9.2) \end{aligned}\right\}, \quad x > y,$$

where A is an arbitrary complex constant. Similarly, the motion of the piston produces upstream propagating pressure and particle velocity fields which are given by

$$\left.\begin{aligned} p(x) &= B\mathrm{e}^{-jk|x-y|} && (5.9.3) \\ u(x) &= \frac{B}{\rho_0 c_0}\mathrm{e}^{-jk|x-y|} && (5.9.4) \end{aligned}\right\}, \quad x < y,$$

where B is again an arbitrary complex constant. Note that we must ensure that the particle velocity on one side of the piston is exactly equal to the particle velocity on the other side of the piston. The only relationship between the arbitrary complex constants that ensures that this is the case is given by $B = -A$. Also note that the net force applied to the massless piston will be given by the difference in pressure across the two piston faces. This can be written as

$$f(y) = S(A - B) = 2AS. \tag{5.9.5}$$

It therefore follows that the net pressure field downstream and upstream of the massless piston can be written in terms of the complex force applied to the piston such that

$$p(x) = \frac{f(y)}{2S}\mathrm{e}^{-jk|x-y|}, \qquad x > y, \tag{5.9.6}$$

$$p(x) = \frac{-f(y)}{2S}\mathrm{e}^{-jk|x-y|}, \qquad x < y. \tag{5.9.7}$$

This form of source which produces antisymmetric radiation of sound waves is known as a *dipole*. Figure 5.18 shows a sketch of the form of the instantaneous pressure and particle velocity fields associated with this type of source.

We shall now show that exactly the same effect can be produced by using two plane monopole sources of equal and opposite strength, the arrangement which leads to the term "dipole" to describe this form of source. Assume that we have two plane monopole sources separated by a distance d as shown in Fig. 5.19. The upstream and downstream plane monopoles have complex strengths $-q$ and $+q$,

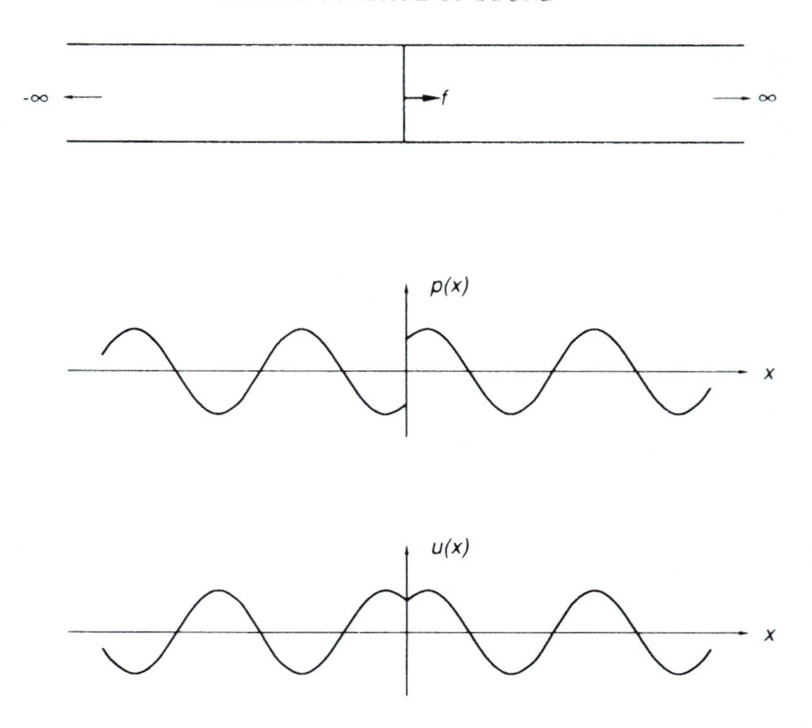

Fig. 5.18 The instantaneous pressure and particle velocity associated with a plane dipole source at one instant during a cycle of harmonic motion. The plane dipole source can be visualised as a massless piston which is forced to oscillate to and fro by the action of the applied force f. Note that this gives rise to a discontinuity in pressure on either side of the piston but that the particle velocity is continuous.

respectively. We can now calculate the net velocities at each of the plane pistons comprising both sources by superposing the velocity fluctuations due to each source. Thus, denoting the velocities as depicted in Fig. 5.19 it can be seen that

$$u_1 = \frac{q}{2S} - \frac{q}{2S}\mathrm{e}^{-jkd}. \tag{5.9.8}$$

We now assume that the wavelength of the sound is much longer than the separation between the two sources, i.e. we have the condition that $kd = 2\pi d/\lambda \ll 1$. Under these circumstances we can use a series expansion of the form $\mathrm{e}^{-jkd} = [1 - jkd - (kd)^2/2! \ldots]$ which shows that when kd is small then the velocity u_1 is given to leading order in the small parameter kd, by

$$u_1 \approx \frac{j(kd)q}{2S}. \tag{5.9.9}$$

The velocity on the other side of the first source, however, is given by

$$u_2 = \frac{-q}{2S} - \frac{q}{2S}\mathrm{e}^{-jkd} \approx \frac{-q}{S}, \tag{5.9.10}$$

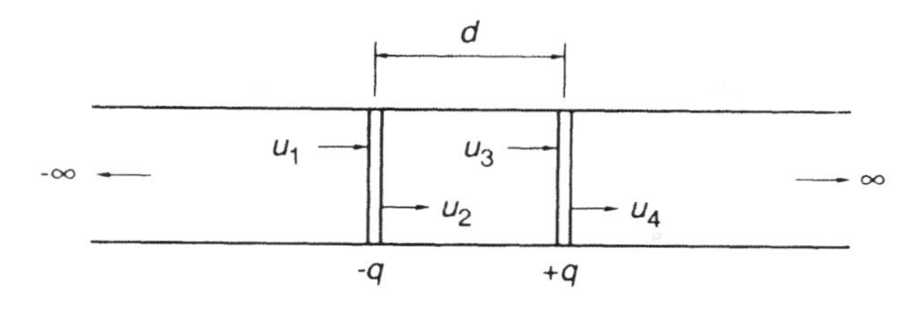

Fig. 5.19 Two harmonic plane monopole sources of equal and opposite strength placed a distance d apart which is small compared to the wavelength of the sound radiated.

and thus has a very large value compared to u_1. Similarly, the velocity at the faces of the pistons associated with the second source are given by

$$u_3 \approx \frac{-q}{S}, \qquad u_4 \approx \frac{j(kd)q}{2S}. \tag{5.9.11a,b}$$

Thus, when these two sources are operating together there is a very large fluctuating volume flow between the two sources, but since the velocity fields upstream and downstream of the two sources *almost* cancel out, there is only a relatively weak velocity fluctuation produced on either side of the source. These velocity fluctuations, u_1 and u_4 are identical (in both modulus and phase) and are equal to $jkdq/2S$. Note that this is a factor (kd) smaller than the piston velocity associated with just a single plane monopole. In fact if we calculate the pressure field radiated downstream as a result of the piston motions associated with the velocities u_1 and u_4 then we must have

$$p(x) = \frac{j\omega\rho_0 qd}{2S}\mathrm{e}^{-jk(x-d)}, \qquad x > d, \tag{5.9.12}$$

and similarly the pressure field radiated upstream will be given by

$$p(x) = \frac{-j\omega\rho_0 qd}{2S}\mathrm{e}^{jkx}, \qquad x < 0. \tag{5.9.13}$$

Note that as $d \to 0$ if we keep the product qd constant, then the pressure fields (5.9.12) and (5.9.13) are directly comparable with those of equations (5.9.6) and (5.9.7) produced by the fluctuating force, where the complex magnitude of the fluctuating force is given by

$$f = j\omega\rho_0 qd. \tag{5.9.14}$$

The physical origin of this force is associated with the fluctuating flow produced between the two monopole sources. The mass of the fluid between the two sources to a first approximation is given by $\rho_0 Sd$, whilst the acceleration of the fluid is given by $(-j\omega q/S)$. Thus the product of this mass and acceleration is equivalent to a force of magnitude $j\omega\rho_0 qd$. Note, however, that this mass times acceleration is acting in the negative x-direction whilst the equivalent force acting on the surrounding fluid is in the positive x-direction. Therefore the force which generates the dipole field is a *reaction force* which balances the inertia force associated with the fluctuating volume

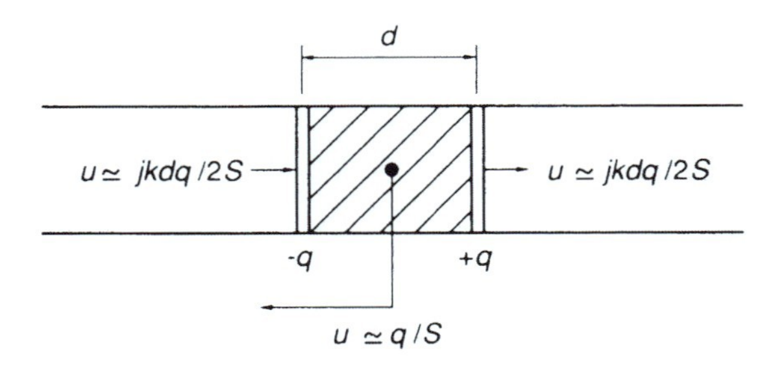

Fig. 5.20 The plane dipole field produced by the superposition of the fields of two plane monopoles of equal and opposite strength placed apart by a distance d which is small compared to the wavelength. There is a large velocity associated with the flow between the two sources, the rate of change of the momentum of which exerts a net reaction force on the fluid in the duct.

flow since the plane monopole sources themselves can supply no net force to the fluid. This is illustrated in Fig. 5.20.

5.10 Unidirectional radiation from a plane monopole/dipole combination

Now we can show that a combination of a monopole and a dipole source can produce perfectly unidirectional radiation at all frequencies. Let us assume that we have a plane monopole and a plane dipole source at a position y in the duct of strengths $q(y)$ and $f(y)$, respectively. The net pressure field downstream and upstream of the source combination can be written as

$$p(x) = \frac{\rho_0 c_0 q(y)}{2S} e^{-jk|x-y|} + \frac{f(y)}{2S} e^{-jk|x-y|}, \qquad x > y, \tag{5.10.1}$$

$$p(x) = \frac{\rho_0 c_0 q(y)}{2S} e^{-jk|x-y|} - \frac{f(y)}{2S} e^{-jk|x-y|}. \qquad x < y. \tag{5.10.2}$$

It is clear that in order to produce $p(x) = 0$ for $x < y$ then we need only specify the relationship between the dipole strength and monopole strength to be given by

$$f(y) = \rho_0 c_0 q(y). \tag{5.10.3}$$

This will then result in a net pressure field downstream and upstream of the sources which is given by

$$p(x) = \frac{f(y)}{S} e^{-jk|x-y|}, \qquad x > y, \tag{5.10.4}$$

$$p(x) = 0, \qquad x < y. \tag{5.10.5}$$

A sketch showing the manner in which a monopole and dipole field can be combined to give zero upstream radiation is illustrated in Fig. 5.21.

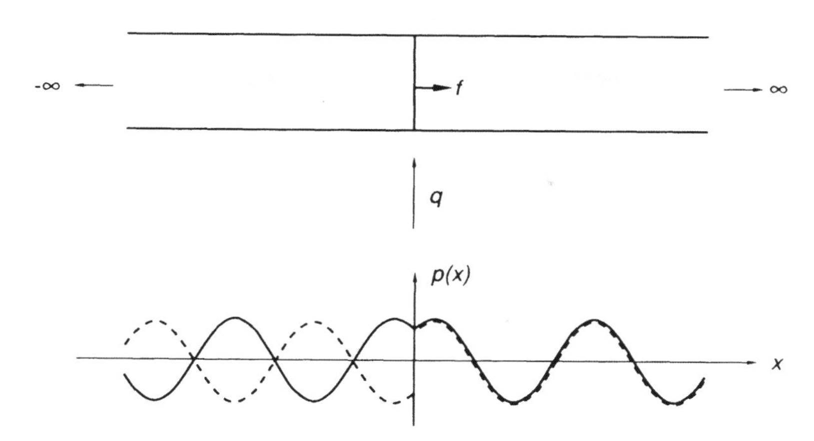

Fig. 5.21 The superposition of the pressure fluctuation due to a plane monopole source (———) with that of a plane dipole source (- - - - -) in order to produce zero upstream radiation and finite downstream radiation.

Now assume that we have a primary plane monopole source at $x = 0$, and our monopole/dipole source combination at $x = L$ with the dipole force related to the monopole source strength by equation (5.10.3). Under these circumstances the pressure field downstream of the secondary source can be written as

$$p(x) = \frac{\rho_0 c_0}{2S} q_p e^{-jkx} + \frac{f}{S} e^{-jk|x - L|}, \qquad x > L. \tag{5.10.6}$$

It therefore follows that in order to provide perfect cancellation of the downstream radiation we need a combination of sources which are specified by complex strengths such that

$$f = \frac{-\rho_0 c_0 q_p}{2} e^{-jkL}, \tag{5.10.7}$$

$$q_s = \frac{-q_p}{2} e^{-jkL}. \tag{5.10.8}$$

The use of a combination of monopole and dipole sources to produce unidirectional radiation was first suggested by Jessel (1968) and Malyuzhinets (1969), and led to the attempted synthesis of such a source combination by using a "tripole". This consisted of two loudspeakers used to form the dipole element and a further loudspeaker acting as a monopole (see Canévet, 1978). However, it is interesting to note that this same combination of source strengths is produced in the low frequency limit when we use just two plane monopole sources. If we return to the analysis of Section 5.7, where we specified the strengths of two plane monopole sources necessary to cancel downstream radiation and produce zero upstream radiation, then the strengths of the two sources were given by

$$q_{s1} = \frac{q_p e^{-jkL}}{2j \sin kd} e^{-jkd}, \qquad q_{s2} = \frac{-q_p e^{-jkL}}{2j \sin kd}. \tag{5.10.9a,b}$$

Again if $kd \ll 1$ or $d \ll \lambda$ then we can use the series expansions

$$e^{-jkd} = \left[1 - jkd - \frac{(kd)^2}{2!} + \ldots\right], \tag{5.10.10}$$

$$\sin kd = kd - \frac{(kd)^3}{3!} + \frac{(kd)^5}{5!} \ldots, \tag{5.10.11}$$

to show that the source strengths to leading order are given by

$$q_{s1} = \frac{q_p e^{-jkL}}{2jkd} - \frac{q_p e^{-jkL}}{2}, \tag{5.10.12}$$

$$q_{s2} = \frac{-q_p e^{-jkL}}{2jkd}. \tag{5.10.13}$$

Thus, the source strength associated with the first secondary source has a component which becomes increasingly large as kd becomes increasingly small, plus a second component which is simply a monopole strength equal to $(-q_p/2)e^{-jkL}$. The second secondary source also has an increasingly large source strength but of opposite sign to that associated with the first secondary source. The large source strengths associated with these two secondary sources combine to give a dipole of strength which is therefore specified by the complex force written as

$$f = j\omega\rho_0 d\left[\frac{-q_p e^{-jkL}}{2jkd}\right] = \frac{-\rho_0 c_0 q_p e^{-jkL}}{2}. \tag{5.10.14}$$

Thus, in principle, at very low frequencies we could use just two plane monopole sources to produce the requisite combination of monopole and dipole strengths in order to perfectly cancel downstream radiation whilst producing zero upstream radiation. However, in order to do this we must have extremely large constituent strengths which are associated with the terms q_p/kd; these strengths clearly becoming larger as kd becomes smaller, as suggested by the lower frequency limit in Fig. 5.13.

5.11 The influence of reflections from the primary source

Now let us consider the influence of reflections which could occur upstream of the primary source. Consider a primary source, shown in Fig. 5.22, which is spaced a distance D from a reflecting surface and a secondary source which is at a distance L. The easiest way to describe the form of the sound field under these circumstances is to use the "method of images". Thus, as shown in Fig. 5.22, the reflections from the surface at $x = 0$ occur as if they were produced by image sources located at $x = -D$ and $x = -L$. The complex pressure in a wave reflected from the surface is equal to the complex pressure in an incident wave multiplied by a complex pressure reflection coefficient R. This complex coefficient accounts for any change in amplitude *and* phase which occurs as a result of the reflecting surface on an incident wave. Now let us calculate the total complex pressure in the region of the duct downstream of the secondary source. First, the pressure which occurs as a direct result of both primary

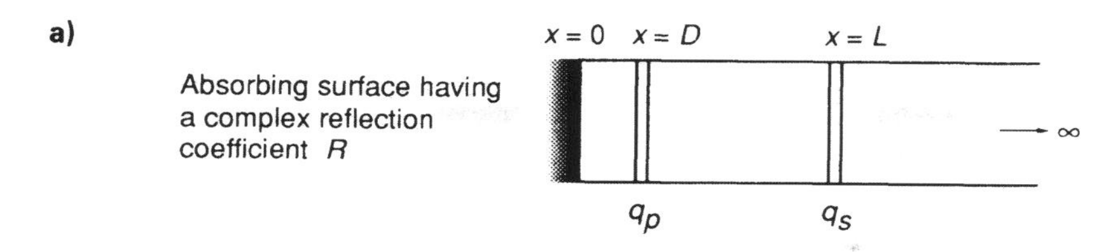

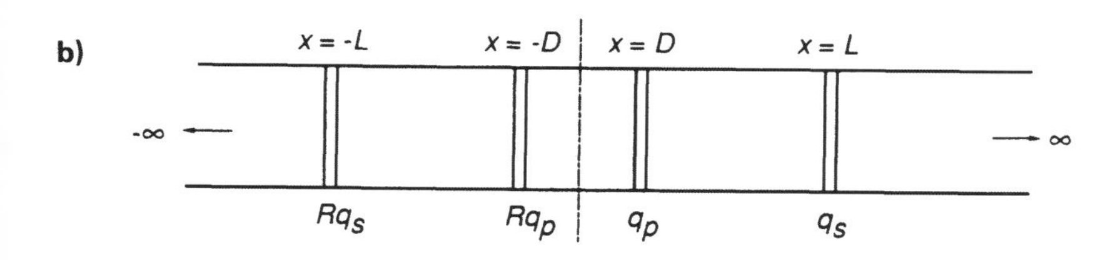

Fig. 5.22 The control of downstream radiation using a secondary source (q_s) when there is a reflecting surface upstream of the primary source (q_p) as shown in (a). The equivalent "image model" is shown in (b).

and secondary sources in the absence of a reflecting surface is given by

$$p_d(x) = \frac{\rho_0 c_0}{2S} q_p \mathrm{e}^{-jk(x-D)} + \frac{\rho_0 c_0}{2S} q_s \mathrm{e}^{-jk(x-L)}, \qquad x \geq L. \tag{5.11.1}$$

Second, the pressure field which results from the image sources whose strengths are reduced by a factor R to account for the properties of the surface is given by

$$p_r(x) = R\left[\frac{\rho_0 c_0}{2S} q_p \mathrm{e}^{-jk(x+D)} + \frac{\rho_0 c_0}{2S} q_s \mathrm{e}^{-jk(x+L)}\right], \qquad x \geq L. \tag{5.11.2}$$

The total pressure field downstream of the secondary source is given by superposing these two components such that

$$p(x) = p_d(x) + p_r(x), \qquad x \geq L. \tag{5.11.3}$$

If we now substitute equations (5.11.1) and (5.11.2) into equation (5.11.3) and set this pressure equal to zero for $x \geq L$, then the required relationship between the complex strengths of the secondary and primary sources is found to be

$$q_s = \frac{-q_p(\mathrm{e}^{jkD} + R\mathrm{e}^{-jkD})}{(\mathrm{e}^{jkL} + R\mathrm{e}^{-jkL})}. \tag{5.11.4}$$

Thus the modulus of the strength of the secondary source is now no longer simply equal to the modulus of the strength of the primary source as it was in the case where no reflections occurred.

It is instructive to examine the case when the primary source is placed close to the reflecting surface so that $D = 0$. This is illustrated in Fig. 5.23. This situation can now be considered to be representative of a primary source which has its own internal impedance: that is, the reflection coefficient R specifies the effect to which

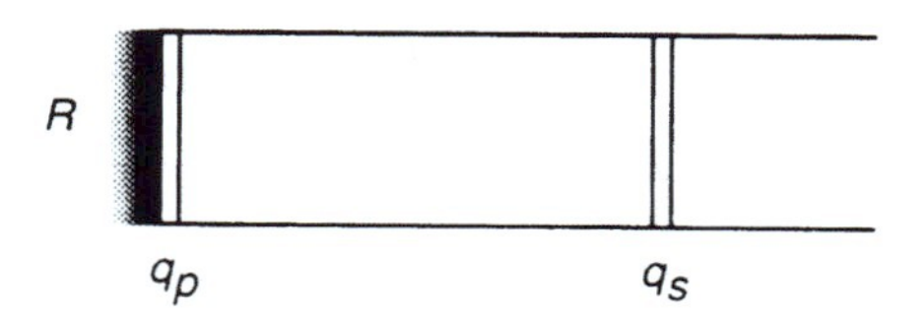

Fig. 5.23 A plane monopole source placed adjacent to an absorbing surface of reflection coefficient R to model the effect of a primary source which reflects incident sound.

incident waves are reflected from the primary source itself. This is relevant, for example, in the case of internal combustion engines (although in this case R varies with time) or where a loudspeaker is mounted in order to drive a duct. First, note that when the reflection coefficient of the primary source is zero, then an identical result is given to that produced by the analysis which excludes the effects of reflections, i.e. $q_s = -q_p e^{-jkL}$. However, when $D = 0$, in general we have

$$q_s = -q_p \frac{(1+R)}{(e^{jkL} + Re^{-jkL})}. \tag{5.11.5}$$

Now note that if $R = 1$, which corresponds to the case when the primary source reflects sound perfectly with no change of phase, then equation (5.11.5) reduces to

$$q_s = -q_p/\cos kL. \tag{5.11.6}$$

This shows that when the value of $kL = n\pi/2$ (where n is an integer) or $L = n\lambda/4$, then the secondary source strength necessary to cancel the field becomes infinite. In practice the value of q_s will be limited by the effects of dissipation at the walls of the duct (see, for example, the analysis presented by Glendinning *et al.*, 1988). These frequencies correspond to the frequencies at which the length of tube between the perfectly reflecting primary source and the secondary source resonates as if it were an organ pipe at very low frequencies. Further insight into this phenomenon is given by rewriting equation (5.11.5) as

$$q_s = \frac{-q_p(1+R)e^{-jkL}}{(1+Re^{-2jkL})}. \tag{5.11.7}$$

If we now use the binomial theorem to expand the denominator of equation (5.11.7) as a series of the form $(1+x)^{-1} = (1 - x + x^2 - x^3 \ldots)$, then it follows that

$$q_s = -q_p(1+R)(e^{-jkL} - Re^{-3jkL} + R^2e^{-5jkL} - R^3e^{-7jkL} \ldots). \tag{5.11.8}$$

An equivalent interpretation of this expression in the time domain can be given by assuming that R is a real frequency independent constant. In this case, transformation of equation (5.11.8) into the time domain yields

$$q_s(t) = -(1+R)\left[q_p\left(t - \frac{L}{c_0}\right) - Rq_p\left(t - \frac{3L}{c_0}\right) + R^2q_p\left(t - \frac{5L}{c_0}\right) \ldots\right]. \tag{5.11.9}$$

This is illustrated in Fig. 5.24 in which it is assumed that $R = 0.5$. A pulse emitted from the primary source is reflected from the secondary source but then suffers a

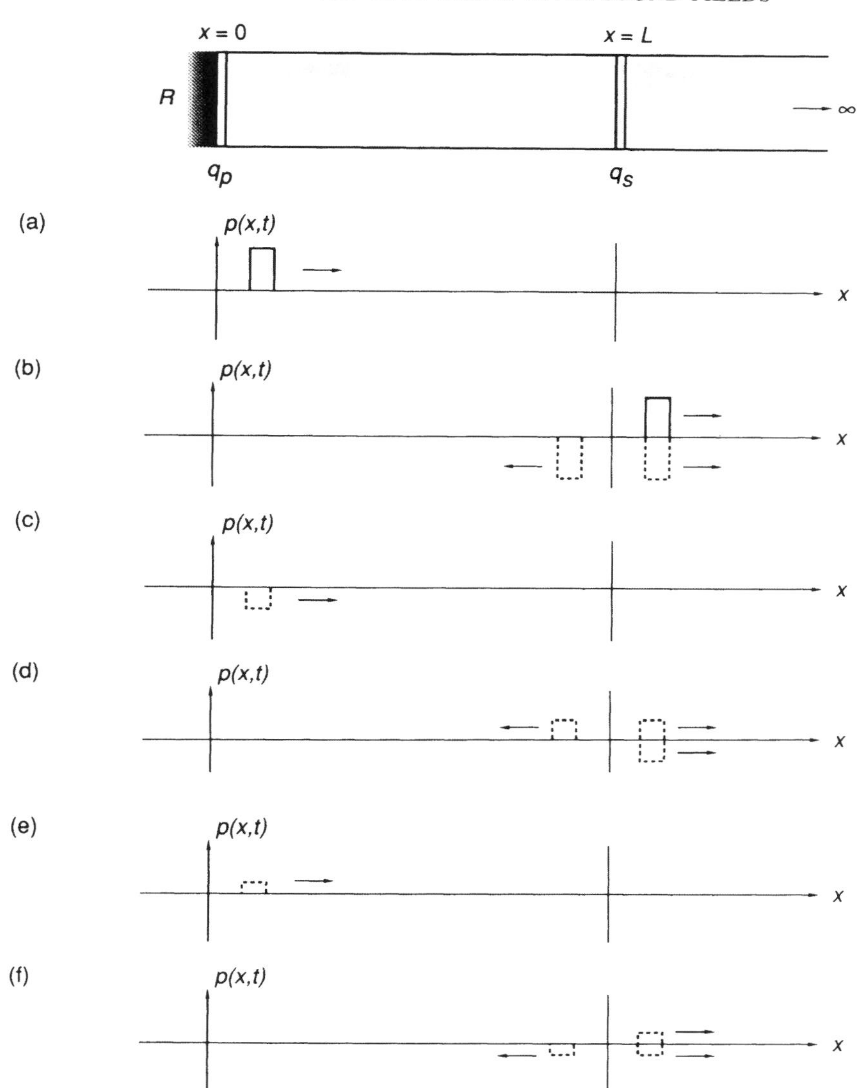

Fig. 5.24 The control of radiation from a reflecting primary source. The secondary source effectively reflects incident sound which is then re-reflected from the primary source with a diminished amplitude. A real value of $R = 0.5$ is assumed in the figure.

reflection at the primary source to return to the position of the secondary source a time $2L/c_0$ later. The secondary source will again act to reflect this pulse which will again be reflected back to the primary source and the process continues *ad infinitum*, with the amplitude of the pulses successively reduced by a factor R on each reflection at the primary source. Thus, we see that considerable demands can be placed on the secondary source strength in dealing with a primary source which strongly reflects incident waves; the secondary source strength required can be very much larger in magnitude than the strength of the primary source.

5.12 A travelling wave model of a one-dimensional enclosed sound field

In order to gain further insight into the possibilities for active control in the presence of reflecting surfaces, let us now consider the simplest possible wave model of an enclosed sound field. Figure 5.25 depicts a rigid wall duct closed at both ends that is excited by a plane monopole source at $x = y$. In the region upstream of the source, the complex pressure and particle velocity fields are given by

$$\left.\begin{aligned} p(x) &= A\mathrm{e}^{-jkx} + B\mathrm{e}^{jkx} \\ u(x) &= \frac{A\mathrm{e}^{-jkx}}{\rho_0 c_0} - \frac{B\mathrm{e}^{jkx}}{\rho_0 c_0} \end{aligned}\right\}, \qquad 0 \le x \le y. \tag{5.12.1, 5.12.2}$$

where the complex particle velocity has been deduced from the complex pressure by using the one-dimensional equation of conservation of momentum, and A and B are arbitrary complex constants. Since the ends of the tube are assumed perfectly rigid and will therefore provide perfect reflection of incident sound, the net particle velocity at $x = 0$ is given by $u(0) = 0$ and it is obvious from equation (5.12.2) that to satisfy this condition we must have $A = B$. Downstream of the monopole source we can again write expressions for the complex pressure and complex particle velocity in terms of the arbitrary complex constants C and D. Thus

$$\left.\begin{aligned} p(x) &= C\mathrm{e}^{-jkx} + D\mathrm{e}^{jkx} \\ u(x) &= \frac{C\mathrm{e}^{-jkx}}{\rho_0 c_0} - \frac{D\mathrm{e}^{jkx}}{\rho_0 c_0} \end{aligned}\right\}, \qquad y \le x \le L. \tag{5.12.3, 5.12.4}$$

Again, since the velocity at $x = L$ must be zero if the end of the tube is perfectly rigid, then we have $u(L) = 0$ and it follows from equation (5.12.4) that $C\mathrm{e}^{-2jkL} = D$. It now remains to evaluate the constants A, B, C and D. There can be no discontinuity of pressure across the source, and we can therefore write

$$p(y) = A(\mathrm{e}^{-jky} + \mathrm{e}^{jky}) = C\mathrm{e}^{-jkL}(\mathrm{e}^{-jk(y-L)} + \mathrm{e}^{jk(y-L)}). \tag{5.12.5}$$

Similarly, any difference in the particle velocity between the positions of the two hypothetical massless pistons representing the plane monopole source must be exactly equal to the source strength divided by the duct's cross-sectional area. This velocity difference can therefore be written as

$$\frac{q(y)}{S} = \frac{C}{\rho_0 c_0}\mathrm{e}^{-jkL}(\mathrm{e}^{-jk(y-L)} - \mathrm{e}^{jk(y-L)}) - \frac{A}{\rho_0 c_0}(\mathrm{e}^{-jky} - \mathrm{e}^{jky}). \tag{5.12.6}$$

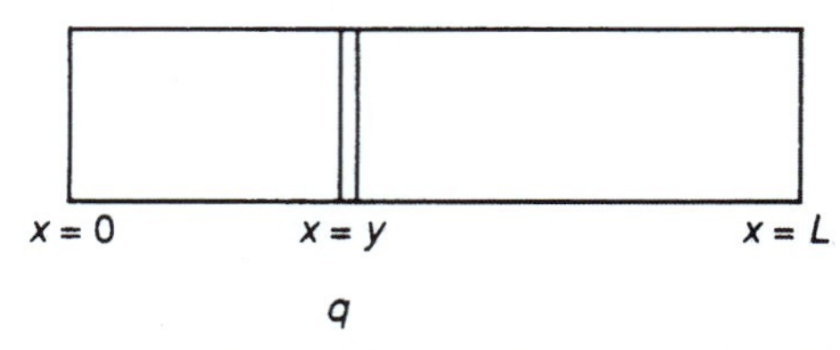

Fig. 5.25 A plane monopole source in a rigid walled tube with rigid terminations.

After some tedious but straightforward algebra we can deduce the constants A and C in terms of the source strength $q(y)$ by using equations (5.12.5) and (5.12.6). It then follows (see, for example, Curtis, 1988) that the expression for the complex pressure in the tube can be written as

$$p(x) = \frac{\rho_0 c_0 q(y)}{jS \sin kL} \cos[k(L-y)] \cos kx, 0 \leq x \leq y, \tag{5.12.7}$$

$$p(x) = \frac{\rho_0 c_0 q(y)}{jS \sin kL} \cos ky \cos[k(L-x)], y \leq x \leq L. \tag{5.12.8}$$

This specification of the sound field as a function of the arbitrary position of the exciting plane monopole source can be used to illuminate several other mechanisms associated with the active control of enclosed sound fields. First, however, let us consider the energy produced inside the enclosure by the action of a given source.

5.13 The time-averaged energy in the enclosure

As we have seen in Section 1.8, the time-averaged acoustic potential and acoustic kinetic energy densities associated with an acoustic disturbance are given by

$$e_p(x) = \frac{|p(x)|^2}{4\rho_0 c_0^2}, \qquad e_k(x) = \frac{\rho_0 |u(x)|^2}{4}. \tag{5.13.1a,b}$$

These quantities, of course, specify the potential and kinetic energies *per unit volume* in the sound field. It will be useful in this and in later chapters to assess the effectiveness of active control by observing the effects of the control on the total energy of the sound field. The total potential energy and kinetic energy associated with the sound field is given by integrating the relevant energy densities over the entire volume considered. These total time-averaged potential and kinetic energies are specified by

$$E_p = \int_V e_p(x) \mathrm{d}V, \qquad E_k = \int_V e_k(x) \mathrm{d}V. \tag{5.13.2a,b}$$

The total energy in the sound field is the sum of these two contributions and can therefore be written as

$$E = E_p + E_k = S \int_0^L \left[\frac{|p(x)|^2}{4\rho_0 c_0^2} + \frac{\rho_0 |u(x)|^2}{4} \right] \mathrm{d}x. \tag{5.13.3}$$

As an example of how this can be quantified consider first the one-dimensional situation above where the termination to the duct at $x = L$ perfectly absorbs incident sound. Thus, a source placed at $x = 0$ having complex strength q will simply generate plane waves which propagate downstream with a complex pressure and particle velocity field specified by

$$p(x) = \frac{\rho_0 c_0 q \mathrm{e}^{-jkx}}{S}, \qquad u(x) = \frac{q \mathrm{e}^{-jkx}}{S}. \tag{5.13.4a,b}$$

If we use equations (5.13.4) (corresponding to a pure travelling wave sound field) in equation (5.13.3) and undertake the necessary integrations, it follows that the total energy (which is equipartitioned between potential and kinetic contributions) is given by

$$E_t = \frac{|q|^2 L \rho_0}{2S}. \tag{5.13.5}$$

This total energy associated with only downstream propagating plane travelling waves will be useful in quantifying a base line level of energy in the duct. Now consider the case where the duct is rigidly terminated at $x = L$ but still excited by a source at $x = 0$. Under these circumstances the complex pressure and particle velocity fields are given by

$$p(x) = \frac{\rho_0 c_0 q \cos k(L-x)}{jS \sin kL} \tag{5.13.6}$$

$$u(x) = \frac{-q \sin k(L-x)}{S \sin kL} \tag{5.13.7}$$

$$0 \leq x \leq L.$$

If we now evaluate the total acoustic potential energy and total acoustic kinetic energy for this sound field by undertaking the appropriate integration for the length L of the tube, then it follows that (see Curtis, 1988)

$$E_p = \frac{E_t}{4 \sin^2 kL}\left(1 + \frac{\sin 2kL}{2kL}\right), \tag{5.13.8}$$

$$E_k = \frac{E_t}{4 \sin^2 kL}\left(1 - \frac{\sin 2kL}{2kL}\right), \tag{5.13.9}$$

where the results have been expressed relative to the energy in the same volume due to a plane travelling wave. Furthermore, the total energy, which is the sum of these two contributions, is given by

$$E = \frac{E_t}{2 \sin^2 kL}. \tag{5.13.10}$$

These expressions are shown graphically in Fig. 5.26. Note that both become infinite at the resonant frequencies of the sound field. Also, above the first resonance frequency both the potential and kinetic energy become approximately equal. At very low frequencies the potential energy becomes very large as the source forces the pressure to increase uniformly in the enclosure, whereas the kinetic energy becomes small as there is little associated particle velocity at very low frequencies.

5.14 The cancellation of the sound field in a one-dimensional enclosure

Now consider the case where we have not only a primary source at $x = 0$, but a secondary source placed at a position $x = d$ along the enclosure length which is still

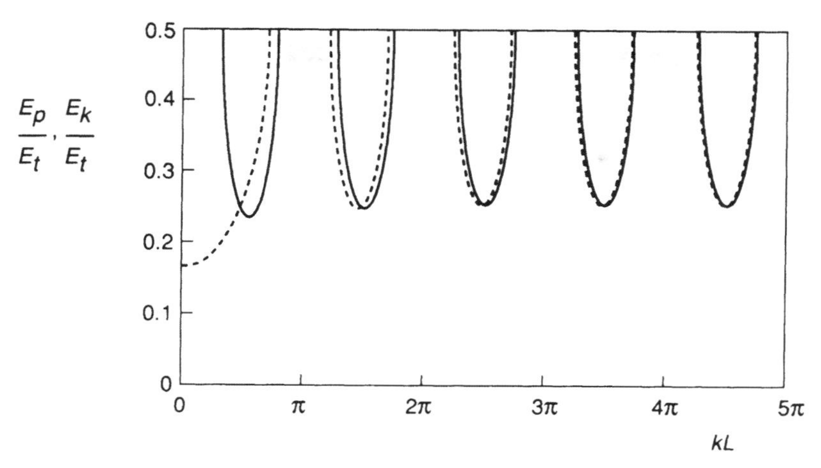

Fig. 5.26 The total time-averaged kinetic (- - - - -) and potential (———) energy in a rigid walled enclosure with rigid terminations excited by a single plane monopole source at one end.

rigidly terminated, as illustrated in Fig. 5.27. The pressure downstream of the secondary source can now be expressed as the contribution due to the primary and secondary sources, respectively, such that

$$\left.\begin{aligned} p_p(x) &= \frac{\rho_0 c_0 q_p \cos k(L-x)}{jS \sin kL} \quad (5.14.1) \\ p_s(x) &= \frac{\rho_0 c_0 q_s \cos kd \cos k(L-x)}{jS \sin kL} \quad (5.14.2) \end{aligned}\right\}, \qquad d \leq x \leq L.$$

As usual we will apply the superposition principle in order to calculate the pressure in the region downstream of the secondary source to ensure that

$$p(x) = p_p(x) + p_s(x) = 0, \qquad d \leq x \leq L. \tag{5.14.3}$$

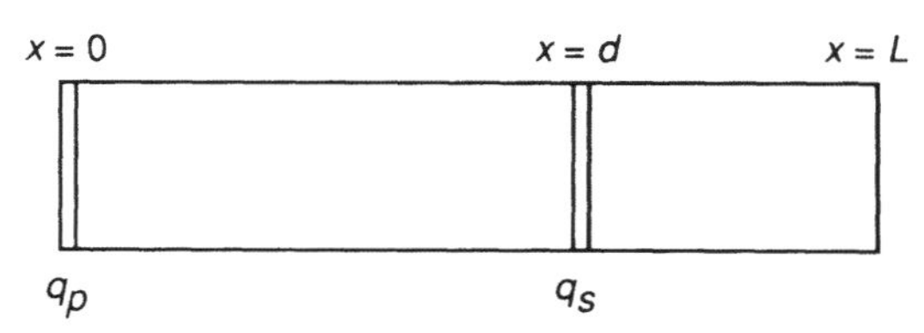

Fig. 5.27 The positions of the primary and secondary sources in a rigid walled tube with rigid terminations when the sound field is cancelled in the region $d \leq x \leq L$.

It is evident from equations (5.14.1) and (5.14.2) that this condition can be satisfied for all values of $x \geq d$ by choosing the secondary source strength such that

$$q_s = \frac{-q_p}{\cos kd}. \tag{5.14.4}$$

Note that this is exactly the relationship that we deduced between the secondary and primary source strengths when we considered the problem described in Section 5.11, where the primary source was placed against a perfectly reflecting surface but where waves were allowed to propagate to infinity downstream. It is therefore evident that the introduction of a further reflecting surface downstream of the primary/secondary source combination has not affected the requirement for the secondary source strength. In fact, as described above, such a secondary source as this will serve to act as a perfectly reflecting open end at $x = d$. This can be illustrated by assuming that the secondary source is placed at the end of the tube at $x = L$ such that the net complex pressure in the tube is described by

$$p(x) = \frac{\rho_0 c_0}{jS \sin kL}[q_p \cos k(L-x) + q_s \cos kx], \qquad 0 \leq x \leq L. \tag{5.14.5}$$

If we now choose the secondary source strength in accordance with equation (5.14.4) such that the pressure at the position of the secondary source is driven to zero, then after some manipulation it can be shown that the resulting pressure field in the tube is given by

$$p(x) = \frac{-\rho_0 c_0 q_p \sin k(L-x)}{jS \cos kL}. \tag{5.14.6}$$

The tube has now become resonant when $\cos kL = 0$, i.e. when $L = n\lambda/4$ where n is an integer. This is shown graphically in Fig. 5.28 which, following Curtis (1988),

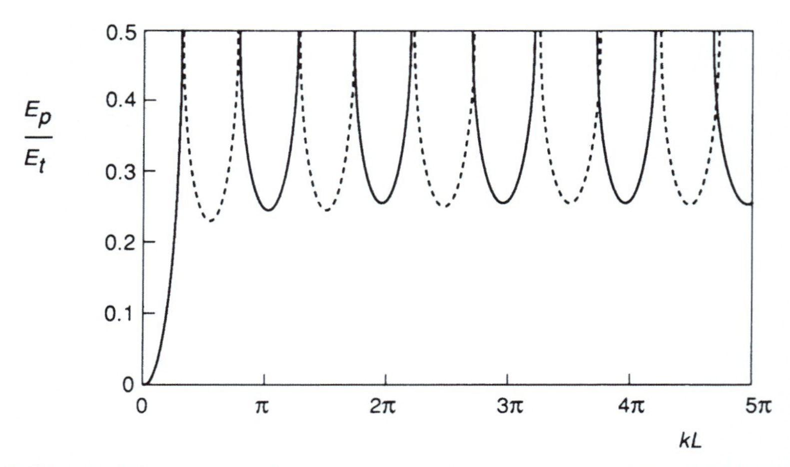

Fig. 5.28 The total time-averaged acoustic potential energy in a rigid walled one-dimensional enclosure when driven by a primary source at $x = 0$ (————). Also shown is the total time-averaged acoustic potential energy (– – – – –) when a secondary source at $x = L$ is used to drive the pressure to zero at $x = L$.

shows a plot of the total acoustic potential energy in the sound field under this condition. This illustrates that adopting this strategy effectively suppresses the original resonances of the sound field but introduces new resonances where previously the sound field was anti-resonant.

5.15 The absorbing termination

Rather than attempting the strategy described in the previous section, it is entirely possible to use the secondary source placed at the end of the duct in order to ensure that no waves are reflected from that end. This can be explained by firstly expanding equation (5.14.5) in terms of the exponential contributions to the cosine functions in the square brackets. Thus using the identity $\cos\theta = (1/2)(e^{j\theta} + e^{-j\theta})$ enables equation (5.14.5) to be written as

$$p(x) = \frac{\rho_0 c_0}{jS \sin kL}[q_p e^{jk(L-x)} + q_p e^{-jk(L-x)} + q_s e^{-jkx} + q_s e^{jkx}], 0 \leq x \leq L. \tag{5.15.1}$$

It can be seen that the second and fourth terms in the square brackets represent waves which are travelling upstream since they both have exponential terms of the form of e^{+jkx}. Thus, to ensure that the upstream travelling wave contribution is zero we need

$$q_p e^{-jk(L-x)} + q_s e^{jkx} = 0, \tag{5.15.2}$$

and it follows that this condition is satisfied if

$$q_s = -q_p e^{-jkL}. \tag{5.15.3}$$

Of course, in the time domain this simply requires that the secondary source strength is a delayed inverted version of the primary source strength. Thus, in the presence of a rigid boundary a simple plane monopole source can act to absorb incident radiation perfectly provided it is related to the primary source strength by

$$q_s(t) = -q_p(t - L/c_0). \tag{5.15.4}$$

A physical interpretation of this process is illustrated in Fig. 5.29. A pulse emitted by the primary source reaches the end of the tube and is reflected perfectly by the end of the tube. At the same time that this reflection occurs, the secondary source emits a negative pulse of the same amplitude, thereby cancelling perfectly the reflection. The net effect is of absorption of the incident pulse.

The action of a secondary source in providing an effective anechoic termination to an otherwise resonant tube was recognised by Bobber (1962) and Beatty (1964) in connection with providing a means for calibrating transducers for underwater acoustic measurements. Considerably more work since that time has been undertaken by Guicking and Karcher (1984) and by Guicking *et al.* (1985) in considering the wider possibilities for the active control of surface acoustic properties. More recently, Orduna-Bustamante and Nelson (1991) have demonstrated

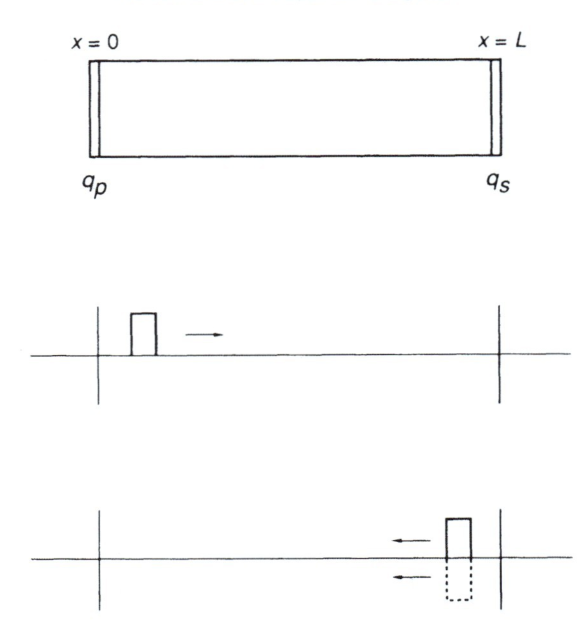

Fig. 5.29 A time domain description of the perfect absorption of a sound pulse incident on a secondary plane monopole source placed adjacent to a rigid surface. The secondary source emits an equal and opposite pulse to that reflected from the surface.

that an absorbing termination can be realised by using a control system based on adaptive signal processing techniques.

5.16 The active minimisation of the total acoustic energy in a one-dimensional enclosure

In addition to absorbing energy, a secondary source placed at the termination of a rigid enclosure of the type discussed above can be made to reduce the time-averaged energy in the enclosure to *less than* that associated with the pure travelling waves generated when perfect absorption is ensured at the termination. The complex pressure field generated by a primary source at $x = 0$ and a secondary source at $x = L$ is given by equation (5.14.5). The associated particle velocity distribution can be deduced by using the momentum conservation equation. If both these expressions are substituted into equation (5.13.4) and the integration over the length of the tube is undertaken, then an equation results which expresses the time-averaged energy in the enclosure as a *quadratic* function of the secondary source complex strength. Thus we have a function of the form

$$E = A|q_s|^2 + b^* q_s + q_s^* b + c, \tag{5.16.1}$$

where A is a positive real number, b is complex and c is a real number which corresponds to the time-averaged energy in the enclosure produced by the primary source alone. As we saw in Sections 5.5 and 5.6, such a function has a unique minimum value associated with an optimal complex secondary source strength q_{s0}. It has been shown by Curtis *et al.* (1987) that the secondary source strength which minimises this function is given by

$$q_{s0} = -q_p \cos kL, \tag{5.16.2}$$

and that the corresponding minimum energy is given by

$$E_0 = E_t/2, \tag{5.16.3}$$

where E_t is the total time-averaged energy associated with only plane travelling waves in the tube (see Section 5.13). Note that E_0 therefore corresponds to the energy in the enclosure at the *anti-resonance* frequencies of the tube. As illustrated in Fig. 5.26, the potential and kinetic contributions to the total energy add to give $E_t/2$ at the higher anti-resonant frequencies.

The means by which the secondary source can make the enclosure appear anti-resonant were illustrated by Curtis *et al.* (1987) using a time domain example. Note that the optimal secondary source strength is related to the primary source strength in the frequency domain by

$$q_{s0} = -q_p \cos kL = -\frac{1}{2} q_p (e^{jkL} + e^{-jkL}), \tag{5.16.4}$$

which can be transformed into the time domain to give

$$q_{s0}(t) = -\frac{1}{2} q_p(t + L/c_0) - \frac{1}{2} q_p(t - L/c_0). \tag{5.16.5}$$

This solution is therefore of an identical form to that given in Section 5.6 for the minimisation of the time-averaged power output of two plane monopole sources in an *infinite* duct. Again, the solution involves a secondary source output which is *non-causal* with respect to the primary source output. As illustrated in Fig. 5.30, the secondary source produces a negative pulse having an amplitude which is one-half that of the pulse produced by the primary source at a time L/c_0 *before* the radiation of the primary pulse. The remaining positive pulse of amplitude one-half is subsequently absorbed by the secondary source at a time L/c_0 *after* its radiation from the primary source. The net effect is that the energy released into the enclosure is proportional to $(-1/2)^2 + (1/2)^2$ as exactly one-half that associated with a pulse of unit amplitude propagating down the tube and being absorbed by the termination. Of course, for single frequency sounds, the non-causal form of the optimal solution in the time domain is of no relevance; under steady-state conditions the secondary source strength can be adjusted to make the tube appear anti-resonant at any single frequency.

We will return to the topic of controlling enclosed sound fields in Chapter 10 and use the idea of minimising the energy of the sound field in order to evaluate the possibilities for active control in more practical three-dimensional enclosures. We shall also show there that minimising the total acoustic energy in the field is a somewhat unrealistic objective, but that minimising the sum of the squared outputs of a number of microphones in the enclosure can, under some circumstances, give a good approximation to minimising the total acoustic potential energy.

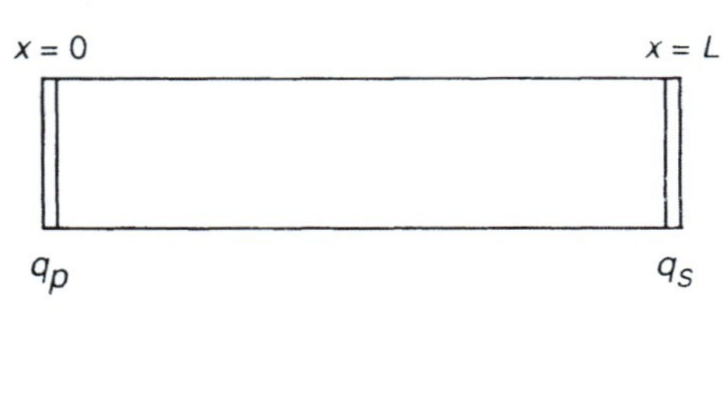

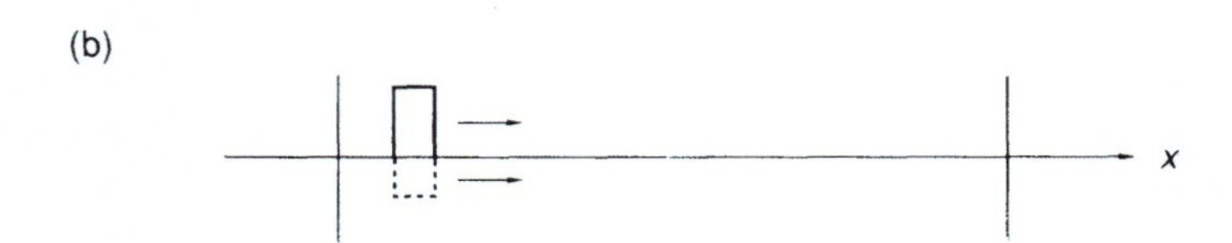

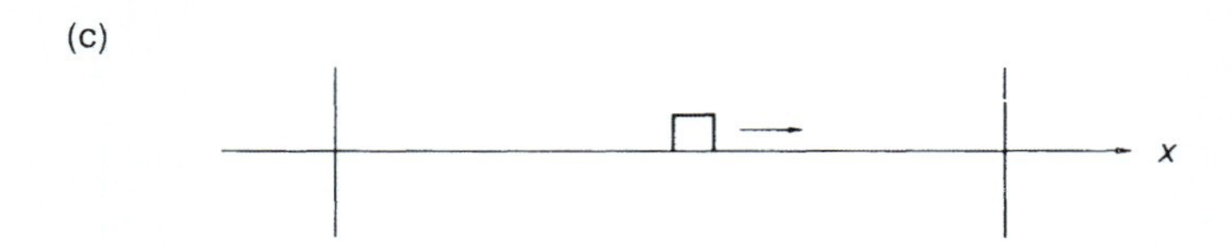

Fig. 5.30 A time domain description of the action of the secondary source in minimising the total time-averaged energy in a one-dimensional rigid walled enclosure. As shown in (a) the secondary source emits a pulse of one-half the amplitude of that emitted by the primary source at a time L/c_0 *before* its emission by the primary source (b). The resulting phase (c) is then absorbed by the secondary source as shown in (d).

6 Single Channel Feedforward Control

6.1 Equivalent block diagram

In this chapter we will consider the design and implementation of the electronic *controller* in a single input, single output active control system. This controller corresponds to the element labelled "V" in Lueg's original patent (1936), which is illustrated in Fig. 5.1 at the start of the previous chapter. We will initially study the behaviour of such a single channel active control system by considering its response in terms of purely electrical inputs and outputs. This allows us to describe its external behaviour completely in terms of an *equivalent electrical block diagram*. It is assumed that the signal used to drive the secondary source is derived from a *detection sensor* placed upstream of the secondary source in a propagating wave field. This sensor may be a microphone, an accelerometer, an optical sensor (in the case of flame noise), a mechanical transducer on a rotating machine, or a purely electrical connection to an electrical machine. This signal is fed through an electrical controller (which is, in practice, some form of electronic filter) before driving the secondary source. The acoustic output of the secondary source can be made to mimic the inverse of the sound from the primary source when it reaches the position of the secondary source, and thus achieve attenuation of the primary field.

We will also introduce the idea of placing an *error sensor* downstream of the secondary source. The output of the error sensor can be used to monitor the performance of the active control system. The objective of the control system is the minimisation of this electrical error signal, whether in the frequency domain, the continuous time domain or the sampled time domain. The design of an optimal electrical controller, and its performance in reducing the error signal, can be assessed in terms of cross-spectral densities, the coherence function or the cross-correlations between the observable electrical signals. The advantage of an equivalent electrical block diagram representation is that it allows the design of the electrical controller to be approached by using the tools of linear systems theory. The approach thus allows the development of practical controller design methods and an assessment of ultimate performance from purely electrical measurements in the system to be controlled.

Before contemplating more detailed design methods, in order to establish the values of the coefficients of the electrical filter being implemented as the controller, some assumptions must be made by the designer regarding the structure and complexity of the filter to be designed. If, for example, a digital filter is being contemplated, the choice must be made of whether an FIR or IIR filter is to be implemented, and of how many coefficients such a filter will probably require in order to achieve a certain performance level. These questions regarding the initial form and complexity of the filter prompt a study of the acoustical interpretation of the electrical block diagrams used above. We will show that the electroacoustic transfer functions of each element in the block diagram can be manipulated to derive the transfer function of the ideal controller in terms of the acoustical characteristics of the sensors, sources and duct in which the controller operates. This enables a physical interpretation to be placed on the form of the electrical controller.

6.2 Single channel control with an independent reference signal

We begin our discussion of single channel controllers by considering the case in which a *reference signal* is available from the detection sensor which is *unaffected* by the action of the secondary source. This assumption considerably simplifies the resulting block diagram, and is valid for a number of problems of practical interest. In particular, many acoustic sources of practical importance produce sound waves which are nearly periodic. Examples are the noise from rotating or reciprocating machines such as internal combustion engines. Direct observation of the mechanical motion of such sources is generally possible, by using a suitable mechanical, electrical or optical transducer, which can be used to provide an electrical reference signal with the same fundamental frequency as the sound emitted. There are a variety of techniques which have been employed in using such an electrical reference as the excitation signal fed to the secondary source, via an electrical controller, in order to implement an active noise control system. The general block diagram of such a single channel control system is illustrated in Fig. 6.1, in which it is assumed, for the sake of illustration, that the sound from the periodic primary source is propagating as plane waves down a duct. This assumption is not necessary for the development of the equivalent block diagram and analysis presented below, but

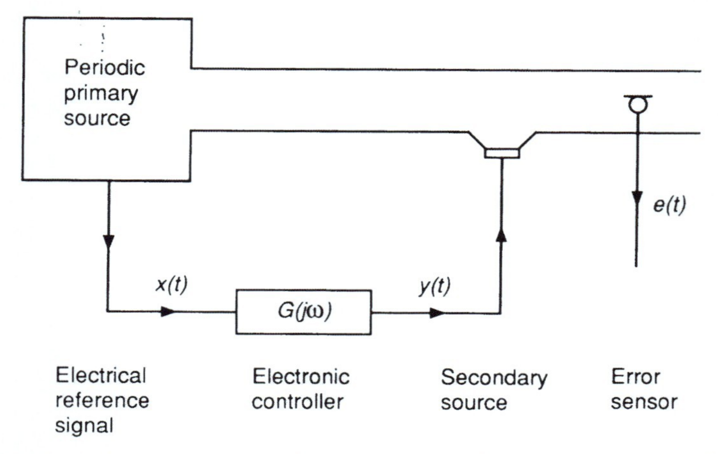

Fig. 6.1 The physical elements of a single channel active control system with an electrical reference signal.

provides a convenient physical interpretation of the acoustical behaviour involved. Similarly, although the secondary source is illustrated as a single loudspeaker in Fig. 6.1, the electrical block diagram would be the same even if the secondary source were a directional array of transducers, provided the array were fed originally from a single electrical signal. We denote the waveform of the reference signal as $x(t)$, that of the signal fed by the electrical controller into the secondary source as $y(t)$, and that from the error sensor as $e(t)$. (Note that in previous chapters we have used the symbols x and y to denote spatial co-ordinates, but in this chapter, and in Chapter 12 when we deal with multi-channel control systems, we will use x and y to denote signals. In these chapters we will have no need to use x and y as spatial co-ordinates.) Before considering any specific realisation of the electrical controller, we can derive a block diagram which completely describes the electrical behaviour of the system illustrated in Fig. 6.1.

A single electrical signal $e(t)$ (usually a voltage) is shown as the output from the error sensor in Fig. 6.1, and a single electrical signal $y(t)$ is fed to the secondary source. Provided these signals drive in to or are driven from electrical components of constant electrical impedance (typically $e(t)$ will drive into a large electrical impedance and $y(t)$ will be supplied from a source of low electrical impedance) it will not be necessary to consider the current supplied to, or driven by, these transducers and the characteristics of the physical system can be described in terms of single input, single output electrical elements rather than the more general electrical two-port network (for a further discussion, see, for example, Kuo, 1966). Indeed, any signal conditioning, impedance matching device or electrical amplifier can be considered as being contained in the physical components of the system illustrated in Fig. 6.1, with the relevant voltages being those which eventually drive or are generated by the control system.

We now make the important assumption that all the electrical, electroacoustic and acoustic elements of the physical system are *linear*. In practice this can only be an approximation, especially for the electroacoustic transducers, but as we have discussed in the introduction to Chapter 3, provided the transducers are good quality units and are operating within their dynamic range constraints, the approximation is generally a good one. The assumption of linearity allows the exact characterisation of the *error path* between the secondary source and error sensor purely in terms of its electrical frequency response function. In general, the transfer function of this error path may be influenced by the operation of the primary source, due to flow down the duct if the primary source is a fan for example (Ross, 1982b). We will return to a discussion of methods of practically estimating the error path transfer function later, but for the time being we will assume that this error path is known and can be characterised by its frequency response, $C(j\omega)$, defined by the equation

$$C(j\omega) = \left. \frac{E(\omega)}{Y(\omega)} \right|_{D(\omega)=0}, \tag{6.2.1}$$

where the contribution of the primary source to the signal from the error sensor is denoted as $D(\omega)$ and assumed to be zero, and $E(\omega)$, $Y(\omega)$ and $D(\omega)$ are the Fourier transforms of the relevant time histories. The assumption of linearity is now again used in order to determine the total output from the error sensor. This output is the

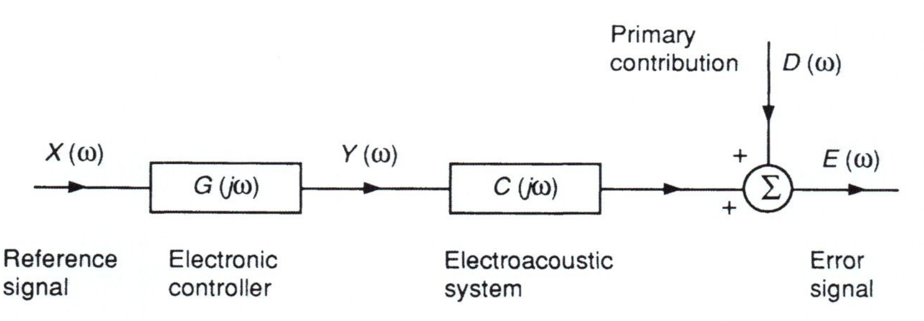

Fig. 6.2 The equivalent electrical block diagram of Fig. 6.1 with linearity assumed.

superposition of the effects of the primary source, $D(\omega)$, and of the secondary source, given by $Y(\omega)$ modified by the error path $C(j\omega)$, so that in general

$$E(\omega) = D(\omega) + C(j\omega)\,Y(\omega). \tag{6.2.2}$$

We now assume that the electrical controller has a frequency response of $G(j\omega)$, so that

$$Y(\omega) = G(j\omega)\,X(\omega). \tag{6.2.3}$$

The total response at the error sensor can thus be written as

$$E(\omega) = D(\omega) + G(j\omega)C(j\omega)\,X(\omega), \tag{6.2.4}$$

and the final equivalent electrical block diagram is shown in Fig. 6.2.

It should be emphasised that the superposition of the electrical signals leading to equation (6.2.4) is a necessary consequence of the *linearity* of the system under consideration, which includes the electrical, electroacoustic and acoustic components. No further assumptions need be made about the connections between these components. The electroacoustic sources could, for example, have finite internal acoustic impedances and so their volume velocities would depend on the pressure they experience in the duct because of acoustic loading effects. The electrical transfer response $C(j\omega)$, and the response of the error sensor with the electrical input to the secondary source set to zero, $D(\omega)$, will both be affected by these loading effects, but the superposition of the electrical signals described by equation (6.2.4) will still be valid.

6.3 Control of a periodic sound at individual harmonics

We now return to the specific problem of a periodic primary source. The most obvious method of designing the electrical controller is to operate on each harmonic of the sound independently. Since the system is assumed linear, there will be no interaction between that part of the controller operating at one harmonic and that part operating at any other. The controller for each harmonic can thus be designed independently. Clearly, in order to implement such a system, the original periodic reference signal waveform must be decomposed into its constituent harmonics and each of these harmonics must be used to generate one harmonic of the output

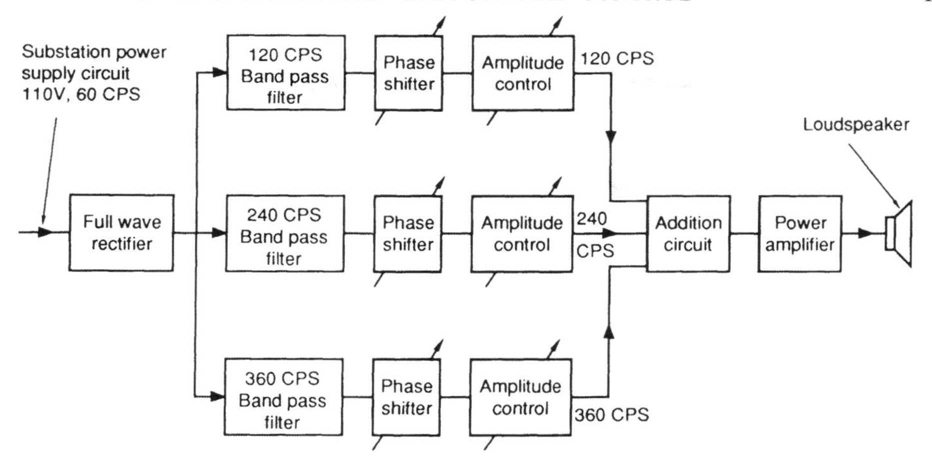

Fig. 6.3 Control of a periodic sound source by using individual harmonics (after Conover, 1956).

waveform $y(t)$. This philosophy was clearly understood by William Conover in the 1950s, who experimented with the active control of the noise radiated by electrical transformers. The noise radiated by transformers is largely at harmonics of twice the "line frequency". Conover's method for generating reference signals at individual harmonics is illustrated in Fig. 6.3, which has been reproduced from Conover (1956). The original reference waveform is a voltage derived from the 60 Hz sinusoidal electrical supply to the transformer substation. This waveform is full-wave rectified in order to double the frequency, and produce a waveform rich in harmonics. Each harmonic is then filtered out with individual bandpass filters. The controller in this case consists of manually adjusted analogue circuits which modify the magnitude and phase of each harmonic before these signals are added together and fed via a power amplifier into the secondary source. Modified forms of such reference waveform generators are still used today. These are implemented, for example, as tracking bandpass filters in applications involving the control of a small number of harmonics whose fundamental frequency changes with time. Another modern manifestation of this control philosophy is used in digital control systems which sample at an integer number of samples per cycle of the primary waveform. The modulus and phase of the individual harmonics can then be calculated by using the discrete Fourier transform which is efficiently implemented by fast Fourier transform (FFT) programs.

The exact method of generating the harmonic reference signals is less important than the independence of the electrical controller design at each harmonic frequency. Figure 6.4 shows the equivalent block diagram at one such harmonic (of angular frequency $n\omega_0$) in which the reference signal is assumed to be a complex phasor of unit amplitude. This allows the controller response and error path response to be characterised by single complex numbers, $G(jn\omega_0)$ and $C(jn\omega_0)$. The component of the complex error signal at this harmonic can now be written as

$$E_n = D_n + G(jn\omega_0)\,C(jn\omega_0). \tag{6.3.1}$$

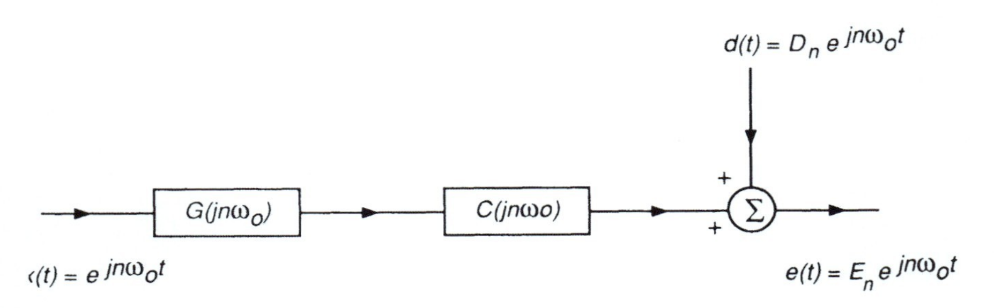

Fig. 6.4 Block diagram at a single harmonic frequency $n\omega_0$.

It has been assumed in this case that no corrupting signal is present in either the reference or error signal. This is equivalent to assuming that the time interval over which $x(t)$ and $e(t)$ are observed is sufficiently long that the bandwidth within which E_n and D_n are measured is small enough to exclude any extraneous background noise. The design of the optimal controller at this harmonic, $G_{\text{opt}}(jn\omega_0)$, is now a trivial task since clearly the harmonic signal at the error microphone can be set to zero if the complex response of the controller at this harmonic is

$$G_{\text{opt}}(jn\omega_0) = -\frac{D_n}{C(jn\omega_0)}. \qquad (6.3.2)$$

Measurement of the error path response at the frequency $n\omega_0$, and of the output at the error sensor due to the primary source at this harmonic are, in principle, sufficient to calculate the optimal controller response in one step. In practice the primary sound field is rarely constant, and measurement errors arise in the estimation of both D_n and $C(jn\omega_0)$. In order to obtain the best performance, therefore, iterative methods are often used to modify continuously $G(jn\omega_0)$ to reduce the error signal. A detailed discussion of these iterative algorithms is left until Chapter 11, where the more general multiple channel problem is considered. The various iterative algorithms which can be applied in the single channel case can be readily deduced from these more general results.

6.4 Control with waveform synthesis

Rather than considering the periodic waveform supplied to the secondary source as the sum, at every instant, of a number of harmonic components, we can approximate the waveform itself by a number of time segments, each having a constant voltage and being of a fixed duration, as illustrated in Fig. 6.5. The adjustment of the levels of each of these segments in order to reduce the mean square value of the final error signal is a control strategy pioneered by Chaplin in the 1970s and called by him "waveform synthesis" (Chaplin *et al.*, 1980). An output waveform such as that illustrated in Fig. 6.5 could be generated, for example, by sending a periodic sequence of numbers from a digital control system to a practical digital-to-analogue converter, which has an inherent "zero order hold" on the output which holds the voltage level constant until the receipt of the next sample. This

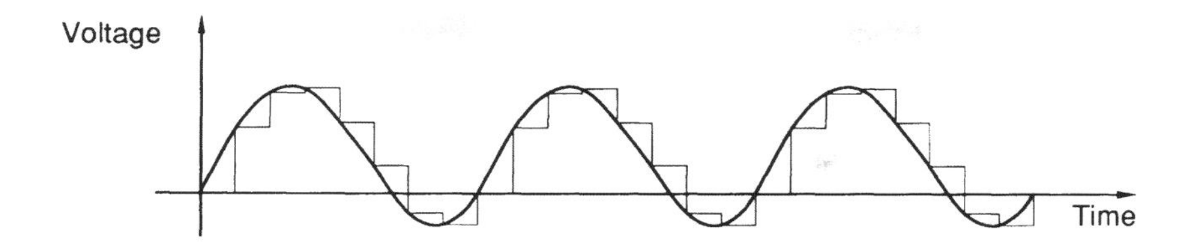

Fig. 6.5 A periodic waveform and its approximation by constant voltage segments as in "waveform synthesis".

suggests that an analysis of such waveform synthesiser techniques could be performed by assuming that the signals in the control system were sampled sequences of the type discussed in Chapter 4.

Such an assumption can be accommodated within the block diagram approach described above, by assuming that the physical system now generates and accepts these *sampled* signals, $x(n)$, $y(n)$ and $e(n)$. The physical system now includes not only the acoustic and electroacoustic elements of the system but also the analogue anti-aliasing and reconstruction filters generally used before and after sampling analogue signals. Indeed, the physical system is assumed to include the analogue-to-digital (ADC) and digital-to-analogue (DAC) converters themselves, as illustrated in Fig. 6.6. Before proceeding to define linear relationships between the sampled variables, and in addition to the assumptions of linearity in the acoustic, electroacoustic and electrical elements of the system, we must now also assume that

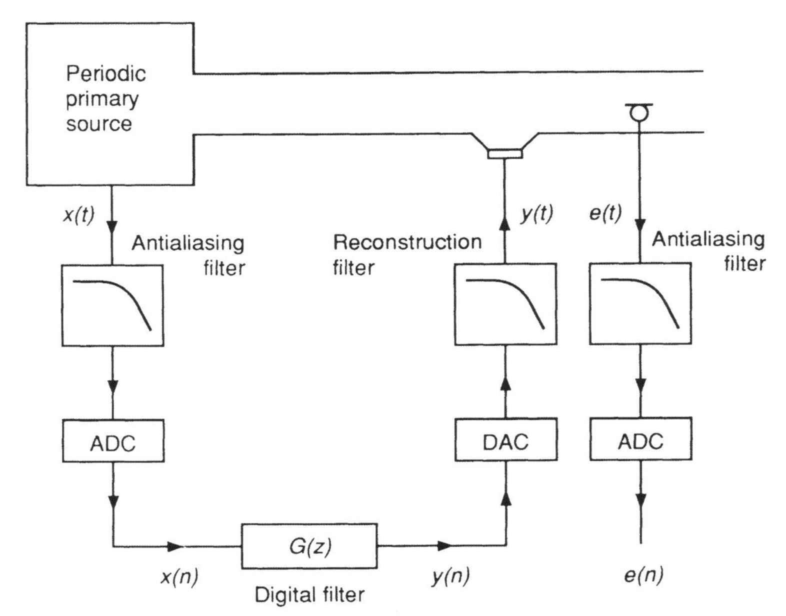

Fig. 6.6 The block diagram of a sampled control system with a digital filter as the controller.

no aliasing is caused by the sampling process. The presence of aliasing (due to, for example, inadequate anti-aliasing filters) causes frequency components above half the sample rate in the continuous error waveform $e(t)$ to appear as components below half the sample rate in the sampled error signal $e(n)$. This would then constitute a form of non-linearity in the loop.

We will find it very convenient to describe the effects of the error path on the sequence $y(n)$ fed to the secondary source by modelling this error path as an FIR digital filter of order J with coefficients c_j. To begin with, J can be made arbitrarily large in order to obtain an arbitrarily close match between the response of the physical path and the response of the digital filter. Thus if the component of the error sequence due to the primary source (which we shall define to be $d(n)$) is zero, then

$$e(n) = \sum_{j=0}^{J-1} c_j y(n-j). \tag{6.4.1}$$

Thus, in general, upon using superposition, the error sequence will be given by

$$e(n) = d(n) + \sum_{j=0}^{J-1} c_j y(n-j). \tag{6.4.2}$$

Regardless of the form of the error path, however, if the signal $y(n)$ fed to the secondary source repeats itself every N samples, then that component of $e(n)$ due to the secondary source will also be periodic and can exactly cancel the periodic signal $d(n)$ (with the same period) due to the primary source. By a similar argument, provided the reference sequence $x(n)$ is of period N samples, any digital filter used as a controller would produce a periodic output, $y(n)$. The relationship between the output, $y(n)$, and input, $x(n)$, of a digital filter has been discussed in Chapter 4. One particularly simple combination of periodic reference signal and digital filter which can be used to generate such a periodic output is illustrated in Fig. 6.7 (Swinbanks, 1985; Elliott and Darlington, 1985; Darlington, 1987). In the case illustrated it has been assumed that the period N is equal to 8 samples. The reference signal is a periodic Kronecker impulse train of period N samples and can be written as

$$x(n) = \sum_{k=-\infty}^{\infty} \delta(n-kN), \tag{6.4.3}$$

where $\delta(n) = 1$, if $n = 0$, otherwise $\delta(n) = 0$. The controller is also an FIR digital filter, with N coefficients, whose ith coefficient is g_i, so that

$$y(n) = \sum_{i=0}^{N-1} g_i x(n-i) = \sum_{k=-\infty}^{\infty} \sum_{i=0}^{N-1} g_i \delta(n-kN-i). \tag{6.4.4}$$

As a result of the sampling properties of the Kronecker delta function, from Section 4.5 it follows that the output of the controller is now just equal to the value of one of the filter coefficients at every sample and thus

$$y(n) = g_p, \tag{6.4.5}$$

where the index p is the smallest positive value of $(n-kN)$ for any k, and may be interpreted as the phase of $y(n)$ relative to $x(n)$. The output sequence $y(n)$ is thus a

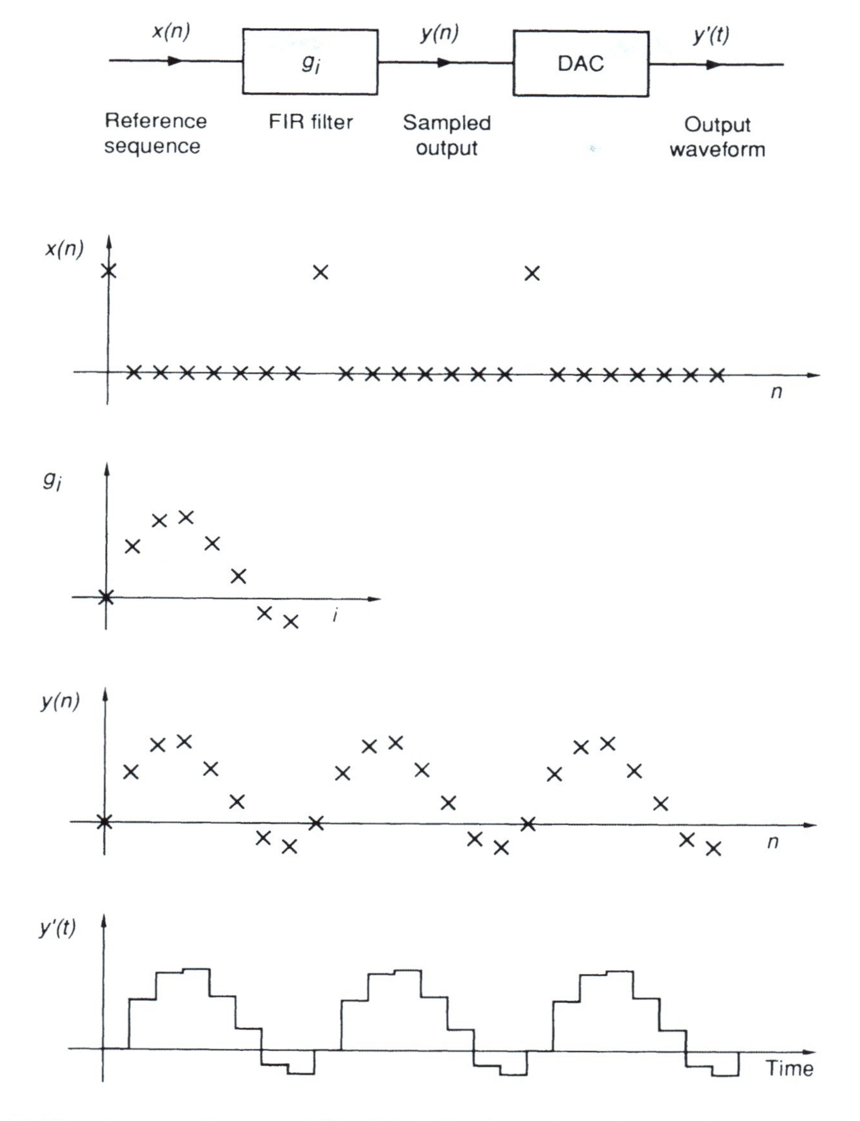

Fig. 6.7 Waveform syntheses as FIR digital filtering of a reference signal consisting of a periodic impulse train.

periodic reproduction of the impulse response of the filter (see Fig. 6.7) and can be generated every sample by looking up one of a set of N filter coefficients, stored for example in a digital memory device. No explicit multiplication is needed in this case to perform the discrete convolution associated with digital filtering. The error sequence may now be written by using equations (6.4.2) and (6.4.4) as

$$e(n) = d(n) + \sum_{j=0}^{J-1} c_j \sum_{i=0}^{N-1} g_i x(n-i-j). \tag{6.4.6}$$

Provided the control filter is time invariant, the two summations may be reordered and the error sequence may be expressed as

$$e(n) = d(n) + \sum_{i=0}^{N-1} g_i r(n-i), \tag{6.4.7}$$

where $r(n)$ is defined as

$$r(n) = \sum_{j=0}^{J-1} c_j x(n-j). \tag{6.4.8}$$

The physical interpretation of $r(n)$ is that it is the sequence which would be generated by passing the reference signal through the error path and it is known as the *filtered reference signal*. Equation (6.4.7) can be rewritten by using the inner product representation of a discrete convolution introduced in Section 4.5. We thus write

$$e(n) = d(n) + \mathbf{r}^{\mathrm{T}}(n)\mathbf{g}, \tag{6.4.9}$$

where the vectors $\mathbf{r}(n)$ and $\mathbf{g}$ are given by

$$\begin{aligned} \mathbf{r}^{\mathrm{T}}(n) &= [r(n), r(n-1), \dots r(n-N+1)], \\ \mathbf{g}^{\mathrm{T}} &= [g_0, g_1, \dots, g_{N-1}]. \end{aligned} \tag{6.4.10}$$

The sum of the mean squared errors taken over one period is given by

$$\sum_{n=0}^{N-1} e^2(n) = \sum_{n=0}^{N-1} (d^2(n) + 2\mathbf{g}^{\mathrm{T}}\mathbf{r}(n)d(n) + \mathbf{g}^{\mathrm{T}}\mathbf{r}(n)\mathbf{r}^{\mathrm{T}}(n)\mathbf{g}). \tag{6.4.11}$$

As follows from the discussion in Chapter 4, the mean squared error is clearly a *quadratic function* of each of the filter coefficients (or waveform segments in the original "waveform synthesis" interpretation), with a unique global minimum if $\mathbf{r}(n)\mathbf{r}^{\mathrm{T}}(n)$, averaged over one period, is positive definite, as in Section 4.7. This property allows a variety of simple adaptive algorithms to be used to adjust each g_i to minimise the mean squared error. Chaplin originally suggested using a trial and error approach (or "power sensing algorithm", Smith and Chaplin, 1983) in which the value of one coefficient is modified, the resulting change in the mean squared error is observed, and the change in the coefficient is retained or abandoned depending on whether the mean squared error rises or falls. More sophisticated algorithms which adjust all the controller coefficients simultaneously in order to minimise the individual sample values of the error sequence have also been suggested (for example the "waveform sensing algorithm" proposed by Smith and Chaplin, 1983). Some of these algorithms have been shown to be variants of the LMS algorithm (see Section 4.8 of Chapter 4), when viewed in terms of the digital filtering formulation presented above (Elliott and Darlington, 1985; Darlington, 1987).

When using this digital filtering approach to active control, care must be taken to ensure that very low frequency components in the measured error signal do not destabilise the system. If some small negative d.c. component is present in the measured error sequence $e(n)$ (due to, for example, offsets in the data converters), an adaptive algorithm may cause the d.c. level of the signal fed to the secondary source to gradually increase in an attempt to cancel the d.c. error component. However, the physical error path will generally have no response at d.c. due to the poor low frequency response of the loudspeaker and microphone. Thus the d.c. output to the secondary source may reach a high level, saturating the controller, without in any way affecting the d.c. offset on the error signal. One solution to this problem is somehow to constrain the filter coefficients so that this uncontrolled increase does not occur by, for example, using a "leak" in the LMS algorithm (Widrow and Stearns, 1985) which subtracts a small percentage off each filter coefficient at every update. Widrow and Stearns (1985) also note that a small amount of random measurement noise in the reference signal has a similar effect. Alternatively a digital high pass filter can be incorporated into the error path to remove the offset in $e(n)$.

6.5 Single channel control of random sound

The active control of periodic sound was discussed above by assuming the availability of a reference signal which contained exactly the same frequency components as the noise to be cancelled, but whose generation was completely independent of any action of the secondary source. This absence of feedback from the secondary source to the reference signal driving the electronic controller allows a very straightforward analysis of the behaviour of such a system which can be regarded as being purely *feedforward* in nature. When designing active control systems for the reduction of random sound, it is rare that such a conveniently independent reference signal is available. Some form of acoustic detector is generally needed to sense the noise coming from the primary source. This detector is usually also influenced by the action of the secondary source. An exception to this rule, in which an independent broadband reference signal can be generated, is in the active control of flame noise, studied by Dines (1984). He found that the instantaneous intensity of the light emitted by the flame in a certain optical waveband was well correlated with the instantaneous acoustic noise generated by the flame. This signal could thus be used as a reference signal in a practical active noise

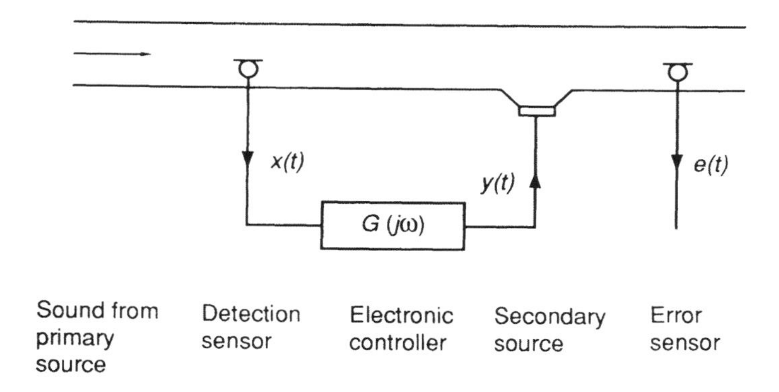

Fig. 6.8 Physical block diagram of a single channel control system for the control of random noise, using an acoustic detection sensor.

control system which was designed on the assumption that there was no feedback between the secondary source and the reference signal.

Figure 6.8 shows the physical arrangement of a more general active control system for random sound which incorporates (1) a detection sensor with output signal $x(t)$, (2) a secondary source, being driven by the signal $y(t)$, and (3) an error sensor with output waveform $e(t)$. The particular arrangement shown again illustrates the control of plane waves propagating in a duct; the general analysis below, however, applies to any single channel active control system in which these three transducers are used. Similarly, although the sensors and sources are shown as single microphones and loudspeakers, they could physically be arrays of transducers connected together to produce a single electrical input or output.

6.6 Frequency domain analysis

In order to derive an equivalent block diagram for the system shown in Fig. 6.8, we recognise that the spectra (Fourier transforms) of the signals from both the detection sensor, $X(\omega)$, and the error sensor, $E(\omega)$, have three components. These components are that due to the primary source, that due to the secondary source, and that due to measurement noise. In the first instance we consider the contributions of these three components at the two sensors independently.

(1) *The component due to the primary source*. We define the spectrum of this component at the detection sensor to be $S(\omega)$, and that at the error sensor to be $P(j\omega)S(\omega)$, where $P(j\omega)$ is the frequency response of the *primary path*. So with the secondary source switched off, and in the absence of measurement noise

$$X(\omega) = S(\omega), \qquad E(\omega) = P(j\omega)S(\omega). \tag{6.6.1a,b}$$

(2) *The component due to the secondary source*. If $Y(\omega)$ is the spectrum of the input to the secondary source, then the contribution of this source at the detection and error sensors is defined to be

$$X(\omega) = F(j\omega)Y(\omega), \qquad E(\omega) = C(j\omega)Y(\omega), \tag{6.6.2a,b}$$

where $C(j\omega)$ is again the frequency response of the *error path*, as used above, and $F(j\omega)$ is the frequency response of the *feedback path*, from secondary source input to detection sensor output.

(3) *The component due to measurement noise*. We define the spectra of the signals at the two sensors due to the noise components as

$$X(\omega) = N_1(\omega), \qquad E(\omega) = N_2(\omega). \tag{6.6.3a,b}$$

If the system is assumed linear, then each of these three contributions will be superposed, to give a total response at the detection sensor of

$$X(\omega) = S(\omega) + F(j\omega)Y(\omega) + N_1(\omega), \tag{6.6.4}$$

and at the error sensor a total response of

$$E(\omega) = P(j\omega)S(\omega) + C(j\omega)Y(\omega) + N_2(\omega). \tag{6.6.5}$$

These relationships are illustrated in the block diagram of Fig. 6.9, in which an

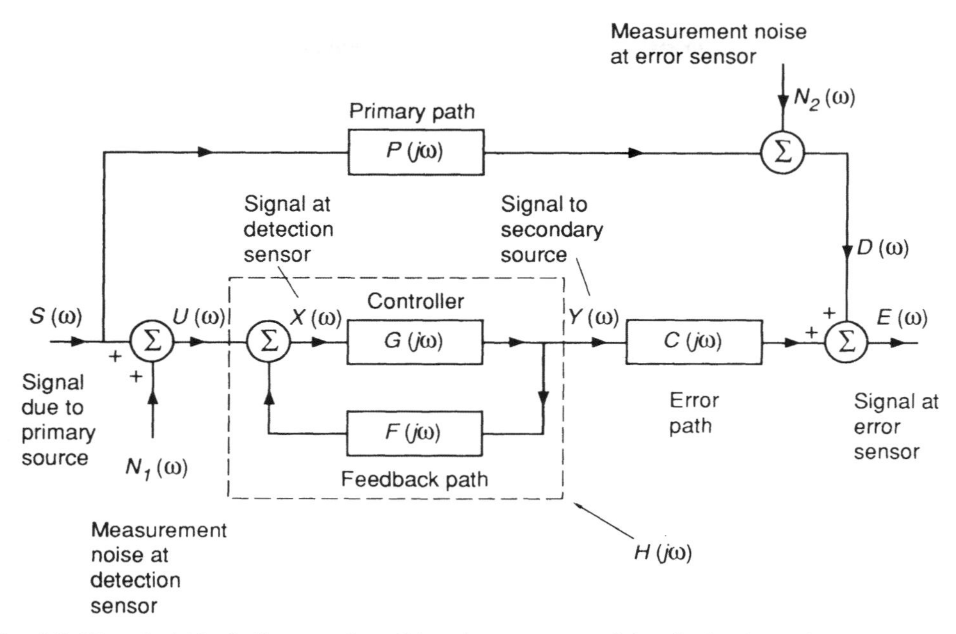

Fig. 6.9 Electrical block diagram describing the response of the single channel control system of Fig. 6.8.

electronic controller, with frequency response $G(j\omega)$, has also been assumed to operate on the detection sensor output in order to produce the secondary source input. Two new signals are defined in this diagram. First we have, at the detection sensor output, the sum of the contributions due to the primary source and measurement noise. This sum is given by

$$U(\omega) = S(\omega) + N_1(\omega). \tag{6.6.6}$$

Second, at the error sensor output, we can also define the sum of the contributions due to the primary source and measurement noise. This is given by

$$D(\omega) = P(j\omega)S(\omega) + N_2(\omega). \tag{6.6.7}$$

In order to simplify this rather complicated block diagram we define the path between $U(\omega)$ and $Y(\omega)$ to be that having the frequency response

$$H(j\omega) = \frac{Y(\omega)}{U(\omega)} = \frac{G(j\omega)}{1 - G(j\omega)F(j\omega)}. \tag{6.6.8}$$

Note that this frequency response function contains the effect of both the controller and the electroacoustic feedback path. The object of the analysis below is to derive procedures for the design of the electrical controller $G(j\omega)$ and to examine the physical significance of this design. The definition of the *dummy controller* having frequency response $H(j\omega)$, which is similar to the "overall model" discussed by Eriksson *et al.* (1987), allows the problem to be again formulated as a purely feedforward control problem with a relatively straightforward design method for

$H(j\omega)$. The philosophy is that once an optimal $H(j\omega)$ has been designed, and the response $F(j\omega)$ is known, the response of the physical controller, $G(j\omega)$, can be calculated from the inverse of equation (6.6.8). Thus $G(j\omega)$ is given by

$$G(j\omega) = \frac{H(j\omega)}{1 + F(j\omega)H(j\omega)}. \tag{6.6.9}$$

With $H(j\omega)$ defined, the block diagram of Fig. 6.9 can be redrawn as the purely feedforward control system illustrated in Fig. 6.10a in which the primary path and measurement noise are not explicitly shown but contained in the definitions of $U(\omega)$ and $D(\omega)$. So the error signal spectrum can be written as

$$E(\omega) = D(\omega) + C(j\omega)H(j\omega)U(\omega). \tag{6.6.10}$$

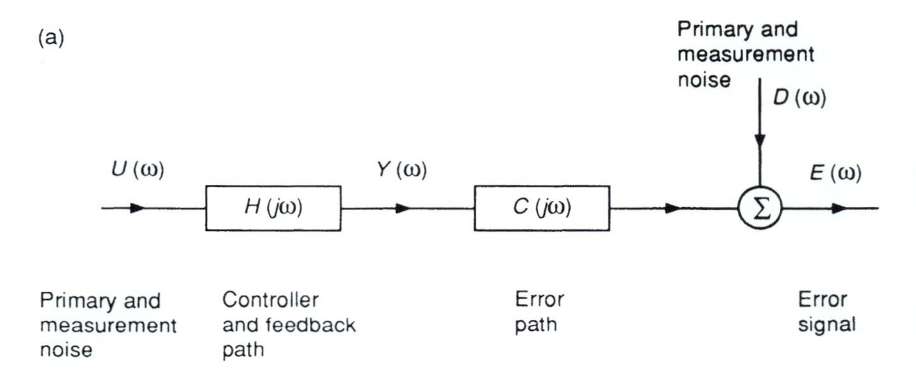

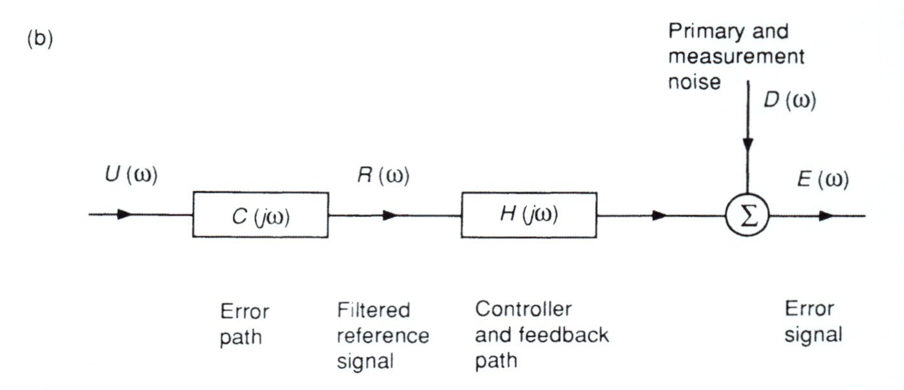

Fig. 6.10 Simplification of the block diagram of Fig. 6.9: (a) as a purely feedforward control problem, (b) as a conventional least squares estimation problem.

The next step is to note that, since the system paths shown in Fig. 6.10a are assumed to be linear and time invariant, they can be transposed with no effect on the error signal. We define the spectrum of the output of the error path in Fig. 6.10b as

$$R(\omega) = C(j\omega)\,U(\omega). \tag{6.6.11}$$

This output signal therefore plays the same role as the *filtered reference signal* introduced in Section 6.4. The whole problem can thus be represented as in Fig. 6.10b which has a topology that is almost identical to the least squares estimation problem discussed in Section 3.8. The error signal may be written simply as

$$E(\omega) = D(\omega) + H(j\omega)R(\omega). \tag{6.6.12}$$

The design problem is now to determine the filter response $H(j\omega)$ which minimises the mean square value of the error signal. Since the signals here are random this can be most easily achieved by minimising the power spectral density of the error signal *at each frequency*. Following the notation introduced in Section 2.7, we may denote the power spectral density of the error signal as

$$S_{ee}(\omega) = E\,[E^*(\omega)\,E(\omega)]. \tag{6.6.13}$$

Expressing this as a function of $H(j\omega)$ by substitution of equation (6.6.12) leads to

$$\begin{aligned} S_{ee}(\omega) = E\,[&D^*(\omega)D(\omega) + D^*(\omega)R(\omega)\,H(j\omega) + H^*(j\omega)R^*(\omega)D(\omega) \\ &+ H^*(j\omega)R^*(\omega)R(\omega)\,H(j\omega)]. \end{aligned} \tag{6.6.14}$$

The various expectations of desired and reference signals can be recognised as power spectral and cross-spectral densities and since $H(j\omega)$ is independent of time the expression can be reduced to

$$\begin{aligned} S_{ee}(\omega) = S_{dd}(\omega) &+ S^*_{rd}(\omega)\,H(j\omega) + H^*(j\omega)S_{rd}(\omega) \\ &+ H^*(j\omega)S_{rr}(\omega)\,H(j\omega), \end{aligned} \tag{6.6.15}$$

where, for example, $S_{rd}(\omega) = E\,[R^*(\omega)D(\omega)]$. This is a scalar complex quadratic form of the type discussed in Section 5.6 and dealt with further in the appendix. The function can be minimised at each frequency ω, since $S_{rr}(\omega)$ is real and positive, to give the optimal value of the frequency response at ω. This is given by

$$H_{\text{opt}}(j\omega) = -\frac{S_{rd}(\omega)}{S_{rr}(\omega)}. \tag{6.6.16}$$

The corresponding minimum value of the power spectral density of the error signal is also given by

$$S_{ee}(\omega)_{\min} = S_{dd}(\omega) - \frac{|S_{rd}(\omega)|^2}{S_{rr}(\omega)}. \tag{6.6.17}$$

Thus the expression for the frequency response function of the optimal dummy controller and the corresponding minimum value of the power spectral density of the error signal are dependent on only the auto- and cross-spectra of the signals $D(\omega)$ and $R(\omega)$. We should emphasise here (as was done in Chapter 5) that frequency domain optimisation does not guarantee the causality of the resulting controller, a point which will be returned to in Section 6.12.

6.7 The effects of measurement noise

Some physical interpretation can be put upon equation (6.6.16) which defines the frequency response of the optimal dummy controller. First, using equations (6.6.6), (6.6.7) and (6.6.11), we can write the cross-spectral density $S_{rd}(\omega)$ in the form

$$\begin{aligned} S_{rd}(\omega) &= E[R^*(\omega)D(\omega)] \\ &= E[(C(j\omega)[S(\omega)+N_1(\omega)])^*(P(j\omega)S(\omega)+N_2(\omega))]. \end{aligned} \quad (6.7.1)$$

On the assumption that the two measurement noise signals are themselves uncorrelated and also uncorrelated with the primary source output signal, the expectation of the cross terms disappear and we are left with

$$S_{rd}(\omega) = C^*(j\omega)P(j\omega)S_{ss}(\omega), \quad (6.7.2)$$

where $S_{ss}(\omega)$ is the power spectral density of the signal at the detection sensor due to the primary source. Similarly, the power spectral density of the reference signal can be written as

$$\begin{aligned} S_{rr}(\omega) &= E[R^*(\omega)R(\omega)] \\ &= E[(C(j\omega)[S(\omega)+N_1(\omega)])^*(C(j\omega)[S(\omega)+N_1(\omega)])], \end{aligned} \quad (6.7.3)$$

which, upon making the same assumptions as above, becomes

$$S_{rr}(\omega) = C^*(j\omega)C(j\omega)[S_{ss}(\omega)+S_{n_1n_1}(\omega)], \quad (6.7.4)$$

where $S_{n_1n_1}(\omega)$ is the power spectral density of the measurement noise at the detection sensor. Hence equation (6.6.16) for the optimal dummy controller can be written in the form

$$H_{\mathrm{opt}}(j\omega) = \frac{-C^*(j\omega)P(j\omega)S_{ss}(\omega)}{C^*(j\omega)C(j\omega)[S_{ss}(\omega)+S_{n_1n_1}(\omega)]}, \quad (6.7.5)$$

which is comparable with the expression for the unconstrained Wiener filter given in Section 4.7. If we define $SNR(\omega) = S_{ss}(\omega)/S_{n_1n_1}(\omega)$ as the signal to noise ratio *at the detection sensor* then the frequency response of the optimal dummy controller can be written as

$$H_{\mathrm{opt}}(j\omega) = -\left[\frac{SNR(\omega)}{1+SNR(\omega)}\right]\frac{P(j\omega)}{C(j\omega)}. \quad (6.7.6)$$

This result is well known in electrical noise cancelling applications (see, for example, Widrow *et al.*, 1975), and in the context of active noise control is similar to those suggested by Ffowcs-Williams *et al.* (1985) and Eriksson and Allie (1988). It is important to note that the optimal electrical controller depends only on the noise level at the detection sensor and *not* on the noise level at the error sensor. In practice it will be more difficult to estimate the individual spectral densities with excessive noise at the error sensor, but assuming these problems of spectral estimation can be overcome, the form of the optimal controller is not affected by measurement noise at the error sensor. The influence of measurement noise at the detection sensor is that as the noise level increases at any frequency, $SNR(\omega)$ will decrease and $H_{\text{opt}}(j\omega)$ will be reduced compared with its noise-free value. The optimal controller is thus performing a balance between cancelling the acoustic noise and reducing the amplification of the measurement noise from the detection sensor. It is clear from these considerations that the design of the optimal controller response is not just a question of system identification in the usual form (discussed by, for example, Åström and Eykhoff, 1971, or Ljung and Söderström, 1983).

In equation (6.6.17), which gives the minimum power spectral density of the error signal, note that the cross and power spectral densities can alternatively be written as

$$S_{rd}(\omega) = E[R^*(\omega)D(\omega)]$$

$$= C^*(j\omega)E[U^*(\omega)D(\omega)] = C^*(j\omega)S_{ud}(\omega), \qquad (6.7.7a)$$

$$S_{rr}(\omega) = |C(j\omega)|^2 S_{uu}(\omega). \qquad (6.7.7b)$$

Equation (6.6.17) for the minimum mean square error thus reduces to

$$S_{ee}(\omega)_{\min} = S_{dd}(\omega) - \frac{|S_{ud}(\omega)|^2}{S_{uu}(\omega)}. \qquad (6.7.8)$$

The power spectral density of the error signal before control is $S_{dd}(\omega)$, so the fractional reduction in the power spectral density of the error signal is given by

$$\frac{S_{ee}(\omega)_{\min}}{S_{dd}(\omega)} = 1 - \frac{|S_{ud}(\omega)|^2}{S_{uu}(\omega)S_{dd}(\omega)} = 1 - \gamma_{ud}^2(\omega), \qquad (6.7.9)$$

where $\gamma_{ud}^2(\omega)$ is the *ordinary coherence function* between the outputs of the detection sensor and error sensor in the absence of control. This result was derived by Ross (1980) and provides a very convenient method for obtaining an estimate of the effect of an active control system from a simple two channel measurement between the outputs of the detection and error sensors. This estimate can be found without the need to calculate the electrical controller response, and even without the need to have a secondary source present at all. If the coherence $\gamma_{ud}^2(\omega) = 0.9$ at some frequency for example, a maximum of 10 dB attenuation can be achieved at the error sensor with an active control system, and if $\gamma_{ud}^2(\omega) = 0.99$, 20 dB will be the maximum attenuation of the error signal at that frequency. This simple result should be treated with some caution, however, for a number of reasons, as follows.

(1) The optimisation of the controller frequency response at each frequency may lead to a controller impulse response which cannot be implemented with a practical filter. At worst, the optimal controller may be required to have a *non-causal* impulse response. Even if the optimal controller is causal, however, in order to obtain the attenuation predicted above, the implementation of the impulse response must be perfectly realised, which may require a very complicated filter design.

(2) In practice the spectral densities above will probably be obtained by averaging across many time histories of the signals. This procedure implicitly assumes that the signals are *stationary*. If this is not the case, then it is possible that at any one time a controller could be designed with the use of short term averaging, with adaptive techniques for example, which would give a larger attenuation than the fixed controller above.

(3) Equation (6.7.9) describes the reduction in the error sensor signal. If *measurement noise* is present in this signal, it cannot be reduced by the action of the

control system, even if the acoustic disturbance from the primary source is perfectly cancelled. Measurement noise in the error microphone will thus cause the predictions of attenuation from sensor coherence measurements to be less than the attenuation which can actually be achieved for the primary sound propagating down the duct.

(4) If *evanescent components* generated by the secondary source are present in the output from the error microphone, due for example to higher order acoustic modes in the duct, the *sum* of the propagating (plane wave) and evanescent (higher order) components will be driven to zero, in the case with no measurement noise, and the coherence formulation above will predict perfect attenuation. In fact, some residual plane wave component will still be left propagating downstream in the duct after control, since the evanescent modes will continue to decay beyond the error microphone and the cancellation achieved at this point will not be sustained. In practice, however, the evanescent modes decay very rapidly and are negligible provided the distance between the error sensor and secondary source is more than about four times the width of the duct and the frequencies are below 0.9 times the cut-on frequency of the higher order mode (Kazakia, 1986).

It is also worth noting that the cross-spectral density between the error signal and the reference signal, $S_{eu}(\omega)$, is zero if $H(j\omega) = H_{\mathrm{opt}}(j\omega)$ (Ffowcs-Williams, 1984). This illustrates the well known fact that the cross-correlation (the Fourier transform of the cross-spectral density) between the error signal and reference signal in an optimally adjusted control system is zero as a result of the principle of orthogonality (see Section 3.7).

6.8 Turbulence as a source of measurement noise

The most important source of measurement noise from acoustic sensors operating in a flow is likely to be that due to turbulence. Since one of the main applications of single channel active control systems is the control of sound propagating in air conditioning systems, the suppression of this turbulence noise becomes a major technological issue. If the sound is propagating in a duct with flow, it is sometimes possible to position the error sensor outside the duct, beyond its exit, so that it still picks up the acoustic signal but is less affected by the flow. This would remove the minor problem associated with measurement noise at the error sensor, but the more severe problems associated with measurement noise at the detection sensor remain. We have made the assumption above that the measurement noise contributions at the detection and error sensors are uncorrelated. If the measurement noise at both sensors is due to turbulence, we can test this assumption fairly easily since it is known that the turbulent pressure fluctuations travel down the duct at a speed of the order of the average flow velocity in the duct, $\bar{u}$, and are correlated over a length scale proportional to the eddy scale, $\bar{u}/f$, where f is the frequency of the turbulent fluctuation. Shepherd *et al.* (1989) suggest that the constant of proportionality is "several times". However, correlation measurements in boundary layers made by Bull (1968) suggest that the constant of proportionality may be more like 20 before the normalised correlation falls below 0.1. If L is the distance between detection and error sensors, the turbulent contributions at these two sensors will be uncorrelated provided $L \gg \bar{u}/f$, i.e.

$$f \gg \frac{\bar{u}}{L}. \tag{6.8.1}$$

For a mean flow speed of 10 m s^{-1} and a distance between sensors of 5 m, the assumption that $N_1(\omega)$ and $N_2(\omega)$ are uncorrelated will thus be valid above about 4 Hz to 40 Hz, depending on the constant of proportionality assumed above.

A number of methods have been proposed for averaging the measured pressure signal over a length scale greater than the eddy scale, in order to average out the turbulent pressure fluctuations and enhance the acoustic signal (Ross, 1982b; Eriksson and Allie, 1987; Shepherd *et al.*, 1989). The most common arrangement consists of a tube with a microphone at one end, and a slit along the length of the tube to allow the pressure in the flow to drive the pressure in the tube. Normally the tube is covered by cloth to maximise turbulence suppression, and some absorptive material is used inside the tube to reduce internal resonances. This type of "turbulence screen" was analysed and developed by Neise and Stahl (1979) but appears to have been originally introduced by Tamm and Kurtze (1954) in order to improve the *directivity* of the microphone and was later used for its turbulent noise suppression qualities by Friedrich (1967). This is an interesting parallel with the work of Shepherd *et al.* (1989), in which various types of turbulent noise suppressor are described in terms of their action as spatial filters. One example given is a multiple microphone arrangement similar to the type originally suggested for use in active control systems for its *directional* characteristics by Swinbanks (1973).

6.9 Practical frequency domain design methods

The object of this section is to discuss methods by which the frequency domain formulation of the optimal controller can be used to design a realisable electronic filter, for use as the controller in a practical active noise control system. The first step is to obtain an expression for the frequency response of the optimal filter to be implemented, $G_{\text{opt}}(j\omega)$, in terms of physically measurable variables. We know from equation (6.6.9) that

$$G_{\text{opt}}(j\omega) = \frac{H_{\text{opt}}(j\omega)}{1 + F(j\omega)\,H_{\text{opt}}(j\omega)}, \tag{6.9.1}$$

where the optimal feedforward transfer function is given by equation (6.6.16), which, using equation (6.7.7), can also be written as

$$H_{\text{opt}}(j\omega) = -\frac{S_{rd}(\omega)}{S_{rr}(\omega)} = \frac{-1}{C(j\omega)}\frac{S_{ud}(\omega)}{S_{uu}(\omega)}. \tag{6.9.2}$$

By injecting into the secondary source a broadband test signal, $v(t)$, uncorrelated with the noise from the primary source, the frequency response functions of the error path and feedback path can be estimated without bias error by using the normal two channel analyser function ($H_1(j\omega)$ in Section 3.6). Thus we can write

$$C(j\omega) = \frac{S_{ve}(\omega)}{S_{vv}(\omega)}, \qquad F(j\omega) = \frac{S_{vx}(\omega)}{S_{vv}(\omega)}. \tag{6.9.3a,b}$$

We now define the function measured by using this estimate of the frequency response function between the error and detection microphone, with no signal fed to the secondary sources, as

$$P'(j\omega) = \left.\frac{S_{xe}(\omega)}{S_{xx}(\omega)}\right|_{Y(\omega)=0} = \frac{S_{ud}(\omega)}{S_{uu}(\omega)}, \tag{6.9.4}$$

where the second form for $P'(j\omega)$ follows from equations (6.6.4) to (6.6.7). This "transfer function" is exactly that required to calculate $H_{opt}(j\omega)$ according to equation (6.9.2), so that

$$H_{opt}(j\omega) = -\frac{P'(j\omega)}{C(j\omega)}. \tag{6.9.5}$$

It should also be noted that since $U(\omega) = S(\omega) + N_1(\omega)$ and again assuming that the signals associated with $S(\omega)$ and $N_1(\omega)$ are uncorrelated, this measured response can also be expressed as

$$P'(j\omega) = \frac{S_{sd}(\omega)}{S_{ss}(\omega) + S_{n_1n_1}(\omega)}. \tag{6.9.6}$$

Now using equation (6.6.7) and the definition given above of the signal to noise ratio at the detection sensor $SNR(\omega)$, it follows that

$$P'(j\omega) = \left[\frac{SNR(\omega)}{1 + SNR(\omega)}\right] P(j\omega), \tag{6.9.7}$$

where $P(j\omega)$ is the frequency response function of the noise-free primary path. The frequency response of the required electronic controller, $G_{opt}(j\omega)$, can now be expressed in terms of the three measured quantities $C(j\omega)$, $F(j\omega)$ and $P'(j\omega)$. Thus, using equations (6.9.1) and (6.9.5) leads to

$$G_{opt}(j\omega) = \frac{-P'(j\omega)}{C(j\omega) - P'(j\omega)F(j\omega)}. \tag{6.9.8}$$

In the case of no measurement noise at the detection sensor, $P'(j\omega) = P(j\omega)$, and the equation for the frequency response of the controller reduces to that given by Ross (1982a) and Elliott and Nelson (1984). The equation which includes the effects of measurement noise was originally presented by Roure (1985).

After having derived the frequency response of the optimal controller, the problem still remains of how to design the coefficients of an electronic filter which is to be used as the controller in a practical realisation of an active control system. Generally speaking the filter design methods outlined in Chapter 4 are not of direct use in this application since they usually start with a specification of the magnitude response in terms of stop bands and pass bands, and do not allow an arbitrary magnitude and phase response to be specified. However, we will discuss one simple method of designing a digital FIR filter from the estimated optimal frequency response. This uses a technique described by Roure (1985) which is a form of a window design method of the type discussed in Section 4.6.

The basic approach used by Roure is to calculate from equation (6.9.8) the frequency response of the optimal controller, $G_{opt}(j\omega)$ by using measurements of $C(j\omega)$, $F(j\omega)$ and $P'(j\omega)$. The discrete Fourier transform is then used to transform $G_{opt}(j\omega)$ (measured at a uniform grid of frequencies, up to half the sample rate of the digital control system) into a sampled impulse response which can be implemented as an FIR filter. The general experimental arrangement used by Roure (1985) is shown in Fig. 6.11. The host microprocessor is used to estimate $C(j\omega)$, $F(j\omega)$ and $P'(j\omega)$, calculate $G_{opt}(j\omega)$ and perform the Fourier transform by using an FFT algorithm. The coefficients of the FIR filters are then loaded into the memory of a fast signal processing microprocessor (Texas Instruments TMS 320) which performs the arithmetic operations which implement the FIR digital filter (convolver). A number of practical difficulties in this simple approach are discussed by Roure and are listed below.

(1) The estimate of the frequency response of the optimal controller calculated above is generally in error at low frequencies due to turbulence and poor loudspeaker response. The estimate generally also tends to large values at high frequencies, so that if all the response over the whole frequency range were Fourier transformed, severe windowing errors would occur. The calculated frequency response is therefore multiplied by a frequency window, which sets the response below 50 Hz to zero and smoothly attenuates the response above the first cut-off frequency of the duct.

(2) After Fourier transformation of the windowed frequency response, the calculated impulse response is found, in general, to have non-causal components. A temporal window is used to set the non-causal part of the calculated impulse

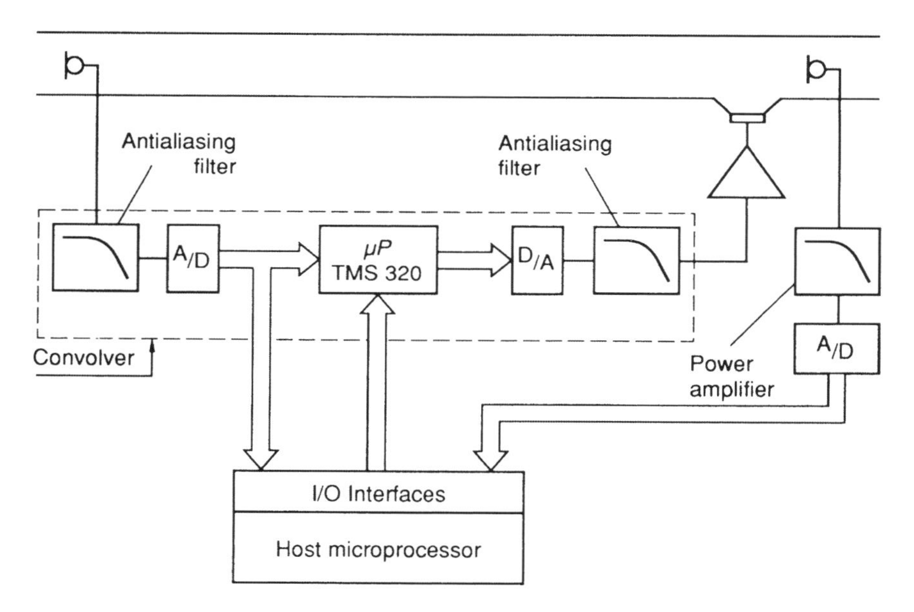

Fig. 6.11 Block diagram of Roure's active noise attenuator (after Roure, 1985).

response to zero and smoothly attenuate the response for long time delays.

(3) As a result of errors in the estimation of the original transfer functions, and errors introduced by the various windowing operations above, the FIR filter originally implemented by the convolver does not give the maximum achievable attenuation. The attenuation is improved by measuring the residual error signal and applying an iterative algorithm which recalculates the required frequency response of the controller, and the procedure detailed above is repeated continuously.

We distinguish between "iterative" algorithms which operate on blocks of data to calculate the new coefficients for the controller over some long period of time compared with the sample rate, and "adaptive" algorithms (discussed later in this chapter) in which each sample of data is used to update the coefficients of the controller on a timescale comparable with the sample rate. Either philosophy of update algorithm will, potentially, allow the filter to adjust itself to correct for errors in its response when the control system is initially switched on, and to track small changes which gradually occur in the operating conditions in the duct. If the operating conditions in the duct change too much, however, the initial estimates of the error path and feedback path will themselves be substantially in error and their use in any algorithm which updates the controller may lead to instability. A number of update algorithms are reasonably robust to these types of error however. In the iterative algorithm used by Roure (1984, 1985) the frequency response of the new controller is calculated as the *product* of the old frequency response and a correction term which depends on the original estimate of the transfer function beween the error and detection microphone, and a new estimate of this transfer function with the original controller operating. In another iterative algorithm suggested by Elliott

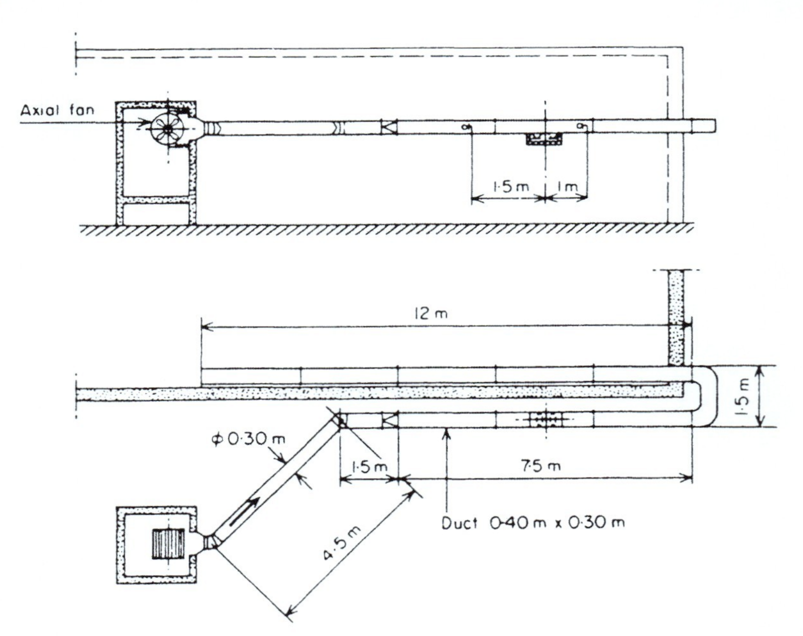

Fig. 6.12 The air conditioning duct installation used in Roure's experiments (after Roure, 1985).

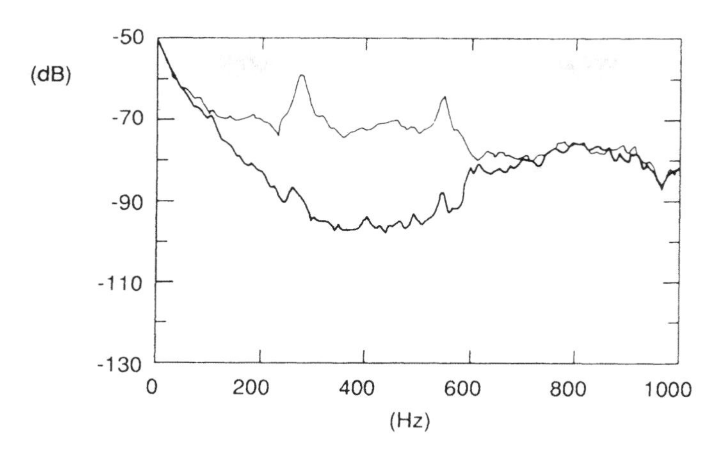

Fig. 6.13 Results of Roure's experiments showing the amplitude spectra of the fan noise at the error microphone with a mean duct velocity of $9\,\mathrm{m\,s^{-1}}$. ———, attenuator off; ———, attenuator on (after Roure, 1985).

and Nelson (1984) the new response is calculated as the *addition* of the old response and a correction term.

The results of applying the controller of Fig. 6.11 to the air conditioning duct illustrated in Fig. 6.12, with a flow rate of $9\,\mathrm{m\,s^{-1}}$, are illustrated in Fig. 6.13, which shows the power spectral density at the error sensor with the active control system switched off (feint) and switched on (dark). A 256 coefficient FIR filter was used as the controller in this experiment, operating at a sample rate of 4096 Hz. Attenuation is achieved between about 100 Hz (below which turbulence noise at the detection microphone was excessive) and about 600 Hz (above which higher order modes could propagate in the duct). Attenuations of about 25 dB are achieved at the fundamental frequency (275 Hz) and second harmonic frequency (550 Hz) of the axial fan driving the duct. An alternative approach to controller design, in which auto- and cross-correlation matrices are used to provide a direct design of the time domain response of a recursive controller, has been used by Ross (1982a), who has also discussed iterative correction algorithms (Ross, 1982b).

6.10 An acoustical interpretation of the optimal controller

In this section we initially consider the simple case of an active control system used to control plane waves of sound propagating in a doubly infinite (or anechoically terminated) duct. We wil! first define the electroacoustic responses of the detection and error sensor (assumed to have no measurement noise) and secondary source in terms of waves propagating in the duct. The individual transfer functions in the equivalent block diagram of Fig. 6.9 are then derived and used to define the optimal controller. The characteristics of this optimal electrical filter are then related to the acoustic properties of the duct. The effect of finite termination impedances in the duct, measurement noise and other practical effects are discussed in the following section.

The geometry of the duct is shown in Fig. 6.14, with the detection sensor at the origin of a co-ordinate system, the secondary source at position l_1 and the error sensor at position $(l_1 + l_2)$. The responses of the detection and error sensors are defined in terms of the complex pressure in a propagating wave at frequency ω travelling downstream in the duct. This results in complex pressures $p^+(0)$ and $p^+(l_1 + l_2)$ at the detection and error sensors respectively. There is also a propagating wave travelling upstream in the duct which produces a complex pressure $p^-(0)$ at the detection sensor and a complex pressure $p^-(l_1 + l_2)$ at the error sensor. We assume that the output from the detection sensor is the complex voltage given by

$$X = M_x p^+(0) + M_x D_x p^-(0), \tag{6.10.1}$$

in which M_x is the response of the detection sensor to downstream propagating pressure wave, in volts/Pascal, and D_x is the dimensionless directivity of the sensor. Both these variables are complex frequency-dependent quantities and may be regarded as frequency response functions, but are shown here simply as complex numbers for notational convenience. Similarly, the output of the error sensor is defined as

$$E = M_e p^+(l_1 + l_2) + M_e D_e p^-(l_1 + l_2), \tag{6.10.2}$$

where M_e (volt/Pa) and D_e (dimensionless) are the electroacoustic response and directivity of the error sensor, defined by analogy with M_x and D_x.

Figure 6.15 shows the propagating waves generated when only the secondary source is in operation. The complex pressures associated with waves travelling in the downstream and upstream directions, are given by

$$p^+(l_1) = L_s Y, \qquad p^-(l_1) = L_s D_s Y, \tag{6.10.3a,b}$$

where L_s is the electroacoustic response of the secondary source (Pa/volt) and D_s is its directivity. Although a single loudspeaker is shown representing the secondary source in these figures, it is again understood that the secondary source could be any array of loudspeakers provided it is driven by the single excitation signal Y. Indeed, as noted in Chapter 5, *only* an array of loudspeakers can give a directivity, D_s, other

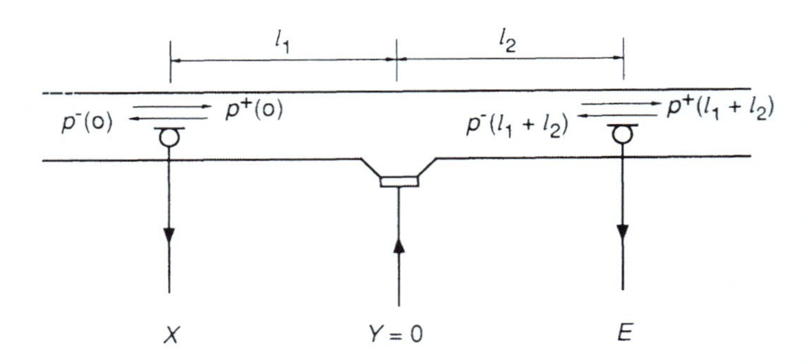

Fig. 6.14 The geometry of the detection and error sensors in the anechoic duct, together with the downstream (p^+) and upstream (p^-) propagating pressure waves.

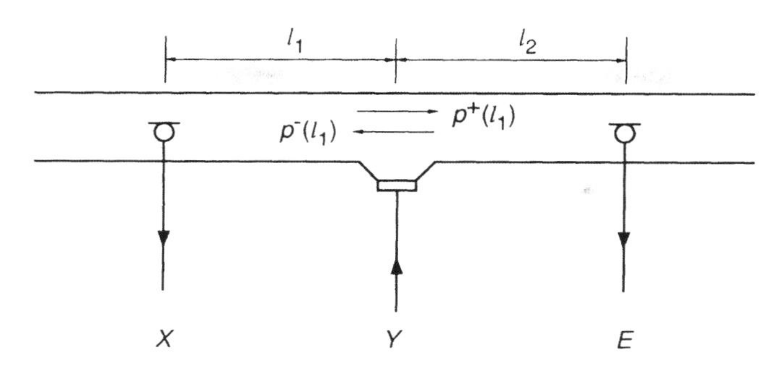

Fig. 6.15 Generation of downstream and upstream propagating waves by the secondary source in the anechoic duct.

than unity. We also assume for the time being that the secondary source(s) have a very high internal acoustic impedance, in other words they act as constant volume velocity sources. This will mean that physically the electroacoustic response L_s will be the product of two parts: (1) the ratio of the volume velocity of the source to the electrical excitation and (2) the ratio of the pressure wave produced in the anechoic duct by the source to its volume velocity. The assumption of high internal acoustic impedance ensures that the first of these two components of L_s will not be disturbed by the presence of any other pressure wave propagating in the duct, since the volume velocities of the sources are unaffected by the pressure "load" they experience in the duct. This assumption will be re-examined later in this section, but for now provides a convenient way of deriving the individual transfer functions in the duct, since if $Y = 0$, the propagation of other pressure waves in the duct will not be affected by the presence of the secondary sources.

It is now a fairly simple matter to deduce the individual transfer functions in the duct, since the propagating components of the pressure are simply related by terms of the form $e^{\pm jkl}$. The error path, $C(j\omega)$, can be deduced by assuming that no pressure wave is propagating downstream from the primary source. Thus the component of pressure propagating downstream at the error sensor, at $l = (l_1 + l_2)$, is due only to that caused by the secondary source and is given by $p^+(l_1) = L_s Y$. It then follows that

$$E = M_e p^+(l_1 + l_2) = L_s M_e e^{-jkl_2} Y, \tag{6.10.4}$$

where k is the wavenumber, which for completeness could have a negative imaginary component to account for dissipation in the duct. There is no upstream travelling wave at the error sensor since the duct is anechoically terminated and thus the frequency response function of the error path is given by

$$C(j\omega) = \frac{E}{Y} = L_s M_e e^{-jkl_2}. \tag{6.10.5}$$

The frequency response $F(j\omega)$ of the feedback path can be similarly calculated. In the absence of any contribution from the primary source, the pressure at the detection sensor is due to the upstream pressure wave defined by equation (6.10.3b) that is generated by the secondary source only. Thus, using equation (6.10.1) leads to the expression

$$X = L_s D_s M_x D_x e^{-jkl_1} Y. \tag{6.10.6}$$

The frequency response of the feedback path thus reduces to

$$F(j\omega) = \frac{X}{Y} = L_s D_s M_x D_x e^{-jkl_1}. \tag{6.10.7}$$

The primary path $P(j\omega)$ is defined as the ratio of the outputs from the error and detection sensors due to the primary source, with $Y = 0$. Since the secondary source is assumed to have a high internal impedance, the downstream propagating pressures at the error and detection sensors are simply related by

$$p^+(l_1 + l_2) = p^+(0)e^{-jk(l_1 + l_2)}. \tag{6.10.8}$$

The primary path frequency response can thus be written as

$$P(j\omega) = \left.\frac{E}{X}\right|_{Y=0} = \frac{M_e}{M_x} e^{-jk(l_1 + l_2)}. \tag{6.10.9}$$

In the absence of any measurement noise at the detection sensor, the frequency response of the optimal controller which will drive $p^+(l_1 + l_2)$ to zero when excited by a primary excitation $p^+(0)$, follows from equations (6.9.7) and (6.9.8), and is given by

$$G_0(j\omega) = \frac{-P(j\omega)}{C(j\omega) - F(j\omega)P(j\omega)}. \tag{6.10.10}$$

Here we have used $G_0(j\omega)$ to denote the value of $G_{opt}(\omega)$ when no measurement noise is present. Substituting into equation (6.10.10) the frequency responses derived above for $C(j\omega)$, $F(j\omega)$ and $P(j\omega)$, we obtain the following important result for the optimal controller in terms of the acoustical characteristics of the system:

$$G_0(j\omega) = \frac{-1}{L_s M_x} \frac{e^{-jkl_1}}{1 - D_s D_x e^{-jk2l_1}}. \tag{6.10.11}$$

The response of the optimal controller does not depend on the properties of the error sensor, since the pressure is driven to zero at this point. The physical significance of the first term in this equation, $1/L_s M_x$, is that the controller must compensate for the electroacoustic response of the detection sensor and secondary source with respect to downstream travelling waves. The numerator of the second term, e^{-jkl_1}, compensates for the delay time of a downstream travelling wave from the detection sensor to the secondary source. The denominator of the second term, $1 - D_s D_x e^{-jk2l_1}$, accounts for any *direct* feedback from the secondary source to the detection sensor by cancelling out its effect. This is illustrated in Fig. 6.16 which shows the transient response of such a controller to a short pulse of pressure propagating downstream from the primary source. In this example it is assumed, for convenience, that the responses of the detection sensor and secondary source, M_x and L_s, are unity, and the directivities of these sources, D_s and D_x, are real constants whose value is less than unity. It is further assumed that $G_0(j\omega)$ is implemented with

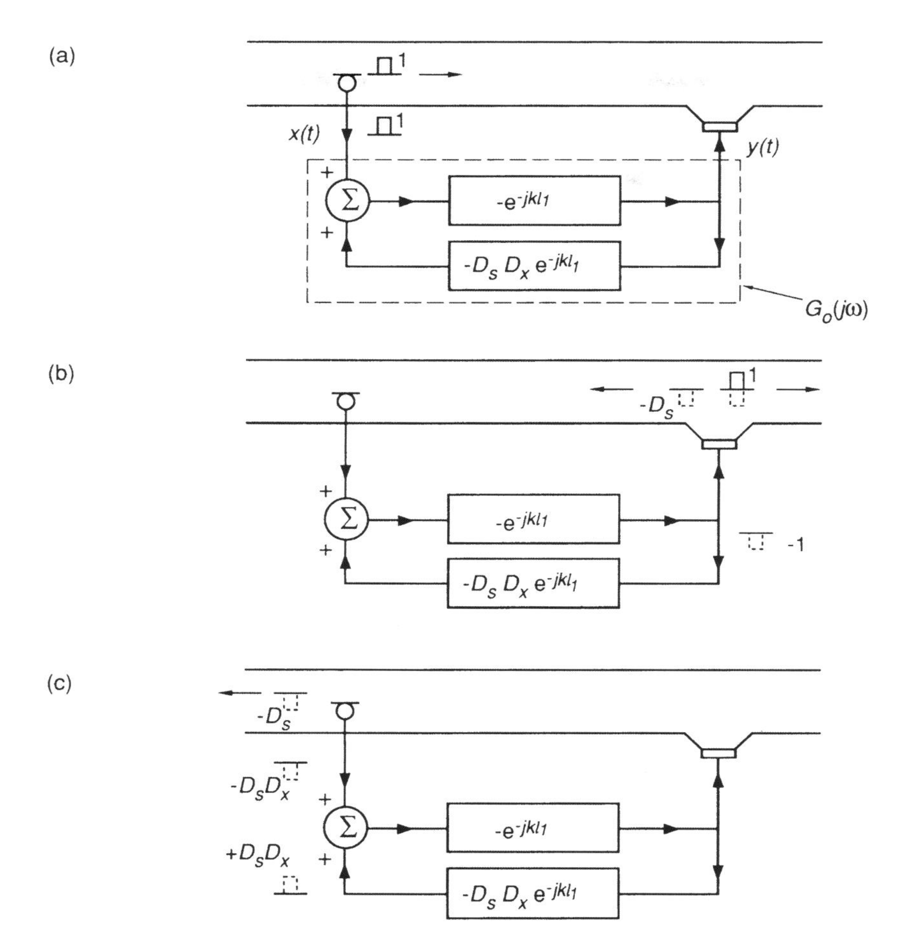

Fig. 6.16 The transient response of the optimal controller having the form $G_0(j\omega) = A(j\omega)/1 - A(j\omega)B(j\omega)$, with $A(j\omega) = -e^{-jkl_1}$, $B(j\omega) = -D_s D_x e^{-jkl_1}$ (a) just after a downstream propagating pulse has passed the detection sensor; (b) just after the downstream propagating pulse has passed the secondary source; (c) just after the upstream propagating reflected pulse has passed the detection sensor.

the recursive structure shown in Fig. 6.16, i.e. it has a frequency response function of the form

$$G_0(j\omega) = \frac{A(j\omega)}{1 - A(j\omega)B(j\omega)}, \tag{6.10.12}$$

where the frequency response functions $A(j\omega)$ and $B(j\omega)$ are given by

$$A(j\omega) = -e^{-jkl_1}, \qquad B(j\omega) = -D_s D_{x0} e^{-jkl_1}, \tag{6.10.13a,b}$$

As the transient pulse of pressure propagates past the detection sensor an electrical pulse of the same shape is generated as shown in Fig. 6.16(a). This pulse passes through the feedforward path $A(j\omega)$ defined above. A time l_1/c_0 later this pulse is fed to the secondary source which generates an inverted downstream wave of the same pulse shape in order to cancel the original pressure pulse due to the primary source. This is shown in Fig. 6.16(b). At the same time, however, the action of the secondary source generates an upstream propagating acoustic pulse of magnitude $-D_s$ times the original pulse. A negative pulse of unit amplitude also begins to pass through the feedback path of the electrical filter $B(j\omega)$ defined above. At a time l_1/c_0 after this the acoustic pulse generates a signal at the detection sensor of $-D_sD_x$ times the original pulse, where the factor D_x results from the directivity of this sensor. However, this is exactly cancelled by the electrical signal given by D_sD_x times the original pulse which has been fed back through the electrical filter $B(j\omega)$. This then ensures no further excitation of the feedforward path $A(j\omega)$ as shown in Fig. 6.16(c). The net result after this time is an upstream travelling pulse of magnitude $-D_s$ times the original pulse, no downstream travelling waves, and no residual electrical excitation of the controller.

It is interesting to note that if *either* the detection sensor, *or* the secondary source were perfectly directional, such that $D_s = 0$ or $D_x = 0$, the controller would have a particularly simple, feedforward, structure. The acoustical effect of these two possibilities are, however, very different. If the secondary source is perfectly directional the acoustic pulse is perfectly *absorbed* by this source, and so no reflected wave is generated to provide further excitation of the detection sensor. If the detection microphone were perfectly directional, but the source was omnidirectional, the pulse would be perfectly *reflected* by the secondary source, in accordance with the theory in Chapter 5. However, this would not affect the detection sensor output, since the pulse would be travelling upstream.

If both the detection sensor and secondary source were omnidirectional (for example if a pressure microphone and a single loudspeaker were used) then $D_s = D_x = 1$ and the response of the controller becomes

$$G_0(j\omega) = \frac{-1}{L_sM_x}\frac{e^{-jkl_1}}{1-e^{-jk2l_1}}, \tag{6.10.14}$$

which can be written in the form

$$G_0(j\omega) = \frac{-1}{L_sM_x}\frac{1}{2j\sin kl_1}, \tag{6.10.15}$$

a result derived by Eghtesadi and Leventhall (1982). This implies that in the absence of any dissipation in the duct the controller has an infinite response at certain frequencies, i.e. it is unstable. This occurs when the distance between the detection sensor and secondary source is exactly an integer number of half wavelengths. Acoustically this is due to the single secondary source reflecting a harmonic primary pressure wave and generating a perfect standing wave of pressure upstream of this source. At frequencies where an integer number of half wavelengths fit between the secondary source and detection sensor, the pressure at the detection sensor will be zero, and an infinite controller response will be necessary to generate any output to the secondary source. In practice this singular condition can be avoided by having some form of acoustic damping between the secondary source and detection sensor which prevents the build-up of a perfect standing wave. Although such a procedure will remove the singularities in $G_0(j\omega)$, the controller will still have a response of large magnitude at certain frequencies, which implies a long impulse response, and this may cause problems with certain implementations of the electrical controller.

6.11 Other factors affecting the optimal controller

Unfortunately, it is unusual in practice to be asked to implement an active control system using perfect transducers in a duct with anechoic terminations, and no flow and no measurement noise! We will consider these important practicalities below.

If arbitrary termination impedances are assumed at the two ends of the duct shown in Fig. 6.14, it is still possible to calculate the response of the error path, feedback path and primary path, although these individual responses are rather more complicated in this case. The terminating impedances introduce multiple paths from the sources to each sensor and generate an overall "reverberant" component in the responses. The individual responses can be shown to be (Roure, 1984, 1985; Elliott and Nelson, 1984)

$$C(j\omega) = \frac{M_e L_s e^{-jkl_2}(1 + D_e R_e)(1 + D_x R_x e^{-jk2l_1})}{1 - R_x R_e e^{-jk2(l_1 + l_2)}}, \tag{6.11.1}$$

$$F(j\omega) = \frac{M_x L_s e^{-jkl_1}(D_x + R_x)(D_s + R_e e^{-jk2l_2})}{1 - R_x R_e e^{-jk2(l_1 + l_2)}}, \tag{6.11.2}$$

$$P(j\omega) = \frac{M_e e^{-jk(l_1 + l_2)}(1 + D_e R_e)}{M_x(1 + D_x R_e e^{-jk2(l_1 + l_2)})}, \tag{6.11.3}$$

where R_x and R_e are the frequency-dependent reflection coefficients experienced by upstream and downstream propagating waves in the planes of the detection and error sensors, respectively. It is sometimes stated that the reason for using directional detection sensors and/or secondary sources is to eliminate feedback from the secondary source to the detection sensor. It is clear, however, from the form of $F(j\omega)$ above, that even if $D_x = 0$ and $D_s = 0$, i.e. both these transducers were perfectly directional, $F(j\omega)$ would not be zero. Physically this is due to the presence of a feedback path corresponding to a sound wave travelling downstream from the secondary source, being reflected off the downstream termination, travelling back upstream until it is reflected off the upstream termination and the resulting downstream propagating sound wave being picked up by the detector microphone. The individual responses defined by equations (6.11.1)–(6.11.3) can now be used in equation (6.10.10) to calculate the optimal controller, $G_0(j\omega)$ in this case. When this tedious procedure is performed, however, it is found that the final form of $G_0(j\omega)$ in terms of the electroacoustic responses of the transducers is *exactly* the same as that derived for the anechoic duct above, equation (6.10.11). A reflective termination at the downstream end of the duct does not affect the form of the optimal controller because the optimal controller completely cancels all downstream propagating components. Thus there can be no acoustic wave which can be reflected from the downstream end and affect the form of the controller. The reason that a reflective termination at the upstream end of the duct does not affect the form of the optimal controller is slightly more subtle. In the general case the secondary source may not be perfectly absorbing and some component of the acoustic travelling waves will be reflected back towards the primary source. Once these reflected components have passed the detection sensor, however, we have seen that their effect is no longer present in the optimal electrical controller. As they are reflected from the upstream termination and travel back down the duct towards the detection sensor they are treated by the active control system as if they were just another component of the primary field to be cancelled, and so require no modification of the electrical controller to cope with their effect.

The effect of the internal acoustic impedance of the secondary source on the form of the controller has been discussed by Silcox and Elliott (1985) and Munjal and Eriksson (1988, 1989). In the equations discussed above, this internal impedance has been assumed to be infinite, but if this condition is relaxed, acoustic waves propagating in the duct will not just be reflected from the duct terminations, but also will be reflected by the impedance discontinuity presented by the secondary source. When the control system is operating optimally, however, the pressure in front of the secondary source will be nearly zero (some pressure components will still exist because of near-field effects, but these will be small, as discussed by Trinder and Nelson, 1983). If the pressure in front of the secondary source is small, the acoustic impedance seen by this source is also small, and the additional shunting effect of the source's internal acoustic impedance will have very little effect on its behaviour. This observation was confirmed experimentally by Silcox and Elliott (1985). It should be noted, however, that both the electroacoustic response of a moving coil loudspeaker as a source, and its internal acoustic impedance, are functions of the mechanical and electrical properties of the loudspeaker. It is thus possible to express the electroacoustic response in terms which include its internal acoustic impedance (Munjal and Eriksson, 1988), although this does not imply that the source is significantly loaded when operating optimally.

Swinbanks (1973), Elliott and Nelson (1984) and Munjal and Eriksson (1989) have also considered the effect of mean flow in the duct on the controller response, and have shown that it affects the solution for the optimal controller only in that the relevant wavenumbers are affected by the Mach number of the flow. Another effect of flow in the duct, however, is to introduce the likelihood that measurement noise will be picked up by the detection sensor. We have seen that the frequency response of the optimal controller is modified by this measurement noise, and by combining equations (6.9.7) and (6.9.8), this can be expressed as

$$G_{\text{opt}}(j\omega) = \frac{-SNR(\omega)P(j\omega)}{C(j\omega) + SNR(\omega)[C(j\omega) - P(j\omega)F(j\omega)]}, \tag{6.11.4}$$

where $SNR(\omega)$ is the acoustic signal to measurement noise ratio at the detection microphone. Equation (6.10.10) for the optimal controller in the *absence of measurement noise*, $G_0(j\omega)$, can be used to express the ratio of the primary path response to the error path response as

$$\frac{P(j\omega)}{C(j\omega)} = \frac{-G_0(j\omega)}{1 - F(j\omega)G_0(j\omega)}. \tag{6.11.5}$$

Equation (6.11.4), for the optimal controller in the presence of noise can now be written using equation (6.11.5) as

$$G_{\text{opt}}(\omega) = \frac{SNR(\omega)G_0(j\omega)}{1 + SNR(\omega) - G_0(j\omega)F(j\omega)}. \tag{6.11.6}$$

Under conditions of high signal to noise ratio, $G_{\text{opt}}(j\omega) \to G_0(j\omega)$ as expected. If the signal to noise ratio is low, however, the optimal controller will not only depend on the electroacoustic responses and directivities of the detection and error sensors that are contained in the acoustical interpretation of the expression for $G_0(j\omega)$ given by equation (6.10.11). The optimal controller will now also depend on the reflection conditions at either end of the duct that are contained in the terms R_e and R_x in the general expression given by equation (6.11.2) for the feedback path $F(j\omega)$. Because the sound pressure at the error microphone is no longer driven to zero when measurement noise is present at the detection microphone, the sound pressure loading the secondary source will also be finite. It is therefore likely that the internal impedance of the loudspeaker will also, now, play a part in determining the frequency response of the optimal controller. All of these effects act to complicate the frequency response of the controller which must be implemented in a real duct. The simple expression given by equation (6.10.11) for the noise-free controller $G_0(j\omega)$ can thus only be used as a guide to the structure and approximate number of coefficients needed for a certain design of active control system.

The acoustical interpretation of the individual transfer functions can also be used to explain the sensitivity of the residual error signal to differences between the frequency response of the implemented controller and the ideal controller. Returning to the case with no measurement noise, the error signal spectrum follows from equation (6.6.5) and is given by

$$E(\omega) = P(j\omega)S(\omega) + C(j\omega)Y(\omega), \tag{6.11.7}$$

where $Y(\omega) = G(j\omega)X(\omega)$ and $X(\omega) = S(\omega) + F(j\omega)Y(\omega)$. Thus the ratio of the error signal spectrum to the primary signal spectrum, $S(\omega)$, can be written for a general controller $G(j\omega)$ in the form

$$\frac{E(\omega)}{S(\omega)} = P(j\omega) + \frac{C(j\omega)G(j\omega)}{1 - G(j\omega)F(j\omega)}. \tag{6.11.8}$$

Equation (6.10.10) for the optimal controller, $G_0(j\omega)$ in this case, can now be used to show that

$$\frac{E(\omega)}{S(\omega)} = P(j\omega)\left[\frac{1 - G(j\omega)/G_0(j\omega)}{1 - G(j\omega)F(j\omega)}\right]. \tag{6.11.9}$$

If $G(j\omega) = G_0(j\omega)$, the term in square brackets is clearly zero. In the more general case this expression shows that a small difference between $G(j\omega)$ and $G_0(j\omega)$ generates a residual error signal proportional to $[1 - G(j\omega)F(j\omega)]^{-1}$ (Swinbanks, 1982; Elliott and Nelson, 1984). The response $G(j\omega)F(j\omega)$ can be identified as the gain around the controller–feedback path loop in Fig. 6.9. When the controller is close to the optimal value, so $G(j\omega) \approx G_0(j\omega)$, the expressions for $G_0(j\omega)$ and $F(j\omega)$ in terms of the acoustic properties of the duct (equations (6.10.11) and (6.11.2)), can be used to express the residual error signal spectrum in terms of the transducer directivities and the reflection coefficients in the duct. The sensitivity of the controller to these parameters, as discussed by La Fontaine and Shepherd (1985), can then be readily deduced.

6.12 Time domain controller design

It was shown in Section 6.6 that the design of the optimal electronic controller in a single channel active noise control system could, by suitable definition of the filtered reference signal $R(\omega)$, be reduced to designing the filter response $H(j\omega)$ which minimised the mean square value of the error signal whose spectrum is given by

$$E(\omega) = D(\omega) + H(j\omega)R(\omega). \tag{6.12.1}$$

In that section, the solution to the least squares design problem was given by minimising the mean square error at each frequency ω, by independent adjustment of the controller response at that frequency to give $H_{\text{opt}}(j\omega)$. The inverse Fourier transform of this complex function of frequency will give the impulse response of the filter which would have to be implemented in order to realise this optimal frequency response. Unfortunately there is no guarantee that this impulse response will be causal. In other words it may require knowledge of the future as well as the past behaviour of the reference signal. Clearly such a filter could not be physically realised. In an active control system for the control of acoustic plane waves in ducts we have seen that the expected form of the optimal controller contains a delay term in order to compensate for the acoustic propagation time in the duct. In this application then, we do not expect to encounter problems with the causality of the optimal filter. In practice, however, several effects serve to complicate this simple interpretation. These include measurement noise, the inversion of the responses of the transducers and analogue filters, errors in spectral estimation and finite precision effects in the numerical calculations. If we have the freedom to move the relative positions of the detection sensor and secondary source we could, in principle, increase the acoustic propagation delay to overcome these practical problems. It has been noted, for example, that Roure (1985) merely windows out the non-causal part of the impulse responses obtained by inverse Fourier transformation of the optimal controller response derived from frequency domain manipulation. There are situations, however, in which the distance between the detection sensor and secondary source is necessarily small, and the unconstrained optimal filter obtained from frequency domain optimisation has a significant non-causal component. Under these situations the optimal causal filter is *not* just the causal part of the optimal unconstrained filter (Hough, 1988). What we must do is to return to the optimisation process for $H(j\omega)$ and impose the constraint of controller causality on the optimisation formulation from the start. This has been discussed in Section 3.8 of Chapter 3 for the continuous time case, and in Section 4.7 of Chapter 4 for a digital filter.

This causally constrained optimal impulse response would still, however, have to be transformed back into the frequency domain, $H(j\omega)$, in order to calculate the physical filter realisation $G(j\omega)$, which accounts for the effects of the feedback term

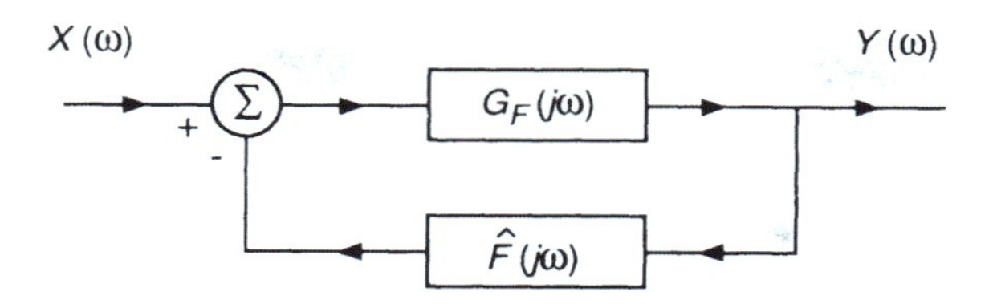

Fig. 6.17 "Feedback cancellation" architecture for the electronic controller.

$F(j\omega)$, using equation (6.6.9). This final procedure cannot be avoided even by using perfectly directional transducers, since in general $F(j\omega)$ is still not zero in this case, as noted above. An alternative suggested by Wanke (1976), Davidson and Robinson (1977) and Chaplin and Smith (1979) is to implement two filters, using the "feedback cancellation" implementation for the controller depicted in Fig. 6.17. In this arrangement, the feedback in the controller $(\hat{F}(j\omega))$ is arranged to cancel out the feedback in the acoustic path $(F(j\omega))$, so that the feedforward path, $G_F(j\omega)$, can be designed by using purely feedforward control methods. This potentially attractive design philosophy can, however, turn out to be very inefficient in terms of the complexities of the filters needed accurately to implement $\hat{F}(j\omega)$ and $G_F(\omega)$ (Elliott and Nelson, 1984; Swinbanks, 1985). This can be demonstrated by comparing the transfer function of this filter architecture with the frequency response of the optimal controller (equation (6.10.10)). The two expressions are given respectively by

$$G(j\omega) = \frac{G_F(j\omega)}{1 + \hat{F}(j\omega)G_F(\omega)}, \qquad G_0(j\omega) = \frac{-P(j\omega)}{C(j\omega) - P(j\omega)\,F(j\omega)}. \qquad (6.12.2)$$

On the assumption that $\hat{F}(j\omega) = F(j\omega)$, then the feedforward path must take the form $G_F(\omega) = -P(j\omega)/C(j\omega)$. We have seen from equation (6.11.2) that in terms of the acoustic variables in the duct $F(j\omega)$ is a complicated function of the directivities of the detection and secondary transducers, and the terminating reflection coefficients at the duct ends. If these reflection coefficients are close to unity, the impulse response of $F(j\omega)$ will be very long. Similarly, the response $P(j\omega)/C(j\omega)$ will also depend on these duct parameters in a way that can be predicted from equations (6.11.1) and (6.11.3). This response will, potentially, have a long impulse response. The form of $G_0(j\omega)$, however, does not depend on these termination properties of the duct and, provided either the detection sensor or the secondary source are reasonably directional, will have a very short impulse response. This argument leads to the conclusion that there are a number of common terms in $G_F(j\omega)$ $(= -P(j\omega)/C(j\omega))$ and $1 + \hat{F}(j\omega)G_F(j\omega)$ $(= 1 - F(j\omega)\,P(j\omega)/C(j\omega))$, which cancel out when the feedback arrangement of Fig. 6.17 is implemented, so that the net response, $G(j\omega)$, is much simpler than the individual filter responses $\hat{F}(j\omega)$ and $G_F(j\omega)$. One advantage of the feedback cancellation controller illustrated in Fig. 6.17 is that the possibility of an unstable loop around the controller and feedback path of Fig. 6.9 will be removed provided that the feedback cancellation path is perfectly adjusted. If, however, $\hat{F}(j\omega)$ does not exactly match $F(j\omega)$ the potential for such an unstable loop still exists. Finally, it is worth noting that Hamada *et al.* (1988) have suggested a simpler arrangement for feedback cancellation, in which a component of the *error* microphone signal, which is placed symmetrically with the detection microphone about the secondary source, is subtracted from the detection microphone signal.

6.13 The effect of electrical delays in the controller

A digital control system of the type illustrated in Fig. 6.6, for example, has a response which can be described in terms of the continuous input signal $x(t)$ and the continuous output signal $y(t)$. This response will be composed of the response of the analogue anti-aliasing and reconstruction filters, the digital-to-analogue and analogue-to-digital converters and the response of the digital filter itself. Provided the analogue filters prevent aliasing so that the controller behaves linearly, the frequency response of the whole controller can be written as

$$G(j\omega) = G_A(j\omega)G_D(e^{j\omega T}), \tag{6.13.1}$$

where $G_A(j\omega)$ is the frequency response of the analogue anti-aliasing and reconstruction filters and accounts for any delays in the data converters and the processing delay in calculating the output $y(n)$ of the digital controller from its input, $x(n)$. $G_D(e^{j\omega T})$ is the response of the digital filter at a normalised frequency of ωT, where T is the sample time. This expression must be used with some care where anti-aliasing and reconstruction filters are *not* present, as noted by Kuo (1980) for example, but it provides a convenient formulation here.

We have seen that in controlling acoustic plane waves in a duct, the response $G_0(j\omega)$ (equation (6.10.11)) has an overall delay of l_1/c_0 seconds, where l_1 is the distance between the detection sensor and the secondary source. Clearly we must ensure that this delay is greater than that inherent in the analogue and processing part of the controller, $G_A(j\omega)$, to ensure that the digital filter is required only to implement a causal response. If we assume that the total delay inherent in $G_A(j\omega)$ is τ_A, then the condition for requiring only a causal digital filter becomes

$$l_1 \geq \tau_A c_0. \tag{6.13.2}$$

In other words, the presence of the practical components necessary to implement a digital controller will place a geometric constraint on the physical design of the active control system (Ffowcs-Williams *et al.*, 1985). As an approximate guide to the magnitude of τ_A, we could allow one sample delay (T seconds) for the processing and converters, and assume that each pole of each analogue filter (which would contribute 6 dB/octave to the cut-off rate) has 45° of phase shift, or 1/8 cycle of delay, at the cut-off frequency f_c. So the total delay for a total of n poles in both the anti-aliasing and reconstruction filters is $n/8f_c$ seconds. The cut-off frequency of the filters (f_c) is typically set at one-third the sample rate, so assuming $f_c = f_s/3$, i.e. $1/f_c = 3T$, then the total delay in the analogue path is approximately

$$\tau_A \approx T\left(1 + \frac{3n}{8}\right). \tag{6.13.3}$$

As a practical example, we assume that the sample rate is 1 kHz, giving a maximum operating frequency of about 300 Hz. This corresponds to using a rectangular duct whose largest dimension is less than about 0.5 m in order to ensure plane wave propagation. Typically the order of the analogue anti-aliasing and reconstruction filters may be 6 poles each, so $n = 12$, $T = 1$ ms in this case and so $\tau_A \approx 5.5$ ms. The minimum spacing between the detection sensor and secondary source in this case is thus about 2 m, i.e. about four times the width of the duct.

6.14 Adaptive FIR digital controllers

In Section 3.8 and Section 4.7 methods of designing optimal analogue and digital filters were described. These were based on minimisation of a mean square error criterion averaged over some long period of time. Such an averaging procedure is implicit in the use of correlation functions or spectral densities. Even when such methods are made responsive to changes (as, for example, in the active control systems described by Ross, 1982b or Roure, 1985) each iteration of the controller is based on some time-averaged measurement of the system under control. Such *iterative* control methods are sometimes characterised as being *off-line*. In other words the control algorithm continues to operate in one device, while any updates are done in a different time frame on a conceptually separate device. In this section we consider methods of adjusting, on a sample by sample basis, the coefficients of a digital filter that is used as the controller in a single-channel active control system. Such procedures are sometimes characterised as being *on-line* although we described them above as being *adaptive*. Since, when using such methods, the coefficients are changed on a timescale equal to the sampling time, the algorithms which can easily be implemented tend to be rather simple. Indeed, the algorithms may be only simple approximations to more mathematically exact and rigorous iteration procedures. It is the fact that the filters are being *continuously* updated, however, which is important in this case, and this feature can make adaptive algorithms *more* responsive to changes than exact methods whose updates take some time to compute.

The simplest such algorithm is a variant of the simple LMS algorithm that was introduced in Section 4.8 for the adaptation of an FIR digital filter. Compared with the simple arrangement for updating the coefficients presented in Chapter 4, the use of an LMS algorithm in active control is complicated by a number of factors. First, the difference between the filter output and the desired signal is no longer directly available to update the coefficients, as it was in the electrical case illustrated in Fig. 4.21. The error signal can now only be observed through the error path, as illustrated in Fig. 6.18(a). In this figure $C(z)$ denotes the response of the error path and $H(z)$ denotes the response of the FIR filter acting as the controller. The reference signal is now the signal $x(n)$ produced directly from the detection sensor. In the similar block diagram described in Section 6.6, the order of the blocks $H(z)$ and $C(z)$ were reversed in order to generate a filtered reference signal and reduce the problem to a more standard form. This is illustrated in Fig. 6.10. In the case of a fixed, time invariant controller, such a reversal is exactly valid. In the case where $H(z)$ is changing with time, however, such a reversal as that shown in Fig. 6.18(b) will produce an error signal, $e_2(n)$, which is not equal to the true signal, $e_1(n)$, except in the case where the adaptation of the coefficients of $H(z)$ takes place on a timescale which is very slow compared with the response time of the error path

$C(z)$. Despite this reservation, Fig. 6.18(b) does suggest a modification of the normal LMS algorithm which could be used in this application. The figure suggests the use of the filtered reference signal, $r(n)$, instead of $x(n)$ in the normal LMS update equation. The update equation then becomes, by analogy with equation (4.8.6),

$$\mathbf{h}(n+1) = \mathbf{h}(n) - \alpha\mathbf{r}(n)e(n), \tag{6.14.1}$$

where α is the convergence coefficient, $\mathbf{h}(n)$ is the vector of controller filter coefficients at the nth sample, and $\mathbf{r}(n)$ is the vector of previous delayed filtered reference signals.

This adaptive filtering algorithm is known as the *filtered-x* LMS algorithm (Widrow and Stearns, 1985), since $r(n)$ is obtained by filtering the reference signal $x(n)$ with the error path response. In practice an electrical filter is often used to *model* the error path, and the filtered reference signal is generated by passing the reference signal through this *estimate* of the error path $\hat{C}(z)$ rather than the error path itself. The algorithm was first proposed by Morgan (1980) and independently for feedforward control by Widrow *et al.* (1981) and for the active control of sound by Burgess (1981). There is also a simplified version of the algorithm, in which the error path is a pure delay of k samples, and in which the adaptive algorithm reduces to

$$\mathbf{h}(n+1) = \mathbf{h}(n) - \alpha\mathbf{x}(n-k)e(n). \tag{6.14.2}$$

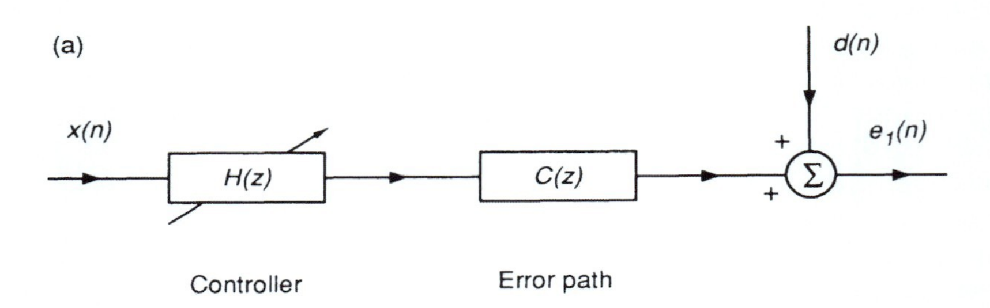

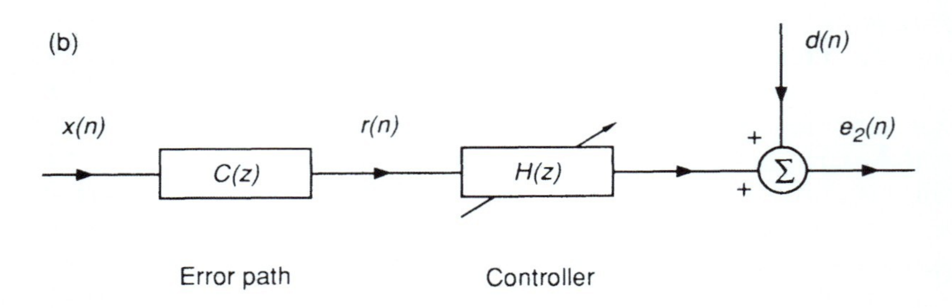

Filtered reference
signal

Fig. 6.18 Block diagram of (a) a simple active control system with an adaptive controller and (b) its *approximate* equivalent.

This is known as the delayed LMS, or DLMS algorithm which was suggested by Widrow (1971), and which has been used in telecommunications applications (Qureshi and Newhall, 1973; Kabal, 1983).

The general properties of the filtered-x LMS algorithm are as follows: (1) it can converge on a timescale comparable with the delay in the error path, and so can rapidly track changes in the primary signal; (2) it is rather robust to errors in the estimate of the error path used to generate the filtered reference, which has a frequency response of $\hat{C}(e^{j\omega T})$; in fact, Morgan (1980) has shown that in the limit of slow adaptation the algorithm will still converge with nearly 90° of phase error in $\hat{C}(e^{j\omega T})$ compared to $C(e^{j\omega T})$; (3) it is relatively easy to implement and, compared with the LMS algorithm, it requires only the additional generation of the filtered reference signal.

A more difficult problem than the effect of the error path on the LMS algorithm is the effect of the feedback path. In general the feedback path can destabilise the LMS algorithm, since it is possible for the adaptive filter to pass through a state in which there is a gain of unity around the feedback path-controller loop. This will cause an instability which can saturate the hardware being used to implement the adaptive filter, and prevent further convergence. A more detailed discussion of instability in such feedback systems will be presented in Chapter 7.

The simplest approach to solving the problems caused by the feedback path is to use a separate "feedback cancellation" filter, as discussed above. Such a filter acts in exactly the same way as the "echo cancellers" used in telephone systems (Sondhi and Berkley, 1980). The incorporation of such a filter into an adaptive digital controller has been investigated by Warnaka *et al.* (1984). These authors suggest designing such a feedback cancellation filter by feeding white noise into the secondary source prior to start-up of the controller, and using an LMS algorithm to minimise the difference between the output of the true feedback path (the signal from the detection microphone) and an FIR model $\hat{F}(z)$ of the feedback path. The adaptation of the main cancellation filter is performed using the delayed LMS algorithm by these authors. Because the physical error path is not a pure delay, however, a third "compensation" filter must be used which when connected in series with the error path, makes the net system appear to be a pure delay. This compensation filter is also designed before the start-up of the controller by feeding the signal from the error sensor through this compensation filter, and comparing its output with a delayed version of the white noise signal being fed to the secondary source. The FIR compensation filter is then adapted by using the LMS algorithm. Two arrangements are suggested for using this compensation filter (of response $\hat{C}^{-1}(z)$) in the adaptation of the cancelling filter, as illustrated in Fig. 6.19(a) and (b). In the first of these (Warnaka *et al.*, 1984) the error signal is passed through the compensation filter before being used to update the cancelling filter. This arrangement has the disadvantage that the DLMS algorithm will minimise the mean square value of the error signal filtered by $\hat{C}^{-1}(z)$, rather than the mean square value of the error signal itself. Because a broadband minimisation is being performed, this may well leave high levels of residual error at the error microphone at some frequencies, if there is a dip in the response of the $\hat{C}^{-1}(z)$ filter at these frequencies. The other arrangement, suggested by Poole *et al.* (1984), is illustrated in Fig. 6.19(b) in which $\hat{C}^{-1}(z)$ is now part of the feedforward part of the controller. Now the signal from

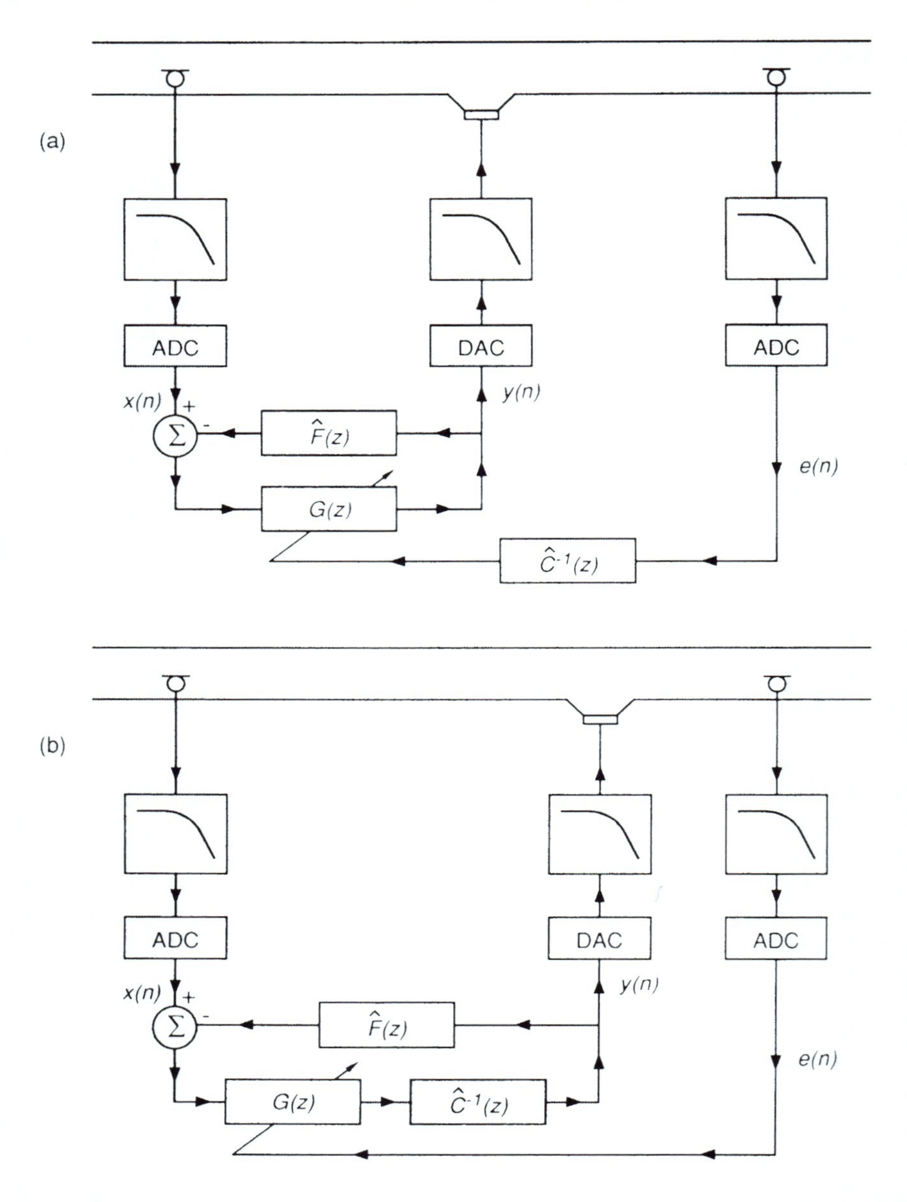

Fig. 6.19 Two methods of using the delayed inverse error path to transform the total error path, seen by the adaptive controller, into a pure delay. The delayed LMS algorithm is then used to update $G(z)$ (after Warnaka *et al.*, 1984, and Poole *et al.*, 1984).

the error sensor is directly minimised. The delay through the feedforward part of the controller is, however, now inevitably increased, which may necessitate a greater separation distance between detection sensor and secondary source and thus a physically larger system. The potential inefficiency of this *feedback cancellation* arrangement has already been discussed above in Section 6.12.

6.15 An adaptive recursive controller

Rather than trying to cancel the feedback path explicitly, the approach taken by Eriksson and his co-workers in a series of papers (1987, 1988, 1989) is to use an adaptive IIR filter as the controller, and use this to model the complete desired controller response, including the effects of the feedback path. The basic configuration is shown in Fig. 6.20 in which $A(z)$ and $B(z)$ are individual FIR filters. These are combined to generate the IIR controller response given by

$$G(z) = \frac{A(z)}{1 - B(z)}. \tag{6.15.1}$$

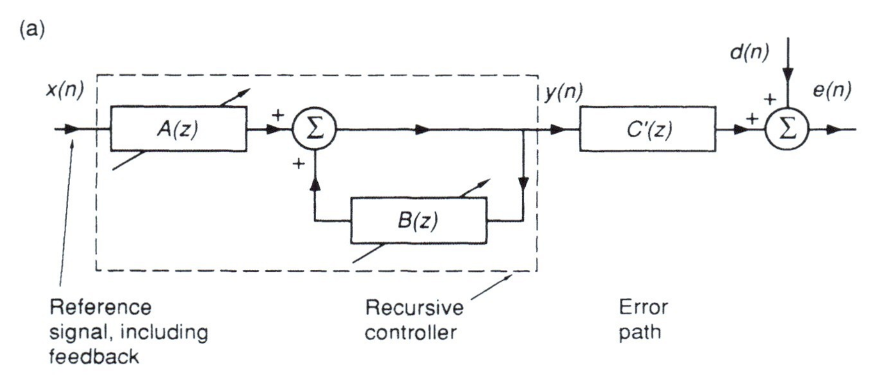

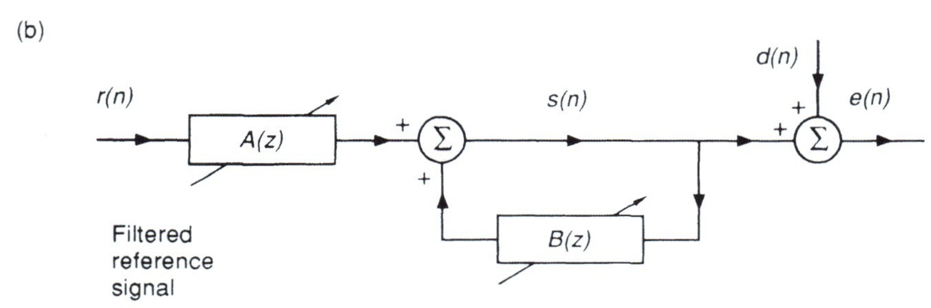

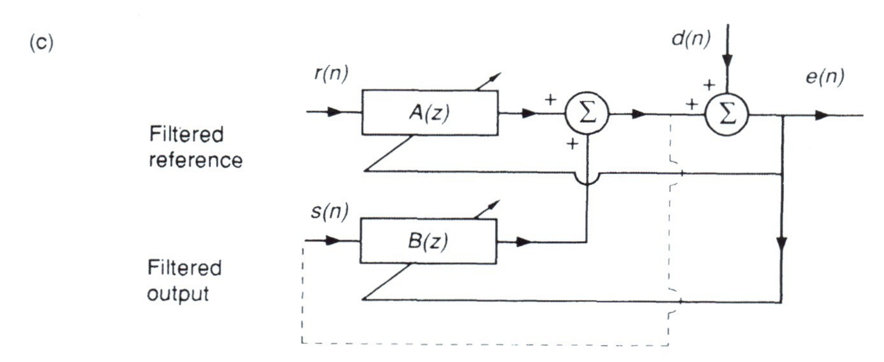

Fig. 6.20 Block diagram of an active control system with a recursive controller: (a) in direct form; (b) in re-ordered form; (c) as a pair of LMS adaptive filters.

Comparing this with the optimal, noise-free filter specified by equation (6.10.11), in terms of the acoustic variables;

$$G_0(j\omega) = \frac{-1}{L_s M_x} \frac{e^{-jkl_1}}{1 - D_x D_s e^{-jk2l_1}}, \qquad (6.15.2)$$

it is clear that if $A(z)$ could model the response $-e^{-jkl_1}/L_s M_x$, i.e. the delayed inverse of the electroacoustic response of the transducers, and $B(z)$ could model the response $D_x D_s e^{jk2l_1}$, i.e. the delayed directivities of the transducers, then the IIR filter could be a very efficient implementation of the controller (as noted by Elliott and Nelson, 1984).

The algorithm used by Eriksson *et al.* (1987) to adapt the coefficients of the $A(z)$ and $B(z)$ paths in the filter is a modification of a simple gradient descent method for recursive filters (Feintuch, 1976). The properties of such adaptive algorithms for IIR filters are still not fully understood, especially when complicated by the physical feedback path present in this case (Flockton, 1989a). In particular, the error surface for the coefficients of an adaptive recursive filter will not generally be quadratic (Elliott and Nelson, 1985b), but may, or may not, be unimodal (Widrow and Stearns, 1985, Ch. 8). We will attempt to describe the action of Eriksson's algorithm here in terms of a block diagram manipulation. Another justification has been presented by Elliott and Nelson (1988), in which the multiple input, multiple output generalisation of this algorithm is presented. Figure 6.20(a) shows the recursive controller being fed by the signal $x(n)$ from the detection microphone with its output $y(n)$ driving through a modified error path, $C'(z)$. This output then combines with the primary field $d(n)$ to produce the error signal, $e(n)$. Note that due to the physical feedback path, the "reference signal", $x(n)$, will depend partly on the controller output, $y(n)$. This dependence is not directly indicated in Fig. 6.20(a). The effect of feedback will be to make the reference signal non-stationary during adaptation (Darlington, 1987). We assume that the adaptive algorithm can cope with this change in the reference signal provided it is adapting slowly. Another effect of the feedback is to complicate the signal path from the secondary source to the error sensor when the controller is in operation. For a non-zero controller response, $G(z)$, this signal path is the combination of the previous error path, $C(z)$, and a contribution from the loop formed by the *feedback* path and controller as can be seen by reference to Fig. 6.9. The net effect is to produce an effective error path $C'(z)$ with the controller operating. This has the transfer function (Elliott and Nelson, 1984)

$$C'(z) = \frac{C(z)}{1 - G(z)F(z)}. \qquad (6.15.3)$$

Provided $A(z)$ and $B(z)$ are only slowly varying, the transfer functions of the error path and controller can be reversed, and the block diagram will look like Fig. 6.20(b), where $r(n)$ is the reference signal, $x(n)$, filtered by $C'(z)$ and $s(n)$ is the controller output, $y(n)$, also filtered by the modified error path $C'(z)$. If the filters are converging sufficiently slowly to ensure that the effects of changes in one filter on the adaptation of the other are not significant, the adaptation of $A(z)$ and $B(z)$ can be considered separately. This enables the block diagram to be redrawn as shown in Fig. 6.20(c). The feedback from the output of the controller to the input of the recursive path, $B(z)$, has deliberately been drawn in dashed to stress the interpretation as two feedforward filters $A(z)$ and $B(z)$. Each of these can be adapted by using LMS algorithms with either $r(n)$ or $s(n)$ as reference signals. The inherent assumption of slowly varying filters, however, means that $s(n)$ is nearly stationary as far as the LMS algorithm used to update $B(z)$ is concerned, provided $1/(1 - B(z))$ is never near instability. The output of the recursive controller is assumed to be

$$y(n) = \sum_{i=0}^{I-1} a_i(n)x(n-i) + \sum_{k=1}^{K} b_k(n)y(n-k), \tag{6.15.4}$$

where $a_i(n)$ and $b_k(n)$ are the ith and kth direct and recursive coefficients in the controller at the nth sample time. The update algorithms for the two sets of filter coefficients which are motivated by Fig. 6.20(c) are those described by Eriksson and are given by

$$a_i(n+1) = a_i(n) - \alpha e(n)r(n-i), \tag{6.15.5a}$$

$$b_k(n+1) = b_k(n) - \alpha e(n)s(n-k), \tag{6.15.5b}$$

where α is a convergence coefficient.

Although a number of assumptions and approximations have been made in justifying this recursive least mean square or "RLMS" algorithm, Eriksson has reported that the algorithm successfully converges in a variety of practical active control applications (Eriksson *et al.*, 1987, 1988, 1989). A development of the RLMS

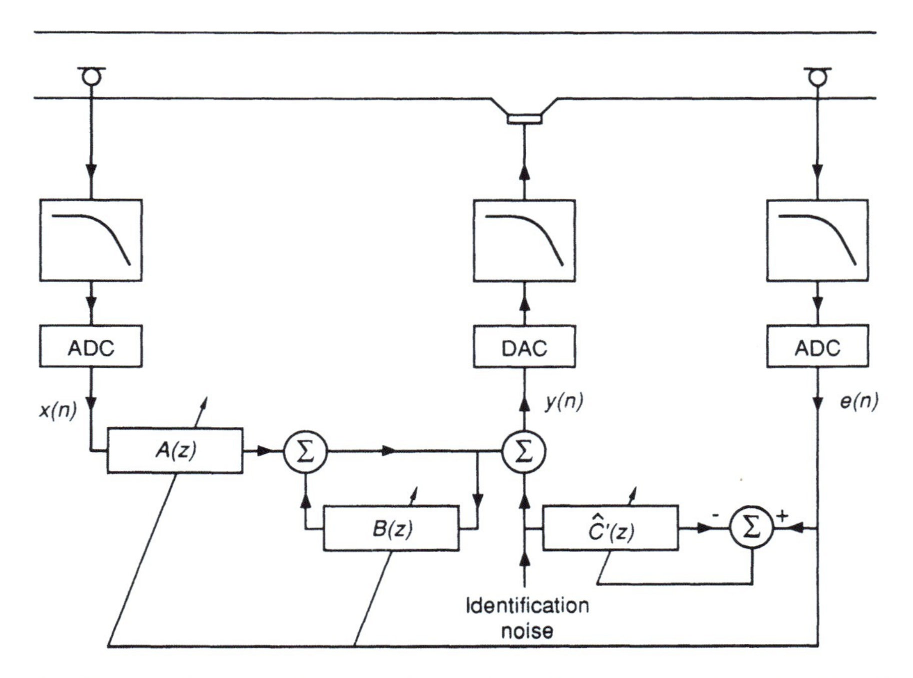

Fig. 6.21 The adaptive recursive controller used by Eriksson *et al.* including identification noise and on-line identification of the error path.

algorithm which explicitly accounts for the presence of the feedback path has been presented by Flockton (1989b).

Another feature of the adaptive controllers described by Eriksson is the fact that although a preliminary estimate of the error path is required (to generate the filtered reference and output signals in the update equation), which is obtained from measurements prior to control, the model of the error path is also continuously updated while control is in progress. In addition to the output from the active controller, a white noise signal generated by using a pseudo random binary sequence is continuously fed to the secondary source as illustrated in Fig. 6.21. A separate LMS algorithm is used to keep the response of the FIR filter, $\hat{C}'(z)$ as close as possible to that of the true error path (Eriksson and Allie, 1989). This continuous identification which proceeds in parallel with the control not only allows tracking of the changing duct response with variations in flow rate etc. but also allows tracking of the way in which the error path changes as the controller converges, since as shown by equation (6.15.3), $\hat{C}'(z)$ is dependent upon $G(z)$. This estimate of the error path response is not biased by the presence of the primary field or the output

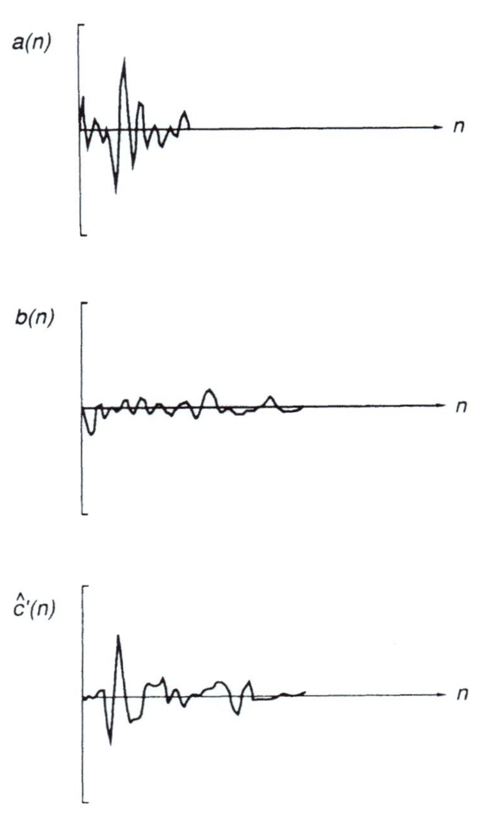

Fig. 6.22 The impulse responses of the converged filters in an adaptive recursive controller; $a(n)$ is that of the feedforward path $A(z)$ (32 coefficients), $b(n)$ is that of the recursive path $B(z)$ (64 coefficients), $\hat{c}'(n)$ is that of the error path $\hat{C}'(z)$ (64 coefficients) (after Eriksson *et al.*, 1989).

from the control system, since these signals are not present at the input to the adaptive modelling filter ($\hat{C}'(z)$). It is pointed out by Eriksson *et al.* (1988), however, that the identification noise can couple into the signal from the detection sensor via the feedback path. This will have the same effect as measurement noise at the detection sensor, and bias the controller response. These authors note that in practice this effect is not found to be a problem, since the final level of identification noise must be relatively low at the error microphone, so that the attenuation of the active controller is not compromised.

Figure 6.22 shows the impulse responses measured by Eriksson and Allie (1989) of the filters $A(z)$, $B(z)$ and $C'(z)$ after convergence of the controller operating in a duct of 0.3 m diameter and length 7.5 m, which is unlined except for a 1.3 m silencer near the primary (electroacoustic) source. The control system is operating at a sample rate of 875 Hz. Figure 6.23 shows the spectrum of the error signal before and after control when the system is used by these authors in a lined supply duct (0.86 m $\times$ 1.12 m) approximately 12 m from a centrifugal fan producing an air speed of about 14 m s^{-1}. The coherence between the detection microphone and error microphones, both fitted with an anti-turbulence screen, is above 0.95 from about 40–140 Hz, where significant reductions in level are observed.

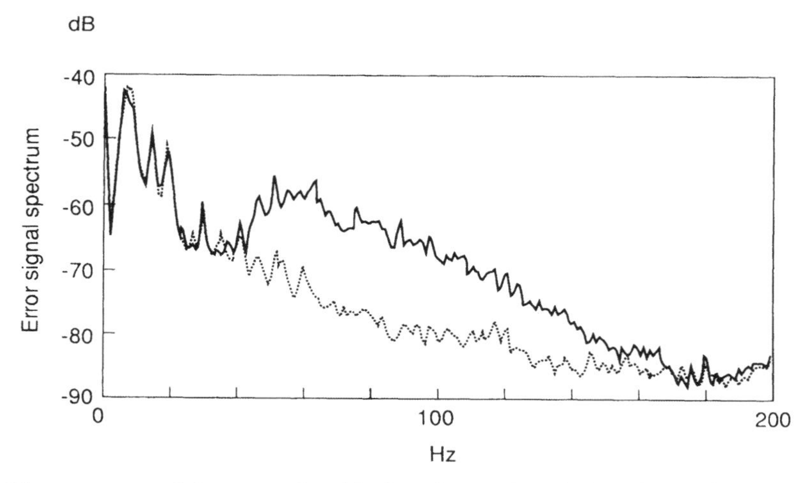

Fig. 6.23 The spectrum of the error signal before (———) and after (·······) the application of a recursive adaptive controlller in an air conditioning duct with an air flow of about 14 m s^{-1} (after Eriksson and Allie, 1989).

7
Single Channel Feedback Control

7.1 Introduction

In what must be regarded as a classic paper in the context of the active control of sound, Olson and May (1953) introduced the "electronic sound absorber". Although the title of the paper is something of a misnomer, Olson and May foresaw many of the applications of active control techniques that are now coming close to practical use. The device described in the paper is illustrated in Fig. 7.1. The "perspective view" of the apparatus illustrates well the state of electronic technology at the time the paper was written. An "electronic microphone" (in which "the impinging sound vibrations directly control the electron stream in a vacuum tube") was used to derive a signal proportional to sound pressure. The microphone had a flat frequency response down to 0 Hz and a phase response that was "less than two degrees from 20 Hz to 400 Hz". The signal was amplified by using a battery powered valve amplifier. The amplifier output was used to drive a loudspeaker with a "high impedance voice coil" which avoided the necessity of a transformer between the vacuum tubes of the amplifier and the loudspeaker. A transformer would introduce a "considerable phase shift" in the low frequency range and this was to be eliminated. In addition, the response of the amplifier was made to roll off at frequencies above about 400 Hz to avoid "positive feedback".

The overall operation of the device was made to ensure that the phase shift could be minimised between the input to the microphone and the output from the loudspeaker. Under these circumstances a simple inversion of the microphone signal could be introduced such that the loudspeaker output was 180° out of phase with the microphone input. If a high gain of the amplifier was used, the device would then act as a "sound pressure reducer"; the detected sound pressure was fed back with a phase inversion and high gain in order to drive the pressure fluctuation towards zero. The operation of the device is thus analogous to the "virtual earth" principle of closed loop feedback (see, for example, Tobey *et al.*, 1971) and is sometimes now referred to as the "acoustical virtual earth" (Trinder and Nelson, 1983) or as the "tight coupled monopole" (Hong *et al.*, 1982).

Driving the sound pressure in the vicinity of the loudspeaker to zero of course ensures that the loudspeaker neither radiates nor absorbs any energy under steady-state conditions. This was recognised by Olson and May but they suggested that the device could also *absorb* sound by "designing the system so the proper phase relations are obtained". They also saw that this was difficult to achieve in practice

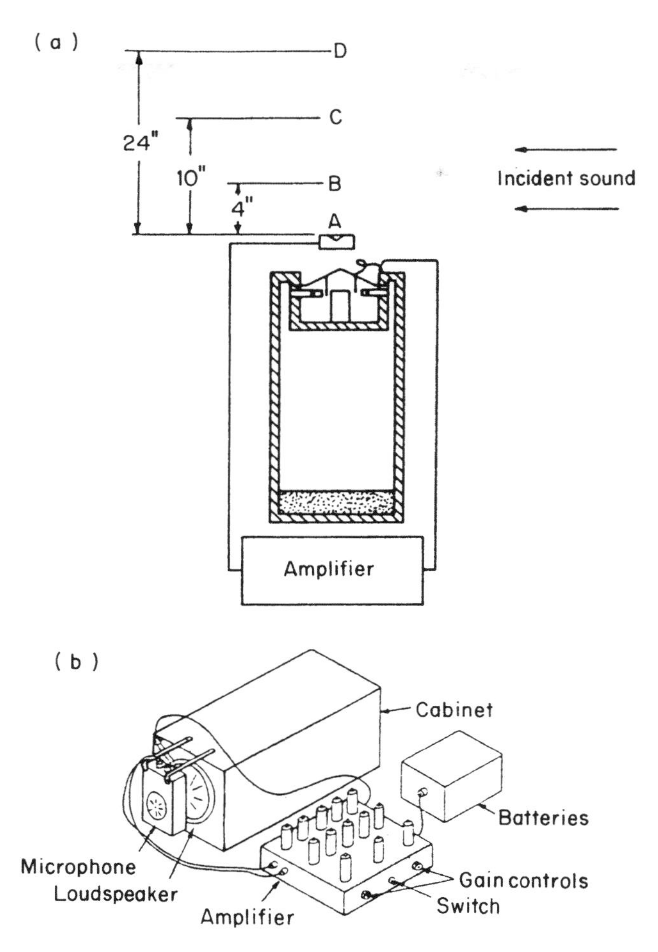

Fig. 7.1 (a) The "electronic sound absorber" of Olson and May reproduced from their 1953 paper together with (b) "a perspective view" of the device.

and suggested that if the region of the loudspeaker and microphone was surrounded by an acoustically resistive screen, then this could be made to dissipate energy. Olson and May suggested that in general "the problem in low frequency sound absorption is to provide an acoustical impedance of relatively small value so that the volume current which introduces the sound absorption will not be limited by high acoustical impedance". Thus, Olson and May reasoned that the feedback system producing low sound pressure could provide a low impedance for terminating the dissipative acoustic impedance provided by the surrounding screen. The active absorption of sound by using a feedforward approach will be further considered in Chapter 9 and in Chapter 11.

However, most of the applications suggested by Olson and May involved the use of the device as a "spot sound reducer". The performance of the apparatus in an "incident plane wave field" is illustrated in Fig. 7.2. Several suggestions for the application of the device in other environments were made, one of which is

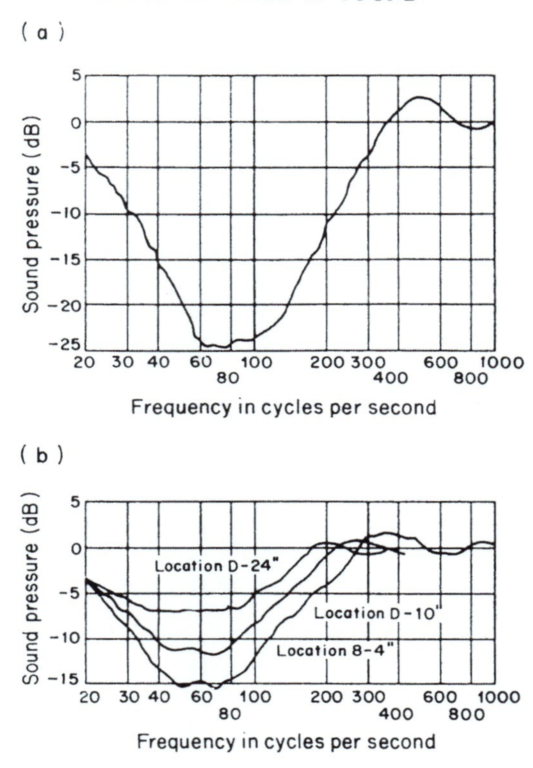

Fig. 7.2 The performance of the "electronic sound absorber" measured by Olson and May (1953). The curves show the "sound pressure reduction frequency characteristic" (a) at location A of Fig. 7.1 and (b) at locations B, C and D of Fig. 7.1.

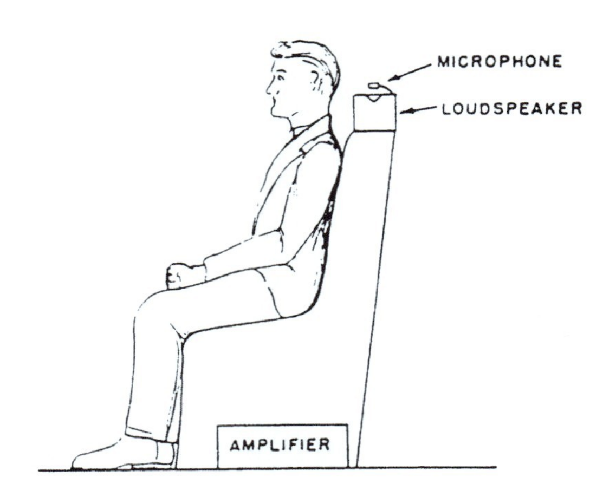

Fig. 7.3 A reproduction of the sketch presented by Olson and May (1953) of the "electronic sound absorber" used in an airplane or automobile to reduce the noise in the vicinity of the occupant's head.

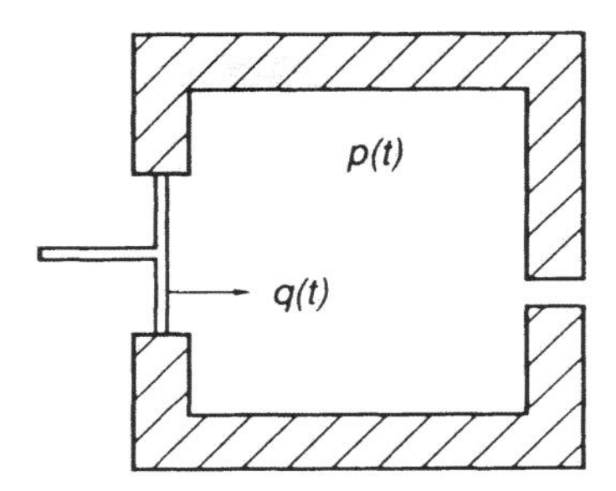

Fig. 7.4 A rigid walled cavity driven by a piston source.

reproduced in Fig. 7.3. This illustrates the application of active control in an "airplane or automobile". Although these applications have since proved feasible (at least for deterministic signals), it has not been accomplished by using the Olson and May technique. This is at least partly due to the problems of designing such a system to avoid the stability problems referred to by Ffowcs-Williams (1984). However, the acoustical properties of the device will be discussed in more detail in Chapter 11, and some reasons suggested as to why the device is not in widespread use. However, Olson and May also foresaw the application of the feedback technique to the control of duct-borne sound and in a later paper, Olson (1956) suggested its use in enhancing the performance of "a noise reducing headset". Both of these applications will be discussed in this chapter. However, we will firstly introduce the ideas of feedback control and present a general framework for the analysis of such feedback systems. This work is described in terms of classical control systems theory since almost all applications of the feedback approach to date have been single channel and implemented with analogue electronics. The application of modern control systems theory and the powerful results of optimal control are only just being brought to bear on the problem (see the papers by Dohner and Shoureshi, 1989 and Costin and Elzinga, 1989). It will be interesting to see whether the applications of modern techniques will improve the possibilities for the more widespread use of the Olson and May device.

7.2 Feedback control of cavity pressure fluctuations

As a preliminary to a more detailed discussion of the feedback control of sound, it will be useful to consider a very simple model problem in the hope of shedding some light on the physical processes involved. The problem that we shall discuss is illustrated in Fig. 7.4. This shows a rigid walled cavity driven by a rigid piston source. The cavity is perfectly sealed apart from a small hole through which air can leak when the air in the cavity is compressed by the piston. As we have seen in Section 3.2, we can describe the dynamics of this system with a first order differential equation of the form

$$C_a \frac{dp(t)}{dt} + \frac{1}{R_a} p(t) = q(t), \tag{7.2.1}$$

where $p(t)$ is the pressure produced in the cavity as a result of the volume flow, $q(t)$, introduced to the cavity by the piston. The term C_a quantifies the acoustic compliance of the cavity and is equal to $V_0/\rho_0 c_0^2$ where V_0 is the cavity volume. The term R_a quantifies the acoustic resistance and defines the rate at which air can escape from the cavity via the "leak" in the cavity wall.

We have seen that the response of this system can be expressed in terms of a transfer function between the rate of volume flow input to the cavity and the resulting pressure fluctuations. Laplace transformation of equation (7.2.1) shows that this transfer function can be written as

$$H(s) = \frac{P(s)}{Q(s)} = \frac{R_a}{1 + sR_a C_a}, \tag{7.2.2}$$

and setting $s = j\omega$ shows that the corresponding frequency response function is given by

$$H(j\omega) = \frac{R_a}{1 + j\omega R_a C_a}. \tag{7.2.3}$$

It has also been demonstrated in Section 3.3 that the pressure response of the system to a unit impulse in volume velocity is given by

$$h(t) = \frac{1}{C_a} e^{-t/R_a C_a}. \tag{7.2.4}$$

Now let us assume that the excitation of the cavity is not in the first instance via motion of the piston but through the vibration of the cavity walls resulting from an externally imposed sound pressure field. Thus, assume that the net rate of volume flow produced by this primary excitation is specified by $q_p(t)$. This fluctuation need not necessarily be deterministic but could in general be a random process. If this is the case, recall from Section 3.6 that if $q_p(t)$ has a power spectral density given by $S_{qq}(\omega)$, then the spectral density of the resulting cavity pressure fluctuations will be given by

$$S_{pp}(\omega) = |H(j\omega)|^2 S_{qq}(\omega). \tag{7.2.5}$$

We can now show that this value of the power spectral density of the pressure fluctuations can be reduced by the action of feedback. Assume that as illustrated in Fig. 7.5, the cavity pressure fluctuation can be perfectly detected and the piston on the cavity wall can be driven such that it produces a secondary volume flow given by

$$q_s(t) = -Kp(t). \tag{7.2.6}$$

Thus the secondary volume flow is proportional to the pressure in the cavity but inverted. It is assumed here, of course, that *ideal* transducers are available. In other words, the cavity pressure is detected by a microphone, amplified by an amplifier and transmitted via loudspeaker (piston) all of which have an infinite, flat, frequency response and produce zero phase shift throughout this infinite frequency range. Naturally, as we shall see later, this situation is far from that encountered in practice. Nevertheless, if the piston *could* be made to move instantaneously in order to produce a volume flow which was a large number (K) times the negative of the

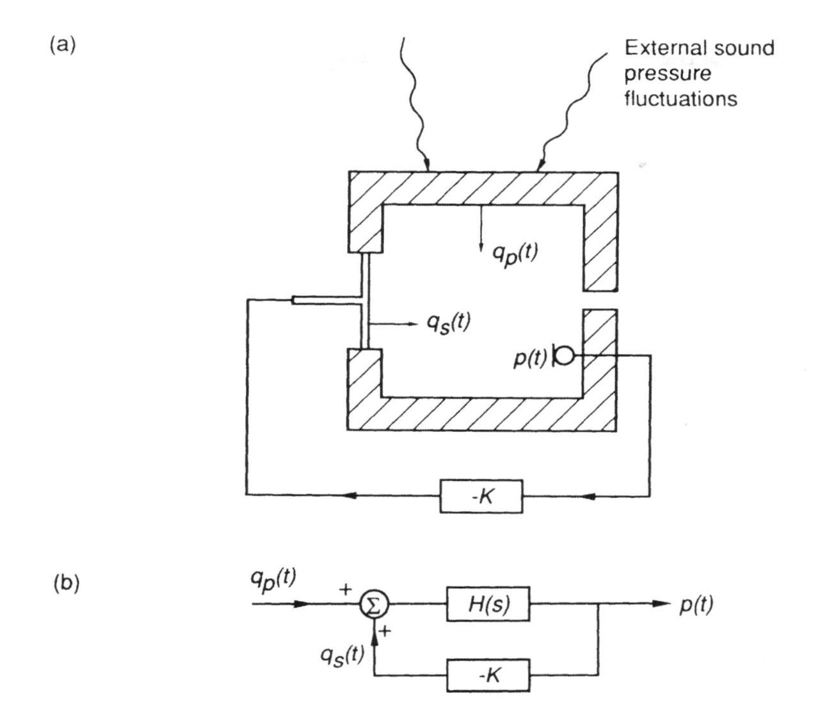

Fig. 7.5 The feedback control of cavity pressure fluctuations. The cavity pressure fluctuations are induced by the vibration of the cavity walls which produce a primary volume velocity of $q_p(t)$. It is assumed that the secondary volume velocity $q_s(t) = -Kp(t)$.

detected pressure, then the pressure fluctuations in the cavity could be reduced substantially. In this case, reductions would be produced even when the primary fluctuation $q_p(t)$ is a random process with a waveform which is not predictable perfectly. In the presence of feedback, the differential equation governing the system can be expressed as

$$C_a \frac{dp(t)}{dt} + \frac{1}{R_a} p(t) = q_p(t) + q_s(t). \tag{7.2.7}$$

Since we are assuming that $q_s(t) = -Kp(t)$, we can rewrite this equation as

$$C_a \frac{dp(t)}{dt} + \left(\frac{1}{R_a} + K \right) p(t) = q_p(t). \tag{7.2.8}$$

If we now define $R_a' = R_a/(1 + KR_a)$, then it follows by analogy with equation (7.2.2) that we can define a modified transfer function between the primary excitation and the resulting cavity pressure fluctuation that is given by

$$H'(s) = \frac{P(s)}{Q_p(s)} = \frac{R_a'}{1 + sR_a'C_a}. \tag{7.2.9}$$

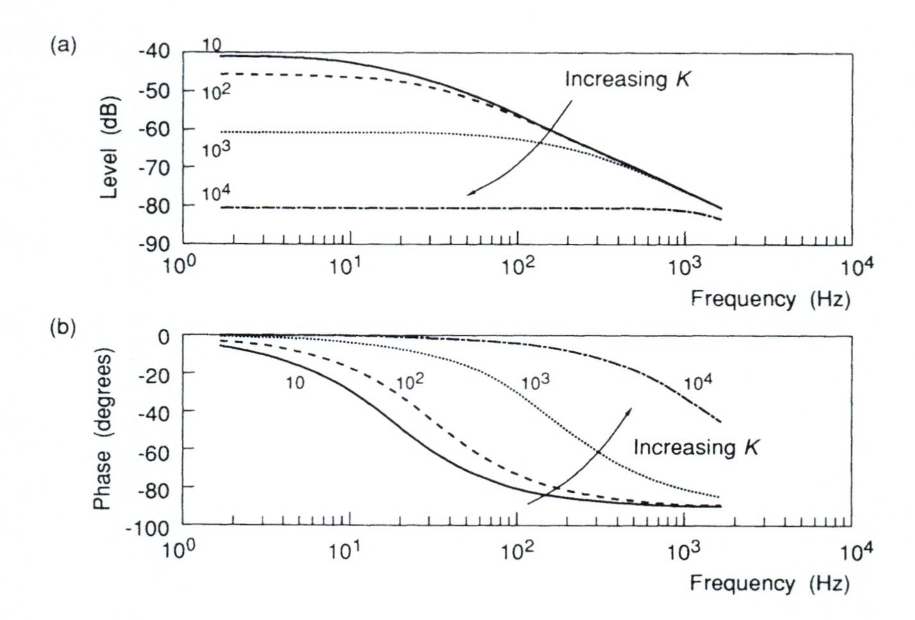

Fig. 7.6 The effect of increasing feedback gain K on the modulus (a) and phase (b) of the frequency response function of the acoustic cavity. Curves show the modulus and phase with values of K of 10, 10^2, 10^3 and 10^4.

The corresponding modified frequency response function and impulse response also follow from equations (7.2.3) and (7.2.4) with R_a replaced by R_a'. Thus R_a' is the modified effective resistance to flow out of the cavity in the presence of feedback. When K is made very large (i.e. such that $KR_a >> 1$), then $R_a' \approx (1/K)$. Thus a large value of feedback gain K produces a very *small* resistance to flow from the cavity. In other words, the build up of pressure in the cavity is prevented by the instantaneous withdrawal of the piston from the cavity. Similarly, if the primary excitation rarefies the air in the cavity, the piston will be moved into the cavity in an attempt to counteract this rarefaction.

The effect of increasing the feedback gain K on the frequency response function is illustrated in Fig. 7.6. Note that, most importantly, reducing the effective resistance to flow from the cavity with a high feedback gain produces a consequent reduction in the modulus of the frequency response function. Thus, in view of equation (7.2.5), the power spectral density of the resulting pressure fluctuations will be reduced by a factor proportional to the reduction in the modulus squared of the frequency response function.

Before leaving this elementary (and very idealised) discussion of the feedback control of sound, it is worth noting that if we make K negative, and in particular less than $-(1/R_a)$, then the effective resistance to flow from the cavity R_a' will become negative, that is the piston will be moved to *reinforce* the cavity pressure fluctuation. Under these circumstances, the impulse response of the system can be written as

$$h(t) = \frac{1}{C_a} e^{+t/|R_a'|C_a}, \tag{7.2.10}$$

which is an exponentially increasing function which grows without bound. The system will be *unstable*, the feedback continuing to reinforce itself and the cavity pressure continuing to build up until some practical limit (such as the capability of the amplifier) prevents any further growth. Such an instability is also directly identified by the migration of the pole of the modified transfer function $H'(s)$ to the right hand half of the s-plane. The question of stability will be returned to using a rather different criterion in Section 7.4.

7.3 A general approach to the feedback control of acoustical systems

The real-life problem of using feedback to control an acoustic system is somewhat removed from the idealised model dealt with in the last section. Figure 7.7 illustrates the components that are combined (for example in the manner described by Olson and May, 1953 and Olson, 1956) in order to produce a reduction in the sound

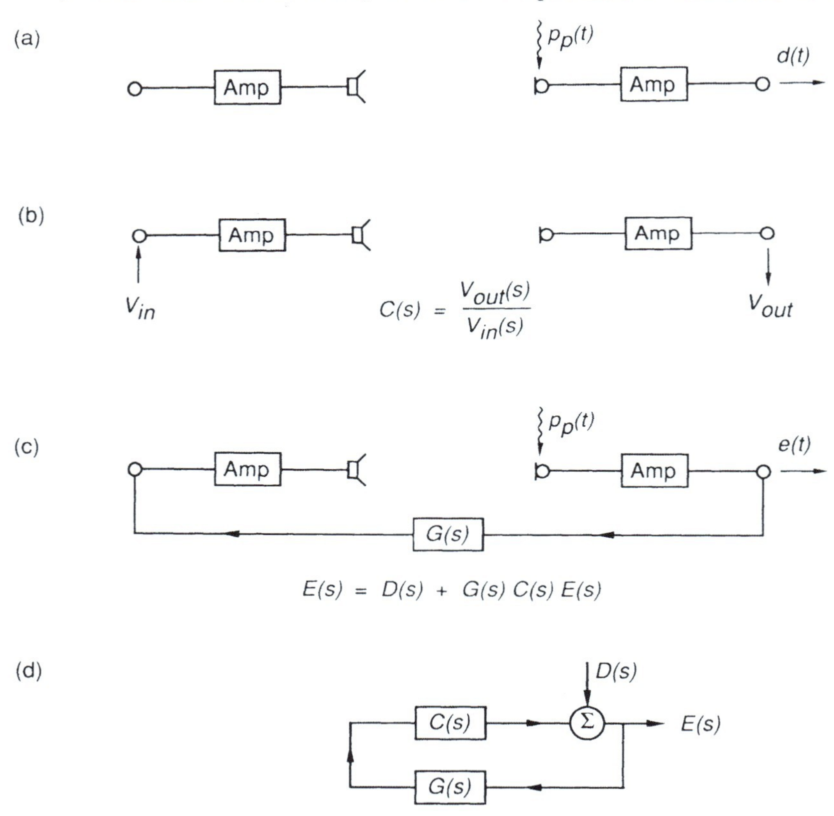

Fig. 7.7 A general approach to the feedback control of sound: (a) the primary field produces a microphone output signal $d(t)$; (b) we define the transfer function of the electroacoustic path between loudspeaker input and microphone output as $C(s)$; (c) the output signal is fed back via a transfer function $G(s)$ and the superposition principle used to define the total error signal produced; (d) the system is illustrated in block diagram form.

pressure fluctuations detected by the microphone. The signal from the microphone is amplified and then fed back to a transducer of some kind (a loudspeaker is shown) which in turn produces an acoustic signal at the microphone which is added to that produced by the primary field. We also have the potential to introduce a filter of some kind into this feedback loop in order to enhance its performance.

With reference to Fig. 7.7, we proceed by adopting the approach used in Chapter 6 and work entirely in terms of the *electrical* variables involved. The analysis that follows is similar in many respects to that presented by Carme (1987). As shown in Fig. 7.7(a), we assume that the primary excitation produces (an amplified) microphone output signal $d(t)$. In order to be consistent with the previous chapter, we now define the electroacoustic transfer function between the loudspeaker input (V_{in}) and the resulting microphone output (V_{out}) as $C(s)$ (see Fig. 7.7(b)). We now have the capacity, as illustrated in Fig. 7.7(c), to feed back the output signal via an electrical network having a transfer function $G(s)$. When feedback is applied in the presence of the primary excitation, the superposition principle (applied to the electrical variables) tells us that the net output signal $e(t)$ will result from the superposition of the primary error signal $d(t)$ with the signal that is produced as a direct result of the feedback. Thus, in terms of Laplace transforms we can write

$$E(s) = D(s) + G(s)C(s)E(s), \tag{7.3.1}$$

which can be expressed in terms of the block diagram illustrated in Fig. 7.7(d). Now note that this expression can be rearranged as

$$E(s)[1 - G(s)C(s)] = D(s), \tag{7.3.2}$$

which can in turn be written in the form

$$E(s) = D(s)\left[\frac{1}{1 - G(s)C(s)}\right]. \tag{7.3.3}$$

We can now consider the primary signal to be the input to a linear system having the transfer function $[1/(1 - G(s)C(s))]$. Therefore it follows from the discussion presented in Chapter 3 (Section 3.5) that if the primary signal is a random process having power spectral density $S_{dd}(\omega)$, then the power spectral density of the error signal in the presence of feedback will be given by

$$S_{ee}(\omega) = S_{dd}(\omega)\left[\frac{1}{|1 - G(j\omega)C(j\omega)|^2}\right]. \tag{7.3.4}$$

Thus in order to minimise, at a given frequency, the power spectral density of the error signal, we clearly have to maximise $|1 - G(j\omega)C(j\omega)|^2$ at that frequency. Assume for the moment that we can express the frequency response $G(j\omega)C(j\omega)$ in terms of its modulus $K(\omega)$ and phase $\phi(\omega)$ such that $G(j\omega)C(j\omega) = K(\omega)e^{-j\phi(\omega)}$. It then follows that

$$|1 - G(j\omega)C(j\omega)|^2 = 1 + K^2(\omega) - 2K(\omega)\cos\phi(\omega). \tag{7.3.5}$$

Thus, broadly speaking, given a value of $C(j\omega)$ we have to manipulate $G(j\omega)$ to ensure that the net gain $K(\omega)$ around the loop is maximised at each frequency and that the phase shift around the loop is in the region of 180°. Note, however, that if

$K(\omega)$ is large, then the value of $\phi(\omega)$ will not be critical (i.e. a value of $90° < \phi(\omega) < 270°$ will at least always ensure that the term $-2K(\omega)\cos\phi(\omega)$ in equation (7.3.5) will be positive).

At first sight, this would appear to be a simple problem to solve. However, we are constrained considerably in our choice of $G(j\omega)$. First, $G(j\omega)$ must be *realisable*, that is, it must have a causal impulse response – we are not at liberty to introduce a filter which produces its output prior to its input (the problem would indeed be simpler if we could!). Second, we have to ensure that the closed loop system that we produce is *stable*. (Note that we only need the *closed* loop to be stable; we could in principle have an unstable "open loop" transfer function $G(s)C(s)$.) We shall now deal with both these constraints, beginning with a discussion of the requirements for closed loop stability.

7.4 The Nyquist stability criterion

Before we proceed to deal formally with the stability of the closed loop system described above, some physical insight into the onset of instability can be gleaned by considering the loop depicted in Fig. 7.8(a). The initial argument presented here is not rigorous and only gives an indication of the conditions leading to instability. First note that we assume no primary excitation is present but we apply a sinusoidal signal to the terminal "A" shown in the figure. If the phase shift around the loop were 360° say, the signal appearing at terminal "B" would be perfectly in phase with the input signal. Now assume that, as illustrated in Fig. 7.8(b), the input signal is disconnected and that a switch connecting terminals A and B is closed. If the gain around the loop is less than unity, the signal in the loop will decay away. However, if the loop gain is greater than unity, the loop will be excited by its own feedback and the response of the loop will grow without bound; the loop will be unstable. In simple terms, if the

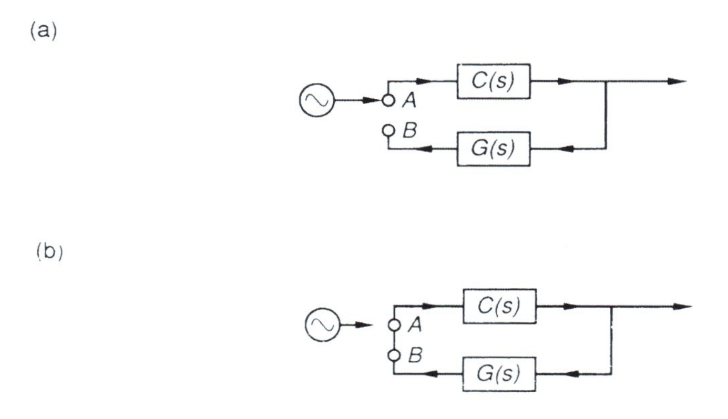

Fig. 7.8 Loop stability: (a) a sinusoidal signal is applied to terminal "A"; (b) the input is disconnected and terminals "A" and "B" are connected; if the phase shift around the loop is 360° and the gain exceeds unity, the system will be excited by its own feedback and will be unstable.

net phase shift $\phi(\omega)$ around the open loop (having the frequency response function $G(j\omega)C(j\omega)$) is 360°, and the magnitude (gain) $K(\omega)$ of this loop frequency response function is greater than unity, then the loop would appear to be unstable. This also appears to be the case if the phase shift is *any* integer multiple (including zero) of 360°.

Although this simple argument for determining the stability of a system holds true in many cases, there are systems which will be stable when the phase shift around the loop is 360° and the gain exceeds unity (see, for example, the discussion presented by Dorf, 1967, Ch.8, of "conditionally stable" systems). The problem of system stability has thus to be dealt with by using the more rigorous techniques of analysis associated with the development of the Nyquist stability criterion in classical control theory. Since there are some differences in approach between that adopted in dealing with classical feedback systems and the feedback problem considered here, it is worth outlining briefly the argument that leads to the Nyquist stability criterion. A fuller description can be found in the many texts on classical control theory; a particularly clear account is presented by Richards (1979, Ch.5).

The essence of the approach is to determine the stability of the *closed loop* system from the properties of the *open loop* transfer function. Thus the closed loop system has a transfer function defined by

$$\frac{D(s)}{E(s)} = \frac{1}{1 - G(s)C(s)} = \frac{1}{F(s)}. \tag{7.4.1}$$

If this function has any poles in the right half of the s-plane, the system will be unstable. Poles of the closed loop transfer function are *zeros* of the denominator $F(s) = 1 - G(s)C(s)$. The argument relies on certain properties that occur when "mapping" the function $F(s)$ from its s-plane representation to the "$F(s)$ plane" where we plot $\mathrm{Im}\{F(s)\}$ against $\mathrm{Re}\{F(s)\}$. In order for these properties to hold, the function $F(s)$ must be *analytic* at all but a finite number of points in the s-plane. This requires that the derivative (at a point $s = s_0$ say) defined by

$$\left.\frac{\mathrm{d}F(s)}{\mathrm{d}s}\right|_{s=s_0} = \lim_{s\to s_0}\left[\frac{F(s) - F(s_0)}{s - s_0}\right], \tag{7.4.2}$$

exists (i.e. the limit has a finite unique value). Points at which $F(s)$ is not analytic are the singularities of $F(s)$; the poles of $F(s)$ are such singular points. In addition, $F(s)$ must be *single valued* and every point on the s-plane must map to one and only one point in the $F(s)$ plane. Finally, a closed contour in the s-plane must map to a closed contour in the $F(s)$ plane.

Now consider a closed contour drawn in the s-plane in a clockwise direction, as illustrated in Fig. 7.9, where we are careful to draw the contour so that it does not pass through any of the singularities in the s-plane. Any points inside the contour are deemed to be *enclosed* by it. It can be argued (see, for example, Richards, 1979, Ch.5) that the corresponding contour in the $F(s)$ plane will encircle (in a clockwise direction) the *origin* of the $F(s)$ plane a number of times given by

$$N = Z - P, \tag{7.4.3}$$

where Z and P are respectively the number of zeros and poles of $F(s)$ enclosed by the contour in the s-plane.

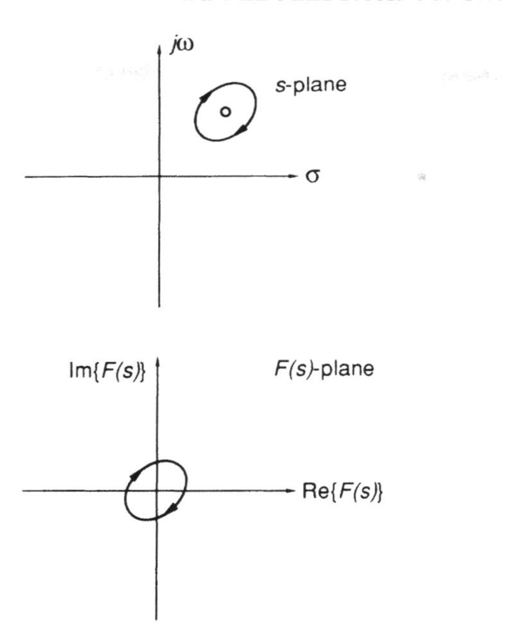

Fig. 7.9 A closed clockwise contour that encircles a zero in the s-plane will map to a closed clockwise contour that encircles the origin in the $F(s)$ plane.

Here, of course, we are concerned specifically with the number of zeros of $F(s)$ in the right half of the s-plane (since these will result in poles of the closed loop system). Thus we choose the contour (the "Nyquist contour") illustrated in Fig. 7.10 which encloses the entire right half of the s-plane. Note that the path is chosen to deviate around any poles of the function $F(s)$ (i.e. poles of $1 - G(s)C(s)$) that lie on the imaginary axis (where $s = j\omega$). Also note that the path along the $j\omega$ axis is exactly that needed to calculate the frequency response function of $F(s)$. This has been illustrated, for example in Fig. 3.4 of Chapter 3. The corresponding path in the $F(s)$ plane is the "Nyquist plot" of the system. If we firstly assume (as is mostly true in practice) that the open loop transfer function $G(s)C(s)$ is stable and has no poles in the right half of the s-plane, then $F(s)$ will have no poles in the right half of the s-plane. (Note that the open loop system $G(s)C(s)$ *must* be stable if we are to be able to plot its open loop frequency response function.) Under these circumstances, $P = 0$ in equation (7.4.3) and the number of zeros Z of $F(s)$ (i.e. poles of the closed loop transfer function) that lie in the right half of the s-plane is given by the number of times the origin is encircled by the Nyquist plot. For stability of the closed loop system we should have *no* zeros of $F(s)$ in the right half of the s-plane. Thus we can state that given a stable open loop system, then the closed loop will be stable, *provided the Nyquist plot of F(s) does not encircle the origin*.

Now we can translate this into the condition that must be met by the Nyquist plot of the open loop transfer function $G(s)C(s)$. Since $G(s)C(s) = 1 - F(s)$ when $F(s) = 0$, then $G(s)C(s) = 1$. Thus, enclosure of the origin by the Nyquist plot of $F(s)$ translates to enclosure of the point $(1, j0)$ by the Nyquist plot of $G(s)C(s)$.

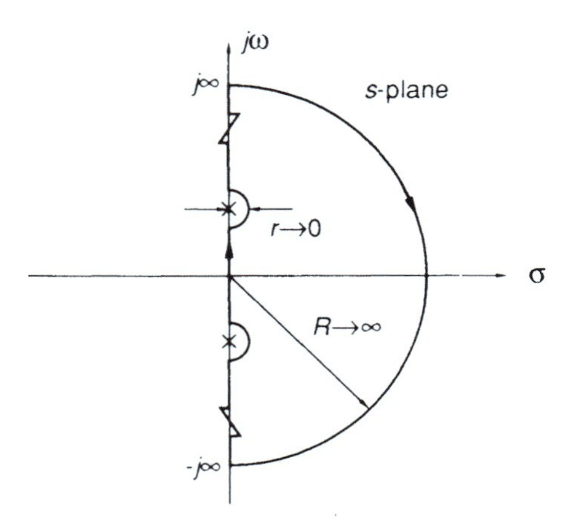

Fig. 7.10 The Nyquist contour. The entire right half of the s-plane is enclosed by assuming the radius "R" tends to infinity in the limit. The contour is drawn to avoid any poles on the $j\omega$ axis by constructing small semicircles whose radius "r" tends to zero in the limit.

Thus we can finally state that, *provided the open loop system $H(s)C(s)$ is stable, the closed loop system having transfer function $[1 - G(s)C(s)]^{-1}$ will be stable provided the Nyquist plot of $G(s)C(s)$ does not encircle the point (1,j0)*. The generalisation of this criterion to deal with *unstable* open loop systems $G(s)C(s)$ can be deduced by following the reasoning presented by, for example, Richards (1979, Ch.5).

We should also point out here that the above reasoning is, in the context of classical control theory, most usually applied to a *negative* feedback system which has a closed loop transfer function whose denominator is of the form $F(s) = 1 + G(s)C(s)$. In this case the Nyquist plot of $G(s)C(s)$ must not encircle the $(-1, j0)$ point if the closed loop is to be stable.

Now consider a simple example of a Nyquist plot illustrated in Fig. 7.11. For a clear description of the techniques used to construct Nyquist plots, the reader is referred to (for example) DiStefano *et al.* (1967). Since $G(s)C(s)$ is a linear time invariant system, the portion of the Nyquist plot corresponding to $s = 0 \to -j\omega\infty$ will be the mirror image in the real axis ($\text{Im}\{G(s)C(s)\} = 0$) of the portion of the plot corresponding to $s = 0 \to +j\omega\infty$. Also, as $s = j\omega \to \pm\infty$, then $G(s)C(s) \to 0$ and the limits of the Nyquist contour corresponding to $j\omega = \pm\infty$ will map to the origin. In view of these properties, in general one only needs to construct the Nyquist plot for the values of s on the Nyquist contour corresponding to $j\omega = 0 \to \infty$. This amounts therefore to evaluating the open loop frequency response function of the system. (There are systems, however, for which it is wise to plot the entire Nyquist plot if the stability properties are to be made clear; see the examples presented by Dorf, 1967, Ch.8.)

7.5 Bode plots and relative stability

A plot is shown in Fig. 7.12 of the modulus and phase of the open loop frequency response function $G(j\omega)C(j\omega)$ corresponding to the Nyquist plot of the system

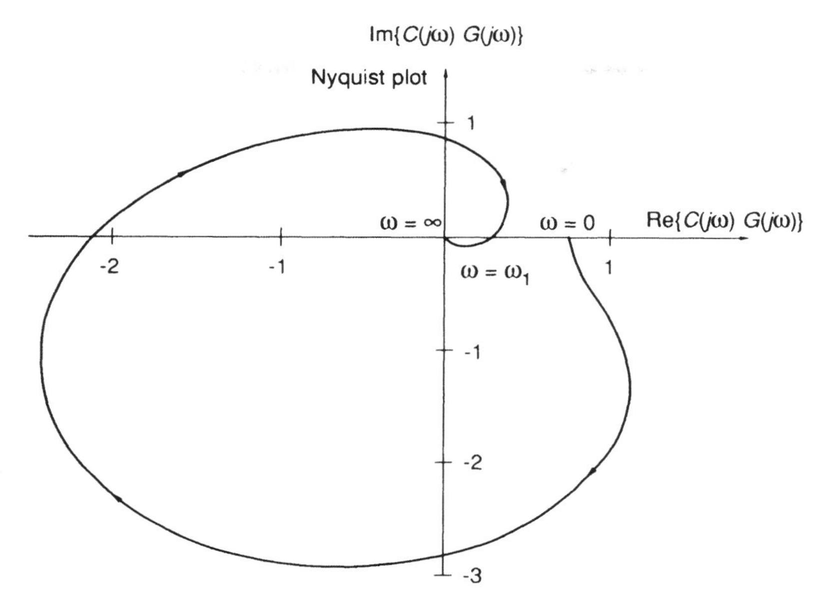

Fig. 7.11 A simple example of a Nyquist plot corresponding to values of ω from zero to $+\infty$. Note that the point (1,*j*0) is not enclosed and therefore the system would be stable when the loop is closed.

illustrated in Fig. 7.11. This frequency response plot is known as a *Bode plot* in the parlance of classical control systems theory, although such plots are usually presented with a logarithmic frequency axis. The proximity of the Nyquist plot to the (1, *j*0) point effectively provides a measure of the *relative* stability of the system. In particular, at the frequencies (ω_i, say) at which the Nyquist plot intersects the positive part of the real axis, and thus the phase shift is an integer multiple of 360° we can define the *gain margin* as

$$G_i = 20 \log_{10} \frac{1}{|G(j\omega)C(j\omega)|_{\omega=\omega_i}}. \tag{7.5.1}$$

This quantifies the level (in decibels) by which the loop gain can be increased before the Nyquist plot encloses the (1, *j*0) point and the loop becomes unstable. This is illustrated on the Bode plot sketched in Fig. 7.12. In this case the first crossing of the real axis occurs at $\omega = \omega_0 = 0$ rad and there is a second at $\omega = \omega_1$ rad. The corresponding gain margins are defined as G_0 and G_1 in Fig. 7.12.

A further, interrelated, measure of relative stability is provided by the *phase margin*. Again, with reference to the Nyquist plot sketched in Fig. 7.11, the phase margin (in this context) represents the amount by which the open loop phase lag may be decreased before the loop becomes unstable. The phase margin is given by the angle ϕ_i defined by the intersection of the Nyquist plot with a circular contour defining an open loop gain of unity. The equivalent representation is illustrated in the Bode plot of Fig. 7.12.

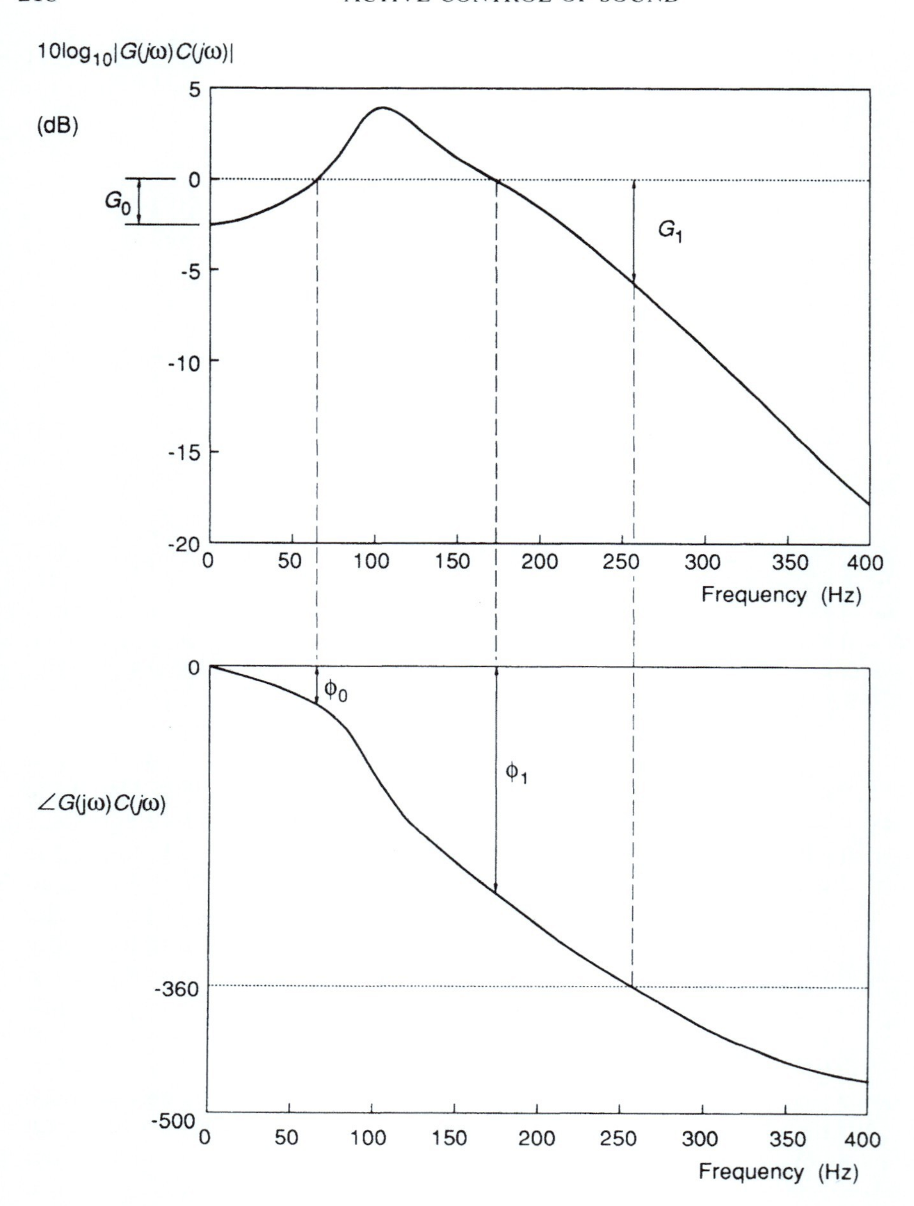

Fig. 7.12 The Bode plot, or frequency response function $G(j\omega)C(j\omega)$, corresponding to the Nyquist plot of Fig. 7.11. The values of G_i and ϕ_i represent the gain and phase margins, respectively.

Finally, it should be emphasised that the gain and phase margins are fairly crude single measures of relative stability and the *detailed form* of the Nyquist plot in the region of the $(1, j0)$ point will determine the relative stability of a given system. In addition, there are no hard and fast rules regarding an acceptable level of gain and phase margin which will be tolerant to changes in the system to be controlled, although as a rough guide, again in the context of classical control, Richards (1979, Ch.5) suggests that a phase margin of between 30° and 60° and a gain margin greater than 6 dB will usually give a sufficient margin for error in order to account for unmodelled changes in the dynamics of the system to be controlled.

7.6 The design of realisable compensation filters

Having established the criteria for both good performance and stability of a feedback control system, we are now in a position to make some general comments on achieving these objectives in practice. The typical form of an electroacoustic frequency response function $C(j\omega)$ in terms of its modulus and phase is, in general, reminiscent of the function depicted in Fig. 7.12. Recall that this frequency response function includes the effects of the loudspeaker, acoustic path and microphone. Note that the phase shift is a steadily falling function of frequency. This will occur as a direct result of the fact that, broadly speaking, in any physical system, there must be a delay between the input to the system and its output. It is clearly advantageous to minimise the magnitude of this delay, and thus the rate of change of phase, in the electroacoustic frequency response function. We have seen that the closed loop will be unstable if the gain exceeds unity when the phase shift reaches −360°. Therefore if, by judicious design of the components in the system, this rate of change of phase can be minimised, and the −360° phase crossing can be shifted to a higher frequency, the system can be made to operate over a potentially broader bandwidth.

However, having done the best that one can to minimise the phase shifts through the system, one is then faced with ensuring that the open loop gain is less than unity at the −360° phase shift (and indeed at 0° phase shift, −720° phase shift, etc.). Olson and May (1953) accomplished this by rolling off their amplifier response above about 400 Hz as described in the introduction to this chapter. Most modern applications of the technique have used a further filter network $G(s)$ to ensure stability. However, we should note that $G(s)$ cannot be chosen to have arbitrary amplitude and phase characteristics in order to stabilise the loop. In particular, one cannot choose an amplitude characteristic which is completely independent of its phase characteristic; both are dictated by a single complex function which is the transfer function $G(s)$ of the system. In addition, we are constrained in the choice of $G(s)$ by the necessity for the filter to be realisable; it must have a causal impulse response.

The problem of causality is overcome provided we restrict our attention to filters having the general transfer function of the form

$$G(s) = \frac{K(s - z_1)(s - z_2)(s - z_3)\ \ldots}{(s - p_1)(s - p_2)(s - p_3)}, \tag{7.6.1}$$

where the poles p_j are in the left half of the s-plane only (see the discussion presented by Kuo, 1966). This of course also implies the stability of $G(s)$, although strictly speaking stability of $G(s)$ is not a necessity in the current context. We are thus at liberty to locate the poles and zeros of $G(s)$ arbitrarily provided we restrict the location of the poles to the left half of the s-plane.

The influence of the location of the poles and zeros on the frequency response function can be deduced using the graphical construction described in Section 3.4 of Chapter 3. It is shown there that the frequency response function can be written as

$$G(j\omega) = \frac{K(N_1N_2N_3\ldots)}{(M_1M_2M_3\ldots)} e^{j(\psi_1+\psi_2+\psi_3\ldots-\theta_1-\theta_2-\theta_3\ldots)}, \tag{7.6.2}$$

where the factors N_i, ψ_i and M_j, θ_j are defined geometrically in Fig. 3.9. Thus the amplitude of the frequency response function at a given frequency will be dictated by the *distance* of the poles and zeros from the point on the $j\omega$ axis corresponding to the frequency ω. The closer are the poles and the further are the zeros, the greater will be the magnitude. The phase of the frequency response function is dictated by the sum of the phase leads ψ_i introduced by the zeros and the lags θ_j introduced by the poles. The larger the number of poles, the larger the net phase lag. However, this lag can be counteracted by the presence of the zeros. It is therefore no accident that in most compensation filters designed to date, as we shall see in the next section, an equal number of poles and zeros have been used so that as $\omega \to \infty$, the net phase shift tends to zero. Furthermore, if the zeros (in addition to the poles) are located exclusively in the left half of the s-plane, then the filter will be *minimum phase* and the net phase shift $[(\psi_1 + \psi_2 + \psi_3 \ldots) - (\theta_1 + \theta_2 + \theta_3 \ldots)]$ will be minimised for a given magnitude response (see Chapter 3).

7.7 Feedback control of the sound field in an ear defender

One of the applications for feedback control suggested by Olson (1956), which has since proved technically feasible, is the use of a feedback system to control the sound field in an ear defender, or headset. In this application, a system of the type illustrated in Fig. 7.13 can be used to reduce the pressure fluctuations in the cavity close to a listener's ear. The two most detailed studies of this problem that have been published to date are recorded in the doctoral theses of Wheeler (1986) and Carme (1987). Wheeler, for example, used a moving coil loudspeaker (referred to as a "telephone" in this context) and a closely spaced microphone to minimise the phase shift introduced into the open loop transfer function as a result of the "acoustic travel time" between telephone and microphone. A similar strategy was adopted by Carme. A typical plot of the electroacoustic frequency response function $C(j\omega)$ measured in Carme's apparatus is illustrated in Fig. 7.14. The Nyquist plot of this transfer function is also shown. The phase of the frequency response function has a steadily falling characteristic which one is at first tempted to associate with the acoustic travel time between telephone and microphone. Recall that a pure delay of duration τ has a frequency response function given by $e^{-j\omega\tau}$; this has unit modulus and a phase shift given by $-\omega\tau = -2\pi fd/c_0$ (or $-360fd/c_0$ degrees), where d is the distance travelled by the sound wave. Thus if d were, say, 10 mm, the phase would fall linearly from zero at 0 Hz to around $-10°$ at 1 kHz (for $c_0 \approx 344$ m s^{-1}). The measured phase shift in Carme's prototype arrangement falls from around $-120°$ at 20 Hz to approximately $-260°$ at 1 kHz and $-360°$ at 1.5 kHz.

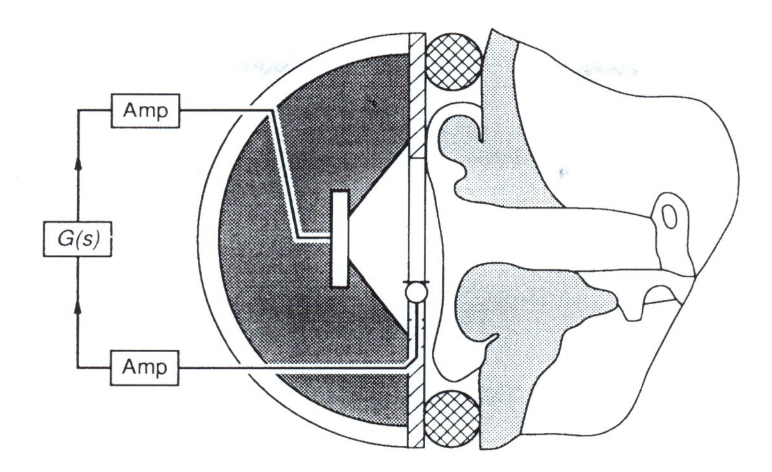

Fig. 7.13 Feedback control of the sound field at the entrance to the ear canal (after Carme, 1987).

It is thus obvious (as was clearly recognised by both Wheeler and Carme) that the phase shift in the electroacoustic response has little to do with the very small acoustic travel time. The phase shift is controlled by the driving (telephone) transducer and its interaction with the acoustic loads provided by the front (auditory) and rear cavities into which the transducer drives. Carme (1987) presents a comprehensive analysis of this interaction and in the design of the prototype apparatus sketched in Fig. 7.13, every effort was made to optimise the design and construction of the ear defender in order to minimise the rate of change of phase over the frequency range shown. Another useful account of the ear defender transfer function is presented by Salloway and Twiney (1985).

However, as the Nyquist plot of the electroacoustic transfer function $C(s)$ illustrates, if the loop is closed (by feeding the microphone output back to the telephone input) then the system is certain to be unstable unless the overall gain of the open loop system is significantly reduced. This is evident from the frequency response function since the modulus of the transfer function exceeds unity when the phase shift reaches $-360°$. The solution to this problem that was devised by Wheeler (1986) was to introduce a further compensation network into the feedback loop which had a transfer function of the general form

$$G(s) = \frac{K(s - z)}{(s - p)}, \tag{7.7.1}$$

where z and p are both real. As we have seen in the last section, provided the zero (z) of this transfer function is placed in the left half plane, this transfer function can be made to be *minimum phase*. This will minimise the contribution of $G(s)$ to the total loop phase shift whilst providing the requisite reduction in gain at higher frequencies (where the phase shift reaches 360°). A plot of the compensating frequency response function of the type used by Wheeler (1986) is shown in Fig. 7.15 together with a sketch of its pole-zero map. Note that for a transfer function of

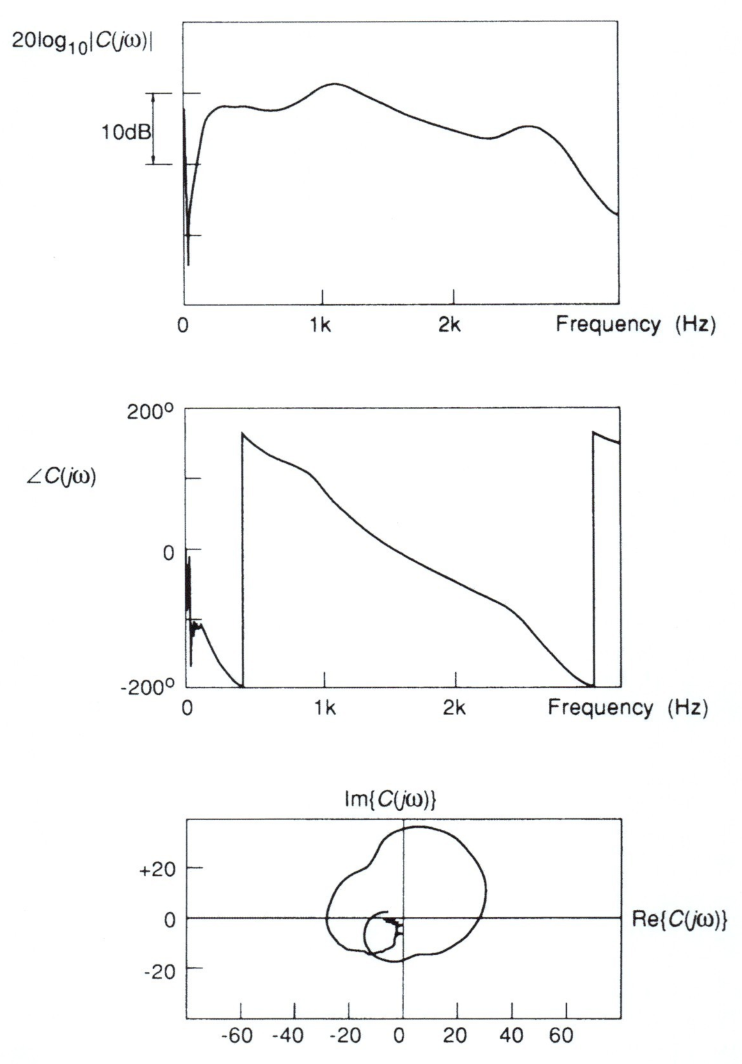

Fig. 7.14 Measurements of the uncompensated electroacoustic frequency response function made by Carme (1987) on a prototype active ear defender.

this type having an equal number of poles and zeros, the phase shift will tend to zero as the frequency $\omega \to \infty$. This is clearly a desirable feature of the compensation filter. A phase characteristic which is firstly *minimum* phase and secondly recovers to zero in the high frequency limit will minimise the addition of further negative phase shift which will cause the phase of $G(j\omega)C(j\omega)$ to pass through $-360°$ at a lower frequency. The means of providing an electrical realisation of this compensating filter are very simple (Wheeler, 1986). The positions of the pole and zero of the filter can be chosen by appropriate selection of the values of the two resistors and the capacitor associated with a simple electrical realisation of this "transient lag network".

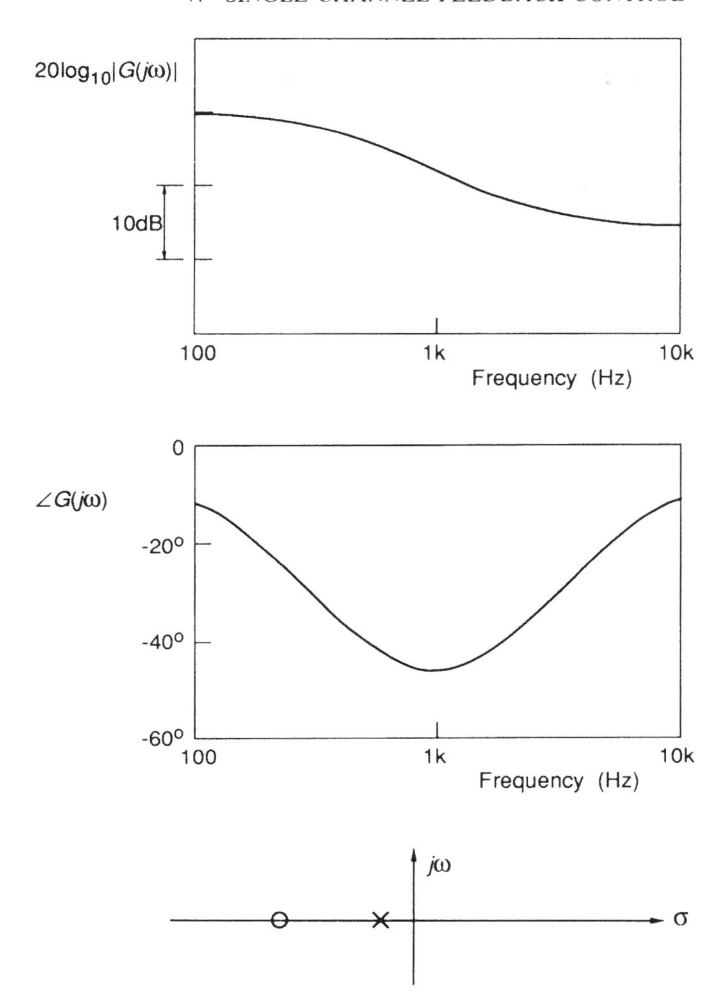

Fig. 7.15 An example of the type of frequency response function of the first order compensating filter used by Wheeler (1986) and a sketch of its pole-zero map.

After a comprehensive study of various compensation filter designs, Carme (1987) concluded that the most appropriate filter for practical purposes had a transfer function of the form

$$G(s) = \frac{K(s - z)(s - z^*)}{(s - p)(s - p^*)}, \tag{7.7.2}$$

where the two poles and two zeros of this transfer function are in complex conjugate pairs. Carme adopted the standard "bi-quad" realisation of this transfer function, where again the pole and zero locations are determined by the appropriate selection of simple electrical components. It is interesting to note that precisely this realisation of compensating filter was also adopted by Saxon (1986) in an attempt to patent a range of pole-zero locations appropriate to the use of feedback in controlling the

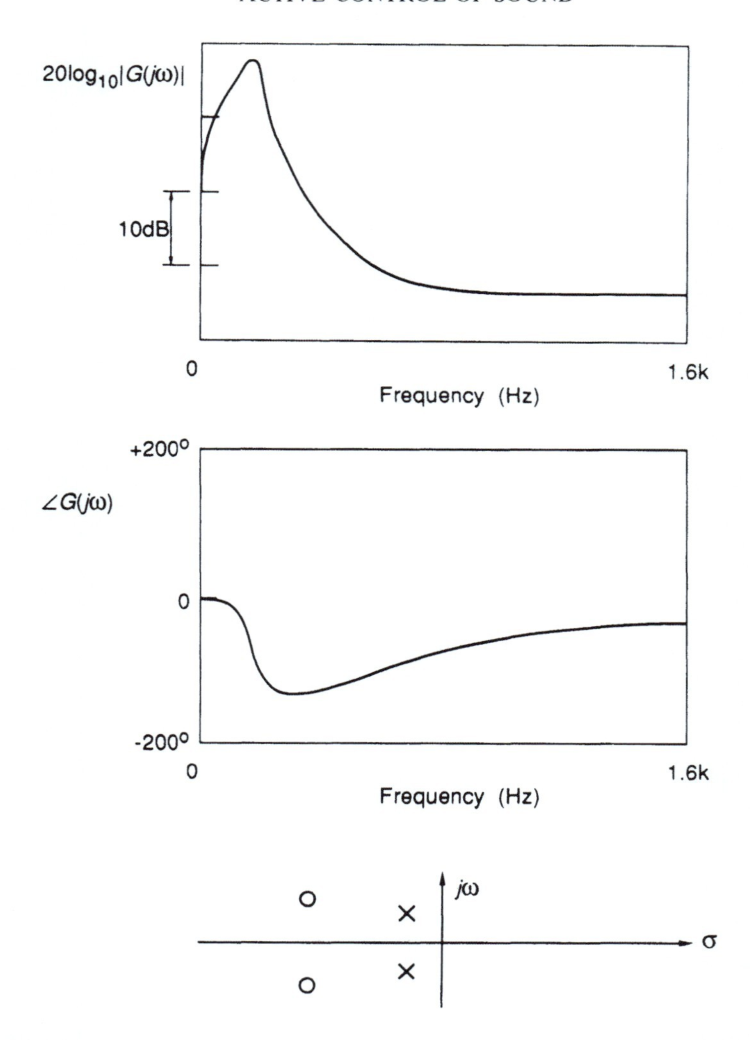

Fig. 7.16 An example of the frequency response function of the second order compensating filter used by Carme (1987) and a sketch of its pole-zero map.

sound field at the entrance to the ear canal. The frequency response function and a sketch of the pole-zero map of the filter used by Carme are illustrated in Fig. 7.16. The recovery of the phase to zero in the high frequency limit is again evident, but in this case the second order nature of the filter enables the gain to be made particularly high in a given frequency range. Thus the open loop gain can be made high in a given bandwidth whilst not producing excessive phase shift outside this bandwidth. These are precisely the requirements for producing efficient stable operation of the closed loop system. Figure 7.17 shows the open loop frequency response function measured by Carme with the addition of a filter $G(s)$ of the type specified by equation (7.7.2). Note that the Nyquist plot has now been modified to ensure that a high value of loop gain can be used without producing encirclement of

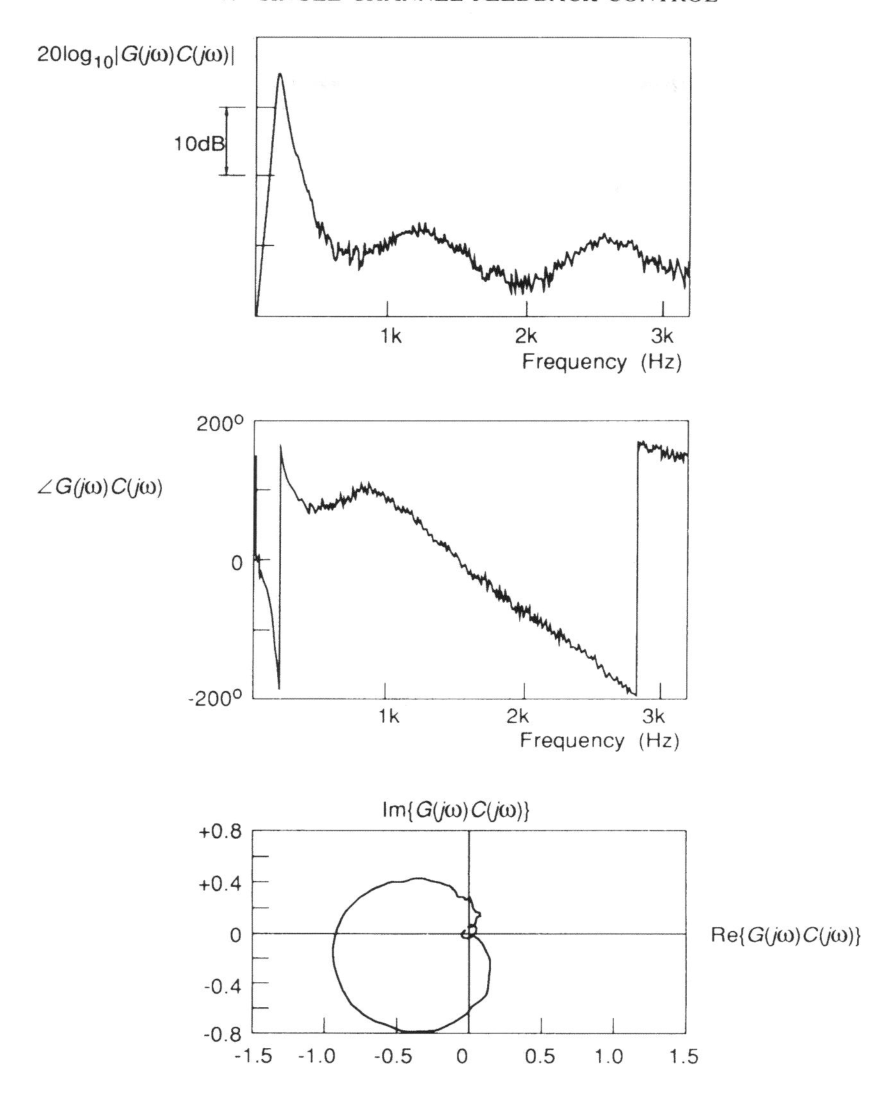

Fig. 7.17 The result of adding a second order compensation filter to the electroacoustic response shown in Fig. 7.14 (after Carme, 1987).

the (1, $j0$) point and that the loop gain is thus very high in the frequency range of approximately 50 Hz to 400 Hz. The result of closing the feedback loop is illustrated in Fig. 7.18 which shows the considerable reduction in the power spectral density of the broadband noise generated by a source external to the ear defender. These impressive results demonstrate clearly the technical feasibility of the technique, and its practical application is currently the subject of considerable attention. This has as much to do with the design of systems to deal with factors such as inter- and intra-subject variability of the electroacoustic response $C(s)$ associated with a range of different wearers of such devices. A useful discussion of these aspects is presented

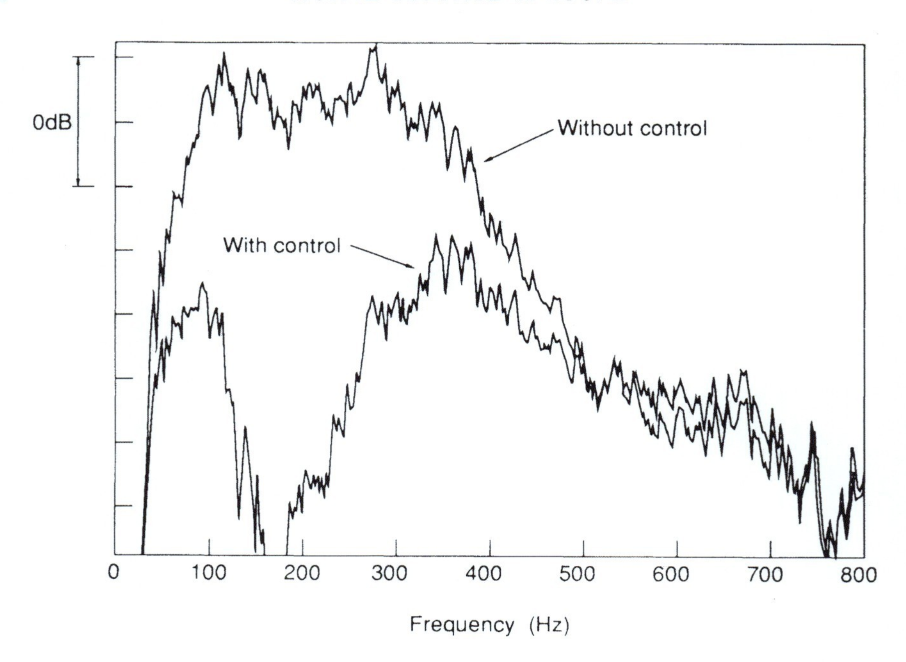

Fig. 7.18 The reduction in the power spectrum of broadband noise produced at the entrance to the ear canal by the closed loop system devised by Carme (1987).

by Wheeler (1986), who also describes a technique for introducing an external communications signal into the loop to be transmitted via the telephone. Suitable pre-emphasis of this signal prevents its reduction by the action of the feedback loop (see Wheeler, 1986 for full details).

7.8 Feedback control of duct-borne sound

One of the applications of feedback control recognised by Olson and May (1953) was the reduction of sound radiated from the outlet of an air conditioning duct or an automobile exhaust system. As we have seen in Chapter 5, the use of a secondary source to produce a condition of zero sound pressure in the region of the secondary source is an effective means of reflecting plane waves propagating downstream in a duct. Olson and May's suggestion for this use of their device appears to have been largely ignored until the early 1980s when several authors re-examined the possibilities offered by this approach. In particular, Leventhall and his co-workers published a series of papers investigating the properties of the "tight coupled monopole" in a duct (Eghtesadi and Leventhall, 1981; Hong *et al.*, 1982; Eghtesadi *et al.*, 1983) and Trinder and Nelson (1983) examined in particular the interaction between the feedback controller and the acoustical properties of the duct in which it is installed.

In the work presented by Eghtesadi *et al.* (1983), the arrangement depicted in Fig. 7.19 was used, with the test duct having dimensions of 450 mm wide, 600 mm high

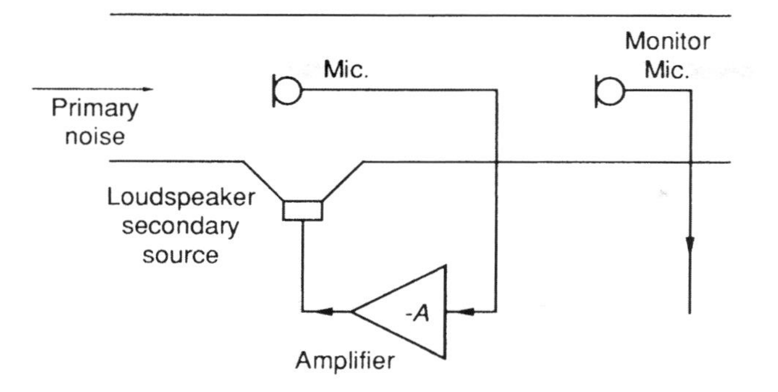

Fig. 7.19 The "tight coupled monopole" in a duct (after Eghtesadi *et al.*, 1983).

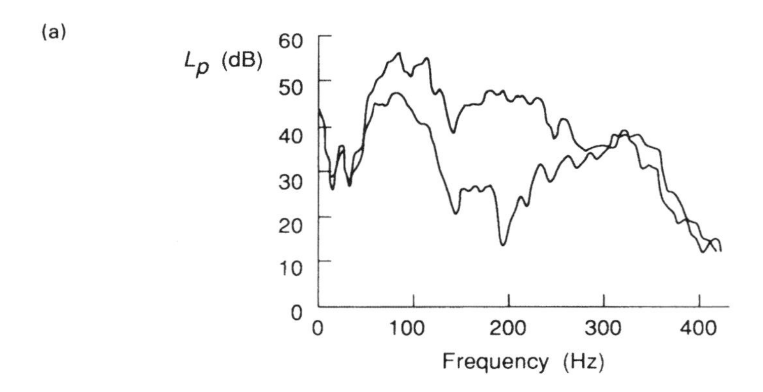

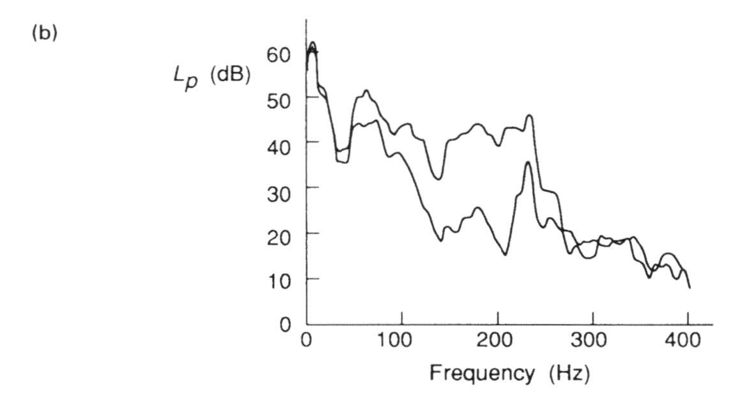

Fig. 7.20 The power spectrum of the sound pressure level measured downstream of the feedback system illustrated in Fig. 7.19 when (a) the primary source was a loudspeaker radiating 500 Hz bandwidth random noise and (b) the primary source was a 2.2 kW axial flow fan producing a flow velocity in the region of 5 m s^{-1}.

and 11.1 m long. Figure 7.20 shows the results presented for the reduction in sound pressure level measured in an absorbently lined termination to the duct. The primary source was firstly a loudspeaker generating random noise over a 500 Hz bandwidth and secondly a 2.2 kW axial fan which produced an airflow in the duct having a velocity in the region of 5 m s^{-1}. The results show convincing reductions in the level of sound propagating downstream. However, no details are presented of the open loop frequency response function and no compensating filter appears to have been used in order to stabilise the system.

In the work presented by Trinder and Nelson (1983) a smaller test duct was used which had a square cross-section of side 230 mm and a length of 2.44 m. The secondary source loudspeaker was mounted on one wall of the duct and a detailed examination of its near field was undertaken. Below the cut-on frequency of the first transverse duct mode, the near field is comprised of a series of exponentially decaying higher order modes. Both the measurements undertaken and computations based on the theory presented by Doak (1973b) showed that the near field contribution to the detected pressure could be minimised by locating the microphone

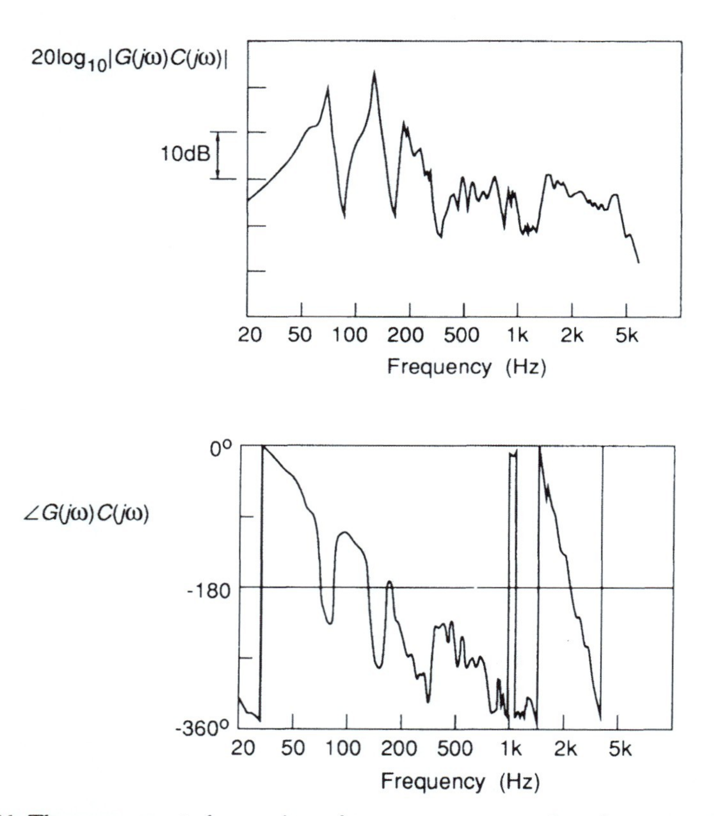

Fig. 7.21 The compensated open loop frequency response function measured by Trinder and Nelson (1983) for a feedback system in an open-ended test duct of cross section 230 mm^3 and 2.44 m in length. (Note that the phase results have been replotted to be consistent with the convention used here.)

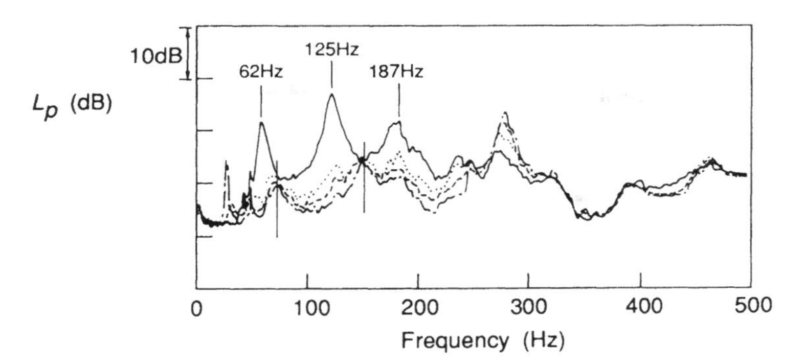

Fig. 7.22 The performance of the feedback controller used by Trinder and Nelson (1983). The results show the power spectrum of the sound pressure level in the centre of the cross-section of the duct exit downstream of the cancelling loudspeaker. ——— Primary source only; ········, --------, —·—·—, with secondary cancelling source, in order of increasing gain.

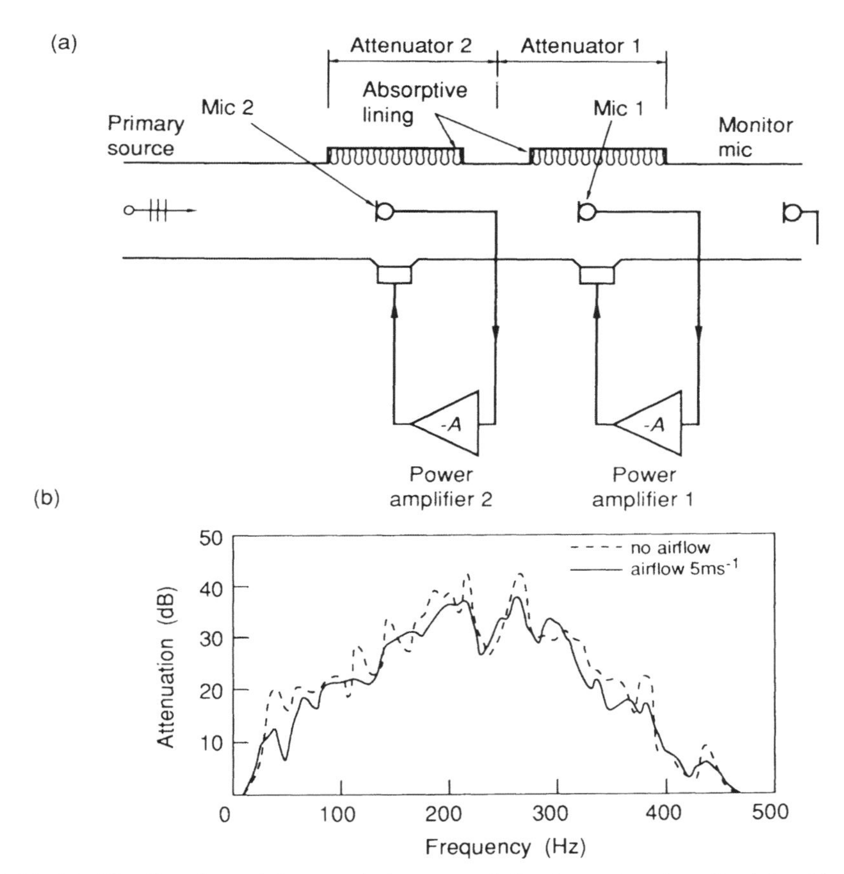

Fig. 7.23 (a) The "tandem tight coupled monopole" attenuator described by Hong *et al.* (1987). (b) The performance of the attenuator as measured by the downstream microphone with a 2.2 kW axial fan source. The system was implemented using "low cost electronics" (see Hong *et al.*, 1987).

on the axis of the loudspeaker at a distance (from the duct wall in which the loudspeaker was mounted) of about 0.4 times the duct width. The contribution of the higher order modes at frequencies above cut-on was minimised by placing absorbent material on the duct wall opposite the loudspeaker. It was also necessary to stabilise the loop by using a low-pass compensating filter which had the effect of further minimising the contribution of higher order modes above cut-on. The compensated open loop frequency response function is illustrated in Fig. 7.21 and the results of applying the control system are shown in Fig. 7.22. It can be seen that the system is particularly effective at the frequencies of the longitudinal resonances of the duct. At these frequencies the open loop gain of the system is high and as demonstrated in Section 7.3, a high open loop gain leads to a good performance of the system. Trinder and Nelson also demonstrated that the introduction of a condition of zero pressure by the action of the controller effectively introduced new resonances in the duct in a manner analogous to that described in Chapter 5. The new resonance frequencies of the duct are evident in the residual sound pressure level detected when the controller was operating (see Fig. 7.22).

Finally, it is worth noting that Hong *et al.* (1982) devised a successful control system comprised of two (electrically) independent feedback systems in cascade in a duct as illustrated in Fig. 7.23. The comprehensive analysis of this system presented by Hong *et al.* (1987) suggests that there is an interaction between the two systems such that the attenuation of downstream propagating plane waves provided by the system is not equal to the sum of the attenuations of the two devices when operating individually. Nevertheless, the additional system does provide worthwhile improvements and an example of the attenuation that can be produced by such a "tandem tight coupled monopole" system is shown in Fig. 7.23. These results were obtained in the duct of cross-section 450 mm $\times$ 600 mm described above, and again the influence of an airflow of 5 m s^{-1} from a fan source was found to be minimal.

8
Point Sources and the Active Suppression of Free Field Radiation

8.1 Introduction

In this chapter we will begin to address the problem of the active control of free field acoustic radiation. Our starting point is an investigation of the nature of the sound field produced when a simple "secondary" point monopole source is introduced in order to control the radiation of an existing "primary" point monopole source, both sources radiating sound at the same single frequency. The form of the resulting sound field has long been understood and in the particular case of the two sources being of equal and opposite strength and separated by a distance that is small compared to the wavelength of the sound radiated, the sound field produced reduces to the well known dipole field. It is well known that to produce substantial reductions of the sound field radiated by a simple source, we have only to introduce in close proximity a simple source of equal and opposite strength. However, a closer examination of the problem reveals that this strategy is not necessarily the best approach, and that to minimise the total sound power radiated by the source combination a subtle variation of the secondary source strength is necessary. The recognition of an *optimal* strategy with regard to the control of single frequency fields stems from the application of the least squares approach first introduced in Chapter 2 and used in Chapter 5 to derive optimal results in a series of one-dimensional acoustical problems. Many of the same physical processes observed in the one-dimensional case are also evident in the equivalent three-dimensional problems. For example, we shall again see the capacity of a secondary source both to absorb radiation from a primary source and also to prevent radiation escaping from a primary source. The latter mechanism of "loading" occurs through the secondary source producing the appropriate pressure on the primary source at the appropriate time. Since there is a finite time taken for sound to propagate from the secondary source to the primary source, the secondary source has to *anticipate* the output of the primary source. As emphasised in Chapter 5, analyses conducted in the frequency domain often reveal such "non-causal" action of the secondary source when the solutions are transformed into the time domain.

An introductory treatment is also given of the calculation of optimal results when the secondary source is constrained to act causally with respect to the primary source. The case examined is of a point primary source whose source strength fluctuation is a stationary random process. A secondary source is introduced in order to control the field. At field points closer to the secondary source than the primary source, the sound pressure fluctuation can be cancelled perfectly through a delayed action of the secondary source. For field points further from the primary source than the secondary source, the secondary source must produce an *optimal prediction* of the primary sound pressure fluctuation in order to minimise the time-averaged value of the pressure fluctuations. This analysis is conducted in the time domain by using the Wiener–Hopf techniques introduced in Chapter 3. It is demonstrated that for stationary random fluctuations in the primary field, the best achievable performance of an active controller will often depend on the statistical properties (and thus predictability) of the primary pressure fluctuation.

The remainder of the work presented in this chapter is conducted in the frequency domain and is thus directly applicable to single frequency (or periodic) primary field fluctuations. Some attention is given to "classical" higher order source distributions of dipole and quadrupole type and the general technique of multipole analysis is introduced. The implications of these classical higher order source distributions for the active control of sound are discussed, but again some effort is made to point out that classical distributions of this type may not be the best source distributions to use in order to produce the optimal global performance of an active control system.

8.2 Interference between the far fields of two point monopole sources

In Chapter 1 we introduced the concept of a point monopole source by observing the behaviour of the sound radiated by a pulsating sphere. We saw that the complex pressure radiated by a point source could be expressed as

$$p(r) = \frac{j\omega\rho_0 q e^{-jkr}}{4\pi r}, \tag{8.2.1}$$

where q is the complex "source strength". Recall that this source strength has the dimensions of volume velocity. In the same way as we have defined the strength associated with one-dimensional sources in Chapter 5, the source strength is given by the product of the velocity of the surface of the pulsating sphere and its surface area. This product is held constant in the limit of the radius of the sphere being reduced to zero. Note that some authors (e.g. Lighthill, 1978) use the *time derivative* of the volume velocity ($j\omega q$) to define the "strength" of a simple source since it is this fluctuation which in three dimensions is directly proportional to the resulting pressure fluctuation. Here, however, we will adopt the definition that is consistent with Chapter 5 and as is used by other contemporary authors (e.g. Pierce, 1981; Morse and Ingard, 1968).

It is a simple matter to apply the principle of linear superposition to the sound fields generated by a primary point monopole source having complex strength q_p and a secondary point monopole source having complex strength q_s. The complex pressure produced by the interference of these two sound fields is given by

$$p(r,\theta) = \frac{j\omega\rho_0 q_p e^{-jkr_p}}{4\pi r_p} + \frac{j\omega\rho_0 q_s e^{-jkr_s}}{4\pi r_s}, \tag{8.2.2}$$

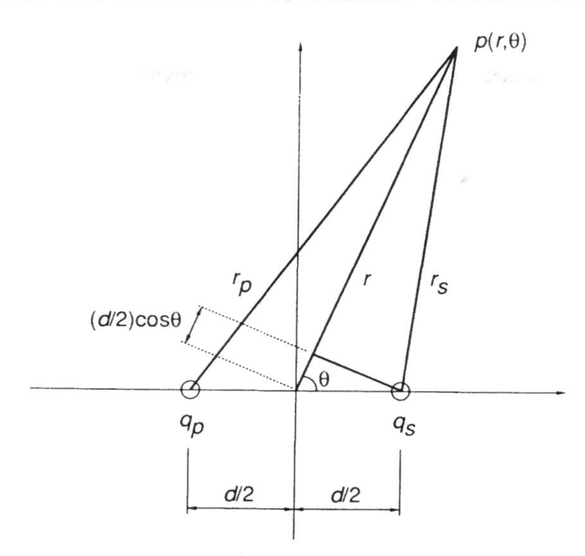

Fig. 8.1 Point primary and secondary sources separated by a distance d.

where r_p and r_s are the radial distances from the source point to the observation point from the primary and secondary sources, respectively, and θ is the angle defined in Fig. 8.1.

If we consider the arrangement of sources depicted in Fig. 8.1 and restrict our attention to observation positions in the *far field*, where the radial distance r is much greater than both the wavelength of the sound radiated and the distance d separating the two sources, then we can make some useful approximations for the distances r_p and r_s. As illustrated in Fig. 8.1, we can write $r_p \approx r + (d/2) \cos \theta$ and $r_s \approx r - (d/2) \cos \theta$. These approximations can be made for the values of r_p and r_s appearing in the complex exponential terms in equation (8.2.2). These terms govern the relative phases of the pressure fluctuations produced by the two sources. The values of r_p and r_s appearing in the denominator of the terms in equation (8.2.2) affect only the amplitude of the pressure in the far field, and as $r \to \infty$, in the denominator we can make the further approximation $r \approx r_p \approx r_s$. By using these well known far field approximations we can easily derive the secondary source strength which ensures that the pressure at some distance r and some angular position θ_0 is driven to zero, i.e. such that

$$p(r,\theta_0) = 0. \tag{8.2.3}$$

Making the far field approximations in equation (8.2.2) together with equation (8.2.3) leads to

$$q_s e^{-jk[r - (d/2) \cos \theta_0]} = -q_p e^{-jk[r + (d/2) \cos \theta_0]}, \tag{8.2.4}$$

which can be written as

$$q_s = -q_p e^{-jkd \cos \theta_0}. \tag{8.2.5}$$

This relationship specifies the complex secondary source strength that ensures that zero pressure is produced in the far field at a particular angular position specified by θ_0. Note that the requisite secondary source strength has a modulus that is equal to that of the primary source and a phase that differs from the primary source by $kd \cos \theta_0$. This result was pointed out by Conover and Ringlee (1955). Also note that in the particular case of $\theta_0 = 90°$, equation (8.2.5) reduces to simply $q_s = -q_p$. Such a combination of source strength has a particular significance when the wavelength of the sound radiated by the sources is much larger than the separation distance between them. In these circumstances the sound field reduces to the well known *dipole* field where zero sound pressure is produced at $\theta = 90°$ and $\theta = 270°$. This source arrangement will be described in more detail later in this chapter.

It is clearly possible to produce zero pressure in the far field of the two sources at a particular location, but of course it is possible that this destructive interference at one particular location may be accompanied by constructive interference at other locations. It is interesting to calculate the form of the sound field when the condition given by equation (8.2.3) is specified. If we substitute equation (8.2.5) into equation (8.2.2), then the resulting pressure field is

$$p(r,\theta) = \frac{j\omega\rho_0 q_p}{4\pi}\left[\frac{e^{-jkr_p}}{r_p} - \frac{e^{-jk(r_s + d\cos\theta_0)}}{r_s}\right]. \tag{8.2.6}$$

Note that this can be written in the form

$$p(r,\theta) = \frac{j\omega\rho_0 q_p e^{-jkr_p}}{4\pi}\left[\frac{1}{r_p} - \frac{e^{-jk(r_s - r_p + d\cos\theta_0)}}{r_s}\right], \tag{8.2.7}$$

and if we again use the far field approximation such that $r_s - r_p = -d \cos \theta$ in the exponential terms and $r \approx r_p \approx r_s$ in the denominator of equation (8.2.7), then it follows that

$$p(r,\theta) = \frac{j\omega\rho_0 q_p e^{-jkr_p}}{4\pi r}\left[1 - e^{-jkd(\cos\theta_0 - \cos\theta)}\right]. \tag{8.2.8}$$

The terms outside the square brackets can be identified as the pressure field $p_p(r,\theta)$ which is due to the primary source in the absence of the secondary source. Therefore, equation (8.2.8) can be written as

$$p(r,\theta) = p_p(r,\theta)\left[1 - e^{-jkd(\cos\theta_0 - \cos\theta)}\right]. \tag{8.2.9}$$

By using the identity $|1 - e^{-jx}|^2 = 2(1 - \cos x)$, we can compare the modulus squared of the far field complex pressure produced by the interference of the two sources to that produced by the primary source alone. It follows that

$$\frac{|p(r,\theta)|^2}{|p_p(r,\theta)|^2} = 2(1 - \cos[kd(\cos\theta_0 - \cos\theta)]). \tag{8.2.10}$$

A simple condition for producing *attenuation* of the primary field at *all* angular locations θ, irrespective of the choice of θ_0, follows from this equation. In order that $|p(r,\theta)|^2 < |p_p(r,\theta)|^2$, then we must have

$$[1 - \cos kd(\cos\theta_0 - \cos\theta)] < (1/2). \tag{8.2.11}$$

Since the maximum value of $\cos \theta_0 - \cos \theta$ is 2 (i.e. when $\theta_0 = 0$ and $\theta = \pi$) then this condition reduces to $[1 - \cos 2kd] < (1/2)$, and it follows that we must have $kd < \pi/6$ or, simply, $d < \lambda/12$. This simple result, derived by Thornton (1988), shows that if a single secondary source is used to drive the pressure to zero in the far field of a single point primary source then, provided the two sources are separated by less than one-twelfth of a wavelength, attenuation of the entire far field will be produced. We will see that ensuring that primary/secondary source separation distances are much less than the wavelength of the sound being radiated is a consistent requirement for the successful global control of sound fields. Before leaving this discussion of the simplest of all three-dimensional active control problems, it is interesting to observe the form of the resulting sound field for various source separation distances relative to the acoustic wavelength and various chosen points of cancellation in the far field. Polar plots of the modulus squared of the complex pressure derived from equation (8.2.10) are shown in Fig. 8.2. The form of the net far field squared pressure is illustrated for separation distances d of $\lambda/8$, $\lambda/4$, $\lambda/2$ and λ and for values of $\theta_0 = 0°$ and $90°$. The most successful attenuation of the field in global terms is illustrated when $d = \lambda/8$ and $\theta_0 = 90°$. The radiation pattern shown is typical of the dipole type field generated by such an arrangement of two sources having equal and opposite strength.

8.3 The power output of two point monopole sources

In determining the global effectiveness of active control, the total power output of given source arrangements is a very useful measure. As we have seen in Chapter 1, the power output of an acoustic source is given by the integral of the acoustic intensity over a surface surrounding the source. In the far field, the mean squared pressure is a measure of the acoustic intensity. Thus, the acoustic power is effectively an addition of all the contributions to the mean squared pressure at all locations in the far field. In order to calculate the total power radiated by an array of sources it is usually necessary to perform an integral of the mean squared pressure over this far field surface. This integral is usually fairly easy to evaluate in relatively simple cases (see, for example, Nelson and Elliott, 1986), but becomes more difficult as the complexity of the source arrangement increases. We will introduce here a technique for computing the sound power output of point sources which will be particularly useful in this chapter.

It is shown in Section 9.11 that we can compute the power output of a harmonic point source of complex strength q from

$$W = \frac{1}{2}\operatorname{Re}\{p^* q\}, \tag{8.3.1}$$

where p is the complex pressure at the position of the source. Thus, in order to compute the sound power radiated by a point source of given strength we need only compute the pressure at the position of the point source that is in phase with the source strength fluctuation. To illuminate this further, consider only a point source when there is no other incident sound field. The expression for the radiated pressure is given by $p = (j\omega\rho_0 q e^{-jkr})/4\pi r$. One might expect that this expression would lead

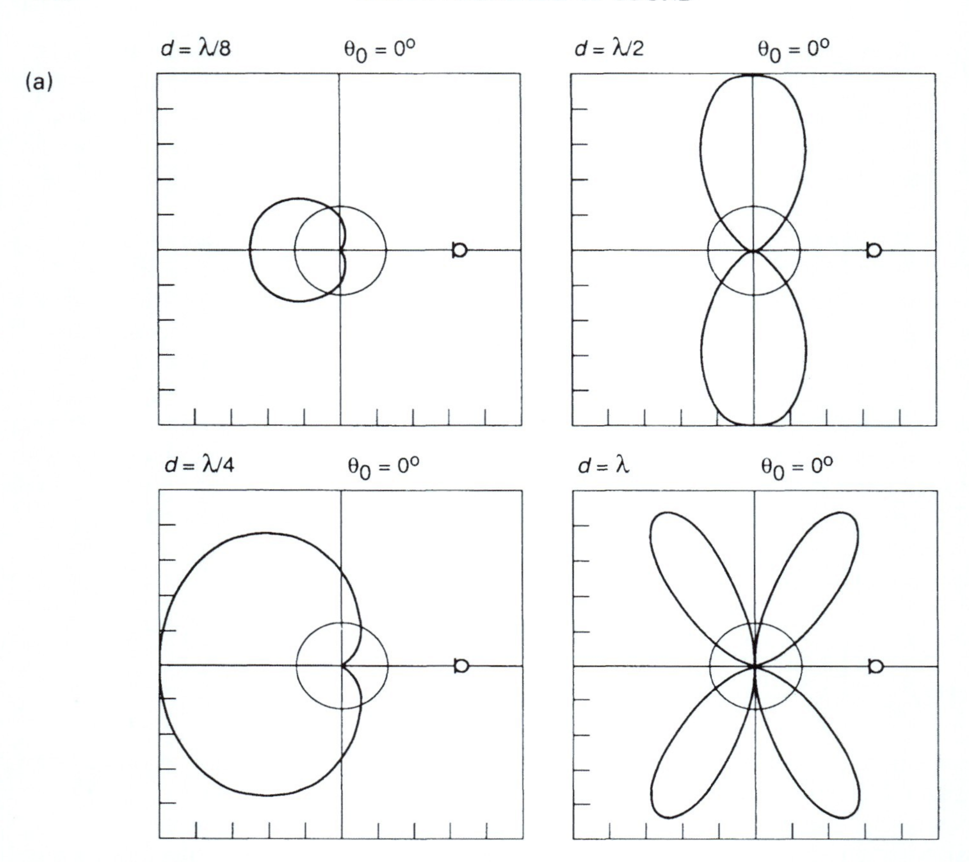

Fig. 8.2 (a) Polar plots of the modulus squared of the far field pressure produced by the interference of the primary and secondary sources when the secondary source strength is adjusted to ensure zero pressure at $\theta_0 = 0°$. The circle shows the modulus squared of the far field pressure due to the primary source alone.

to difficulties in the use of equation (8.3.1) since p becomes infinite as $r \to 0$. However, if we substitute this expression for p into equation (8.3.1) we can write the result in the form given by

$$W = \lim_{kr \to 0} \frac{\omega^2 \rho_0}{8\pi c_0} |q|^2 \operatorname{Re}\left\{\frac{je^{-jkr}}{kr}\right\}. \tag{8.3.2}$$

The complex term in the curly brackets can be written as

$$\frac{je^{-jkr}}{kr} = \frac{\sin kr}{kr} + j\frac{\cos kr}{kr}. \tag{8.3.3}$$

Although the imaginary part of this expression becomes infinite as $r \to 0$, the real part remains bounded, since $(\sin kr)/kr \to 1$ as $kr \to 0$. The expression for the source power output thus reduces to

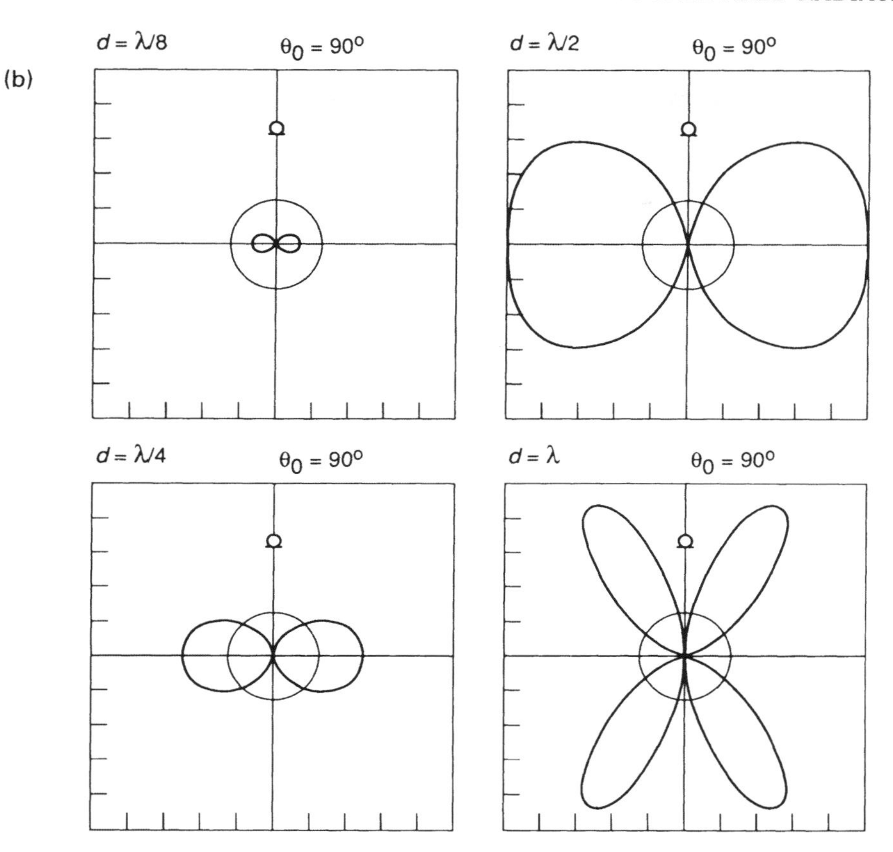

Fig. 8.2 (b) Polar plots of the modulus squared of the far field pressure produced by the interference of the primary and secondary sources when the secondary source strength is adjusted to ensure zero pressure at $\theta_0 = 90°$. The circle shows the modulus squared of the far field pressure due to the primary source alone.

$$W = \frac{\omega^2 \rho_0}{8\pi c_0}|q|^2 = \frac{1}{2} Z_0 |q|^2, \tag{8.3.4}$$

which is the result deduced in Section 1.13 of Chapter 1, and where Z_0 is equal to $\omega^2 \rho_0 / 4\pi c_0$.

Now let us apply this result to the problem of the two point monopoles discussed in the last section. We can compute the power output of the combination of the sources by adding the power outputs of the individual sources (see, for example, Pierce, 1981, Ch.1). If we denote p_{pp} as the pressure at the position of the primary source due to the primary source and p_{sp} as the pressure due to the secondary source at the position of the primary source and denote similarly p_{ps} and p_{ss} as the pressures at the position of the secondary source, then the total source power output can be written as

$$W = \frac{1}{2} \mathrm{Re}\{(p_{pp} + p_{sp})^* q_p\} + \frac{1}{2} \mathrm{Re}\{(p_{ps} + p_{ss})^* q_s\}. \tag{8.3.5}$$

If we use the relationship between pressure and source strength specified by equation (8.2.1) this expression reduces to

$$W = \left[\frac{1}{2} Z_0 \mathrm{Re}\left\{ |q_p|^2 - \frac{q_p q_s{}^* j e^{jkd}}{kd} \right\}\right] + \left[\frac{1}{2} Z_0 \mathrm{Re}\left\{ |q_s|^2 - \frac{q_s q_p{}^* j e^{jkd}}{kd} \right\}\right], \tag{8.3.6}$$

where the two terms in square brackets are the power outputs of the primary and secondary sources, respectively. The terms $Z_0|q_p|^2/2$ and $Z_0|q_s|^2/2$ represent the respective power outputs of the primary and secondary sources that would be produced if each source were radiating sound in the absence of the other. The remaining terms quantify the influence that one source has on the power output of the other. A particular aspect of this interaction was recognised by Ffowcs-Williams (1984), who calculated the source power outputs when the secondary source strength was of the same amplitude as that of the primary source but differed in phase by an angle ϕ. Thus, the secondary source strength was related to that of the primary source by $q_s = q_p e^{j\phi}$. Under these circumstances the expression for the total power output reduces to

$$W = \frac{1}{2} Z_0 |q_p|^2 \left[\left(1 + \frac{\sin(kd - \phi)}{kd}\right) + \left(1 + \frac{\sin(kd + \phi)}{kd}\right)\right]. \tag{8.3.7}$$

Again, the terms in round brackets represent the respective power outputs of the primary and secondary sources. Both these power outputs and the total power output are shown plotted in Fig. 8.3 as a function of ϕ for values of kd such that $d = \lambda/64$, $\lambda/32$ and $\lambda/16$. It is evident from these graphs that at some phase angles ($\phi \approx 90°$, $270°$) then there is a very large flow of power between the two sources; thus, for example, at $\phi \approx 270°$, the secondary source *absorbs* a considerable amount of power, whilst the primary source is induced to increase considerably its power output. The imbalance between this large respective inflow and outflow of power accounts for the radiation to the far field. We have noted the capacity of secondary sources to absorb energy in Chapter 5; here we see that for closely spaced sources adjusting the strength of the secondary source to absorb energy is successful only in inducing a larger flow of energy from the primary source and is not a useful strategy for controlling the field. We will return to the topic of the active absorption of sound in Chapter 9.

It is also interesting to note the total power output of the sources when the secondary source strength is optimally adjusted in order to cancel the pressure at an angular position θ_0 in the far field. As we have seen in equation (8.2.5), this condition is given by $q_s = -q_p e^{-jkd \cos \theta_0}$. Substitution of $\phi = -kd \cos \theta_0$ into equation (8.3.7) shows that

$$W = \frac{1}{2} Z_0 |q_p|^2 \left[2 - \frac{2 \sin kd}{kd} \cos (kd \cos \theta_0) \right], \tag{8.3.8}$$

where $Z_0|q_p|^2/2$ is the power output of the primary source in the absence of the secondary source. The condition for the power output to be reduced can thus be written as

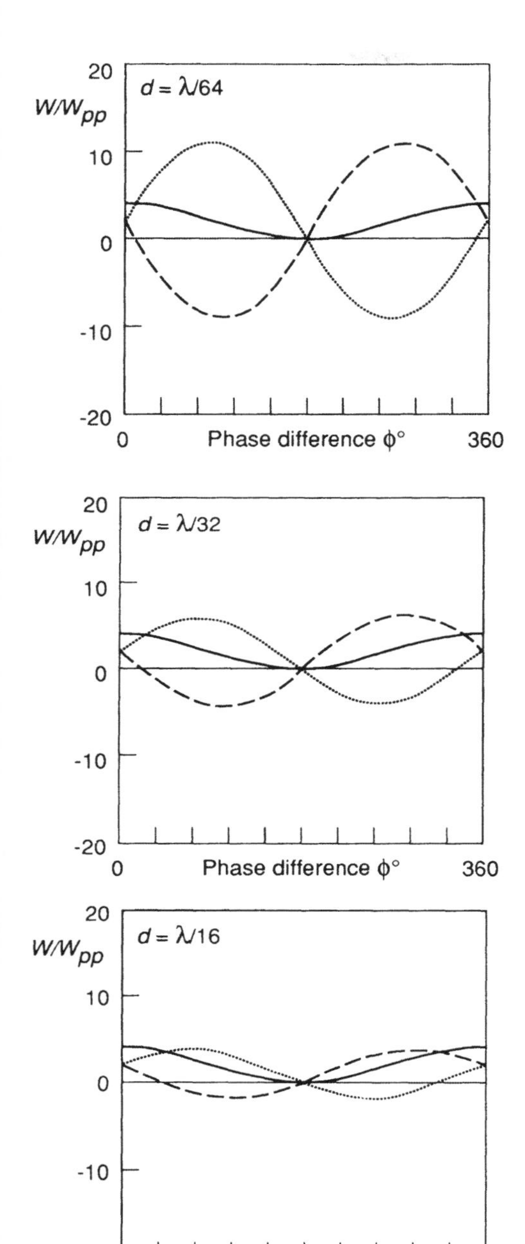

Fig. 8.3 The power output of the primary and secondary sources of the same source strength amplitude. Results are shown as a function of the phase difference between the two sources and are plotted on linear scales relative to the power output $W_{pp} = \omega^2 \rho_o |q_p|^2 / 8\pi c_o$ which is that radiated by the primary source acting alone. — — – Primary source output, ······· secondary source output, ——— total power output.

$$\frac{\sin kd \cos(kd \cos\theta_0)}{kd} > (1/2), \tag{8.3.9}$$

which reduces to

$$2 \sin kd \cos(kd \cos\theta_0) > kd. \tag{8.3.10}$$

The most stringent criterion occurs when $\theta_0 = 0°$ ($\cos\theta_0 = 1$) and since $2 \sin kd \cos kd = \sin 2kd$, the inequality can be written as

$$\sin 2kd > kd. \tag{8.3.11}$$

This criterion is satisfied when kd is less than approximately $\pi/3$. This is satisfied by $d < \lambda/6$ which is a less stringent criterion than that deduced in Section 8.2 for the far field pressure to be everywhere reduced by producing zero pressure at one angular location.

Note also that if the pressure is driven to zero at right angles to the source pair ($\theta_0 = 90°$), which represents the least stringent criterion for the power output to be reduced, then the condition reduces to

$$\frac{\sin kd}{kd} > \frac{1}{2}. \tag{8.3.12}$$

This criterion is satisfied when kd is less than approximately $2\pi/3$ or $d < \lambda/3$. So even for the best choice of θ_0, driving the far field pressure to zero will produce an *increase* in net power output when the source separation distance is greater than about one-third of a wavelength.

8.4 The minimum power output of the sources

Now we shall see that the source separation distance criterion for producing global reductions in far field sound pressure is far less stringent if we do not apply control by forcing the pressure to zero at one angular location. Equation (8.3.5) for the total power output of the source combination can be reduced to the complex quadratic form

$$W = A|q_s|^2 + q_s^* b + b^* q_s + c, \tag{8.4.1}$$

where $A = Z_0/2$, $b = (1/2)Z_0 \operatorname{sinc} kdq_p$ (using the abbreviation $(\sin x)/x = \operatorname{sinc} x$) and $c = (1/2)Z_0|q_p|^2$ (see Nelson *et al.*, 1987a). This last term is, of course, equal to power output due to the primary source alone, which we shall denote by W_{pp}. Since A is greater than zero, this quadratic function has a unique minimum specified by the optimal secondary source strength given by $A^{-1}b$ and the minimum power output given by $c - b^*A^{-1}b$ (see Section 5.5). It therefore follows that the optimal secondary source strength and resulting minimum power output can be written, respectively, as

$$q_{s0} = -q_p \operatorname{sinc} kd, \tag{8.4.2}$$

$$W_0 = W_{pp}[1 - \operatorname{sinc}^2 kd]. \tag{8.4.3}$$

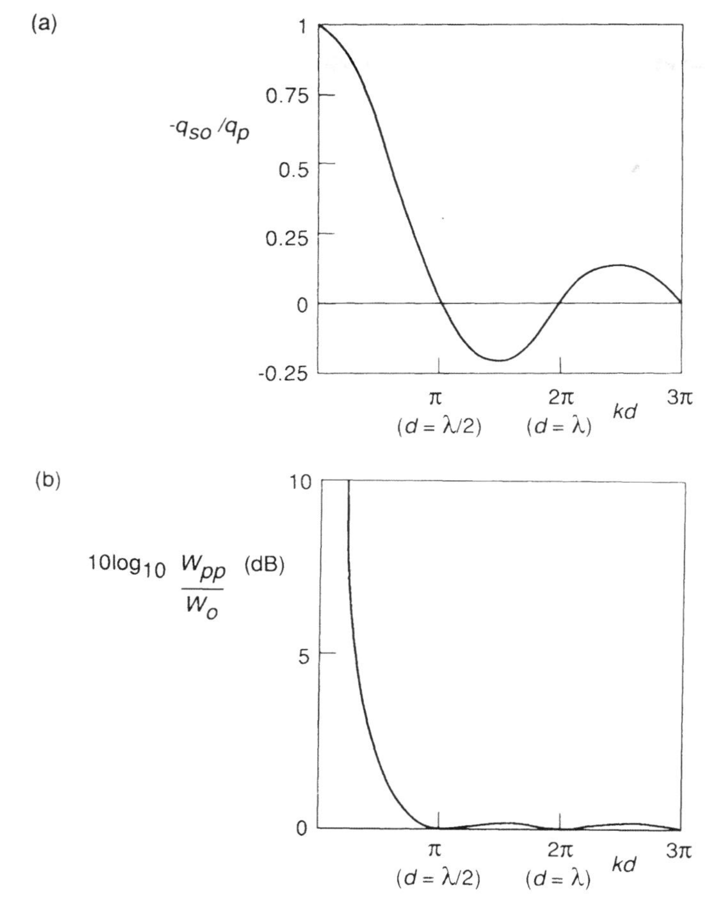

Fig. 8.4 (a) The ratio of the optimal secondary source strength to the primary source strength. Note that for $0 < kd < \pi$ the secondary source is out of phase with the primary source, but for $\pi < kd < 2\pi$ the secondary source is in phase with the primary source. (b) The maximum achievable reduction in total sound power radiated.

Graphs showing the behaviour of these two functions are shown in Fig. 8.4. As illustrated in this figure, when the separation distance between the two sources is increased relative to the acoustic wavelength, the strength of the secondary source must be progressively decreased. Thus, in contrast to the case discussed in Section 8.2, where the secondary source was used to cancel the far field pressure at a point, the secondary source strength in this case does not have the same modulus as the primary source. Note also that the secondary source always remains *180° out of phase* with the primary source and only its amplitude is reduced as d is increased until $d = \lambda/2$. For source separations such that $\lambda/2 < d < \lambda$, in order to achieve optimal results, the secondary source must be *in phase* with the primary source but of a considerably reduced amplitude. As the two sources become increasingly separated, the secondary source strength asymptotes to zero. In fact, once the separation distance of $\lambda/2$ is exceeded, there is very little that can be done with the secondary source to reduce the net power radiated to the far field. Thus, as we have seen in Section 8.2, although it is always possible to produce zero pressure at some angular position in the far field by using a secondary source, if the sources become separated by a distance that is large compared to the wavelength then this control *must* be achieved at the expense of a net increase in the total acoustic energy radiated.

It is also interesting to calculate the form of the far field complex pressure when the secondary source strength is optimally adjusted. Substitution of equation (8.4.2) into equation (8.2.2) and use of the usual far field approximations shows that

$$p(r,\theta) = p_p(r,\theta)(1 - \operatorname{sinc} kd\, \mathrm{e}^{jkd\cos\theta}), \tag{8.4.4}$$

where again p_p is the complex pressure radiated by the primary source alone. The expression for the modulus squared of the far field pressure relative to the modulus squared of the primary pressure field is thus

$$\frac{|p(r,\theta)|^2}{|p_p(r,\theta)|^2} = [1 + \operatorname{sinc}^2 kd - 2\operatorname{sinc} kd \cos(kd\cos\theta)]. \tag{8.4.5}$$

The form of this pressure field is illustrated in the polar diagrams shown in Fig. 8.5 for various values of the separation distance d relative to the acoustic wavelength. Note that the radiation pattern produced has some similarities with that of a dipole source when $d = 2\lambda/16$ for example, but as the source separation is increased relative to the acoustic wavelength the directivity pattern becomes distinctly different with a finite value of far field pressure produced at $\theta = 90°$ and $\theta = 270°$.

We can also calculate the power outputs of the primary and secondary sources individually when the secondary source strength is optimally adjusted to minimise the total power radiated. The first term in square brackets in equation (8.3.6) represents the power radiated by the primary source. This can be written as

$$W_{p0} = \frac{\omega^2\rho_0}{8\pi c_0}(|q_p|^2 + q_{s0}^* q_p \operatorname{sinc} kd). \tag{8.4.6}$$

Substitution of the optimal value of the secondary source strength q_{s0} then shows that this expression reduces to

$$W_{p0} = W_{pp}(1 - \operatorname{sinc}^2 kd). \tag{8.4.7}$$

Note that this is exactly equal to the minimum power output and therefore *all* the power flowing must come from the primary source. This is confirmed if we calculate the power output due to the secondary source which will be given by the second term in the square brackets in equation (8.3.6). Substitution of the optimal value q_{s0} shows that

$$W_{s0} = 0. \tag{8.4.8}$$

Thus, under optimal conditions the secondary source *neither radiates nor absorbs any net acoustic power*.

Before leaving the discussion of the minimisation of the power output of the sources, it is important to note that these results are achieved at the expense of the secondary source which acts non-causally with respect to the primary source. Equation (8.4.2) specifies the frequency response function of the optimal filter which

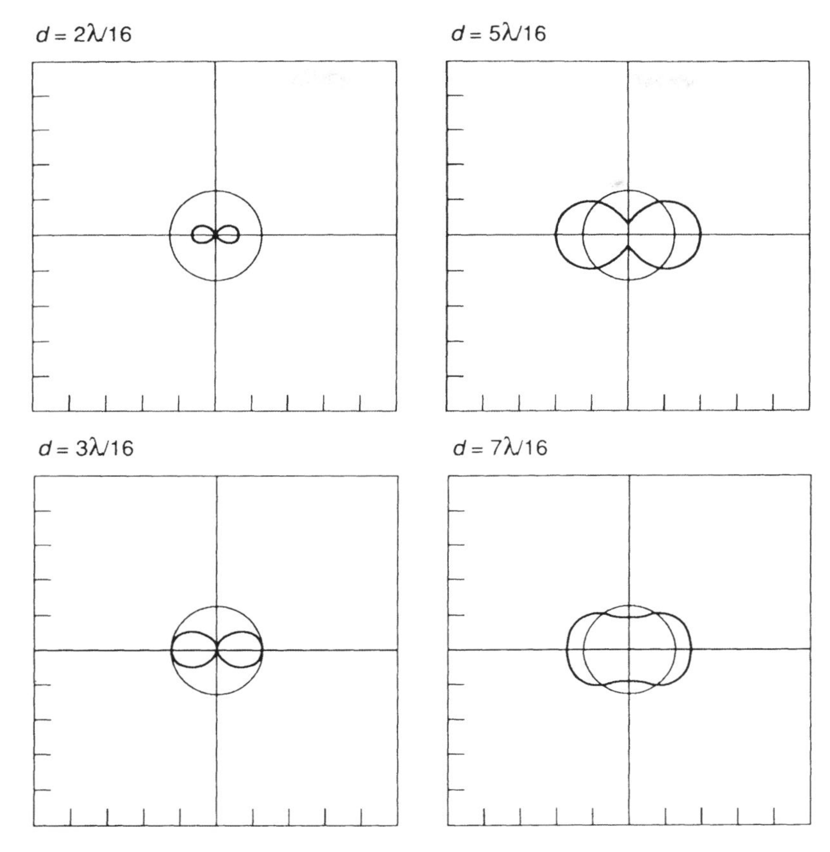

Fig. 8.5 Polar plots of the modulus squared of the far field pressure when the secondary source strength is adjusted to minimise the total power radiated. The circle shows the modulus squared of the far field pressure produced by the primary source alone.

relates the secondary source strength to the primary source strength in order to minimise the power radiated. This frequency response function is given by

$$\frac{q_{s0}}{q_p} = H(j\omega) = -\frac{\sin(\omega d/c_0)}{(\omega d/c_0)}. \tag{8.4.9}$$

It can be shown via inverse Fourier transformation (see, for example, Papoulis, 1981, Ch. 1) that the corresponding impulse response of this filter is given by

$$h(t) = -\frac{[H(t + d/c_0) - H(t - d/c_0)]}{(2d/c_0)} \tag{8.4.10}$$

where $H(x)$ is the Heaviside unit step function which is specified by $H(x) = 1, x > 1$ and $H(x) = 0, x < 1$. This form of impulse response is shown in Fig. 8.6. It is evident that the optimal filter relating the secondary and primary source strengths has a *non-causal* impulse response. Thus, if optimal results are to be achieved, the secondary source must emit signals a well-defined time d/c_0 in advance of the primary source. Not surprisingly this time difference corresponds to the time for an

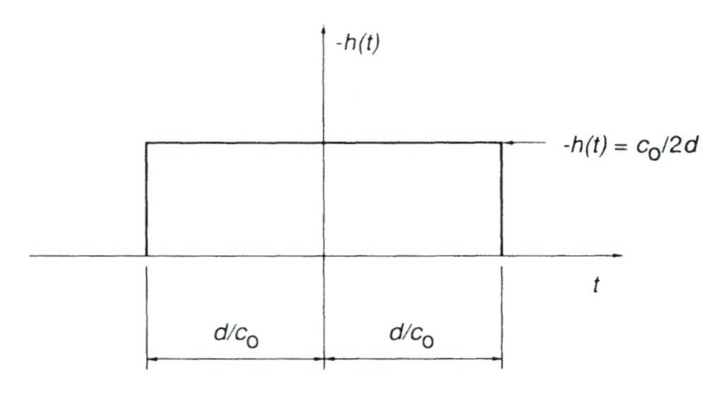

Fig. 8.6 The impulse response function with which the primary source strength time history must be convolved to deduce the optimal secondary source strength time history.

acoustic signal to propagate from one source to the other. There are two ways of physically interpreting this necessity for advance information. First, it can be seen that if cancellation of a given signal from the primary source is to be achieved for far field positions on the axis of the two sources to the left of the primary source in Fig. 8.1, then in order to compensate for the difference in travel times from the two sources to the field point the secondary source must emit the inverse signal at a time d/c_0 before it is emitted by the primary source. Similarly, for points on the axis of the two sources which are to the right of the secondary source in Fig. 8.1, the output of the secondary source must be delayed by a time d/c_0. The second explanation for this behaviour lies in the two mechanisms by which the power output of the source combination can be modified. First, and as we have described in detail in the one-dimensional case in Section 5.6, the primary source power output can be modified by the addition of the secondary source pressure field which changes its ability to radiate power. If the appropriate pressure is to be produced on the primary source then the secondary source must clearly emit its signals at a time d/c_0 prior to those emitted by the primary source. Similarly, the second possible mechanism of power output reduction is by sound power absorption. If the output from a primary source is to be absorbed by the secondary source then the secondary source must clearly act at a time d/c_0 after the emission of the relevant signal from the primary source. This aspect of the production of optimal results is of little consequence when one is dealing with purely periodic signals. However, when the output of the primary source is, for example, random in nature, then the frequency domain results derived here cannot be readily applied. We will return to this problem in Section 8.6.

8.5 The minimisation of the sum of the squared pressures at a number of far field error sensors

In our consideration of the simple combination of point primary and secondary monopole sources we have shown that there is a well-defined global optimum for the minimisation of the total source power output. Of course, to monitor this total power output requires measurement in the far field of the modulus of the squared pressure at all locations. In practice one is restricted to the use of a finite number of measurement positions or "error detectors" in order to estimate the total far field power radiated. It is interesting to see to what extent the use of a finite number of sensors can enable the globally optimal reduction to be approximated. Let us assume that we have a number L of sensors all located in the same plane as the two sources in the far field and specify a vector $\mathbf{p}$ of complex far field pressures

$$\mathbf{p}^{\mathrm{T}} = [p(r,\theta_1)\ p(r,\theta_2)\ \dots\ p(r,\theta_L)], \tag{8.5.1}$$

where θ_l denotes the angular location of the lth sensor and T denotes the vector transpose. The total complex pressure is a superposition of the fields due to the primary and secondary sources, and we can write this as

$$\mathbf{p} = \mathbf{z}_p q_p + \mathbf{z}_s q_s, \tag{8.5.2}$$

where $\mathbf{z}_p$ and $\mathbf{z}_s$ are the vectors of complex acoustic transfer impedances given by

$$\mathbf{z}_p^{\mathrm{T}} = \frac{j\omega\rho_0}{4\pi}\left[\frac{\mathrm{e}^{-jkr_{p1}}}{r_{p1}}\ \frac{\mathrm{e}^{-jkr_{p2}}}{r_{p2}}\ \dots\ \frac{\mathrm{e}^{-jkr_{pL}}}{r_{pL}}\right], \tag{8.5.3a}$$

$$\mathbf{z}_s^{\mathrm{T}} = \frac{j\omega\rho_0}{4\pi}\left[\frac{\mathrm{e}^{-jkr_{s1}}}{r_{s1}}\ \frac{\mathrm{e}^{-jkr_{s2}}}{r_{s2}}\ \dots\ \frac{\mathrm{e}^{-jkr_{sL}}}{r_{sL}}\right], \tag{8.5.3b}$$

and the terms r_{pl} and r_{sl} denote the radial distances from the primary and secondary sources, respectively, to the lth sensor. If we define a cost function equal to the sum of the squared moduli of the pressures at these L sensors, then we can write this as

$$J_p = \sum_{l=1}^{L} |p(r,\theta_l)|^2 = \mathbf{p}^{\mathrm{H}}\mathbf{p}, \tag{8.5.4}$$

where the symbol H denotes the complex conjugate of the vector transpose. Substituting equation (8.5.2) into this expression shows that

$$J_p = (\mathbf{z}_p q_p + \mathbf{z}_s q_s)^{\mathrm{H}}(\mathbf{z}_p q_p + \mathbf{z}_s q_s). \tag{8.5.5}$$

Expansion of this expression results in

$$J_p = |q_p|^2\mathbf{z}_p^{\mathrm{H}}\mathbf{z}_p + q_p^*\mathbf{z}_p^{\mathrm{H}}\mathbf{z}_s q_s + q_s^*\mathbf{z}_s^{\mathrm{H}}\mathbf{z}_p q_p + |q_s|^2\mathbf{z}_s^{\mathrm{H}}\mathbf{z}_s, \tag{8.5.6}$$

which can be recognised as a quadratic function of the complex secondary source strength of the type defined by equation (8.4.1), where in this case $A = \mathbf{z}_s^{\mathrm{H}}\mathbf{z}_s$, $b = \mathbf{z}_s^{\mathrm{H}}\mathbf{z}_p q_p$ and $c = |q_p|^2\mathbf{z}_p^{\mathrm{H}}\mathbf{z}_p$. It follows that the secondary source strength which minimises the sum of the squared pressures in the far field can be written as

$$q_{s0} = -\frac{\mathbf{z}_s^{\mathrm{H}}\mathbf{z}_p}{\mathbf{z}_s^{\mathrm{H}}\mathbf{z}_s} q_p. \tag{8.5.7}$$

Expansion of the vector products appearing in this equation shows that

$$\mathbf{z}_s^{\mathrm{H}}\mathbf{z}_p = \left(\frac{\omega\rho_0}{4\pi}\right)^2\left[\frac{\mathrm{e}^{-jk(r_{p1}-r_{s1})}}{r_{s1}r_{p1}} + \frac{\mathrm{e}^{-jk(r_{p2}-r_{s2})}}{r_{s2}r_{p2}}\ \dots\ \frac{\mathrm{e}^{-jk(r_{pL}-r_{sL})}}{r_{sL}r_{pL}}\right], \tag{8.5.8}$$

$$\mathbf{z}_s^{\mathrm{H}}\mathbf{z}_s = \left(\frac{\omega\rho_0}{4\pi}\right)^2\left[\frac{1}{r_{s1}^2} + \frac{1}{r_{s2}^2} + \dots\ \frac{1}{r_{sL}^2}\right]. \tag{8.5.9}$$

If we adopt the usual far field approximations for the terms for both the exponential and denominator terms in these expressions (see Section 8.2), equation (8.5.7) reduces to

$$q_{s0} = -\frac{q_p}{L}\sum_{l=1}^{L} e^{-jkd\cos\theta_l}. \tag{8.5.10}$$

The optimal secondary source strength that minimises the sum of the squared pressures at the L far field sensors is therefore relatively simply defined. The effect on the total source power output of adopting this strategy can be assessed by substitution of this relationship into the expression for the total power output of the two point sources given by equation (8.3.5). After some algebra the expression for the total power output that results is given by

$$W = W_{pp}\left\{1 + \left[\frac{1}{L}\sum_{l=1}^{L}\cos(kd\cos\theta_l)\right]^2 + \left[\frac{1}{L}\sum_{l=1}^{L}\cos(kd\sin\theta_l)\right]^2 - 2\,\mathrm{sinc}\,kd\left[\frac{1}{L}\sum_{l=1}^{L}\cos(kd\cos\theta_l)\right]\right\} \tag{8.5.11}$$

where again W_{pp} is the power output due to the primary source alone.

The results of evaluating the expression (8.5.11) are shown in Table 8.1 when there is one sensor at $\theta_1 = 0°$, two sensors at locations $\theta_1 = 0°$, $\theta_2 = 45°$, three sensors at $\theta_1 = 0°$, $\theta_2 = 45°$, $\theta_3 = 90°$ and four sensors at $\theta_1 = 0°$, $\theta_2 = 45°$, $\theta_3 = 90°$, $\theta_4 = 135°$. Results are shown for various source separation distances relative to the acoustic wavelength and also shown are the source power outputs when the secondary source is optimally adjusted to minimise the total power radiated.

It is evident that close to optimal results can be achieved with a relatively small number of sensors (four, for example) and adjusting the secondary source strength to minimise the sum of their squared outputs. This appears an effective strategy if we wish to avoid the increase in power output that often occurs when we adjust the

Table 8.1 The total power output of a combination of a point primary source and a point secondary source when the secondary source strength is adjusted to minimise the sum of the squared pressures at L angular locations in the far field. Results are shown relative to the power output of the primary source.

Source separation	Minimum power output $10\log_{10}(W_0/W_{pp})$ (dB)	Power output produced by minimising sum of squared pressures $10\log_{10}(W/W_{pp})$ (dB)			
		$L = 1$	$L = 2$	$L = 3$	$L = 4$
$\lambda/8$	−7.2	−1.3	−2.3	−4.4	−6.5
$3\lambda/16$	−4.1	+1.5	+0.6	−1.5	−3.5
$\lambda/4$	−2.3	+3.0	+2.2	0.0	−1.6
$\lambda/2$	0.0	+3.0	+2.5	+0.5	0.0
λ	0.0	+3.0	+1.4	+1.6	+0.5
2λ	0.0	+3.0	0.0	+0.7	0.0

secondary source to produce zero pressure at one location. A more extensive study of this approach has been undertaken by Thornton (1988) who examined a number of primary/secondary source geometries and consistently found that close to optimal results could often be achieved with a relatively small number of sensors. Thornton found for the case considered here that negligible further reductions in power output were produced by increasing the number of sensors from four to twenty.

8.6 The constraint of causality

Thus far in our discussion of the active control of free field sound we have limited our attention to single frequency radiation. This is largely due to the ease with which calculations are undertaken in the frequency domain. Now, however, we will demonstrate how equivalent calculations can be undertaken in the time domain. The necessity of performing these types of calculation arises when the excitation with which we are dealing is *unpredictable*. This applies, for example, to primary sources whose output time history is a random process. The need for an approach in the time domain can again be simply illustrated by considering the cancellation of the pressure in the far field of two point monopole sources. The frequency domain condition for far field cancellation at a point follows from equation (8.2.2) and can be written as

$$q_s = -q_p e^{-j\omega(r_p - r_s)/c_0}, \tag{8.6.1}$$

where, as illustrated in Fig. 8.7(a), r_p and r_s are the distances from the primary and secondary sources, respectively, to the far field point considered. We can assume that the secondary source strength to achieve cancellation is related to the primary source strength via an "optimal filter" which has the frequency response function $H_0(j\omega) = -e^{-j\omega(r_p - r_s)/c_0}$. Inverse Fourier transformation shows that this frequency response function has the impulse response $h_0(t) = -\delta[t - (r_p - r_s)/c_0]$. For field points which are at a greater distance from the primary source than from the secondary source, such that $r_p > r_s$, then this impulse response is causal. However, for positions which are closer to the primary source than the secondary source, such that $r_p < r_s$, then $h_0(t) = -\delta[t + (r_s - r_p)/c_0]$ and this impulse response is clearly *non-causal*. We see that in order to achieve cancellation at such a field point the secondary source must anticipate the output of the primary source.

Now consider the case where we do not have any advance information regarding the primary source output but that we can detect the output of the primary source at the instant it is produced. This output waveform is assumed to have a stationary random time history. Also assume that we can pass the signal through a filter, whose impulse response is constrained to be causal, in order to generate the secondary source strength output which, in some sense, produces the best results in the far field. One way of treating this problem is to cast it in the form of a Wiener filtering problem of the type discussed in Section 3.8. This is most easily explained by regarding the far field pressure as the output of a system whose block diagram is depicted in Fig. 8.7(b). Since the systems with which we are dealing are linear and time invariant, this block diagram can be rewritten as Fig. 8.7(c). Now we can see that the determination of the optimal filter reduces to a Wiener filtering problem if

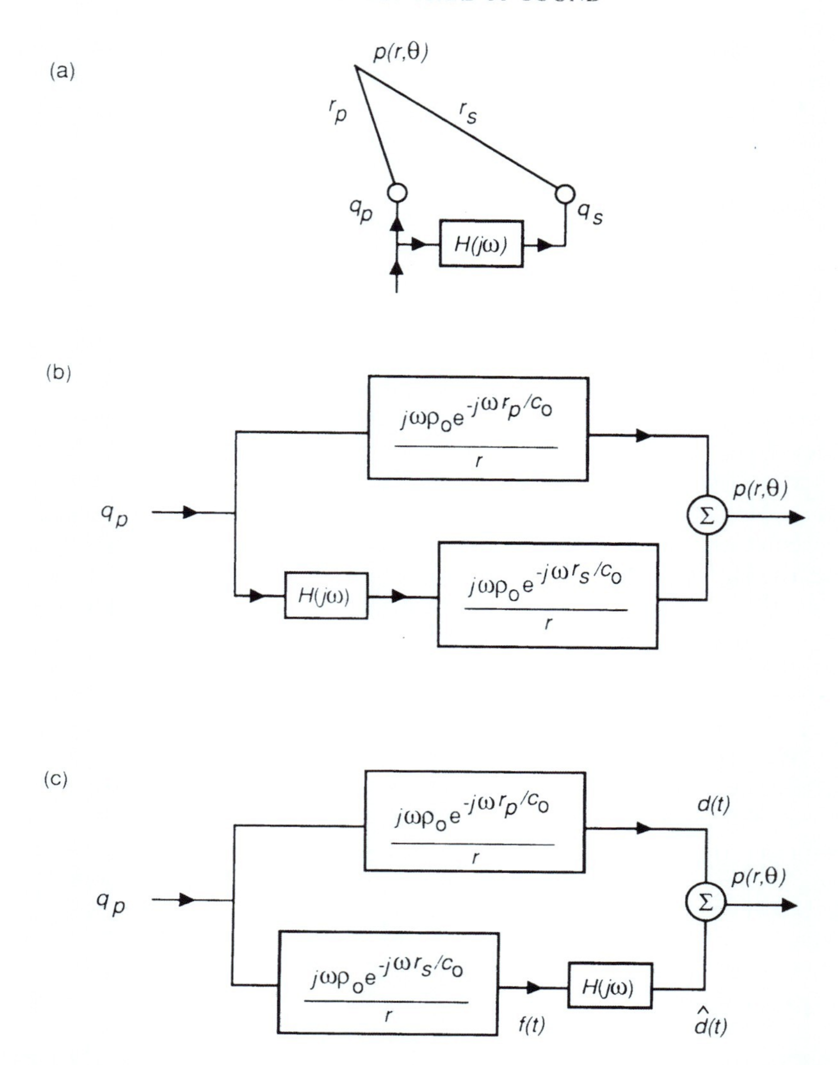

Fig. 8.7 The free field radiation problem is shown in (a) and in block diagram form in (b). The equivalent relationship is shown in (c) with transfer functions reversed. Note that the observation point is assumed to be in the far field such that $r_p \simeq r_s \simeq r$ in the denominator of the acoustic frequency response functions.

we assume that the objective is to minimise the mean square pressure in the far field. Thus, using the terminology of optimal filtering theory (see, for example, Van Trees, 1965, Ch. 6) the "received" or "reference" signal $f(t)$ and the "desired" signal $d(t)$ can be interpreted as shown in the block diagram of Fig. 8.7c. The only difference between this problem and that of a classical Wiener filtering problem is that we wish to minimise the *sum* of the signal $d(t)$ and its estimate $\hat{d}(t)$, rather than the difference between them. Thus, we assume that the signal $\hat{d}(t)$ is given by

$$\hat{d}(t) = \int_0^\infty h(\tau_1)f(t - \tau_1)\mathrm{d}\tau_1 \quad (h(\tau_1) = 0,\ \tau_1 < 0), \tag{8.6.2}$$

and we wish to minimise the function given by

$$J_p = E[p^2(t)] = E[(d(t) + \hat{d}(t))^2]. \tag{8.6.3}$$

It follows from the reasoning presented in Section 3.8 that a necessary and sufficient condition for the impulse response $h(\tau_1)$ to minimise this function is given by a Wiener–Hopf integral equation of the form

$$R_{fd}(\tau_2) + \int_0^\infty h_0(\tau_1)R_{ff}(\tau_2 - \tau_1)\mathrm{d}\tau_1 = 0, \qquad 0 < \tau_2 < \infty, \tag{8.6.4}$$

where $R_{fd}(\tau) = E[f(t)d(t + \tau)]$, $R_{ff}(\tau) = E[f(t)f(t + \tau)]$ and $h_o(\tau_1)$ represents the optimal value of $h(\tau_1)$. (Note that this equation does not include a minus sign before the integral and thus differs from equation (3.8.5) since we are minimising the expectation of the *sum* of $d(t)$ and $\hat{d}(t)$ rather than the difference between them.)

In our acoustical problem it follows from the block diagram illustrated in Fig. 8.7(c) that the signals $d(t)$ and $f(t)$ are respectively

$$d(t) = \frac{\rho_0 \dot{q}_p(t - r_p/c_0)}{4\pi r}, \quad f(t) = \frac{\rho_0 \dot{q}_p(t - r_s/c_0)}{4\pi r}, \tag{8.6.5a,b}$$

where the dot symbolises differentiation with respect to time and replaces the factor $j\omega$ in the frequency domain. Thus, we can write the integral equation in the form

$$R_{\dot{q}\dot{q}}\left(\tau_2 - \frac{r_p - r_s}{c_0}\right) + \int_0^\infty h_0(\tau_1)R_{\dot{q}\dot{q}}(\tau_2 - \tau_1)\mathrm{d}\tau_1 = 0, \qquad 0 < \tau_2 < \infty, \tag{8.6.6}$$

where $R_{\dot{q}\dot{q}} = E[\dot{q}_p(t)\dot{q}_p(t + \tau)]$ is the autocorrelation function associated with the time derivative of the primary source strength fluctuation. It is clear that for values $r_p > r_s$ the equation has the solution

$$h_0(\tau_1) = -\delta\left(\tau_1 - \frac{r_p - r_s}{c_0}\right), \tag{8.6.7}$$

which is equivalent to the solution derived above from an analysis in the frequency domain. However, for far field points closer to the primary source than to the secondary source, the integral equation can be written as

$$R_{\dot{q}\dot{q}}(\tau_2 + \eta) + \int_0^\infty h_0(\tau_1)R_{\dot{q}\dot{q}}(\tau_2 - \tau_1)\mathrm{d}\tau_1 = 0, \qquad \left\{\begin{matrix} r_s > r_p \\ 0 < \tau_2 < \infty \end{matrix}\right\}, \tag{8.6.8.}$$

where $\eta = (r_s - r_p)/c_0$. This form of the integral equation is well known in the field of signal processing and has already been referred to in Section 3.8. The equation is usually written in the form

$$R_{ff}(\tau_2 + \eta) + \int_0^\infty h_0(\tau_1)R_{ff}(\tau_2 - \tau_1)\mathrm{d}\tau_1 = 0, \qquad 0 < \tau_2 < \infty. \tag{8.6.9}$$

The equation specifies the conditions necessary for a solution to the *pure prediction problem* where the signal $d(t)$ is related to the signal $f(t)$ by $d(t) = f(t + \eta)$. That is, we wish to operate on the signal $f(t)$ in order to produce an optimal estimate of its value at a time η in the future. The solution to this problem is also well known and will be outlined only briefly here.

First, note that if the received signal $f(t)$ is white noise, $w(t)$ say, then we can write the autocorrelation function of the received signal as

$$R_{ff}(\tau_2 - \tau_1) = R_{ww}(\tau_2 - \tau_1) = \delta(\tau_2 - \tau_1), \tag{8.6.10}$$

and it therefore follows from equation (8.6.4) that for a white noise signal $f(t)$, the impulse response of the optimal filter is specified by

$$h_{0w}(\tau_2) = \begin{Bmatrix} -R_{wd}(\tau_2), & \tau_2 \geqslant 0 \\ 0, & \tau_2 < 0 \end{Bmatrix}, \tag{8.6.11}$$

where $R_{wd}(\tau_2) = E[w(t)d(t + \tau_2)]$ is the cross-correlation function between white noise and the desired signal $d(t)$. In addition, provided that $f(t)$ has a rational power spectrum, it can be shown that this signal can be considered to be the output of a "shaping filter" which is excited by a white noise input (see, for example, Van Trees, 1965, Ch. 6). It can also be shown that such a shaping filter has a realisable inverse and that passing $f(t)$ through the inverse of the shaping filter produces white noise. This is illustrated in Fig. 8.8. We can relate the impulse response of this shaping filter to the impulse response of the optimal filter which is the solution to the pure prediction problem. Expressing the desired signal as $d(t) = f(t + \eta)$, we can write the cross-correlation function between white noise and the desired signal as

$$R_{wd}(\tau_2) = E[w(t)f(t + \eta + \tau_2)] = E[w(t - \eta - \tau_2)f(t)]. \tag{8.6.12}$$

If we now substitute for the value of $f(t)$ which is given by the convolution of the white noise input and the impulse response $y(\tau)$ of the shaping filter, this reduces to

$$R_{wd}(\tau_2) = E[w(t - \eta - \tau_2) \int_0^\infty y(\tau)w(t - \tau)\mathrm{d}\tau] =$$

$$\int_0^\infty y(\tau)E[w(t - \eta - \tau_2)w(t - \tau)]\mathrm{d}\tau. \tag{8.6.13}$$

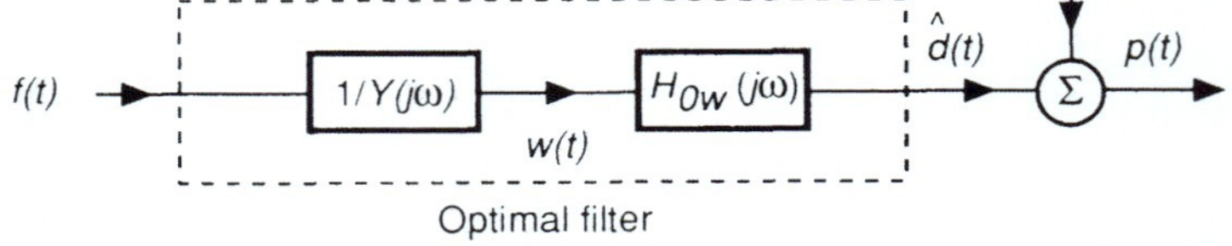

Fig. 8.8 The solution to the Wiener filtering problem for received signals $f(t)$ having rational spectra. The impulse response of the filter H_{0w} $(j\omega)$ is given by h_{0w} $(\tau_1) = -R_{fd}$ (τ_1) for $\tau_1 \geqslant 0$ and h_{0w} $(\tau_1) = 0$ for $\tau_1 < 0$.

The term $E[w(t - \eta - \tau_2)w(t - \tau)]$ can be written as $\delta(\tau - (\tau_2 + \eta))$ and therefore this equation reduces to

$$R_{wd}(\tau_2) = y(\tau_2 + \eta). \tag{8.6.14}$$

The impulse response of the optimal filter is therefore related to that of the shaping filter by

$$h_{0w}(\tau_2) = -y(\tau_2 + \eta). \tag{8.6.15}$$

Thus, provided that we can express the form of the received signal $f(t)$ as the output of a white noise excited shaping filter then the solution to the pure prediction problem is straightforward.

Nelson *et al.* (1990a) considered the case where the primary source strength is a stationary random fluctuation which is the result of driving the source output from a white noise excited second order system. The time derivative of the source strength output is therefore given by exciting with white noise the system whose transfer function can be written as

$$Y(s) = \frac{s\omega_n^2}{s^2 + 2\zeta\omega_n^s + \omega_n^2}, \tag{8.6.16}$$

where ζ is the damping ratio and ω_n the natural frequency of the second order system. This system has the impulse response given by

$$y(\tau) = -\frac{\omega_n^3}{\omega_0}\,\mathrm{e}^{-\zeta\omega_n\tau}\sin(\omega_0\tau - \phi), \tag{8.6.17}$$

where $\omega_0 = \omega_n(1 - \zeta^2)^{1/2}$ and $\phi = \tan^{-1}[(1 - \zeta^2)^{1/2}]/\zeta$. Although the algebra is a little tedious, we can deduce the transfer function of the optimal filter specified by equation (8.6.15) from this form of impulse response. A full description of the procedure is given by Nelson *et al.* (1990a) (see also Nelson *et al.*, 1988a). In addition to specifying the form of the optimal filter, one can also calculate the corresponding reduction in mean square pressure produced by implementing this filter. In general, it can be shown that the minimum value of J_p produced by the implementation of the optimal filter is given by

$$J_{p0} = E[d^2(t)] + E[d(t)\hat{d}_0(t)], \tag{8.6.18}$$

where $\hat{d}_0(t)$ is the output of the optimal filter. Upon noting that $E[d^2(t)] = J_{pp}$, the mean square pressure due to the primary field alone, it can be shown that for the case considered here

$$\frac{J_{p0}}{J_{pp}} = 1 - \frac{\mathrm{e}^{-2\zeta\omega_n\eta}}{\omega_0^2}\,[\omega_n^2\sin^2\omega_0\eta + (\omega_0\cos\omega_0\eta - \zeta\omega_n\sin\omega_0\eta)^2]. \tag{8.6.19}$$

The reduction in pressure is a function of both the natural frequency and the damping ratio of the second order filter. More particularly, the reduction appears as a function of the parameter $\omega_n\eta = \omega_n|r_p - r_s|/c_0$. This parameter can also be written as $2\pi|r_p - r_s|/\lambda_n$, which specifies the ratio of the difference in path length between the primary and secondary sources to the field point to the wavelength of the sound at the natural frequency of the shaping filter. A plot of the function J_{p0}/J_{pp} is shown

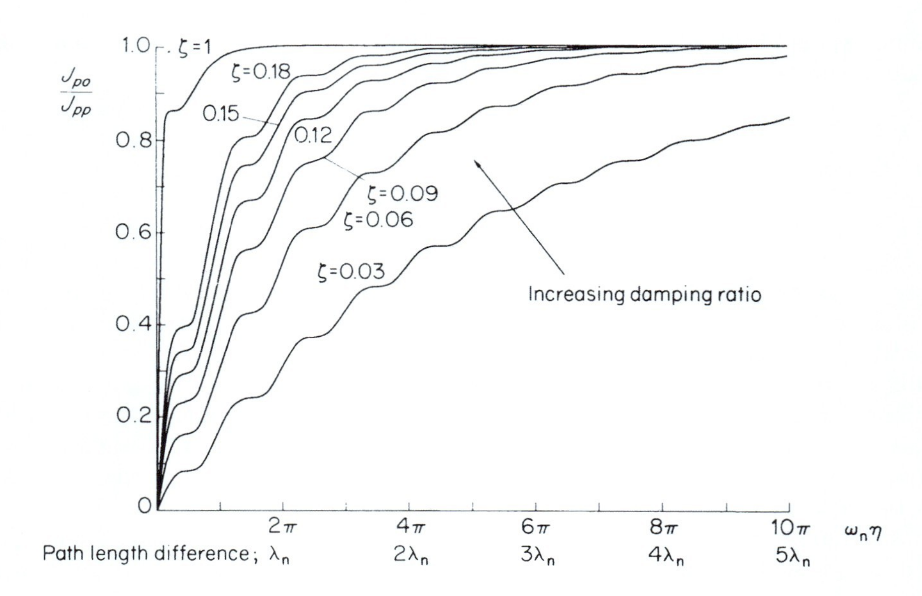

Fig. 8.9 The normalised minimum mean squared pressure in the arc $r_p < r_s$ as a function of the damping ratio of the second order shaping filter and the parameter $\omega_n\eta$. The latter quantifies the difference in path length between the primary and secondary source and the field point relative to the acoustic wavelength at the natural frequency of the shaping filter.

in Fig. 8.9 as a function of the path length difference parameter $\omega_n\eta$ and the damping ratio ζ of the second order filter. As one would expect, the largest reductions in mean squared pressure are produced when the shaping filter is lightly damped and when the path length difference of the field point is small compared to the acoustic wavelength at the natural frequency of the shaping filter. Also note that if either the damping ratio $\zeta \to 0$ (and the primary signal tends to a pure tone) or the path length difference $\omega_n\eta \to 0$ then $J_{p0}/J_{pp} \to 0$ and perfect cancellation will be produced.

The equivalent time domain calculation has been undertaken by Nelson *et al.* (1990a), which allows the specification of the minimum *power output* of a primary/secondary source pair in a free field when the primary source output is considered to be produced by a white noise excited shaping filter. The solution to this problem has been shown to consist of an optimal filter which has two well defined components. The first of these components is again a result of the need to predict future outputs of the primary source from current information in order that the secondary source may produce the appropriate "loading" on the primary source in order to reduce the primary source's ability to radiate sound. The second component of the optimal filter appears as a simple delay related to the travel time between the two sources. This component can be physically related to the action of the secondary source in absorbing sound from the primary source output. Thus, as we have seen in Section 5.6 and in Section 8.4, the mechanisms of active control are the prevention of energy escape from the primary source (through loading with a pressure produced by the secondary source) and the absorption of energy by the secondary source. These mechanisms can also be clearly identified when dealing with random source outputs.

8.7 The point dipole source

We have already noted in Section 8.2 the particular form of far field sound pressure produced by the interference of two sources of equal and opposite strength when these are spaced a small distance apart relative to the acoustic wavelength at the frequency of interest. We will now consider this "point dipole" source in more detail. We have seen in Chapter 5 that the dipole source has a particular significance in the active control of one-dimensional sound fields, where in combination with a monopole source it can act to produce unidirectional radiation. We shall see in Chapter 9 that this is also the case when a continuous layer of monopoles and dipoles is used to control three-dimensional radiation. As a prerequisite to this discussion it will be useful to describe the nature of the point dipole field. The point dipole results when the product of the strength of the component sources and their separation distance is held constant as the separation distance is allowed to vanish. With reference to Fig. 8.10 the net complex pressure produced by this arrangement of two equal and opposite sources can be written as

$$p(r,\theta) = j\omega\rho_0 q\left(\frac{e^{-jkr_1}}{4\pi r_1} - \frac{e^{-jkr_2}}{4\pi r_2}\right). \tag{8.7.1}$$

If we assume that the vector $\mathbf{x}$ (corresponding to co-ordinates x_1, x_2, x_3) specifies the position at which we wish to calculate the sound field and that the vector $\mathbf{y}$ (corresponding to co-ordinates y_1, y_2, y_3) specifies the position of the source, then for the particular arrangement illustrated in Fig. 8.10, we can write the radial distances r_1 and r_2 from the two component monopole sources as

$$r_1 = \sqrt{[(x_1 - (y_1 + \varepsilon_1))^2 + x_2^2 + x_3^2]}\,, \tag{8.7.2}$$

$$r_2 = \sqrt{[(x_1 - (y_1 - \varepsilon_1))^2 + x_2^2 + x_3^2]}\,, \tag{8.7.3}$$

where ε_1 is a distance which is small compared to the acoustic wavelength. The terms in the round brackets of equation (8.7.1) can now be regarded as functions of the type $f(y_1 \pm \varepsilon_1)$, such that we can use a Taylor series expansion of the form

$$f(y_1 + \varepsilon_1) = f(y_1) + \varepsilon_1 \frac{\partial}{\partial y_1} f(y_1) + \frac{1}{2!}\varepsilon_1^2 \frac{\partial^2}{\partial y_1^2} f(y_1) \;\ldots, \tag{8.7.4}$$

where ε_1 is the small expansion parameter. The partial derivatives in the series are evaluated with (x_1, x_2, x_3) kept constant. Thus we can write

$$\frac{e^{-jkr_1}}{4\pi r_1} \approx \frac{e^{-jkr}}{4\pi r} + \varepsilon_1 \frac{\partial}{\partial y_1}\left(\frac{e^{-jkr}}{4\pi r}\right) + \frac{1}{2!}\varepsilon_1^2 \frac{\partial^2}{\partial y_1^2}\left(\frac{e^{-jkr}}{4\pi r}\right) \;\ldots, \tag{8.7.5}$$

$$\frac{e^{-jkr_2}}{4\pi r_2} \approx \frac{e^{-jkr}}{4\pi r} - \varepsilon_1 \frac{\partial}{\partial y_1}\left(\frac{e^{-jkr}}{4\pi r}\right) + \frac{1}{2!}\varepsilon_1^2 \frac{\partial^2}{\partial y_1^2}\left(\frac{e^{-jkr}}{4\pi r}\right) \;\ldots. \tag{8.7.6}$$

In the limit of $\varepsilon_1 \to 0$, we need consider only the terms to order ε_1 in the series, and the expression for the radiated pressure reduces to

$$p(r,\theta) = j\omega\rho_0 q 2\varepsilon_1 \frac{\partial}{\partial y_1}\left(\frac{e^{-jkr}}{4\pi r}\right). \tag{8.7.7}$$

Denoting $d = 2\varepsilon_1$ as the distance between the two sources we can also write this expression as

$$p(r,\theta) = j\omega\rho_0 qd \frac{\partial r}{\partial y_1}\cdot\frac{\partial}{\partial r}\left(\frac{e^{-jkr}}{4\pi r}\right). \tag{8.7.8}$$

Since $r = \sqrt{[(x_1 - y_1)^2 + x_2^2 + x_3^2]}$, then $\partial r/\partial y_1 = (y_1 - x_1)/r = -\cos\theta$, where the angle θ is defined in Fig. 8.10. In addition, we have

$$\frac{\partial}{\partial r}\left(\frac{e^{-jkr}}{4\pi r}\right) = -\frac{jke^{-jkr}}{4\pi r}\left(1 - \frac{j}{kr}\right), \tag{8.7.9}$$

and therefore the net radiated field can be written as

$$p(r,\theta) = -\frac{\omega\rho_0 qe^{-jkr}}{4\pi r}(kd\cos\theta)\left(1 + \frac{1}{jkr}\right). \tag{8.7.10}$$

This is the sound field produced by a *point dipole* source. It is the sound field radiated in the limit when the distance d tends to zero but the product (qd) is held constant. Thus the strengths of the component monopoles are increased as the distance between them is reduced. The product (qd) is known as the *strength* or *moment* of the dipole. In the far field, when one is many wavelengths from the source, such that $kr \to \infty$, then this expression reduces to $p = p_M(jkd\cos\theta)$, where

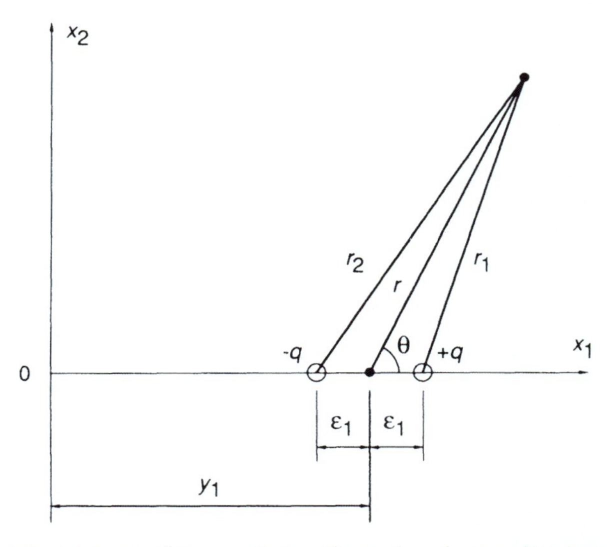

Fig. 8.10 Geometrical arrangement of two sources of equal and opposite strength separated by a distance $d = 2\varepsilon_1$.

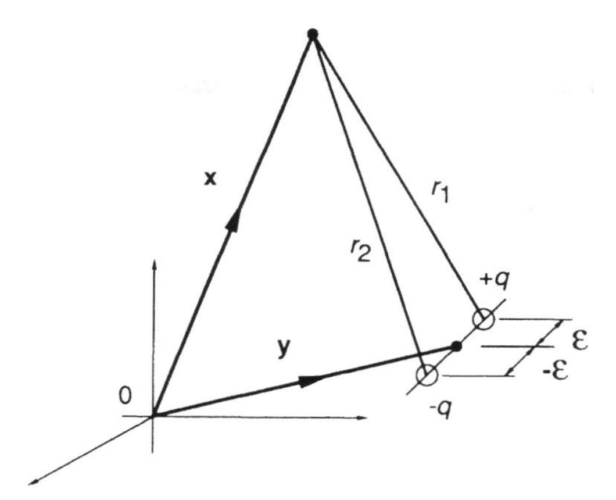

Fig. 8.11 Co-ordinate system for the analysis of the dipole field when the component monopole sources have arbitrary orientation.

p_M represents the far field pressure radiated by a point monopole source of strength q. Thus, the far field pressure radiated differs from that of a monopole in two respects. First, at $\theta = 0$, it is reduced in amplitude by a factor kd and second, it has a directional dependence specified by the term cos θ. In addition, close to the source, then the term $1/jkr$ becomes important and characterises the more intense near field of the dipole source. This near field has large velocity fluctuations associated with the oscillatory flow to and fro between the two component sources of the dipole. We can calculate the form of the velocity fluctuation by evaluating the radial component of particle velocity u_r. We know from the momentum conservation equation (1.11.9) that for harmonic fluctuations, the complex radial particle velocity is given by $u_r(r,\theta) = (j/\omega\rho_0)\partial p/\partial r$. It follows from differentiation of equation (8.7.10) that the radial component of the particle velocity due to the dipole can be written as

$$u_r(r,\theta) = -\frac{q\mathrm{e}^{-jkr}}{4\pi r}k^2 d \cos\theta\left(1 - \frac{2j}{kr} - \frac{2}{(kr)^2}\right). \tag{8.7.11}$$

Similarly we can calculate the *tangential* component of particle velocity u_θ which is given by (see Morse and Ingard, 1968, Ch. 7)

$$u_\theta(r,\theta) = -\frac{q\mathrm{e}^{-jkr}}{4\pi r}k^2 d \sin\theta\left(\frac{j}{kr} + \frac{1}{(kr)^2}\right). \tag{8.7.12}$$

We shall show in Chapter 9 that an identical sound field is produced by the application of a point force to the medium. The origin of this force is associated with the rate of change of momentum of the fluid flowing between the two component monopoles. The relationship between the fluctuating flow produced between the two sources and the net force on the fluid is most easily described in the one-dimensional case considered in Section 5.9.

It is also important to note that in the three-dimensional case, the sound field produced depends on the orientation of the axis on which the two component monopole sources lie. This can be explained with reference to Fig. 8.11, where the radial distance from the two component monopole sources to the field point are given by $r_1 = |\mathbf{x} - (\mathbf{y} + \boldsymbol{\varepsilon})|$ and $r_2 = |\mathbf{x} - (\mathbf{y} - \boldsymbol{\varepsilon})|$. The vector $\boldsymbol{\varepsilon}$ has components $(\varepsilon_1\mathbf{i}, \varepsilon_2\mathbf{j}, \varepsilon_3\mathbf{k})$ where ε_1, ε_2 and ε_3 are all small and $\mathbf{i}$, $\mathbf{j}$, $\mathbf{k}$ specify the unit vectors in the three Cartesian co-ordinate directions. In this case each of the terms in the square brackets of equation (8.7.1) can be regarded as a function of the three variables y_1, y_2, y_3 and can be expanded in the multi-dimensional Taylor series given by

$$\begin{aligned} f(y_1 + \varepsilon_1, y_2 + \varepsilon_2, y_3 + \varepsilon_3) &= f(y_1, y_2, y_3) \\ &+ \left(\varepsilon_1 \frac{\partial}{\partial y_1} + \varepsilon_2 \frac{\partial}{\partial y_2} + \varepsilon_3 \frac{\partial}{\partial y_3}\right) f(y_1, y_2, y_3) \\ &+ \frac{1}{2!}\left(\varepsilon_1 \frac{\partial}{\partial y_1} + \varepsilon_2 \frac{\partial}{\partial y_2} + \varepsilon_3 \frac{\partial}{\partial y_3}\right)^2 f(y_1, y_2, y_3) + \ldots \end{aligned} \tag{8.7.13}$$

See, for example, Spencer *et al.* (1977) for a description of this form of series expansion. Note that we can also write this series expansion more compactly as

$$f(\mathbf{y} + \boldsymbol{\varepsilon}) = f(\mathbf{y}) + \boldsymbol{\varepsilon}.\boldsymbol{\nabla}_y f(\mathbf{y}) + \frac{1}{2!}(\boldsymbol{\varepsilon}.\boldsymbol{\nabla}_y)^2 f(\mathbf{y}) \ldots, \tag{8.7.14}$$

where $\boldsymbol{\nabla}_y f = \partial f/\partial y_1 \mathbf{i} + \partial f/\partial y_2 \mathbf{j} + \partial f/\partial y_3 \mathbf{k}$. Thus we have used $\boldsymbol{\nabla}_y$ to denote the del operator with respect to the y co-ordinates. We shall continue to use simply $\boldsymbol{\nabla}$ to denote the del operator with respect to the x co-ordinates. Thus to leading order in the small parameter vector $\boldsymbol{\varepsilon}$ we can write the terms in the brackets of equation (8.7.1) as

$$\frac{e^{-jkr_1}}{4\pi r_1} \approx \frac{e^{-jkr}}{4\pi r} + \boldsymbol{\varepsilon}.\boldsymbol{\nabla}_y \left(\frac{e^{-jkr}}{4\pi r}\right), \tag{8.7.15}$$

$$\frac{e^{-jkr_2}}{4\pi r_2} \approx \frac{e^{-jkr}}{4\pi r} - \boldsymbol{\varepsilon}.\boldsymbol{\nabla}_y \left(\frac{e^{-jkr}}{4\pi r}\right). \tag{8.7.16}$$

Since $r = |\mathbf{x} - \mathbf{y}|$ then the expression for the pressure field can be written in the form

$$p(r) = j\omega\rho_0 q\mathbf{d}.\boldsymbol{\nabla}_y \left(\frac{e^{-jkr}}{4\pi r}\right). \tag{8.7.17}$$

The product $q\mathbf{d}$ is the *vector dipole strength* which, as we shall see in Chapter 9, can be directly related to the force vector applied to the fluid. Finally, it is sometimes useful to note that

$$\boldsymbol{\nabla}_y \left(\frac{e^{-jkr}}{4\pi r}\right) = (\boldsymbol{\nabla}_y r)\frac{\partial}{\partial r}\left(\frac{e^{-jkr}}{4\pi r}\right), \tag{8.7.18}$$

and also that

$$\boldsymbol{\nabla}_y r = -\boldsymbol{\nabla} r = -\frac{1}{r}(\mathbf{x} - \mathbf{y}). \tag{8.7.19}$$

Since $\mathbf{n}_r = (\mathbf{x} - \mathbf{y})/r$ is the unit vector pointing from the source towards the field point, then the expression for the pressure field can be written as

$$p(r) = -j\omega\rho_0 q\left[\mathbf{d}.\mathbf{n}_r \frac{\partial}{\partial r}\left(\frac{e^{-jkr}}{4\pi r}\right)\right]. \tag{8.7.20}$$

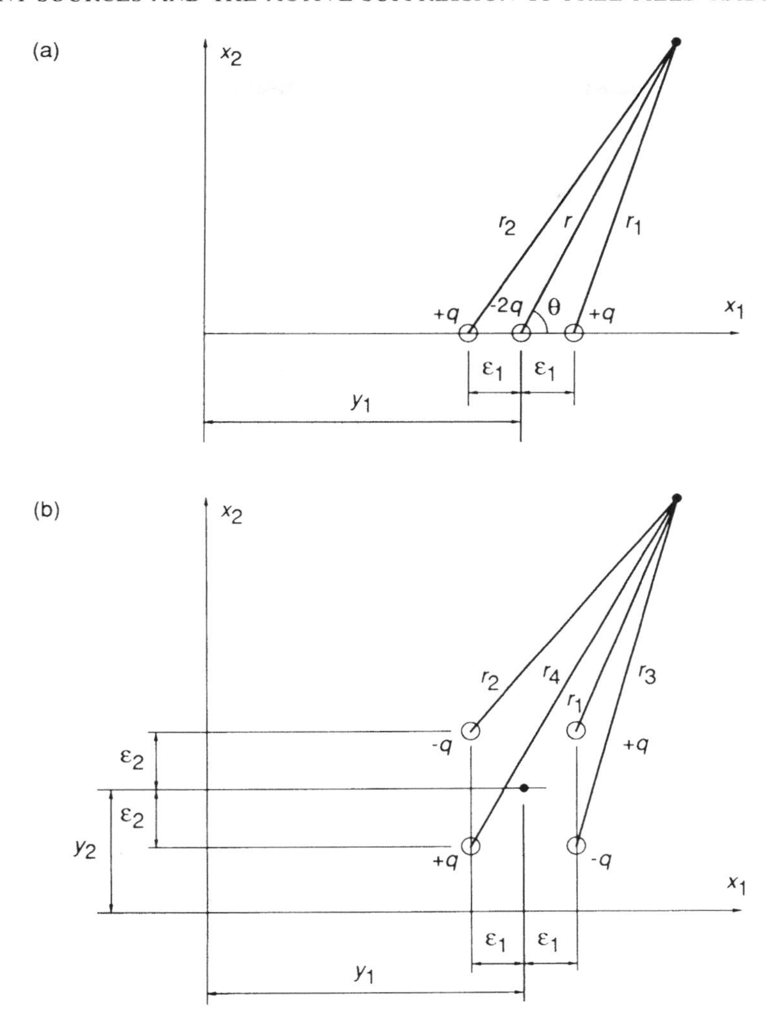

Fig. 8.12 Co-ordinate systems used for the analysis of (a) a longitudinal quadrupole and (b) a lateral quadrupole.

8.8 Point quadrupole sources

A further sound field of particular interest is that produced by the source arrangement depicted in Fig. 8.12(a), where we have a source of strength $-2q$ at the position $(y_1,0,0)$ and further sources of strength $+q$ at positions $(y_1 + \varepsilon_1,0,0)$ and $(y_1 - \varepsilon_1,0,0)$. This arrangement, known as a *longitudinal quadrupole source*, corresponds to two opposed dipoles positioned a distance ε_1 apart; each dipole consists of two monopole sources again a distance ε_1 apart. The pressure field radiated is

$$p(r,\theta) = j\omega\rho_0 q \left[\frac{e^{-jkr_1}}{4\pi r_1} + \frac{e^{-jkr_2}}{4\pi r_2} - \frac{2e^{-jkr}}{4\pi r} \right]. \tag{8.8.1}$$

If we use the series expansions given by equations (8.7.5) and (8.7.6) then the terms of order ε_1 ($O(\varepsilon_1)$) add to zero and only the terms $O(\varepsilon_1^2)$ remain. The expression for the pressure field that results is thus given by

$$p(r,\theta) = j\omega\rho_0 q d^2 \frac{\partial^2}{\partial y_1^2}\left(\frac{e^{-jkr}}{4\pi r}\right), \tag{8.8.2}$$

where we have replaced ε_1 by d. The product qd^2 is known as the strength of the quadrupole. An explicit expression for the sound field radiated can be derived by evaluating the double derivative with respect to y_1 in equation (8.8.2). After some algebra, the expression that results is given by

$$p(r,\theta) = -\frac{j\omega\rho_0 q e^{-jkr}}{4\pi r}(kd)^2 \left[\cos^2\theta \left(1 + \frac{3}{jkr} + \frac{3}{(jkr)^2} \right) - \frac{1}{jkr} - \frac{1}{(jkr)^2} \right]. \tag{8.8.3}$$

Note the limit of this expression in the far field: when $kr \to \infty$, the expression takes the form $p = -p_M(kd \cos\theta)^2$, where p_M is again the pressure radiated by a point monopole source of strength q. Thus the amplitude of the pressure at $\theta = 0$ is now a factor $(kd)^2$ less than that of a monopole, and we now have a directivity which is dependent on $\cos^2\theta$. In addition, as kr becomes small, then terms $O(1/kr)^2$ dominate the near field of the quadrupole. The directivity of a longitudinal quadrupole is sketched in Fig. 8.13.

Another form of quadrupole field is that produced by the source arrangement depicted in Fig. 8.13(b). This type of arrangement is known as a *lateral quadrupole* and has the pressure field given by

$$p(r) = j\omega\rho_0 q \left(\frac{e^{-jkr_1}}{4\pi r_1} - \frac{e^{-jkr_2}}{4\pi r_2} - \frac{e^{-jkr_3}}{4\pi r_3} + \frac{e^{-jkr_4}}{4\pi r_4} \right). \tag{8.8.4}$$

If, as illustrated in Fig. 8.13(b), we assign to the centre of the quadrupole source array the co-ordinates (y_1, y_2, 0), then the expression for the radial distance r_1, for example, can be written as

$$r_1 = \sqrt{[(x_1 - (y_1 + \varepsilon_1))^2 + (x_2 - (y_2 + \varepsilon_2))^2 + x_3^2]}. \tag{8.8.5}$$

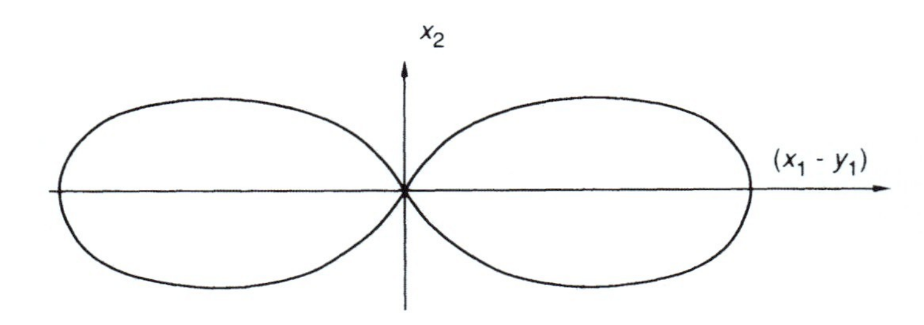

Fig. 8.13 Directivity of a longitudinal quadrupole.

Similar expressions can be written for the other distances r_2, r_3 and r_4. Use of the Taylor series expansion specified in equation (8.7.13) shows that we can write

$$\frac{e^{-jkr_1}}{4\pi r_1} \approx \frac{e^{-jkr}}{4\pi r} + \left(\varepsilon_1 \frac{\partial}{\partial y_1} + \varepsilon_2 \frac{\partial}{\partial y_2} \right) \frac{e^{-jkr}}{4\pi r} + \frac{1}{2!} \left(\varepsilon_1 \frac{\partial}{\partial y_1} + \varepsilon_2 \frac{\partial}{\partial y_2} \right)^2 \frac{e^{-jkr}}{4\pi r}. \quad (8.8.6)$$

Similar series expansions can be derived for the other terms in the brackets of equation (8.8.4). The summation of these series expansions again shows that terms $O(\varepsilon)$ vanish and the expression for the pressure field to $O(\varepsilon^2)$ reduces to

$$p(r) = j\omega\rho_0 q d^2 \frac{\partial^2}{\partial y_1 \partial y_2} \left(\frac{e^{-jkr}}{4\pi r} \right), \quad (8.8.7)$$

for a "square" quadrupole with $d = 2\varepsilon_1 = 2\varepsilon_2$. Evaluation of the partial derivatives with respect to y_1 and y_2 shows, again after some effort, that the pressure field can be written as

$$p(r) = -\frac{j\omega\rho_0 q e^{-jkr}}{4\pi r} (kd)^2 \left[\frac{(x_1 - y_1)(x_2 - y_2)}{r^2} \left(1 + \frac{3}{jkr} + \frac{3}{(jkr)^2} \right) \right] \quad (8.8.8)$$

Note again that the far field pressure due to this source array is a factor $(kd)^2$ less than that due to a monopole source of strength q. The directivity of the far field radiation from a lateral quadrupole of this type is shown in Fig. 8.14.

In addition to the particular quadrupole arrangements considered here, one can also conceive of longitudinal quadrupoles which are oriented with their axes parallel with the y_2 and y_3 axes. Thus, there are three types of longitudinal quadrupole, each of which have the pressure field specified by equation (8.8.2) but containing one of the differential operators $\partial^2/\partial y_1^2$, $\partial^2/\partial y_2^2$, $\partial^2/\partial y_3^2$, Similarly in addition to the particular lateral quadrupole arrangement considered above that consists of a

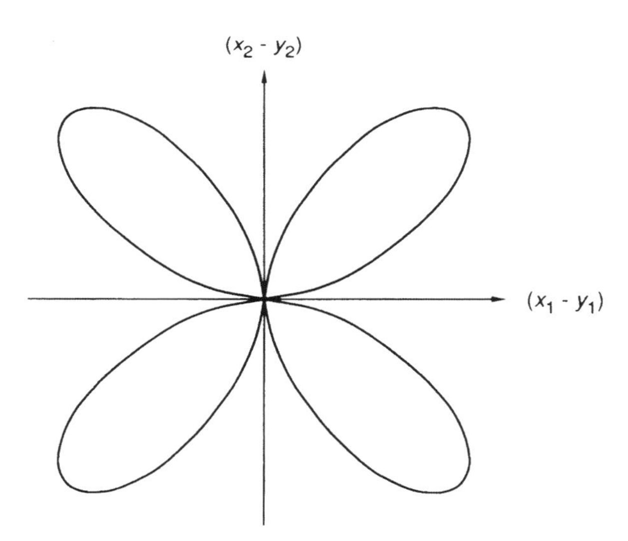

Fig. 8.14 Directivity of a lateral quadrupole.

distribution of monopole sources in the y_1, y_2 plane, one can equally conceive of identical source arrangements but in the y_1, y_3 plane, and the y_2, y_3 plane. Thus, there are in all three types of lateral quadrupole corresponding to the differential operators given by $\partial^2/\partial y_1 \partial y_2$, $\partial^2/\partial y_1 \partial y_3$, and $\partial^2/\partial y_2 \partial y_3$. (Note that these three operators are also equivalent to the operators $\partial^2/\partial y_2 \partial y_1$, $\partial^2/\partial y_3 \partial y_1$ and $\partial^2/\partial y_3 \partial y_2$.)

Finally it is interesting to note that the combination of three orthogonal longitudinal quadrupoles having the same source strength produces a monopole-like radiation field. If we combine the three fields due to the three component longitudinal quadrupoles (see equation (8.8.2)), the expression for the radiated pressure can be written as

$$p_Q = d^2 \left(\frac{\partial^2}{\partial y_1^2} + \frac{\partial^2}{\partial y_2^2} + \frac{\partial^2}{\partial y_3^2} \right) \frac{j\omega\rho_0 q e^{-jkr}}{4\pi r}. \tag{8.8.9}$$

This can be written more compactly as

$$p_Q = d^2 \nabla_y^2 p_M, \tag{8.8.10}$$

where p_M is again the pressure field radiated by a single point monopole. We know, however, from the Helmholtz equation that $(\nabla_y^2 + k^2)p_M = 0$ and therefore it follows that

$$p_Q = -(kd)^2 p_M. \tag{8.8.11}$$

Thus the field of the three orthogonal longitudinal quadrupoles is identical to that of a single point monopole source but reduced in amplitude by the factor $(kd)^2$.

8.9 Multipole analysis

If we have a collection of point monopole sources clustered around a given position specified by the vector $\mathbf{y}$, and the region occupied by the sources is small compared to the acoustic wavelength, we can again use a series expansion technique to express the net field of the source distribution in terms of an equivalent series of *point multipoles* located at the position $\mathbf{y}$. The net field due to the combination of monopoles each having strength q_n is given by

$$p(r) = \sum_{n=1}^{N} \frac{j\omega\rho_0 q_n e^{-jkr_n}}{4\pi r_n}. \tag{8.9.1}$$

where r_n represents the radial distance from the nth monopole source to the field point (see Fig. 8.15) and $r = |\mathbf{x} - \mathbf{y}|$. The spatial dependence of the pressure field due to the nth monopole can be expanded by using equation (8.7.14) as

$$\frac{e^{-jkr_n}}{4\pi r_n} = \frac{e^{-jkr}}{4\pi r} + \mathbf{d}_n . \boldsymbol{\nabla}_y \left(\frac{e^{-jkr}}{4\pi r} \right) + \frac{1}{2!} (\mathbf{d}_n . \boldsymbol{\nabla}_y)^2 \left(\frac{e^{-jkr}}{4\pi r} \right) + \ldots, \tag{8.9.2}$$

where $\mathbf{d}_n$ is the position vector of the nth monopole relative to the point specified by $\mathbf{y}$. This is illustrated in Fig. 8.15. Thus the net field can be expressed as

$$p(r) = \left[\sum_{n=1}^{N} q_n + \left(\sum_{n=1}^{N} q_n \mathbf{d}_n \right) . \boldsymbol{\nabla}_y + \frac{1}{2} \sum_{n=1}^{N} q_n (\mathbf{d}_n . \boldsymbol{\nabla}_y)^2 \ldots \right] \frac{j\omega\rho_0 e^{-jkr}}{4\pi r}. \tag{8.9.3}$$

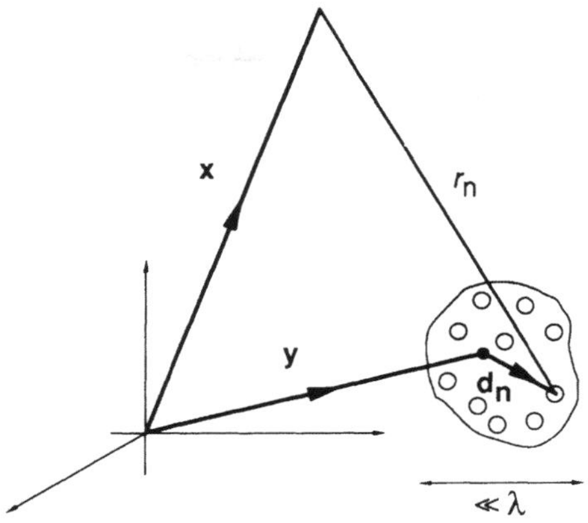

Fig. 8.15 Co-ordinate system used to express the field at **x** due to a number of point monopoles in a compact region as a series of multipoles at **y**.

Following the nomenclature adopted by Pierce (1981, Ch. 4), we can now define respectively the monopole and dipole strengths of the source distribution as

$$Q = \sum_{n=1}^{N} q_n, \qquad \mathbf{D} = \sum_{n=1}^{N} q_n \mathbf{d}_n. \tag{8.9.4}$$

Note that the third term in the series results from the operator

$$(\mathbf{d}_n.\boldsymbol{\nabla}_y)^2 = \left(d_{n1}\frac{\partial}{\partial y_1} + d_{n2}\frac{\partial}{\partial y_2} + d_{n3}\frac{\partial}{\partial y_3}\right)\left(d_{n1}\frac{\partial}{\partial y_1} + d_{n2}\frac{\partial}{\partial y_2} + d_{n3}\frac{\partial}{\partial y_3}\right), \tag{8.9.5}$$

and that there are nine possible combinations of differential operator which result from this product. Six of these terms express the fields of lateral quadrupoles (i.e. the operators $\partial^2/\partial y_1 \partial y_2 = \partial^2/\partial y_2 \partial y_1$, $\partial^2/\partial y_1 \partial y_3 = \partial^2/\partial y_3 \partial y_1$ and $\partial^2/\partial y_2 \partial y_3 = \partial^2/\partial y_3 \partial y_2$) whilst the other three terms are associated with longitudinal quadrupoles (i.e. the operators $\partial^2/\partial y_1^2$, $\partial^2/\partial y_2^2$, $\partial^2/\partial y_3^2$). Thus, if we define the quadrupole strengths of the source distribution as

$$Q_{\mu\nu} = \frac{1}{2}\sum_{n=1}^{N} d_{n\mu} d_{n\nu} q_n, \tag{8.9.6}$$

where we have again adopted the notation of Pierce (1981), the net field can be written in the form

$$p(r) = \left[Q + \mathbf{D}.\boldsymbol{\nabla}_y + Q_{\mu\nu} \sum_{\mu=1,\ \nu=1}^{\mu=3,\ \nu=3} \frac{\partial^2}{\partial y_\mu \partial y_\nu} \cdots \right] \frac{j\omega\rho_0 e^{-jkr}}{4\pi r}. \tag{8.9.7}$$

It is also useful to note that when applied to a function $f(r)$ then $\boldsymbol{\nabla} f(r) = -\boldsymbol{\nabla}_y f(r)$ and the form of the radiated field can be expressed as

$$p(r) = \left[Q - \mathbf{D}.\boldsymbol{\nabla} + Q_{\mu\nu} \sum_{\mu=1,\ \nu=1}^{\mu=3,\ \nu=3} \frac{\partial^2}{\partial x_\mu \partial x_\nu} \cdots \right] \frac{j\omega\rho_0 \mathrm{e}^{-jkr}}{4\pi r}. \tag{8.9.8}$$

Finally, we should point out that the technique of multipole analysis can be extended to deal with even more complex compact regions having an arbitrary distribution of source strength, and the formulation presented here can be made more general (see, for example, the work of Doak, 1973a).

8.10 Kempton's suggestion

Inspection of equation (8.9.8) suggests that, given a collection of sources occupying a compact region, it might be possible to achieve a cancellation of the far field by evaluating the multipole moments of the source distribution and then introduce a set of multipole sources having strengths equal in amplitude but opposite in sign to those of the source distribution. Kempton (1976) recognised this possibility and investigated it by doing some numerical calculations for a simple case, that of attempting to cancel the far field of a point monopole at the co-ordinate position $(-d,0,0)$ by an appropriate set of multipoles at the origin. By using computer algebra, he evaluated the degree of far field cancellation as a function of the finite number M of terms used in the multipole expansion of the point monopole field, which is

$$p(r) = \frac{j\omega\rho_0 q}{4\pi}\left[1 - d\frac{\partial}{\partial x_1} + \frac{d^2}{2!}\frac{\partial^2}{\partial x_1^2} \cdots \frac{d^M}{M!}\frac{\partial^M}{\partial x_1^M} \right] \frac{\mathrm{e}^{-jkr}}{r}, \tag{8.10.1}$$

where $r=(x_1^2 + x_2^2 + x_3^2)^{1/2}$ is distance from the origin. The calculations showed that the reduction achieved with the series truncated after the Mth term was

$$20 \log_{10} \left[\frac{M!}{(kd)^M \cos^M \theta} \right] \text{dB}. \tag{8.10.2}$$

Thus, for example, if $kd = 1$ a far field reduction of almost 30 dB is attained for $\theta = 0°$ and 180° with only four terms used in the series. If $kd = 5$, however, 14 terms are needed to achieve the same cancellation. Figure 8.16 shows some typical results obtained by Kempton for $kd = 1,3,5$ at different angles θ. It can be seen that large sound pressures are predicted at positions near the origin. This is related to the fact, as Kempton pointed out, that the multipole expansion does not converge for values of $r < d$. Further examples of Kempton's results are shown in a more graphical form in Fig. 8.17(a) for $kd = 3$ and Fig. 8.17(b) for $kd = 5$. The ability of the multipole series to match the far field when an increasing number of terms is used is well illustrated in both these cases.

In evaluating Kempton's results, it must be recognised at the outset that although with $M \rightarrow \infty$ the multipole expression (8.10.1) converges uniformly to the field produced by the point monopole at $(-d,0,0)$ for all positions outside the sphere $r = d$, it does not do so for any points inside this sphere. At $r = 0$ the point multipole field is simply $j\omega\rho_0 q\mathrm{e}^{-jkd}/4\pi d$ but the multipole expression (8.10.1) blows up completely: its Mth term goes to infinity like $1/r^{(M+1)}$ as $r \rightarrow 0$. As Kempton

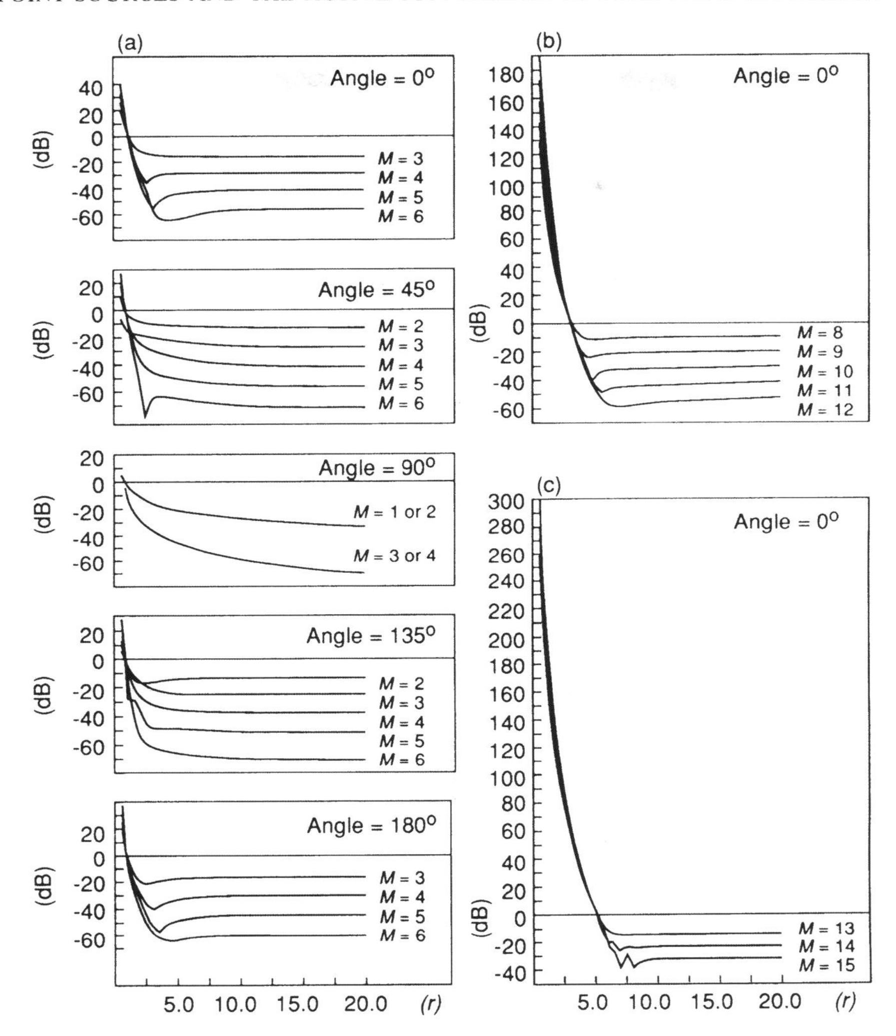

Fig. 8.16 The interference of the fields of the monopole and the series of multipoles at different distances. Number of multipoles in the series = M. Far field limits indicated to the right. (a) $kd = 1$; angles 0°, 45°, 90°, 135°, 180°. (b) $kd = 3$; angle 0°. (c) $kd = 5$; angle 0° (after Kempton, 1976).

mentioned (p. 476 of his paper), "it is clear that such an infinite collection of multipoles is not sensible". Equally clearly, no point multipole, of any order including monopole, is a physical reality, because of this singular behaviour at the point where it is located. The difficulty in approximating higher order point multipoles by an array of real transducers is all too evident.

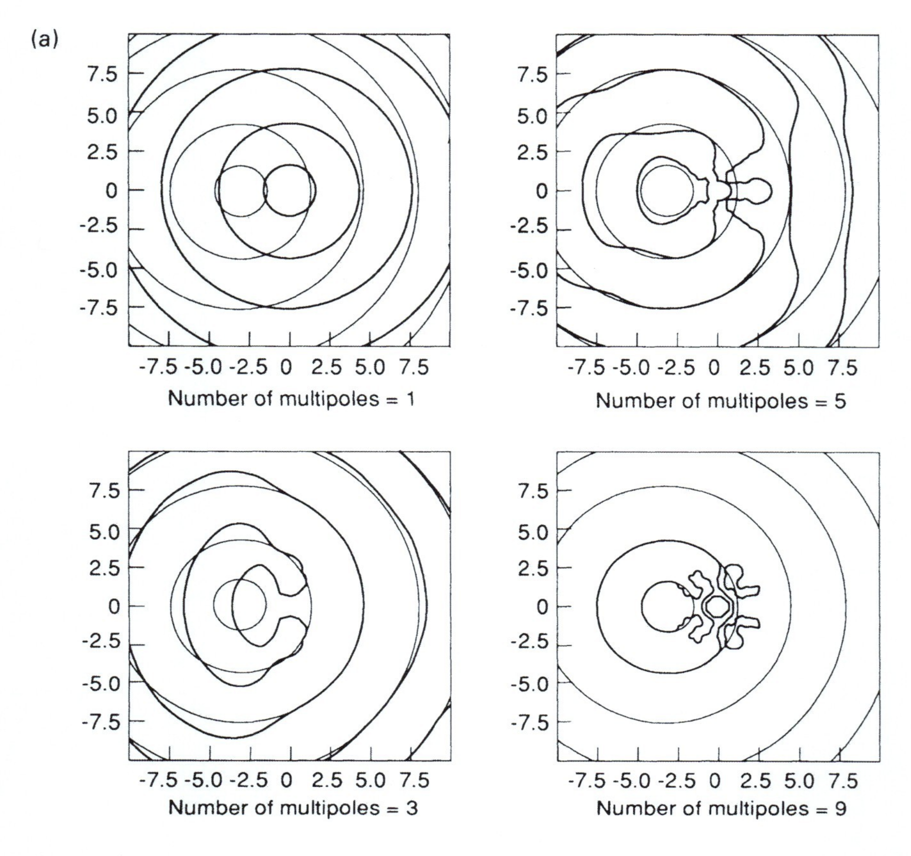

Fig. 8.17 (a) Contours of constant phase of the fields due to the point monopole at $(-d,0,0)$ and a series of multipoles at the origin. Results after Kempton (1976) for $kd = 3$.

8.11 Control of periodic free field sound using multiple sources and sensors

It follows from the discussion of the preceding sections that increasingly large cancellation of the far field from a compact source can be produced by using increasingly sophisticated distributions of control sources. As a simple example, given a point monopole source, then a secondary source placed close to it will produce a reduction in the far field pressure of $O(kd)$ as the monopole field degenerates into a dipole-type field. The use of two secondary sources appropriately arranged close to a primary source can produce a reduction $O(kd)^2$ as a quadrupole type field is produced. A simple example of the practical use of this technique is illustrated in Fig. 8.18. This shows the results of some experiments undertaken by Hesselmann (1978) who used two secondary loudspeaker sources on either side of an electrical transformer (which he assumed to radiate as a monopole source at 100 Hz)

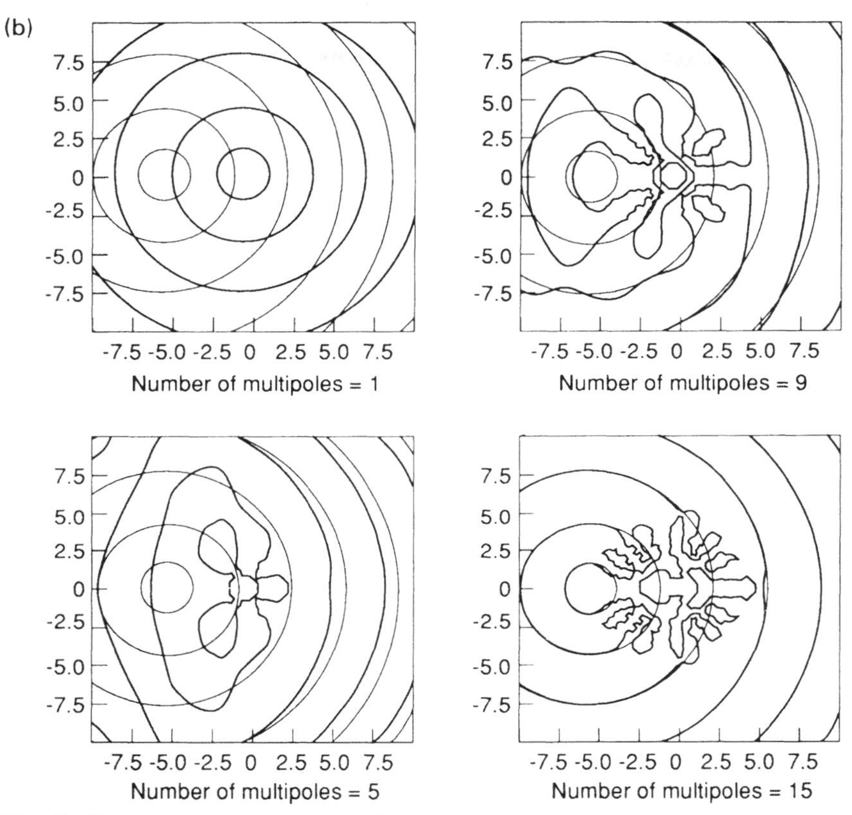

Fig. 17 (b) Contours of constant phase of the fields due to the point monopole at $(-d,0,0)$ and a series of multipoles at the origin. Results after Kempton (1976) for $kd = 5$.

and attempted to produce a net longitudinal quadrupole-type source distribution. Note that the reductions in far field pressure are accompanied by increased near field pressures. However, in practical situations, to extend this approach to the extreme illustrated by Kempton is formidably difficult for the reasons suggested at the end of Section 8.10. One recognises that the use of multiple secondary sources must enable a greater degree of cancellation to be produced but that we also need to "sense" the far field pressure produced by the interference of the primary and secondary source distributions A logical practical approach to the problem is that presented here. This is to use the sum of the squared pressures at a number of far field locations as a measure of the total field radiated and then to determine the complex strengths of a number of point secondary sources required in order to minimise that sum. The analysis is again conducted in the frequency domain and the constraint of causality described in Section 8.6 should always be borne in mind. However, for deterministic sources, such as electrical transformers, the analysis presented below provides an ideal theoretical framework. Thus, we define a vector $\mathbf{p}_p$ of complex acoustic pressures produced by the primary source at L locations in the sound field. This is specified by

$$\mathbf{p}_p^{\mathrm{T}} = [p_p(\mathbf{x}_1)\ p_p(\mathbf{x}_2)\ \ldots.\ p_p(\mathbf{x}_L)]. \tag{8.11.1}$$

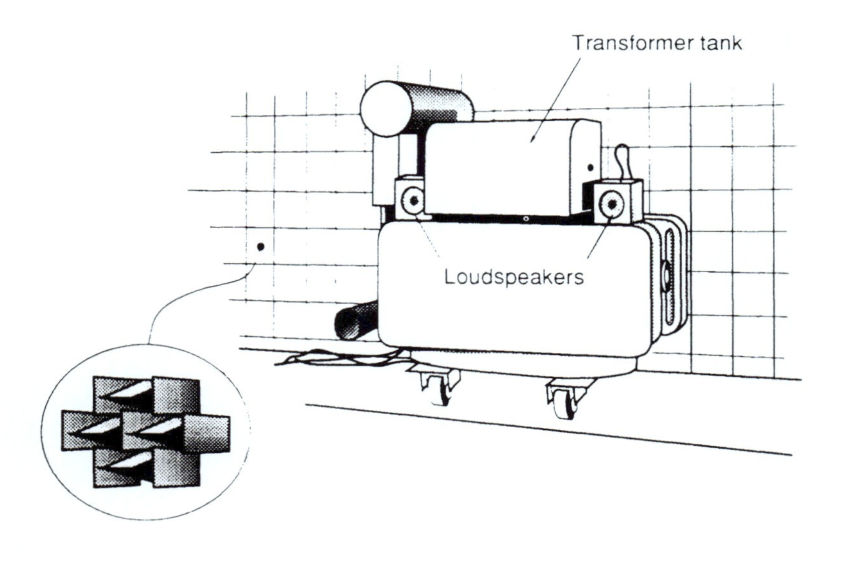

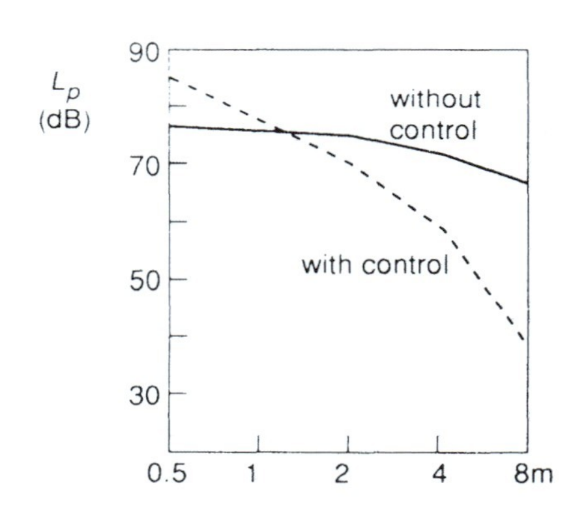

Fig. 8.18 Results of the experiment reported by Hesselmann (1978), using two loudspeakers to control the sound radiated by a 100 kVA transformer tank. A sketch of the experimental arrangement is shown together with the variation of sound pressure level at 100 Hz (averaged over several directions) with distance from the transformer tank. Note the increase in sound pressure in the near field of the tank.

We now wish to synthesise a source distribution which interferes with the pressures at these locations and which produces a vector of pressures

$$\mathbf{p}_s^{\mathrm{T}} = [p_s(\mathbf{x}_1)\ p_s(\mathbf{x}_2)\ \ldots.\ p_s(\mathbf{x}_L)]. \tag{8.11.2}$$

If we further assume that M secondary sources are used, then we can define a complex vector of source strengths

$$\mathbf{q}_s^{\mathrm{T}} = [q_s(\mathbf{y}_1)\ q_s(\mathbf{y}_2)\ \ldots.\ q_s(\mathbf{y}_M)]. \tag{8.11.3}$$

We will further assume that the complex pressures produced by the secondary sources are related to the strengths of the sources via a matrix $\mathbf{Z}$ of complex acoustic transfer impedances (see Section 1.12) such that

$$\mathbf{p}_s = \mathbf{Z}\mathbf{q}_s, \tag{8.11.4}$$

where the matrix $\mathbf{Z}$ is specified by

$$\mathbf{Z} = \begin{bmatrix} Z(\mathbf{x}_1|\mathbf{y}_1) & Z(\mathbf{x}_1|\mathbf{y}_2) & \ldots & Z(\mathbf{x}_1|\mathbf{y}_M) \\ Z(\mathbf{x}_2|\mathbf{y}_1) & Z(\mathbf{x}_2|\mathbf{y}_2) & \ldots & Z(\mathbf{x}_2|\mathbf{y}_M) \\ \cdot & & & \\ \cdot & & & \\ Z(\mathbf{x}_L|\mathbf{y}_1) & Z(\mathbf{x}_L|\mathbf{y}_2) & \ldots & Z(\mathbf{x}_L|\mathbf{y}_M) \end{bmatrix}. \tag{8.11.5}$$

We now use the principle of superposition to write the total vector of pressures produced at the L locations as

$$\mathbf{p} = \mathbf{p}_p + \mathbf{p}_s, \tag{8.11.6}$$

which by using equation (8.11.4) can be written in the form

$$\mathbf{p} = \mathbf{p}_p + \mathbf{Z}\mathbf{q}_s. \tag{8.11.7}$$

If we choose the number of secondary sources M to be equal to the number of measurement locations L, then we can ensure that the vector $\mathbf{p} = 0$, so that

$$\mathbf{p}_p + \mathbf{Z}\mathbf{q}_s = 0, \tag{8.11.8}$$

where the matrix $\mathbf{Z}$ is square. This system of linear equations has the solution given by

$$\mathbf{q}_s = -\mathbf{Z}^{-1}\mathbf{p}_p, \tag{8.11.9}$$

provided that $\mathbf{Z}$ is non-singular (see the appendix). It is therefore always possible by using L secondary sources to produce zero pressure in a harmonic sound field at L locations. This property was exploited by Kido and Onoda (1972) who used up to three loudspeakers and three microphones to control the far field radiation produced by an electrical transformer. Figure 8.19 shows their results for the 100 Hz component of the transformer noise spectrum. It can be seen that concentrating the three sensing microphones in a certain arc enabled substantial reductions to be produced over a reasonably wide angle. Note, however, that increases in the sound pressure were also produced at other angular locations (see, for example, Fig. 8.19(c)). This is very often the case with active control. If an equal number of sources and sensors are used to produce zero pressure at a certain number of locations, whilst being very effective at those locations, substantial increases can be produced at other positions in the sound field. As we have seen in Section 8.5, a technique for producing reductions which are smaller but which can be more widespread is given by minimising the sum of the squares of the pressures in the sound field at a greater number of locations L than there are secondary sources M. Thus, we define a cost function specified by

$$J_p = \sum_{l=1}^{L} |p(\mathbf{x}_l)|^2, \tag{8.11.10}$$

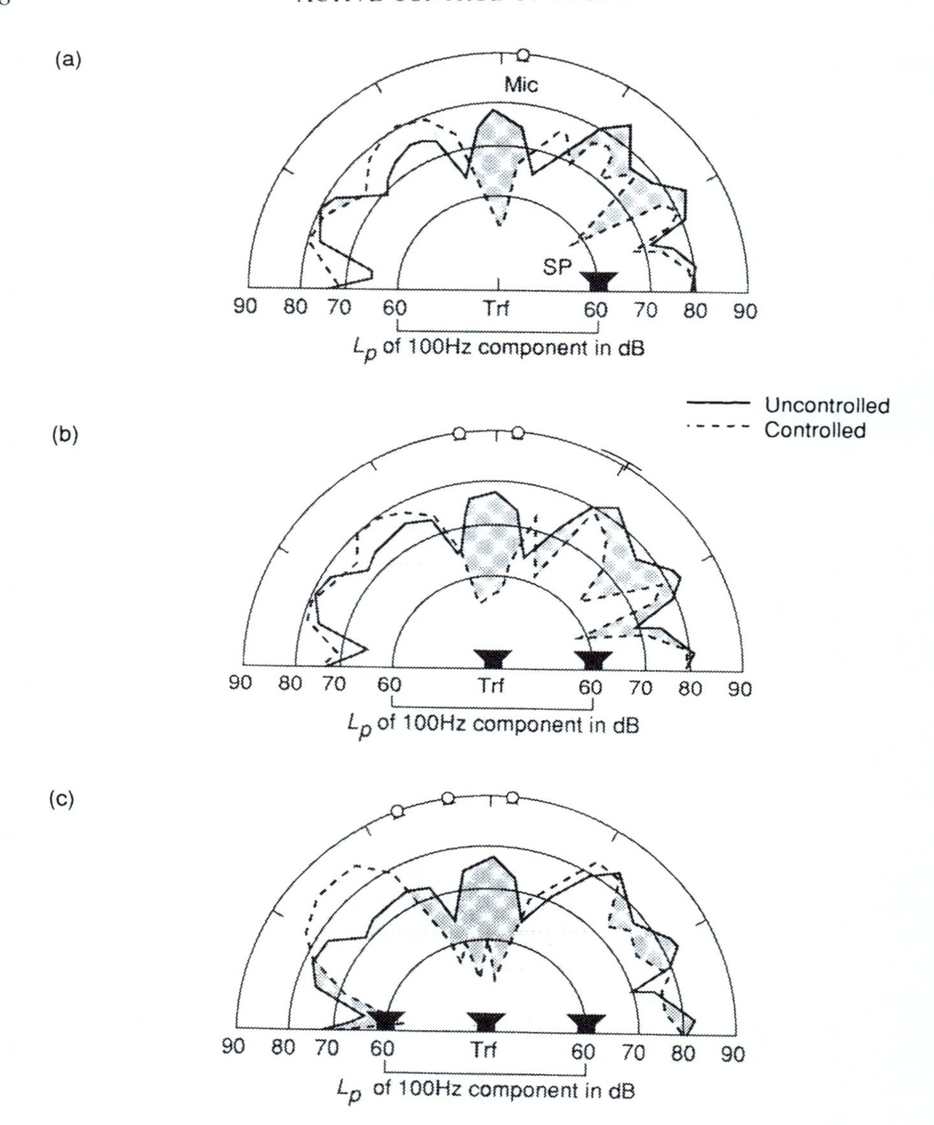

Fig. 8.19 Practical experimental results of field trials of a three-source/three-sensor active control system devised by Kido and Onoda (1972) to control the noise from a sub-station transformer. Results are shown in the form of directivity plots in the horizontal plane using (a) one source/one sensor (b) two sources/two sensors and (c) three sources/three sensors. Thick lines represent uncontrolled conditions and broken lines represent controlled conditions.

which, in vector notation, can be written as

$$J_p = \mathbf{p}^{\mathrm{H}}\mathbf{p}, \tag{8.11.11}$$

where the superscript H denotes the Hermitian transpose or complex conjugate of the vector transpose. Substitution of equation (8.11.7) into this expression shows that the cost function can be written as

$$J_p = (\mathbf{p}_p + \mathbf{Z}\mathbf{q}_s)^{\mathrm{H}}\,(\mathbf{p}_p + \mathbf{Z}\mathbf{q}_s). \tag{8.11.12}$$

Expansion of this expression gives

$$J_p = \mathbf{p}_p{}^{\mathrm{H}}\mathbf{p}_p + \mathbf{p}_p{}^{\mathrm{H}}\mathbf{Z}\mathbf{q}_s + \mathbf{q}_s{}^{\mathrm{H}}\mathbf{Z}^{\mathrm{H}}\mathbf{p}_p + \mathbf{q}_s{}^{\mathrm{H}}\mathbf{Z}^{\mathrm{H}}\mathbf{Z}\mathbf{q}_s. \tag{8.11.13}$$

and it is evident that this is a quadratic function of the secondary source strength vector $\mathbf{q}_s$ which has the general form

$$J_p = \mathbf{q}_s{}^{\mathrm{H}}\mathbf{A}\mathbf{q}_s + \mathbf{q}_s{}^{\mathrm{H}}\mathbf{b} + \mathbf{b}^{\mathrm{H}}\mathbf{q}_s + c, \tag{8.11.14}$$

where the matrix $\mathbf{A} = \mathbf{Z}^{\mathrm{H}}\mathbf{Z}$, the vector $\mathbf{b} = \mathbf{Z}^{\mathrm{H}}\mathbf{p}_p$ and the scalor $c = \mathbf{p}_p{}^{\mathrm{H}}\mathbf{p}_p$. The latter is the sum of the squared pressures due to the primary source alone. A comprehensive discussion of the properties of this quadratic form is presented in the appendix. The important properties of the function are that, provided $\mathbf{A}$ is positive definite, it has an optimal vector of secondary source strengths and a unique minimum value which are specified by

$$\mathbf{q}_{s0} = \mathbf{A}^{-1}\mathbf{b}, \qquad J_{p0} = c - \mathbf{b}^{\mathrm{H}}\mathbf{A}^{-1}\mathbf{b}. \tag{8.11.15a,b}$$

Thus, one approach to the problem of suppressing far field radiation is to introduce a multiplicity of secondary sources whose complex strengths are adjusted in order to minimise the pressure at a number of far field locations. Although the use of this technique has not yet been reported within the context of free field radiation, its application to enclosed sound fields has been well researched and will be discussed in detail in Chapter 10.

8.12 The power output of arrays of point sources

In Section 8.3 we used the property of point sources that the power output can be computed from a knowledge of their complex strength and the total complex pressure produced upon them. Here we extend this principle to deal with a multiplicity of monopole point sources and establish some properties of monopole source distributions which are likely to be useful in controlling free field radiation. First, note that if we specify a complex vector $\mathbf{q}$ of source strengths and a complex vector $\mathbf{p}$ of total pressures produced at the positions of the secondary sources (such that the order of $\mathbf{p}$ and $\mathbf{q}$ are equal) then in vector notation we can write the total power output of the source array as

$$W = (1/2)\mathrm{Re}\{\mathbf{p}^{\mathrm{H}}\mathbf{q}\}. \tag{8.12.1}$$

This expression effectively adds the individual power outputs of all the point sources. Now note that the complex vector of pressures produced at the positions of the sources is related to the complex source strengths by $\mathbf{p} = \mathbf{Zq}$ where $\mathbf{Z}$ is a matrix of complex acoustic transfer and input impedances which relates the pressures at all the source points to the strengths of all the sources. Thus, each source will produce a pressure on each of the other sources in addition to producing a pressure "on itself". Substitution of this relationship into equation (8.12.1) shows that

$$W = (1/2)\mathrm{Re}\{\mathbf{q}^{\mathrm{H}}\mathbf{Z}^{\mathrm{H}}\mathbf{q}\}. \tag{8.12.2}$$

Writing the real part of this expression as half the sum of the term in curly brackets and its complex conjugate yields

$$W = (1/4)\{\mathbf{q}^{\mathrm{H}}\mathbf{Z}^{\mathrm{H}}\mathbf{q} + \mathbf{q}^{\mathrm{T}}\mathbf{Z}^{\mathrm{T}}\mathbf{q}^*\}. \tag{8.12.3}$$

Since both terms are scalars, then they will be equal to their own transpose. Thus, we can write the second term in the curly brackets as $\mathbf{q}^{\mathrm{T}}\mathbf{Z}^{\mathrm{T}}\mathbf{q}^* = \mathbf{q}^{\mathrm{H}}\mathbf{Zq}$. In addition the transfer impedance matrix will be symmetric, so that $\mathbf{Z} = \mathbf{Z}^{\mathrm{T}}$ (see, for example, Nelson *et al.*, 1987a). This follows from the principle of reciprocity, which will be discussed in more detail in Chapter 9. Therefore equation (8.12.3) can be reduced to

$$W = (1/2)\mathbf{q}^{\mathrm{H}}\mathrm{Re}\{\mathbf{Z}\}\mathbf{q}. \tag{8.12.4}$$

This represents a compact means of evaluating the power output of a collection of point sources without having to undertake potentially difficult integrations of the far field squared pressure.

If we now specify a complex vector $\mathbf{q}_p$ of primary sources and a complex vector $\mathbf{q}_s$ of secondary sources, we can also write the power outputs due to these two source distributions as

$$W_p = (1/2)\mathrm{Re}\{[\mathbf{p}_p(\mathbf{y}_p) + \mathbf{p}_s(\mathbf{y}_p)]^{\mathrm{H}}\mathbf{q}_p\}, \tag{8.12.5}$$

$$W_s = (1/2)\mathrm{Re}\{[\mathbf{p}_p(\mathbf{y}_s) + \mathbf{p}_s(\mathbf{y}_s)]^{\mathrm{H}}\mathbf{q}_s\}. \tag{8.12.6}$$

In equation (8.12.5), $\mathbf{p}_p(\mathbf{y}_p)$ and $\mathbf{p}_s(\mathbf{y}_p)$ are the pressures produced at the positions of the primary sources by the primary and secondary sources, respectively. Similarly, in equation (8.12.6), $\mathbf{p}_p(\mathbf{y}_s)$ and $\mathbf{p}_p(\mathbf{y}_s)$ are those pressures produced at the positions of the secondary sources by the primary and secondary sources, respectively. We can further relate these pressures to the source strengths that produce them via transfer impedance matrices. We can thus define

$$\mathbf{p}_p(\mathbf{y}_p) = \mathbf{Z}(\mathbf{y}_p|\mathbf{y}_p)\mathbf{q}_p, \quad \mathbf{p}_s(\mathbf{y}_p) = \mathbf{Z}(\mathbf{y}_p|\mathbf{y}_s)\mathbf{q}_s, \tag{8.12.7a,b}$$

$$\mathbf{p}_p(\mathbf{y}_s) = \mathbf{Z}(\mathbf{y}_s|\mathbf{y}_p)\mathbf{q}_p, \quad \mathbf{p}_s(\mathbf{y}_s) = \mathbf{Z}(\mathbf{y}_s|\mathbf{y}_s)\mathbf{q}_s. \tag{8.12.8a,b}$$

Use of these relationships shows that we can write the total power output $W_p + W_s$ as

$$W = (1/2)\mathrm{Re}\{\mathbf{q}_p{}^{\mathrm{H}}\mathbf{Z}^{\mathrm{H}}(\mathbf{y}_p|\mathbf{y}_p)\mathbf{q}_p + \mathbf{q}_s{}^{\mathrm{H}}\mathbf{Z}^{\mathrm{H}}(\mathbf{y}_p|\mathbf{y}_s)\mathbf{q}_p$$
$$+ \mathbf{q}_p{}^{\mathrm{H}}\mathbf{Z}^{\mathrm{H}}(\mathbf{y}_s|\mathbf{y}_p)\mathbf{q}_s + \mathbf{q}_s{}^{\mathrm{H}}\mathbf{Z}^{\mathrm{H}}(\mathbf{y}_s|\mathbf{y}_s)\mathbf{q}_s\}. \tag{8.12.9}$$

Using manipulations similar to those used in deducing equation (8.12.4) from (8.12.2) and also noting that the principle of reciprocity ensures that $\mathbf{Z}(\mathbf{y}_s|\mathbf{y}_p) =$

$\mathbf{Z}^{\mathrm{T}}(\mathbf{y}_p|\mathbf{y}_s)$, enables this expression to be written as

$$W = (1/2)[\mathbf{q}_p^{\mathrm{H}}\mathrm{Re}\{\mathbf{Z}(\mathbf{y}_p|\mathbf{y}_p)\}\mathbf{q}_p + \mathbf{q}_s^{\mathrm{H}}\mathrm{Re}\{\mathbf{Z}(\mathbf{y}_p|\mathbf{y}_s)\}\mathbf{q}_p + \mathbf{q}_p^{\mathrm{H}}\mathrm{Re}\{\mathbf{Z}^{\mathrm{T}}(\mathbf{y}_p|\mathbf{y}_s)\}\mathbf{q}_s + \mathbf{q}_s^{\mathrm{H}}\mathrm{Re}\{\mathbf{Z}(\mathbf{y}_s|\mathbf{y}_s)\}\mathbf{q}_s]. \quad (8.12.10)$$

An example of the use of this expression in evaluating the power output of a distribution of point primary and point secondary sources is given by considering the particular case where we have a point primary source of strength $+q$ at the origin and two secondary sources on the same axis as the primary source, both at a distance d from the primary source and having complex strength $-(1/2)q$. When $kd << 1$, this corresponds to the arrangement associated with a longitudinal quadrupole. Thus, the terms in equation (8.12.10) are defined by

$$\mathbf{q}_s = -(1/2)q\begin{bmatrix}1\\1\end{bmatrix}, \qquad \mathrm{Re}\{\mathbf{Z}(\mathbf{y}_p|\mathbf{y}_s)\} = Z_0\begin{bmatrix}\mathrm{sinc}\ kd\\ \mathrm{sinc}\ kd\end{bmatrix},$$

$$\mathrm{Re}\{\mathbf{Z}(\mathbf{y}_s|\mathbf{y}_s)\} = Z_0\begin{bmatrix}1 & \mathrm{sinc}\ 2kd\\ \mathrm{sinc}\ 2kd & 1\end{bmatrix}, \quad (8.12.11a,b,c)$$

where we have used equation (1.12.7) to define the elements of the impedance matrices. Also note that $\mathrm{Re}\{\mathbf{Z}(\mathbf{y}_p|\mathbf{y}_p)\} = \omega^2\rho_0/4\pi c_0 = Z_0$. Substitution of these relationships into equation (8.12.10) produces an expression for the power output given by

$$W = \frac{1}{2}Z_0|q|^2[(1/2)(3 - 4\ \mathrm{sinc}\ kd + \mathrm{sinc}\ 2\ kd)]. \quad (8.12.12)$$

Note that this can also be written in terms of the power output W_{pp} produced by the primary source in the absence of the secondary sources. Thus

$$\frac{W}{W_{pp}} = (1/2)(3 - 4\ \mathrm{sinc}\ kd + \mathrm{sinc}\ 2\ kd). \quad (8.12.13)$$

It will prove useful to undertake a Taylor series expansion of this expression in the limit of $kd \to 0$. This yields, to leading order,

$$\frac{W}{W_{pp}} = \frac{(kd)^4}{20} + \ldots, \quad (8.12.14)$$

which shows that the power output of the primary source is reduced by a significant factor by using this quadrupole type of arrangement of secondary sources.

8.13 The minimum power output of point source arrays

Whilst using the source distribution specified in the last section produces a significant reduction in the power output of the primary source, this is not necessarily the best reduction in power output that can be produced when using two secondary sources to control the field of a single primary source. This can be demonstrated by writing equation (8.12.10) for the power output of a distribution of primary and secondary sources as the quadratic function

$$W = \mathbf{q}_s^{\mathrm{H}}\mathbf{A}\mathbf{q}_s + \mathbf{q}_s^{\mathrm{H}}\mathbf{b} + \mathbf{b}^{\mathrm{H}}\mathbf{q}_s + c. \tag{8.13.1}$$

The terms in this expression are

$$\mathbf{A} = (1/2)\mathrm{Re}\{\mathbf{Z}(\mathbf{y}_s|\mathbf{y}_s)\}, \qquad \mathbf{b} = (1/2)\mathrm{Re}\{\mathbf{Z}(\mathbf{y}_p|\mathbf{y}_s)\}\mathbf{q}_p,$$
$$c = (1/2)\mathbf{q}_p^{\mathrm{H}}\mathrm{Re}\{\mathbf{Z}(\mathbf{y}_p|\mathbf{y}_p)\}\mathbf{q}_p. \tag{8.13.2a,b,c}$$

As we have already seen, provided $\mathbf{A}$ is positive definite, this function has the optimal value of secondary source strengths given by $\mathbf{q}_{s0} = -\mathbf{A}^{-1}\mathbf{b}$, and corresponding minimum power output specified by $W_0 = c - \mathbf{b}^{\mathrm{H}}\mathbf{A}^{-1}\mathbf{b}$. Thus, if we again choose a linear array consisting of a single primary source with two secondary sources on the same axis, both separated from it by a distance d, then the matrix $\mathbf{A}$ and complex vector $\mathbf{b}$ associated with equation (8.13.1) are

$$\mathbf{A} = \frac{1}{2}Z_0\begin{bmatrix} 1 & \mathrm{sinc}\ 2kd \\ \mathrm{sinc}\ 2kd & 1 \end{bmatrix}, \quad \mathbf{b} = \frac{1}{2}Z_0\, q_p\begin{bmatrix} \mathrm{sinc}\ kd \\ \mathrm{sinc}\ kd \end{bmatrix}. \tag{8.13.3a,b}$$

Inversion of this matrix is relatively straightforward and the solution for the optimal secondary source strengths reduces to

$$q_{s10} = q_{s20} = -q_p\left(\frac{\mathrm{sinc}\ kd}{1 + \mathrm{sinc}\ 2kd}\right). \tag{8.13.4}$$

Similarly, the minimum power output can be expressed relative to the power output produced by the primary source alone and is given by

$$\frac{W_0}{W_{pp}} = 1 - \left(\frac{2\ \mathrm{sinc}^2\ kd}{1 + \mathrm{sinc}\ 2kd}\right). \tag{8.13.5}$$

It is interesting to undertake series expansions of these terms in the low frequency limit when $kd << 1$. These yield, to leading order

$$q_{s10} = q_{s20} = -q_p\left(\frac{1}{2} - \frac{1}{12}(kd)^2\ \ldots\right), \tag{8.13.6}$$

$$\frac{W_0}{W_{pp}} = \frac{(kd)^4}{45}. \tag{8.13.7}$$

In the low frequency limit the minimum power output produced by optimally adjusting the secondary source strengths is significantly *less* than the power output produced by the quadrupole-like arrangement discussed in Section 8.12. The explanation for this finding lies in a multipole analysis of the source distribution produced by this optimal arrangement of secondary sources. With reference to Section 8.9, which specifies the multipole moments associated with an arbitrary collection of point sources, for this particular arrangement it follows that the monopole moment of the source distribution is given by

$$Q = \sum_{n=1}^{N} q_n = q_p - 2q_p\left(\frac{1}{2} - \frac{(kd)^2}{12}\ldots\right), \tag{8.13.8}$$

which therefore reduces to

$$Q = q_p\frac{(kd)^2}{6}. \tag{8.13.9}$$

This optimal source arrangement has a non-vanishing monopole source moment, whereas for the quadrupole-like arrangement there is zero associated monopole moment. However, the strength of this monopole moment is very small, being $O(kd)^2 q_p$. In fact it produces a pressure field of the same order as the quadrupole moment associated with the distribution. Application of equation (8.9.4) shows that the dipole moment of the optimal source distribution vanishes and equation (8.9.6) shows that there is one quadrupole moment specified by

$$Q_{11} = \frac{1}{2} \sum_{n=1}^{N} d_{n1} d_{n1} q_n = - \frac{q_p d^2}{2}. \tag{8.13.10}$$

The far field pressures produced by the monopole and quadrupole moments of the optimal source distribution are given respectively by

$$p_M(r,\theta) = \frac{j\omega\rho_0 q_p \mathrm{e}^{-jkr}}{4\pi r} \frac{(kd)^2}{6}, \tag{8.13.11}$$

$$p_Q(r,\theta) = \frac{j\omega\rho_0 q_p \mathrm{e}^{-jkr}}{4\pi r} \left[- \frac{(kd)^2}{2} \cos^2 \theta \right]. \tag{8.13.12}$$

Thus, the monopole moment is of the same order of magnitude as the quadrupole moment and the interference between these two pressure fields produces a further reduction in the power output over that which could be produced by the quadrupole source alone. This type of behaviour has been further illustrated by Nelson *et al.* (1987a), who also investigated the other source arrangements depicted in Fig. 8.20 in detail. Again, significant reductions in power output were produced over that which would be expected by simply using classical multipole arrangements. The optimal secondary source strengths and sound power output reductions for the cases shown are plotted in Fig. 8.21. In the particular case of a tetrahedral array of secondary sources depicted in Fig. 8.20(c), then it was found that an *octopole* source distribution could be produced since the secondary sources and primary sources combine to form a weak monopole field which exactly cancels the field associated with three orthogonal longitudinal quadrupoles of equal strength (see Nelson *et al.*, 1986a, 1987a for further details).

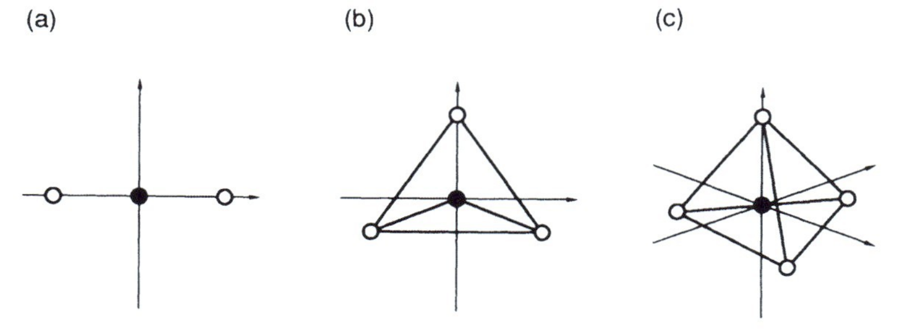

Fig. 8.20 Geometrical arrangements showing a single primary source and two, three and four secondary sources, respectively. In each case, the secondary sources are separated from the primary source by a distance d and the secondary sources are separated from each other by a distance D. In the cases shown: (a) for two secondary sources $D = 2d$, (b) for three secondary sources $D = \sqrt{3}d$ and (c) for four secondary sources $D = (4/\sqrt{6})d$.

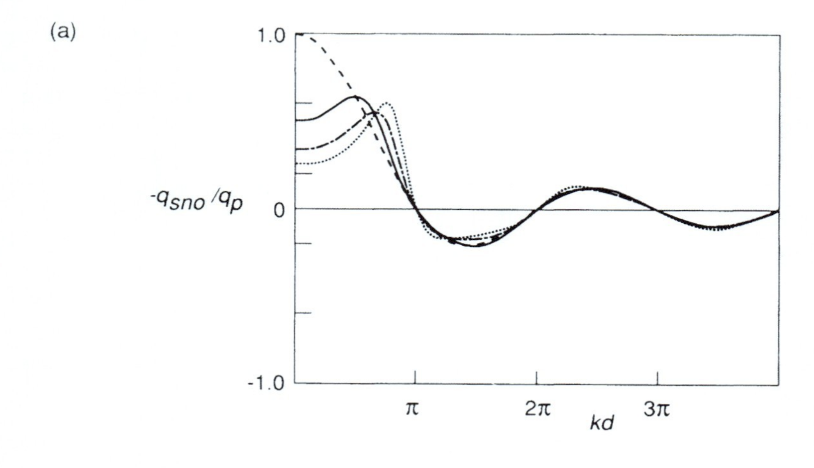

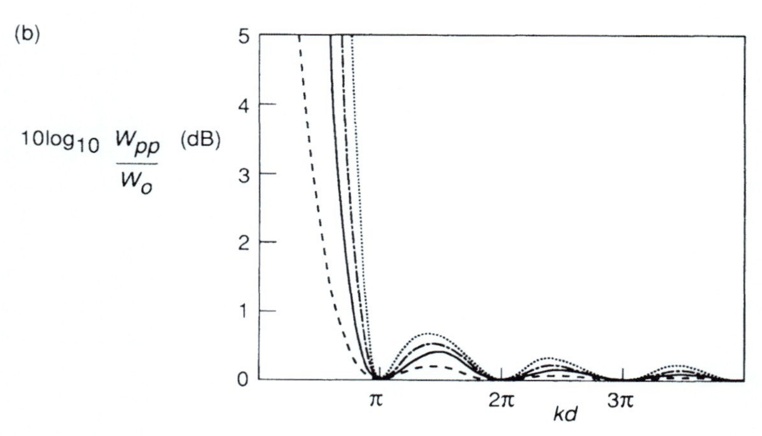

Fig. 8.21 (a) The ratio of the optimal secondary source strengths to the primary source strength. Number of secondary sources: – – – – one; ——— two; —·—·— three; ······· four (after Nelson *et al.*, 1987a). (b) The maximum achievable reduction in total sound power radiated when using a number of secondary sources to control the field of a single primary source: – – – – one; ——— two; —·—·— three; ······· four (after Nelson *et al.*, 1987a).

9
Continuous Source Distributions and the Active Absorption of Free Field Radiation

9.1 Introduction

In this chapter we introduce a more general treatment of the active control of free field radiation. First we will discuss the possibilities for using *continuous source layers* to provide perfect control of free field sound. The theoretical basis for the discussion follows directly from the process of solving the wave equation, and is a natural consequence of the classical relationships between the sound field in a given spatial region and the field properties on the surfaces bounding the region. However, the application of what amounts to Huygens' principle to the active control problem was not pointed out until the 1960s by both Jessel (in France) and Malyuzhinets (in the Soviet Union). As we shall see, the direct application of this theoretical approach is still far from a technological reality. This is due to the requirement for a continuous distribution of both monopole and dipole sources which act to absorb incident acoustic radiation. Control in this case is achieved by using the unidirectional radiation produced by an appropriate monopole/dipole combination of the type described in the one-dimensional case dealt with in Chapter 5. It is also shown in Chapter 5 that a monopole type secondary source is capable of producing perfect reflection of incident radiation in one dimension. In this chapter we show that this mechanism of control can also be produced through the use of a continuous monopole layer in three dimensions. The theoretical possibilities offered by continuous distributions of secondary sources include the perfect suppression of radiation from vibrating bodies and the perfect suppression of radiation scattered by a body; both these possibilities will be considered in outline in this chapter.

Following the discussion of continuous layers of secondary sources it is natural to ask to what extent these layers may be replaced by distributions of discrete sources. This question has been the focus for much research (again particularly in France and the Soviet Union) and this is reviewed in this chapter. A question which also arises when considering the discretisation of continuous source layers is the extent to which individual point sources can absorb incident sound. Here we begin by calculating the maximum sound power that can be absorbed by individual point sources of both monopole and dipole type. We also demonstrate that the superposition of a point monopole source and a point dipole source can absorb the sum of the powers absorbed by the sources individually. In addition we consider the absorption of sound by higher order sources and in particular deduce the maximum sound power that can be absorbed by quadrupole sources. A discussion is also presented of sound power absorption by arbitrary arrays of point sources and we show that some compact arrays of point sources have the potential to produce very large absorptions of sound power.

9.2 The inhomogeneous wave equation

Thus far in our treatment of free field acoustic radiation we have dealt only with point monopole type sources and the particular combinations of point monopole sources which reduce to higher order multipole sources. Now we will discuss continuous distributions of source strength. A source distribution of monopole type can be visualised by dividing a region of the medium into a number of tiny three-dimensional elements and assuming that each element contains an inhomogeneity (such as a sphere or a bubble) which pulsates and thereby introduces a fluctuating volume flow into the element. The magnitude of the additional volume flow will vary from element to element. This fluctuating volume flow *per unit volume* is specified by $q_{vol}(\mathbf{x},t)$ and in the limit (when the elements in the medium are infinitesimally small) this fluctuating rate of addition of volume flow is a continuous function of position, $\mathbf{x}$, and time, t. Under these circumstances the mass conservation equation described in Section 1.10 will be modified by the addition of a further term which quantifies this local rate of addition of volume. The mass conservation equation can be written to first order in the fluctuating quantities as

$$\frac{\partial \rho(\mathbf{x},t)}{\partial t} + \rho_0 \boldsymbol{\nabla}.\mathbf{u}(\mathbf{x},t) = \rho_0 q_{vol}(\mathbf{x},t). \tag{9.2.1}$$

In a similar way we can conceive of a distribution of fluctuating force per unit volume acting on the medium, specified by the vector $\mathbf{f}_{vol}(\mathbf{x},t)$. This can be visualised, for example, by assuming that the fluid elements referred to above contain tiny spheres which oscillate to and fro and thereby apply a net fluctuating force to each element. The linearised momentum conservation equation under these circumstances reduces to

$$\rho_0 \frac{\partial \mathbf{u}(\mathbf{x},t)}{\partial t} + \boldsymbol{\nabla} p(\mathbf{x},t) = \mathbf{f}_{vol}(\mathbf{x},t). \tag{9.2.2}$$

If we use the adiabatic relationship between the pressure and density fluctuations given by $p(\mathbf{x},t) = c_0^2 \rho(\mathbf{x},t)$ and operate on the momentum conservation equation with the divergence operator ($\boldsymbol{\nabla}.$) and then subtract the time derivative of the mass conservation equation from it, we can derive the inhomogeneous wave equation

$$\left(\nabla^2 - \frac{1}{c_0^2}\frac{\partial}{\partial t^2}\right) p(\mathbf{x},t) = \boldsymbol{\nabla}.\mathbf{f}_{vol}(\mathbf{x},t) - \frac{\rho_0 \partial q_{vol}(\mathbf{x},t)}{\partial t}. \tag{9.2.3}$$

The terms on the right side of this equation can be interpreted as sources of sound (see, for example, the discussions by Doak, 1973a and Dowling and Ffowcs-Williams, 1983, Ch. 7). In the particular case of harmonic fluctuations in the field quantities, this equation reduces to the inhomogeneous Helmholtz equation

$$(\nabla^2 + k^2)p(\mathbf{x}) = \nabla.\mathbf{f}_{\text{vol}}(\mathbf{x}) - j\omega\rho_0 q_{\text{vol}}(\mathbf{x}). \tag{9.2.4}$$

Now $\mathbf{f}_{\text{vol}}(\mathbf{x})$ and $q_{\text{vol}}(\mathbf{x})$ are interpreted as the distribution of complex force per unit volume and complex volume velocity per unit volume applied to the medium. We shall demonstrate that these source terms may be respectively identified as dipole and monopole source distributions. First, however, we will discuss briefly a well-established technique for solving the inhomogeneous scalar Helmholtz equation.

9.3 The Green function and the principle of reciprocity

The solution to equation (9.2.4) is most readily dealt with by using a Green function. In general, the equation that we wish to solve is of the form

$$(\nabla^2 + k^2)p(\mathbf{x}) = -Q_{\text{vol}}(\mathbf{x}), \tag{9.3.1}$$

where Q_{vol} is the source strength per unit volume, which in terms of the analysis presented in the last section, may be of monopole or dipole type. We should also note in passing that in analysing the problem of sound generated aerodynamically, Lighthill (1952) showed that the general relevant mass and momentum conservation equations may be combined to give an inhomogeneous wave equation for the fluctuating mass density in a mass conserving continuum with a right side which can be interpreted as a *quadrupole* type source. Irrespective of the multipole order of the source distribution, the solution to equation (9.3.1) can be deduced in terms of the field produced by a monopole source at a point. This field is proportional to the *Green function* which is denoted by $G(\mathbf{x}|\mathbf{y})$ and which satisfies the equation

$$(\nabla^2 + k^2)G(\mathbf{x}|\mathbf{y}) = -\delta(\mathbf{x} - \mathbf{y}). \tag{9.3.2}$$

Physically speaking, the Green function describes the spatial dependence of the complex pressure field produced by a harmonic point monopole source at $\mathbf{y}$. Thus the complex pressure at $\mathbf{x}$ due to a harmonic point monopole source of unit strength at $\mathbf{y}$ is given by $p(\mathbf{x}) = j\omega\rho_0 G(\mathbf{x}|\mathbf{y})$. The presence of a point monopole at $\mathbf{y}$ is represented by the three-dimensional Dirac delta function

$$\delta(\mathbf{x} - \mathbf{y}) = \delta(x_1 - y_1)\delta(x_2 - y_2)\delta(x_3 - y_3).$$

This function has properties which are analogous to the one-dimensional Dirac delta function $\delta(x - y)$ introduced in Chapter 2. Most importantly, if we have some continuous function $f(\mathbf{x})$ in three-dimensional space, then the three-dimensional delta function exhibits the "sifting" property

$$\int_V f(\mathbf{x})\delta(\mathbf{x} - \mathbf{y})\mathrm{d}V = \begin{Bmatrix} f(\mathbf{y}), & \mathbf{y} \text{ within } V \\ 0, & \mathbf{y} \text{ outside } V \end{Bmatrix}, \tag{9.3.3}$$

where V is a volume in the medium bounded by the surface S. Note that we can interpret this volume and its bounding surface in the alternative ways depicted in Figs 9.1(a) and 9.1(b). Thus the volume V could simply be enclosed by the surface S as illustrated in Fig. 9.1(a). In the case illustrated in Fig. 9.1(b), the outer part of the bounding surface S may be considered to lie at infinity. The volume V is then of infinite extent but excludes the region bounded by the inner part of the surface S.

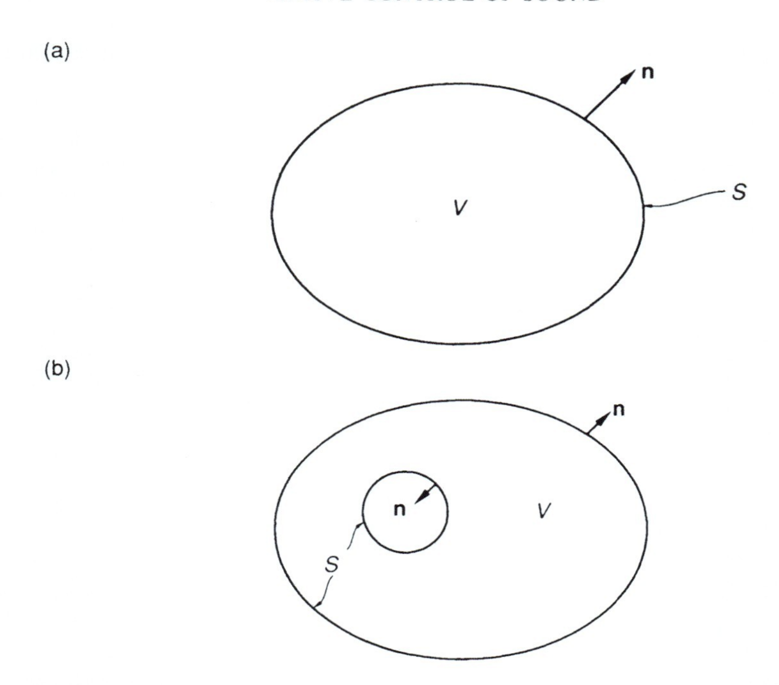

Fig. 9.1 (a) The volume V in the medium and its bounding surface S. (b) An alternative interpretation where the outer part of the bounding surface S can be considered to be at infinity.

We should also note the value of the volume integral when $\mathbf{y}$ lies exactly on the surface S. To some extent this is a matter of convention and depends on the definition adopted for the value of a singular integral such as (9.3.3) when $\mathbf{y}$ is coincident with one of the limits of the integration. For example, Morse and Feshbach (1953, Ch. 7) use the convention that the volume integral (9.3.3) yields $f(\mathbf{y})$ for $\mathbf{y}$ on S. However, another often used definition is that the integral yields $f(\mathbf{y})/2$ for $\mathbf{y}$ on S (see, for example, Junger and Feit, 1986, Ch. 4).

Before using the Green function to derive the solution to the inhomogeneous scalar wave equation, it is useful to establish formally that the Green function satisfies the principle of reciprocity. That is, the complex pressure at $\mathbf{y}_1$ (say) that is produced by a point monopole source at $\mathbf{y}_2$ is identical to the complex pressure at $\mathbf{y}_2$ that is produced by a point monopole source at $\mathbf{y}_1$. The spatial dependence of the two pressure fields due to these sources may be written as $G(\mathbf{x}|\mathbf{y}_1)$ and $G(\mathbf{x}|\mathbf{y}_2)$ respectively and they must satisfy

$$(\nabla^2 + k^2)\, G(\mathbf{x}|\mathbf{y}_1) = -\delta(\mathbf{x} - \mathbf{y}_1), \tag{9.3.4}$$

$$(\nabla^2 + k^2)\, G(\mathbf{x}|\mathbf{y}_2) = -\delta(\mathbf{x} - \mathbf{y}_2). \tag{9.3.5}$$

If we multiply the first of these equations by $G(\mathbf{x}|\mathbf{y}_2)$, and subtract from it the second equation multiplied by $G(\mathbf{x}|\mathbf{y}_1)$, then it follows that

$$G(\mathbf{x}|\mathbf{y}_2)\nabla^2 G(\mathbf{x}|\mathbf{y}_1) - G(\mathbf{x}|\mathbf{y}_1)\nabla^2 G(\mathbf{x}|\mathbf{y}_2) = G(\mathbf{x}|\mathbf{y}_1)\delta(\mathbf{x} - \mathbf{y}_2) - G(\mathbf{x}|\mathbf{y}_2)\delta(\mathbf{x} - \mathbf{y}_1). \tag{9.3.6}$$

Now we integrate both sides of equation (9.3.6) over a volume V that contains the point sources at $\mathbf{y}_1$ and $\mathbf{y}_2$. If S is the surface that encloses the volume V, then Green's theorem states that

$$\int_V (a\nabla^2 b - b\nabla^2 a)\mathrm{d}V = \int_S (a\boldsymbol{\nabla} b - b\boldsymbol{\nabla} a).\mathbf{n}\ \mathrm{d}S, \tag{9.3.7}$$

where a and b are scalar functions and $\mathbf{n}$ is the unit vector normal to the surface S pointing *outwards* from the volume V (see Fig. 9.1). Integrating equation (9.3.6) over the volume V thus shows that

$$\int_S [G(\mathbf{x}|\mathbf{y}_2)\boldsymbol{\nabla} G(\mathbf{x}|\mathbf{y}_1) - G(\mathbf{x}|\mathbf{y}_1)\boldsymbol{\nabla} G(\mathbf{x}|\mathbf{y}_2)].\mathbf{n}\ \mathrm{d}S$$

$$= G(\mathbf{y}_2|\mathbf{y}_1) - G(\mathbf{y}_1|\mathbf{y}_2) \tag{9.3.8}$$

where the property of the delta function (9.3.3) has been used in integrating the right side of equation (9.3.6). We can now show that provided certain boundary conditions are satisfied on S, then the left side of equation (9.3.8) is zero. For example, assume that the pressure and the normal component of particle velocity at a point on S are linearly related by some specific impedance $z(\mathbf{x})$. Remembering that $G(\mathbf{x}|\mathbf{y}_1)$ and $G(\mathbf{x}|\mathbf{y}_2)$ are proportional to the complex pressures in the two sound fields, it follows from the linearised momentum conservation equation that at this point on the surface S

$$\boldsymbol{\nabla} G(\mathbf{x}|\mathbf{y}_1).\mathbf{n} = -\frac{j\omega\rho_0 G(\mathbf{x}|\mathbf{y}_1)}{z(\mathbf{x})}, \quad \boldsymbol{\nabla} G(\mathbf{x}|\mathbf{y}_2).\mathbf{n} = -\frac{j\omega\rho_0 G(\mathbf{x}|\mathbf{y}_2)}{z(\mathbf{x})}. \tag{9.3.9a,b}$$

Substitution of these expressions into the integral of equation (9.3.8) shows that the left side vanishes for all points on S and thus

$$G(\mathbf{y}_2|\mathbf{y}_1) = G(\mathbf{y}_1|\mathbf{y}_2). \tag{9.3.10}$$

The principle of reciprocity therefore holds for two points inside a volume V in which the Helmholtz equation is satisfied and on whose bounding surface S there is some prescribed acoustic impedance. Conditions of zero normal particle velocity (a rigid surface) or zero pressure (a pressure release surface) are clearly special cases corresponding to an infinite or a zero impedance respectively. Note also that in the case of either a part or a whole of the bounding surface being at infinity, then the relevant impedance is the characteristic acoustic impedance of the medium. This is illustrated in Fig. 9.1(b). This is due to the fact that as the distance from the sources of sound approaches infinity then the waves which they emit will become increasingly like plane waves. This amounts to the application of the *Sommerfeld radiation condition*, which requires that, as the radial distance r from a harmonic source of sound approaches infinity,

$$\lim_{r\to\infty} \{r[p(\mathbf{x}) - \rho_0 c_0 u_r(\mathbf{x})]\} = 0, \tag{9.3.11}$$

where u_r is the radial component of the particle velocity field. This condition requires that as $r \to \infty$, then only *outgoing* waves can exist. (See, for example, Pierce, 1981, Ch. 4, for fuller discussions of both the reciprocity principle and the Sommerfeld radiation condition.)

9.4 The solution to the inhomogeneous wave equation

Having established the principle of reciprocity, we can now use the Green function to derive the solution to the inhomogeneous Helmholtz equation in the form of an integral equation for the complex acoustic pressure. In general terms we wish to solve an equation of the form

$$\nabla_y^2 p(\mathbf{y}) + k^2 p(\mathbf{y}) = -Q_{\text{vol}}(\mathbf{y}), \qquad (9.4.1)$$

where, for example, the source strength distribution could consist of a combination of monopole- and dipole-type sources where $Q_{\text{vol}}(\mathbf{y}) = j\omega\rho_0 q_{\text{vol}}(\mathbf{y}) - \boldsymbol{\nabla}_y.\mathbf{f}_{\text{vol}}(\mathbf{y})$. Note that we have used the co-ordinate $\mathbf{y}$ instead of the co-ordinate $\mathbf{x}$ in the wave equation. This substitution has been made deliberately in order to enable the solution to be easily derived. Also note that we have again used the convention introduced in Chapter 8 that $\boldsymbol{\nabla}_y$ is the del operator in y co-ordinates, i.e. $\mathbf{i}\partial/\partial y_1 + \mathbf{j}\partial/\partial y_2 + \mathbf{k}\partial/\partial y_3$ where $\mathbf{i}$, $\mathbf{j}$, $\mathbf{k}$ are the appropriate unit vectors. We shall now use the Green function $G(\mathbf{y}|\mathbf{x})$ which satisfies

$$\nabla_y^2 G(\mathbf{y}|\mathbf{x}) + k^2 G(\mathbf{y}|\mathbf{x}) = -\delta(\mathbf{y} - \mathbf{x}), \qquad (9.4.2)$$

to help us find the solution to equation (9.4.1). Thus $G(\mathbf{y}|\mathbf{x})$ is proportional to the complex pressure produced at $\mathbf{y}$ due to a harmonic point source at $\mathbf{x}$ and, as we have seen, an important property of the Green function here is that it satisfies the principle of reciprocity, and therefore that $G(\mathbf{y}|\mathbf{x}) = G(\mathbf{x}|\mathbf{y})$. In order to derive the solution to equation (9.4.1) we first multiply the equation by $G(\mathbf{y}|\mathbf{x})$ and subtract equation (9.4.2) multiplied by $p(\mathbf{y})$. This yields

$$G(\mathbf{y}|\mathbf{x})\nabla_y^2 p(\mathbf{y}) - p(\mathbf{y})\nabla_y^2 G(\mathbf{y}|\mathbf{x}) = -Q_{\text{vol}}(\mathbf{y})G(\mathbf{y}|\mathbf{x}) + p(\mathbf{y})\delta(\mathbf{y} - \mathbf{x}). \quad (9.4.3)$$

We now integrate both sides of this equation over a volume V that is bounded by the surface S. Using the properties of the delta function specified by equation (9.3.3) shows that

$$\int_V G(\mathbf{y}|\mathbf{x})\boldsymbol{\nabla}_y^2 p(\mathbf{y}) - p(\mathbf{y})\boldsymbol{\nabla}_y^2 G(\mathbf{y}|\mathbf{x})\mathrm{d}V + \int_V Q_{\text{vol}}(\mathbf{y})G(\mathbf{y}|\mathbf{x})\mathrm{d}V = \begin{Bmatrix} \mathrm{p}(\mathbf{x}), & \mathbf{x} \text{ within } V \\ 0, & \mathbf{x} \text{ outside } V \end{Bmatrix}. \qquad (9.4.4)$$

We now use Green's theorem as specified by equation (9.3.7). This enables the first volume integral on the left side of equation (9.4.4) to be transformed into an integral over the surface S. Also using the property $G(\mathbf{y}|\mathbf{x}) = G(\mathbf{x}|\mathbf{y})$ then enables equation (9.4.4) for points $\mathbf{x}$ inside V to be written as

$$p(\mathbf{x}) = \int_V Q_{\text{vol}}(\mathbf{y})G(\mathbf{x}|\mathbf{y})\mathrm{d}V + \int_S [G(\mathbf{x}|\mathbf{y})\boldsymbol{\nabla}_y p(\mathbf{y}) - p(\mathbf{y})\boldsymbol{\nabla}_y G(\mathbf{x}|\mathbf{y})].\mathbf{n}\mathrm{d}S. \quad (9.4.5)$$

Thus, given a distribution of source strength $Q_{\text{vol}}(\mathbf{y})$ in the volume V, we can compute the contribution to the complex pressure $p(\mathbf{x})$ produced by this source distribution by performing the volume integral on the right side of equation (9.4.5). However, there will also be a contribution from the surface integral term in equation (9.4.5). The solution to this equation thus appears to require a specification of the pressure $p(\mathbf{y})$ and pressure gradient $\boldsymbol{\nabla}_y p(\mathbf{y})$ on the surface S and a knowledge of the

Green function $G(\mathbf{x}|\mathbf{y})$. Note that there is (in principle) considerable latitude in the choice of Green function that we use to solve a given problem. The only assumptions that we have made is that $G(\mathbf{x}|\mathbf{y})$ satisfies equation (9.3.2) and that the principle of reciprocity holds. Thus the Green function can, in principle, be chosen to be that which satisfies any impedance boundary condition on S and not necessarily the boundary condition of the particular physical problem at hand. The Green function for example can always be chosen to satisfy $\nabla_y G(\mathbf{x}|\mathbf{y}) = 0$, in which case we only need specify $\nabla_y p(\mathbf{y})$ in order to compute the resulting pressure field. This problem will be returned to in Section 9.9. A comprehensive discussion of Green function techniques is given by Morse and Feshbach (1953, Ch. 7).

9.5 The solution of the inhomogeneous wave equation in an unbounded medium

If the medium is unbounded, we can apply the Sommerfeld radiation condition discussed in Section 9.3 to show that the surface integral is zero on the right side of equation (9.4.5). Thus the bounding surface S is at infinity and since the Sommerfeld radiation condition requires that the field there will consist of only *outgoing* plane waves, then the two terms in the surface integral will add to zero at all points on the surface S. The solution to the inhomogeneous wave equation thus becomes

$$p(\mathbf{x}) = \int_V Q_{\mathrm{vol}}(\mathbf{y}) G(\mathbf{x}|\mathbf{y}) \mathrm{d}V. \tag{9.5.1}$$

In addition, in an unbounded medium, the Green function $G(\mathbf{x}|\mathbf{y})$ takes a particularly simple form; it is known as the *free space Green function* $g(\mathbf{x}|\mathbf{y})$ where

$$G(\mathbf{x}|\mathbf{y}) = g(\mathbf{x}|\mathbf{y}) = \frac{\mathrm{e}^{-jk|\mathbf{x}-\mathbf{y}|}}{4\pi|\mathbf{x}-\mathbf{y}|}. \tag{9.5.2}$$

The proof that this represents a solution to equation (9.3.2) has been presented in outline by, for example, Morse and Ingard (1968). Now consider the case where the "source" term on the right side of the inhomogeneous scalar Helmholtz equation (9.2.4) is due to a distribution of volume sources such that $Q_{\mathrm{vol}}(\mathbf{y}) = j\omega\rho_0 q_{\mathrm{vol}}(\mathbf{y})$. Then the solution to the equation becomes

$$p(\mathbf{x}) = \int_V j\omega\rho_0 q_{\mathrm{vol}}(\mathbf{y}) g(\mathbf{x}|\mathbf{y}) \mathrm{d}V. \tag{9.5.3}$$

The physical interpretation of this equation should now become clear; the volume integral simply represents the superposition of all the contributions of the elemental monopole sources of strength $q_{\mathrm{vol}}(\mathbf{y})$ per unit volume. In the particular case of a point monopole source at $\mathbf{y}_1$ where $q_{\mathrm{vol}}(\mathbf{y}) = q\delta(\mathbf{y} - \mathbf{y}_1)$, performing the volume integral yields

$$p(\mathbf{x}) = \frac{j\omega\rho_0 q \mathrm{e}^{-jk|\mathbf{x}-\mathbf{y}_1|}}{4\pi|\mathbf{x}-\mathbf{y}_1|}, \tag{9.5.4}$$

which is the familiar expression for the complex pressure produced by a point monopole source.

In the case of a distribution of *force* per unit volume applied to the medium, then $Q_{vol}(\mathbf{y}) = -\nabla_y.\mathbf{f}_{vol}(\mathbf{y})$ and the solution can be written as

$$p(\mathbf{x}) = -\int_V \nabla_y.\mathbf{f}_{vol}(\mathbf{y})g(\mathbf{x}|\mathbf{y})\mathrm{d}V. \qquad (9.5.5)$$

We can use the vector identity $a\nabla_y.\mathbf{b} = \nabla_y.(a\mathbf{b}) - (\nabla_y a).\mathbf{b}$ to rewrite this expression as

$$p(\mathbf{x}) = \int_V \mathbf{f}_{vol}(\mathbf{y}).\nabla_y g(\mathbf{x}|\mathbf{y}) - \nabla_y.[g(\mathbf{x}|\mathbf{y})\mathbf{f}_{vol}(\mathbf{y})]\mathrm{d}V. \qquad (9.5.6)$$

The divergence theorem can now be used to show that the contribution of the second term in the volume integral vanishes. Thus since

$$\int_V \nabla_y.[g(\mathbf{x}|\mathbf{y})\mathbf{f}_{vol}(\mathbf{y})]\mathrm{d}V = \int_S g(\mathbf{x}|\mathbf{y})\mathbf{f}_{vol}(\mathbf{y}).\mathbf{n}\ \mathrm{d}S, \qquad (9.5.7)$$

and if we assume that $\mathbf{f}_{vol}(\mathbf{y})$ is non-zero over a restricted region of the medium, then if we choose the bounding surface S to be just outside this restricted region, it follows that the surface integral will be zero. Thus the solution for the complex pressure can be written as

$$p(\mathbf{x}) = \int_V \mathbf{f}_{vol}(\mathbf{y}).\nabla_y g(\mathbf{x}|\mathbf{y})\mathrm{d}V. \qquad (9.5.8)$$

We can now make explicit that the distribution of force per unit volume acts as a distribution of *dipole* sources. Consider the particular case of a point force applied to *the medium at* $\mathbf{y}_1$ where $\mathbf{f}_{vol}(\mathbf{y}) = \mathbf{f}\delta(\mathbf{y} - \mathbf{y}_1)$. In this case, performing the volume integral yields

$$p(\mathbf{x}) = \mathbf{f}.\nabla_y\left(\frac{\mathrm{e}^{-jk|\mathbf{x}-\mathbf{y}_1|}}{4\pi|\mathbf{x}-\mathbf{y}_1|}\right). \qquad (9.5.9)$$

Comparison of this expression with that given by equation (8.7.17) demonstrates that in the limit that $\mathbf{d} \to 0$ (but $q\mathbf{d}$ remains finite) the sound field produced by two closely spaced monopoles of strength q and $-q$ and separated by a vector distance $\mathbf{d}$ is exactly equivalent to the sound field produced by a point force $\mathbf{f}$ where $\mathbf{f} = j\omega\rho_0 q\mathbf{d}$.The integral over the volume in equation (9.5.8) thus evaluates the superposition of the fields of a distribution of such sources.

9.6 The Kirchhoff–Helmholtz integral equation

If in a given volume V there is zero source strength, $Q_{vol}(\mathbf{y})$, and if we choose the Green function $G(\mathbf{x}|\mathbf{y})$ to be the free space Green function $g(\mathbf{x}|\mathbf{y})$, the integral form of solution given by equation (9.4.5) reduces to the *Kirchhoff–Helmholtz integral equation* (see, for example, Pierce, 1981, Ch. 4). This is given by

$$\int_S [g(\mathbf{x}|\mathbf{y})\nabla_y p(\mathbf{y}) - p(\mathbf{y})\nabla_y g(\mathbf{x}|\mathbf{y})].\mathbf{n}\mathrm{d}S = \begin{Bmatrix} p(\mathbf{x}), & \mathbf{x} \text{ within } V \\ 0, & \mathbf{x} \text{ outside } V \end{Bmatrix}. \qquad (9.6.1)$$

The significance of this equation within the context of active noise control was first pointed out by both Jessel and Malyhuzinets during the 1960s. In the next section we shall demonstrate the significance of this integral relationship to the active absorption of sound. First, however, note that the terms appearing on the left side of this equation have a well defined physical interpretation. Since the equation of conservation of momentum for harmonic excitation shows that $\nabla_y p(\mathbf{y}) = -j\omega\rho_0\mathbf{u}(\mathbf{y})$, then the first term in the integral of the left side of equation (9.6.1) can be written as

$$-\int_S g(\mathbf{x}|\mathbf{y})j\omega\rho_0\mathbf{u}(\mathbf{y}).\mathbf{n}\mathrm{d}S = \int_S g(\mathbf{x}|\mathbf{y})j\omega\rho_0 q_{\mathrm{surf}}(\mathbf{y})\mathrm{d}S, \qquad (9.6.2)$$

where $q_{\mathrm{surf}}(\mathbf{y}) = -\mathbf{u}(\mathbf{y}).\mathbf{n}$ is a volume velocity per unit surface area, and therefore represents a distribution of monopole source strength. (Note that since $\mathbf{n}$ points out from the volume considered, then q_{surf} represents a positive rate of introduction of volume into the volume V.) Similarly, the second term in the integral of equation (9.5.1) can be written as

$$-\int_S p(\mathbf{y})\nabla_y g(\mathbf{x}|\mathbf{y}).\mathbf{n}\mathrm{d}S = -\int_S \mathbf{f}_{\mathrm{surf}}(\mathbf{y}).\nabla_y g(\mathbf{x}|\mathbf{y})\mathrm{d}S, \qquad (9.6.3)$$

where $\mathbf{f}_{\mathrm{surf}}(\mathbf{y}) = p(\mathbf{y})\mathbf{n}$ is a force per unit area acting at the surface S. It can be seen by comparison of this equation with equation (9.5.8) that this term has a clear interpretation as the sound field generated by a surface distribution of dipole sources. Thus, in the absence of any volume sources, the pressure at a given position inside the volume V can be specified by the equivalent monopole and dipole source layers on the bounding surface S.

The physical interpretation of this monopole/dipole source combination is most easily understood with respect to the discussion of the one-dimensional case introduced in Chapter 5 (see Section 5.10). In that case we saw that a plane monopole source which produced the same radiation in both directions could be combined with an anti-symmetrically radiating plane dipole source to give unidirectional radiation. In this case, we see that if we distribute elementary plane monopole and dipole sources over the bounding surface S, then provided their relative strengths are appropriately defined, we can produce the sound field $p(\mathbf{x})$ within the volume V whilst producing zero acoustic pressure outside the volume V. An elementary plane dipole source specified by the effective force per unit surface area $\mathbf{f}_{\mathrm{surf}}$ will produce antisymmetric radiation (with the "cos θ" directivity pattern described in Section 8.7) whilst an elementary plane monopole source will produce omnidirectional radiation. These elementary dipole sources can be visualised as small massless pistons which oscillate to and fro and thus produce force per unit area on the fluid. In the case of an elementary monopole source, two small massless pistons are forced to oscillate in anti-phase by the introduction and withdrawal of volume flow. In the particular case of plane wave radiation, when $p(\mathbf{y}) = \rho_0 c_0\mathbf{u}(\mathbf{y}).\mathbf{n}$, then the resulting directivity of the combined radiation from a single elementary monopole/dipole source can be shown to take a cardioid form (see, for example, Mangiante, 1977). Equation (9.6.1) shows that the integrated effect of such elementary contributions is again to produce unidirectional radiation. A further description of this principle and details of its application to problems in both acoustic and electromagnetic radiation is given by Baker and Copson (1950) who also describe the relationship of this formulation to Huygens' Principle and the contributions of both Helmholtz and Kirchhoff to its development.

9.7 Active absorption and reflection with continuous source Layers

The use of the Kirchhoff–Helmholtz integral equation (9.6.1) can be demonstrated by assuming that in a given volume V we wish to provide perfect suppression of a primary sound field $p_p(\mathbf{x})$ within the volume whilst leaving the sound field outside the volume unchanged. The principle is illustrated in Fig. 9.2, where in Fig. 9.2(a) the volume V is an interior volume completely enclosed by the bounding surface S. Under these circumstances we generate a secondary complex pressure field by

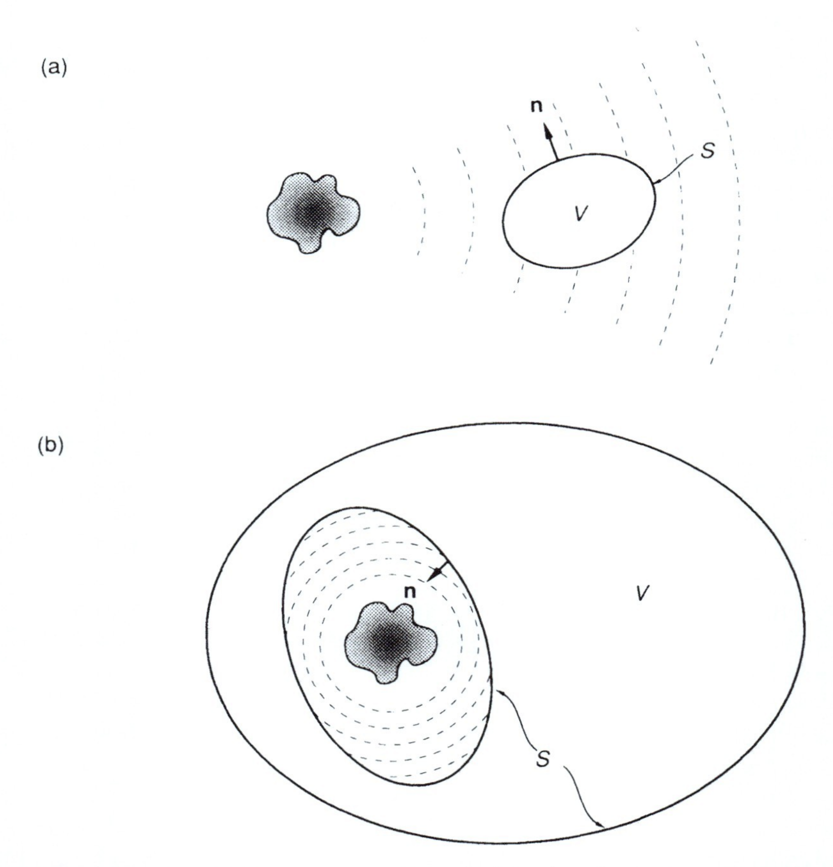

Fig. 9.2 (a) A monopole/dipole source layer is introduced on the surface S to completely cancel the field in V whilst leaving the field outside V completely unchanged. (b) An alternative interpretation where the outer part of the bounding surface S is considered to lie at infinity.

introducing a monopole/dipole layer on the surface S which produces the sound field specified by

$$\int_S g(\mathbf{x}|\mathbf{y})j\omega\rho_0 q_{\text{surf}}(\mathbf{y}) - \mathbf{f}_{\text{surf}}(\mathbf{y}).\nabla_y g(\mathbf{x}|\mathbf{y})].\mathbf{n}\mathrm{d}S = \begin{Bmatrix} p_s(\mathbf{x}), & \mathbf{x} \text{ within } V \\ 0, & \mathbf{x} \text{ outside } V \end{Bmatrix}. \quad (9.7.1)$$

We now choose the strengths of the secondary sources on S such that

$$j\omega\rho_0 q_{\text{surf}}(\mathbf{y}) = -\nabla_y p_p(\mathbf{y}).\mathbf{n}, \qquad \mathbf{f}_{\text{surf}}(\mathbf{y}) = -p_p(\mathbf{y})\mathbf{n}, \quad (9.7.2\text{a,b})$$

where $p_p(\mathbf{y})$ is the complex pressure on the surface S produced by the primary field. This choice of secondary source strengths thus ensures that in the volume V, $p_s(\mathbf{x}) = -p_p(\mathbf{x})$ by virtue of equation (9.6.1). Thus, upon using the superposition principle, the total pressure field reduces to

$$p(\mathbf{x}) = p_p(\mathbf{x}) + p_s(\mathbf{x}) = \begin{Bmatrix} 0, & \mathbf{x} \text{ within } V \\ p_p(\mathbf{x}), & \mathbf{x} \text{ outside } V \end{Bmatrix}. \quad (9.7.3)$$

Hence, as shown in Fig. 9.2(a), we could introduce a layer of monopole and dipole sources on the surface S which completely and perfectly cancelled the sound field within V, whilst leaving the sound field outside V completely unchanged. This principle applies identically to the situation depicted in Fig. 9.2(b). In this case the volume V is an exterior volume bounded by the surface S shown in Fig. 9.2(b) where part of the surface S lies at infinity. Under these circumstances we can again use precisely the reasoning outlined above and introduce a layer of monopole and dipole sources on the surface S in order to provide complete suppression of the radiation from a source surrounded by the surface S.

The capacity of a continuous monopole/dipole layer to radiate sound undirectionally can be used to demonstrate that we can also use continuous layers to *reflect* incident radiation and thereby maintain silence in a given closed volume V. Here we follow the argument presented by Curtis (1988). Consider the situation depicted in Fig. 9.3(a). First we can generate a secondary pressure field $p_s(\mathbf{x})$ exactly as that specified by equation (9.7.1) above. This will again ensure that the sound field within the volume V is cancelled. If we now define the volume V' which is bounded by a surface S' lying just outside the surface S and the primary source, we can place a source distribution $q'_{\text{surf}}(\mathbf{y}')$ and $\mathbf{f}'_{\text{surf}}(\mathbf{y}')$ on the surface S' which produces a field specified by

$$\int_S [g(\mathbf{x}|\mathbf{y}')j\omega\rho_0 q'_{\text{surf}}(\mathbf{y}') - \mathbf{f}'_{\text{surf}}(\mathbf{y}').\ \nabla_y g(\mathbf{x}|\mathbf{y}')]\mathrm{d}S' = \begin{Bmatrix} p'_s(\mathbf{x}), & \mathbf{x} \text{ within } V' \\ 0, & \mathbf{x} \text{ outside } V' \end{Bmatrix} \quad (9.7.4)$$

where we note that $\mathbf{n}'$ is the unit normal vector pointing outwards from the volume V'. The total complex sound pressure field is now

$$p(\mathbf{x}) = \begin{Bmatrix} p_p(\mathbf{x}) + p_s(\mathbf{x}) = 0, & \text{within } V \\ p_p(\mathbf{x}) + p'_s(\mathbf{x}), & \text{within } V' \end{Bmatrix}, \quad (9.7.5)$$

and we note that in the volume between S and S', the sound field remains equal to the primary field $p_p(\mathbf{x})$. Since the choice of $q'_{\text{surf}}(\mathbf{y}')$ and $\mathbf{f}'_{\text{surf}}(\mathbf{y}')$ can be made entirely arbitrarily without producing any additional pressure field inside V then we

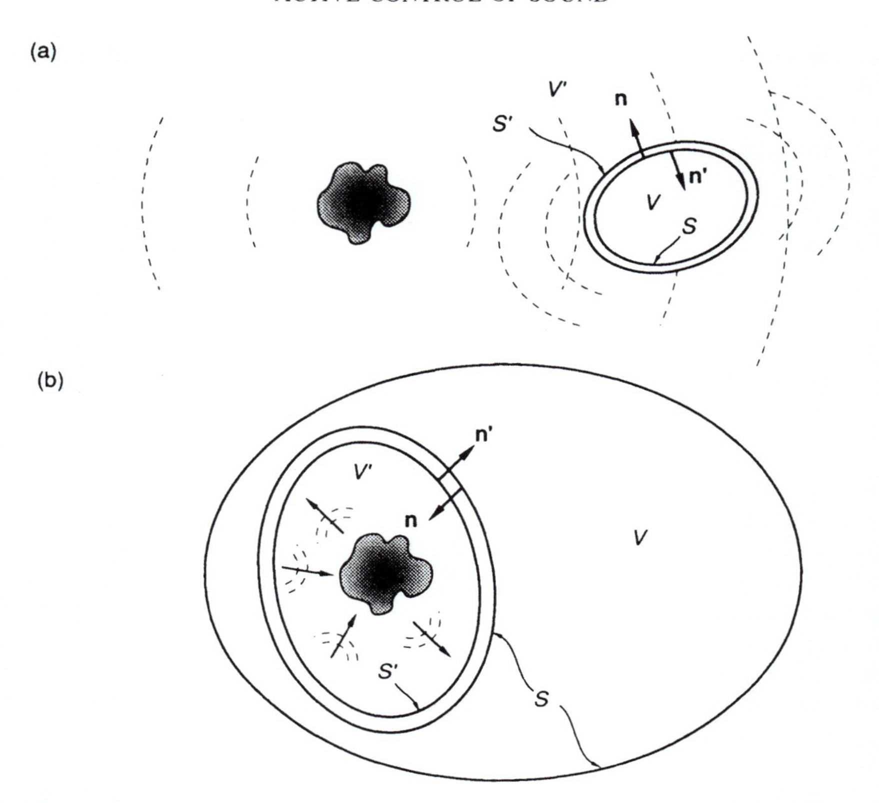

Fig. 9.3 Reflection of sound using continuous source layers. (a) A continuous monopole/dipole layer is introduced onto the surface S' which produces a field V' but not V. When the surfaces S and S' are brought together, the source distribution on S' can be chosen to, for example, cancel the dipole strength of the source distribution on S. (b) An alternative interpretation where the outer part of the bounding surface S can be considered to lie at infinity.

are at liberty to define these source strengths on S' as we wish. These define respectively the equivalent monopole or dipole strengths of the continuous source layer. For example, if we let the part of the surface S' that surrounds the surface S become coincident with the surface S, then we could choose the dipole strength on that part of S' to cancel exactly the dipole strength of the secondary source layer on S. That is, we ensure that

$$\mathbf{f}_{\text{surf}}(\mathbf{y}).\mathbf{n} + \mathbf{f}'_{\text{surf}}(\mathbf{y}')\ .\mathbf{n}' = 0, \tag{9.7.6}$$

when the surfaces S' and S are brought together. The cancelling secondary source distribution then consists only of a monopole layer. Since there is no dipole strength associated with this secondary source distribution, there is no net force applied to the fluid by the secondary source. There can therefore be no discontinuity in the pressure at the surface and since $p(\mathbf{x}) = 0$ inside V, then a zero pressure boundary condition must be imposed by the secondary source distribution. The continuous layer of monopoles consisting of $q_{\text{surf}}(\mathbf{y})$ and $q'_{\text{surf}}(\mathbf{y}')$ therefore *reflects* incident sound in a manner which is exactly analogous to the action of the single plane monopole source in a duct that is described in Chapter 5. The pressure $p_s'(\mathbf{x})$ specifies this reflected sound field. Note also that the same argument applies to the case depicted in Fig. 9.3(b). In this case, a standing wave field will be produced inside the volume V' as the surrounding monopole layer reflects sound back to the primary source distribution. Again, analogous to the one-dimensional case, the form of this sound field will be determined by the impedance (reflection coefficient) of the primary source distribution (see Section 5.11).

Finally, note that we could also choose $q_{\mathrm{surf}}'(\mathbf{y}')$ to ensure that the net secondary source distribution has zero *monopole* strength. Thus we ensure that when the surfaces S and S' are brought together

$$q_{\mathrm{surf}}(\mathbf{y}) - q_{\mathrm{surf}}'(\mathbf{y}') = 0, \tag{9.7.7}$$

where the minus sign is necessary since $q_{\mathrm{surf}}'(\mathbf{y}')$ is defined as being positive when a positive rate of flow of volume is introduced into V'. The sound field in V is thus cancelled using only a dipole layer. With no monopole secondary source strength, there can be no discontinuity in normal pressure gradient (particle velocity) at the surface. Thus, since in V the pressure gradient must be zero, the secondary source distribution must apply a boundary condition of zero normal particle velocity. In this case the field $p_s'(\mathbf{x})$ is the field reflected from the source layer as if it were a rigid boundary. The choice of zero dipole and zero monopole strength represents the two extremes of the range of acoustic impedance from zero to infinity that can be provided by a continuous secondary source distribution which still maintains perfect cancellation of the sound field in a given volume V.

9.8 Active suppression of radiation from vibrating bodies

The sound field radiated by a vibrating body in an otherwise unbounded medium can be expressed in terms of the integral form of solution to the wave equation given by equation (9.4.5). In the absence of any "volume sources" $Q_{\mathrm{vol}}(\mathbf{y})$, the solution for such a "primary" field can be written in the form

$$p_p(\mathbf{x}) = \int_S [G(\mathbf{x}|\mathbf{y})\boldsymbol{\nabla}_y p_p(\mathbf{y}) - p_p(\mathbf{y})\boldsymbol{\nabla}_y G(\mathbf{x}|\mathbf{y})].\mathbf{n}\,\mathrm{d}S, \tag{9.8.1}$$

where S now defines the surface of the vibrating body and $p_p(\mathbf{y})$ and $\boldsymbol{\nabla}_y p_p(\mathbf{y})$ define the complex pressure and its gradient on the surface of the body. We are again at liberty to choose the free space Green function and substitute $g(\mathbf{x}|\mathbf{y})$ for $G(\mathbf{x}|\mathbf{y})$ in equation (9.8.1) in which case the problem reduces to the solution of the Kirchhoff–Helmholtz integral equation. It is also obvious from this that if we place a continuous distribution of secondary sources on the surface S of the body which produce a pressure field $p_s(\mathbf{x})$ in V, we can ensure that $p_s(\mathbf{x}) = -p_p(\mathbf{x})$ and that the entire radiated field is cancelled. Thus the monopole and dipole strengths of the secondary source distribution as are defined by $j\omega\rho_0 q_{\mathrm{surf}}(\mathbf{y}) = -\boldsymbol{\nabla} p_p(\mathbf{y}).\mathbf{n}$ and $\mathbf{f}_{\mathrm{surf}}(\mathbf{y}) = -p_p(\mathbf{y})\mathbf{n}$. This is simply an expression of the fact that the absorbing layer depicted in Fig. 9.2b can be made coincident with the surface of the body radiating the primary field.

In general for arbitrary vibrating bodies, the solution of the Kirchhoff–Helmholtz integral equation must be accomplished numerically if the radiated primary field $p_p(\mathbf{x})$ is to be predicted. One is usually interested in the computation of the sound field radiated when the surface velocity (or acceleration) of the vibrating body is specified; that is, we know the value of $\nabla_y p_p(\mathbf{y}) = -j\omega\rho_0\mathbf{u}_p(\mathbf{y})$ on the surface S. If we specify $G(\mathbf{x}|\mathbf{y}) = g(\mathbf{x}|\mathbf{y})$, we then first have to solve the integral equation (9.8.1) in order to compute $p_p(\mathbf{y})$. A useful early review of numerical techniques for accomplishing this is given by Schenck (1968). Recent work by Cunefare and Koopmann (1991a, 1991b) describes a numerical technique for the solution to the Kirchhoff–Helmholtz integral equation using the boundary element method (see, for example, Brebbia *et al.*, 1984). By using this approach it is possible to calculate numerically the complex vector of pressures ($\mathbf{p}_p$ say) produced at the number of surface elements into which the radiating surface is divided. The complex normal velocity and thus monopole source strength associated with each of these elements is specified *a priori*. The numerical technique calculates the complex transfer impedance matrix $\mathbf{Z}$ where $\mathbf{p}_p = \mathbf{Z}\mathbf{q}_p$ and $\mathbf{q}_p$ is the vector of complex monopole strengths associated with the surface elements. Cunefare and Koopmann then consider the introduction of (monopole) secondary source elements onto the vibrating surface and determine the optimal vector of complex secondary source strengths which minimise the total power radiated by the body. (The appropriate theoretical technique for this computation can be deduced by following the analysis presented in Sections 8.12 and 8.13 of Chapter 8 and treating each source element on the vibrating body as a point source.) Cunefare and Koopmann concluded that global reductions in power output can be achieved by using a single small active area when the primary velocity distribution is in phase over the entire surface of the body and the acoustic wavelength is large compared to the dimensions of the body. When the body considered was a sphere with a "dipole-like" surface vibration (with vibrations of opposite phase on two sides of the sphere), it was observed that one secondary source element alone was not capable of producing substantial reductions in total power output, even at low wave numbers. However, two diametrically opposite active elements could achieve substantial reductions in power output, the two elements together being capable of producing a field which could closely match the primary field.

9.9 Active suppression of radiation from vibrating plane surfaces

It is also possible, under some circumstances, to derive an analytical solution to the Kirchhoff–Helmholtz integral equation. The conditions for which this is possible include the important case in which we can find a Green function $G(\mathbf{x}|\mathbf{y})$ which satisfies the condition $\nabla_y G(\mathbf{x}|\mathbf{y}).\mathbf{n} = 0$ on the surface S. In this case the radiated pressure field given by equation (9.8.1) reduces to

$$p_p(\mathbf{x}) = \int_S G(\mathbf{x}|\mathbf{y})\nabla_y p_p(\mathbf{y}).\mathbf{n}\,\mathrm{d}S, \tag{9.9.1}$$

and the need to solve an integral equation for the pressure on S is circumvented. Note that in general we can choose the Green function $G(\mathbf{x}|\mathbf{y})$ to be of the form

$$G(\mathbf{x}|\mathbf{y}) = g(\mathbf{x}|\mathbf{y}) + F(\mathbf{x}|\mathbf{y}), \tag{9.9.2}$$

where $F(\mathbf{x}|\mathbf{y})$ is the solution of the homogeneous Helmholtz equation $(\nabla^2 + k^2)F(\mathbf{x}|\mathbf{y}) = 0$ inside the volume V considered (see, for example, Morse and Ingard, 1968, Ch. 7). The task is thus to find a function $F(\mathbf{x}|\mathbf{y})$ which when added to the free space Green function $g(\mathbf{x}|\mathbf{y})$ will ensure that $\boldsymbol{\nabla}_y G(\mathbf{x}|\mathbf{y}).\mathbf{n} = 0$ at all points on the surface S. The conditions for which this is easily possible analytically have been described by, for example, Junger and Feit (1986, Ch. 4). First, the boundary S must be completely defined by specifying the value of a single co-ordinate of a system of three orthogonal co-ordinates (this is possible, for example, in the case of an infinite plane boundary, an infinite cylinder or a sphere). Second, the Helmholtz equation must be *separable* in the given co-ordinate system. This is also possible for the geometries described above (see Junger and Feit, 1986).

The simplest of these cases is the infinite plane boundary. If the boundary is rigid, a Green function whose normal derivative on the surface of the boundary is zero is readily constructed by using the "method of images". Thus the function $F(\mathbf{x}|\mathbf{y})$ is proportional to the complex pressure field produced in the volume V by the "image" of a point monopole source in the solid surface S. If the rigid plane were defined by $y_3 = 0$, for example, and a point monopole were placed just above the surface at the position $(y_1, y_2, y_3 + \varepsilon)$, the total pressure field would be described by

$$G(\mathbf{x}|\mathbf{y}) = g(\mathbf{x}|y_1, y_2, y_3 + \varepsilon) + g(\mathbf{x}|y_1, y_2, y_3 - \varepsilon), \tag{9.9.3}$$

where the second term is the complex pressure field due to the image source at the position $(y_1, y_2, y_3 - \varepsilon)$. The interference between these two fields will ensure that the component of particle velocity normal to the surface is always zero and therefore the condition $\boldsymbol{\nabla}_y G(\mathbf{x}|\mathbf{y}).\mathbf{n} = 0$ will be satisfied. In the limit of the distance ε tending to zero (i.e. the source and its image are brought together on the surface) then the Green function becomes the "hard walled" Green function given by

$$G(\mathbf{x}|\mathbf{y}) = 2g(\mathbf{x}|\mathbf{y}), \tag{9.9.4}$$

and the Kirchhoff–Helmholtz integral equation reduces to the *Rayleigh equation* which can be written as

$$p_p(\mathbf{x}) = 2j\omega\rho_0 \int_S g(\mathbf{x}|\mathbf{y})\mathbf{u}_p(\mathbf{y}).\mathbf{n}\, \mathrm{d}S. \tag{9.9.5}$$

This form of solution to the radiation problem was used by Deffayet and Nelson (1988), who studied the problem of the active suppression of sound radiation from a vibrating rectangular panel in an infinite rigid baffle. The velocity distribution on the surface $y_3 = 0$ is defined by

$$u_{p3} = U \sin\left(\frac{n\pi x_1}{a}\right) \sin\left(\frac{m\pi x_2}{b}\right) \quad 0 < x_1 < a, \quad 0 < x_2 < b, \tag{9.9.6}$$

where a and b are the dimensions of the panel. At low frequencies (ka, $kb << 1$) the radiation efficiency of the panel is most easily described by using the "corner monopole" model of the acoustic source distribution. When the integers n and m are both odd numbers (an "odd–odd" mode) the panel radiates as if there were four in-phase monopole sources at the corners. For n even and m odd or n odd and m even (an "odd–even" mode), the radiation efficiency is dipole-like. Finally, for n and m both even numbers (an "even–even" mode), the radiation is of quadrupole order. Deffayet and Nelson studied the degree to which the sound power radiated by the panel could be minimised by the introduction of point secondary monopole sources placed coincident with the panel surface. Again, substantial reductions were limited to wavelengths which were large compared to the panel dimensions, but one of the most important conclusions of this work was that the secondary source distribution had to be appropriately matched to the primary source distribution. For odd–odd (monopole type) modes one secondary source was sufficient but for odd–even (dipole type) modes at least two secondary sources were required. In addition, in the latter case, the location of the secondary sources was found to be critical (the axis of the "secondary" dipole had to be aligned with that of the primary dipole type source). Finally, it was demonstrated that four secondary sources could reduce further the power output of even–even (quadrupole type) modes.

More recent work on this problem has concentrated on controlling the *vibrations* of the surface with the objective of reducing the ability of the surface to radiate sound rather than adding additional secondary acoustic sources. For example, Fuller (1988) has argued that fewer actuators (acting on the structure itself) will be required if one is able to control the efficiently radiating modes of the vibrating surface. Fuller and his co-workers (Fuller, 1990; Fuller *et al.*, 1989) have concentrated on problems of sound transmission and radiation from finite vibrating panels and demonstrated both theoretically and experimentally that such a panel may be forced to respond in a manner which is inefficient in radiating sound and significant global reductions in sound power output may be achieved (see also the experiments of Thomas *et al.*, 1990, the theoretical work of Vyalyshev *et al.*, 1986, and the earlier work of Walker and Yaneske, 1976a,b, on the feedback control of panel vibrations).

9.10 Active suppression of scattered acoustic radiation

It is also possible in principle to render a scattering body invisible to incident acoustic radiation by surrounding the body with a continuous layer of secondary sources which will provide complete suppression of the radiation scattered by the body whilst leaving any incident sound field completely unchanged. One approach to this is to consider the volume V depicted in Fig. 9.2(a) to contain within it a scattering body (of arbitrary acoustic impedance) as illustrated in Fig. 9.4. The sound field within the remainder of the volume can still be made equal to zero by using the secondary source layer on the surrounding surface S that was specified by equation (9.7.1). This source distribution ensures the perfect absorption of the primary (incident) sound field. Note that, strictly speaking, we should also include a source layer on the surface of the body itself, but since the solution we are seeking is zero pressure (and pressure gradient) inside V, then this secondary source layer will have zero source strength. We can thus argue that as we let the surface S collapse onto

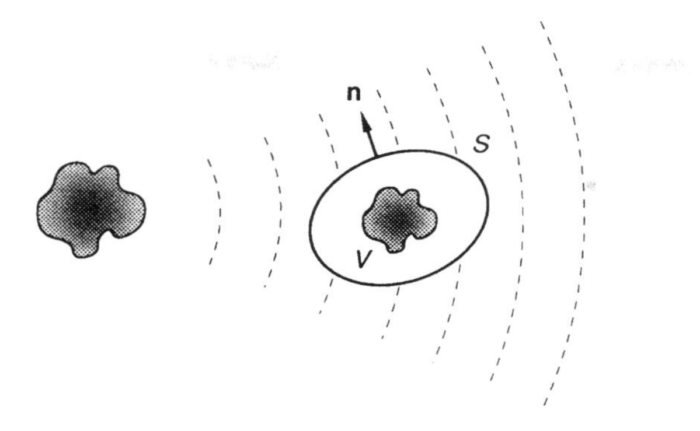

Fig. 9.4 The active suppression of scattered radiation by a secondary source layer which provides perfect absorption of the incident sound field.

the surface of the body, then the effect on the sound field outside V will remain the same as that depicted in Fig. 9.4. The monopole/dipole source layer on the surface of the body is thus specified in terms of the incident pressure field.

Another approach to the active suppression of scattered radiation is to specify the secondary source distribution in terms of the *scattered* pressure field. Thus the complex pressure field $p(\mathbf{x})$ is assumed to be that produced by the primary source in the absence of the scattering body ($p_p(\mathbf{x})$ say) plus a contribution $p_{\text{scat}}(\mathbf{x})$ which is attributed to the presence of the body and thus defined by $p_{\text{scat}}(\mathbf{x}) = p(\mathbf{x}) - p_p(\mathbf{x})$. The sum of $p_{\text{scat}}(\mathbf{x})$ and $p_p(\mathbf{x})$ must therefore satisfy the boundary conditions on the surface of the scattering body. Such boundary conditions could include a prescribed distribution of surface velocity on the body (in the case of a body which both "radiates" and scatters sound). Irrespective of these boundary conditions we can always ensure that the field $p_{\text{scat}}(\mathbf{x})$ is perfectly cancelled in the volume V depicted in Fig. 9.5 by including on the surface S a distribution of monopole and dipole sources whose strengths are defined by

$$j\omega\rho_0 q_{\text{surf}}(\mathbf{y}) = -\nabla_y p_{\text{scat}}(\mathbf{y}).\mathbf{n}', \quad \mathbf{f}_{\text{surf}}(\mathbf{y}) = -p_{\text{scat}}(\mathbf{y})\mathbf{n}'. \tag{9.10.1}$$

The result of adopting this approach is illustrated in Fig. 9.5(a). The primary field is allowed to penetrate S but the scattered field is prevented from escaping by virtue of the secondary source distribution on S. Clearly, as the surface S is allowed to collapse onto the scattering body, then this source distribution will produce identical results to that source distribution specified in terms of the primary field only and whose effect is illustrated in Fig. 9.4. In addition to these two approaches to the suppression of scattered radiation, it is also possible to conceive of a superposition on S of both a source distribution which absorbs incident primary radiation *and* one which cancels any scattered radiation. This approach may be considered useful when $p_{\text{scat}}(\mathbf{x})$ defined above includes a contribution due to sound "radiated" by surface motion of the body (i.e. the body is not just a passive scatterer). Such a case is illustrated in Fig. 9.5(b).

In the particular case of a *rigid* scatterer, the incident primary field and the resulting scattered field are related by the need to satisfy the boundary condition of

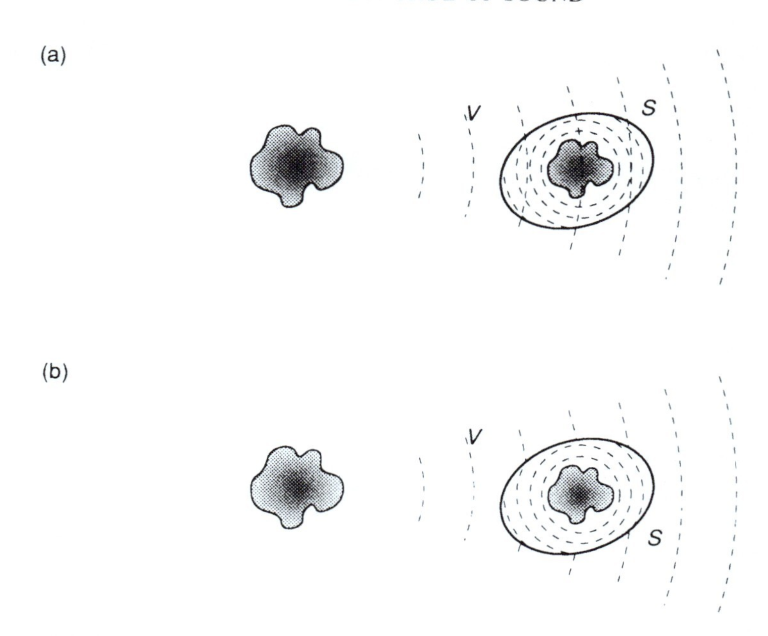

Fig. 9.5 (a) A secondary source layer on S is used to cancel the radiation scattered by body enclosed by S. (b) A secondary source layer on S is used to both absorb incident radiation (as illustrated in Fig. 9.4) and cancel radiation from the body.

zero normal particle velocity on the body and therefore $\nabla_y p_{\text{scat}}(\mathbf{y}) = -\nabla_y p_p(\mathbf{y})$ in a direction normal to the surface of the body. If in addition we can find, for the body considered, a Green function which satisfies $\nabla_y G(\mathbf{x}|\mathbf{y}).\mathbf{n} = 0$ on the surface of the body, then the scattered sound field in the volume V is specified by

$$p_{\text{scat}}(\mathbf{x}) = \int_S G(\mathbf{x}|\mathbf{y})\nabla_y p_{\text{scat}}(\mathbf{y}).\mathbf{n}\,\mathrm{d}S = -\int_S G(\mathbf{x}|\mathbf{y})\nabla_y p_p(\mathbf{y}).\mathbf{n}\,\mathrm{d}S. \quad (9.10.6)$$

Thus the surface of the body must be surrounded by a layer of monopole sources whose strength is directly related to the particle velocity distribution of the incident sound field, i.e. $j\omega\rho_0 q_{\text{surf}}(\mathbf{y}) = -\nabla_y p_p(\mathbf{y}).\mathbf{n}$. The action of the secondary sources in this case can be interpreted physically as the three-dimensional generalisation of the "absorbing termination" dealt with in Section 5.15. In that case, we saw that a plane monopole source placed adjacent to a rigid duct termination could provide perfect absorption of any incident radiation. Such an absorbing termination can be realised physically by causing the rigid surface itself to vibrate with the appropriate particle velocity; it is thus clear that in the case described here, if the surface of the initially rigid scatterer can be made to vibrate with the normal velocity proportional to $\nabla_y p_p(\mathbf{y})$, then zero scattered radiation will result. The realisation of the active control of scattered radiation by using a finite number of discrete secondary monopole sources has been considered by a number of Soviet authors. For example, Kleshchev and Klyukin (1974, 1977) have undertaken a theoretical examination of the problem of compensating the diffracted field of spheroidal scatterers in the field of a harmonic point source. Similar problems have been examined by Zavadskaya *et al.* (1975) and Urusovskii (1986).

9.11 The acoustic energy balance equation

Having described the mechanism of sound absorption by continuous source layers, we shall derive an energy balance equation which specifically accounts for the power radiated by distributions of monopole and dipole type sources. Here we follow the approach adopted by Levine (1980a,b). The equation can be derived by first taking the dot product of the velocity $\mathbf{u}(\mathbf{x},t)$ with the linearised equation of conservation of momentum given by equation (9.2.2). This yields

$$\mathbf{u}(\mathbf{x},t).\left[\rho_0 \frac{\partial \mathbf{u}(\mathbf{x},t)}{\partial t} + \nabla p(\mathbf{x},t)\right] = \mathbf{u}(\mathbf{x},t).\mathbf{f}_{\text{vol}}(\mathbf{x},t). \qquad (9.11.1)$$

If we now use the vector identity $\nabla.(a\mathbf{b}) = (\nabla a).\mathbf{b} + a\nabla.\mathbf{b}$ we can show that

$$\mathbf{u}(\mathbf{x},t).\nabla p(\mathbf{x},t) = \nabla.p(\mathbf{x},t)\mathbf{u}(\mathbf{x},t) - p(\mathbf{x},t)\nabla.\mathbf{u}(\mathbf{x},t). \qquad (9.11.2)$$

Furthermore it follows from the mass conservation equation (see equation (9.2.1)) that

$$\nabla.\mathbf{u}(\mathbf{x},t) = q_{\text{vol}}(\mathbf{x},t) - \frac{1}{\rho_0}\frac{\partial \rho(\mathbf{x},t)}{\partial t}. \qquad (9.11.3)$$

Using these relationships enables equation (9.11.1) to be written as

$$\mathbf{u}(\mathbf{x},t).\rho_0 \frac{\partial \mathbf{u}(x,t)}{\partial t} + \nabla.[p(\mathbf{x},t)\mathbf{u}(\mathbf{x},t)] + \frac{p(\mathbf{x},t)}{\rho_0}\frac{\partial \rho(\mathbf{x},t)}{\partial t}$$
$$= \mathbf{u}(\mathbf{x},t).\mathbf{f}_{\text{vol}}(\mathbf{x},t) + p(\mathbf{x},t)q_{\text{vol}}(\mathbf{x},t). \qquad (9.11.4)$$

Upon assuming the adiabatic relationship between the pressure and density given by $p(\mathbf{x},t) = c_0^2\rho(\mathbf{x},t)$ and also noting that $\partial/\partial t(\mathbf{u}.\mathbf{u}) = 2\mathbf{u}.(\partial\mathbf{u}/\partial t)$ shows that this equation can be written in the form

$$\frac{\partial}{\partial t}\left[\frac{1}{2}\rho_0 u^2(\mathbf{x},t) + \frac{1}{2}\frac{p^2(\mathbf{x},t)}{\rho_0 c_0^2}\right] + \nabla.[p(\mathbf{x},t)\mathbf{u}(\mathbf{x},t)]$$
$$= \mathbf{u}(\mathbf{x},t).\mathbf{f}_{\text{vol}}(\mathbf{x},t) + p(\mathbf{x},t)q_{\text{vol}}(\mathbf{x},t). \qquad (9.11.5)$$

This is an energy balance equation, where the first term in the square brackets on the left side is the time rate of change of the energy density in the medium, whilst the second term on the left side is the divergence of the acoustic intensity. Thus, when applied to a given volume of the medium, the first term represents the rate of change of energy within a volume and the second term represents the rate of outflow of energy from the volume. The terms on the right side of the equation can clearly be interpreted as sources of energy and result directly from dipole sources doing work on the medium, as shown by the product $\mathbf{f}_{\text{vol}}.\mathbf{u}$, and through monopole sources imparting energy as represented by the product $q_{\text{vol}}\text{p}$. In the particular case of harmonic fluctuations then time averaging this equation shows that

$$\nabla.[(1/2)\mathrm{Re}\{p(\mathbf{x})\mathbf{u}(\mathbf{x})\}] = (1/2)\mathrm{Re}\{\mathbf{u}^*(\mathbf{x}).\mathbf{f}_{\mathrm{vol}}(\mathbf{x}) + p^*(\mathbf{x})q_{\mathrm{vol}}(\mathbf{x})\}, \quad (9.11.6)$$

where the relevant variables are now complex. The time-averaged product of pressure and particle velocity gives the time-averaged intensity vector **I**. The divergence theorem shows that

$$\int_V \nabla.\mathbf{I}\mathrm{d}V = \int_S \mathbf{I}.\mathbf{n}\ \mathrm{d}S, \quad (9.11.7)$$

where S is the surface enclosing a given volume V and **n** is the outward normal unit vector from the surface. Therefore, integrating equation (9.11.6) over the volume V enclosing the sources yields the power output

$$W = \int_V (1/2)\mathrm{Re}\{\mathbf{u}^*(\mathbf{x}).\mathbf{f}_{\mathrm{vol}}(\mathbf{x}) + p^*(\mathbf{x})q_{\mathrm{vol}}(\mathbf{x})\}\mathrm{d}V. \quad (9.11.8)$$

In the particular case of point sources, where $\mathbf{f}_{\mathrm{vol}}(\mathbf{x}) = \mathbf{f}\delta(\mathbf{x} - \mathbf{y})$ and $q_{\mathrm{vol}}(\mathbf{x}) = q\delta(\mathbf{x} - \mathbf{y})$ then the expression for the power output of the sources reduces to

$$W = (1/2)\mathrm{Re}\{\mathbf{u}^*(\mathbf{y}).\mathbf{f} + p^*(\mathbf{y})q\}. \quad (9.11.9)$$

This is a formal expression that the power output of a point monopole source can be computed from the time-averaged product of its strength and the pressure fluctuation at the position of the source. We also see that the power output of a dipole type source can be computed from the time-averaged product of the force applied, **f**, and the particle velocity, **u**, in the direction of the applied force.

9.12 Sound power absorption by monopole and dipole sources

In considering the possibility for constructing the type of continuous absorbing layer described in Section 9.7, we shall evaluate to what extent elementary sources of monopole and dipole type can absorb incident radiation. The approach adopted here follows that presented by Nelson *et al.* (1986a,b), in which the optimisation theory introduced in Chapter 5 and described in the appendix has been used to find the maximum sound power absorbed by the basic multipole source types. Let us consider the sound power absorbed by a simple point monopole from an incident plane wave. In undertaking such a calculation (and the calculations that follow) one has, of course, to bear in mind that no account is taken of the influence of the absorbing source on the output of the source from which the plane wave originates. It follows from the acoustic energy balance equation presented in Section 9.11 that the *sound power output* of such a monopole source situated at the position **y** is given by

$$W = (1/2)\mathrm{Re}\{[p(\mathbf{y}) + p_p(\mathbf{y})]^*q\}. \quad (9.12.1)$$

Here $p(\mathbf{y}) = Z(\mathbf{y}|\mathbf{y})q$ is the pressure produced by the monopole upon itself. The pressure $p_p(\mathbf{y})$ is the complex pressure associated with the incident plane wave at the position of the monopole source. It follows from this and the discussion of Section 5.5 that this expression can be written as

$$W = q^*Aq + q^*b + b^*q, \tag{9.12.2}$$

where $A = (1/2)\text{Re}\{Z(\mathbf{y}|\mathbf{y})\}$ and $b = (1/4)p_p(\mathbf{y})$. Note the absence of the scalar constant c usually associated with the quadratic function of this form. The optimal source strength which minimises the power output of the monopole is given by $q_0 = -A^{-1}b$, and the corresponding minimum power output is given by $W_0 = -b^*A^{-1}b$. Thus, in general, the values of q_0 and W_0 are given by

$$q_0 = -\frac{p_p(\mathbf{y})}{2\text{Re}\{Z(\mathbf{y}|\mathbf{y})\}}, \qquad W_0 = \frac{-|p_p(\mathbf{y})|^2}{8\text{Re}\{Z(\mathbf{y}|\mathbf{y})\}}. \tag{9.12.3a,b}$$

Note that these expressions are generally applicable when $p_p(\mathbf{y})$ is interpreted as the pressure produced on the source by some external means (not just an incident plane wave) and $Z(\mathbf{y}|\mathbf{y})$ quantifies the pressure produced on the source by itself. Since $\text{Re}\{Z(\mathbf{y}|\mathbf{y})\} = (\omega^2\rho_0/4\pi c_0)$ as described in Section 8.3, then it follows that

$$q_0 = -\frac{2\pi c_0}{\omega^2\rho_0}p_p(\mathbf{y}), \qquad W_0 = -\frac{\pi c_0}{2\omega^2\rho_0}|p_p(\mathbf{y})|^2. \tag{9.12.4a,b}$$

First note that the optimal monopole strength is 180° out of phase with the incident plane wave pressure. Thus, in order to minimise the source power output, the source will produce a volume inflow at the same time as the incident wave produces a compression. Of most significance is that the minimum power output has a *negative* value. This corresponds to the *maximum power absorbed* by the monopole. Note that this expression can be written in the form $[|p_p(\mathbf{y})|^2/2\rho_0c_0](\lambda^2/4\pi)$, and since $[|p_p(\mathbf{y})|^2/2\rho_0c_0]$ is the incident intensity associated with the plane wave, then it follows that the action of the monopole is to produce an equivalent absorbing area given by $(\lambda^2/4\pi)$. This area is described by a circle of one wavelength in circumference, which demonstrates that *low* frequency sound is more readily absorbed by active means. Figure 9.6 shows the distribution of sound pressure amplitude and the acoustic intensity in the region of a monopole source whose strength has been adjusted to maximise its absorption of energy from an incident plane wave.

An equivalent calculation can also be undertaken for a dipole source. In this instance it has been demonstrated in Section 9.11 that the dot product of the particle velocity vector with the applied force vector when time averaged will yield the power output of a point dipole. Thus, for example, if we have a plane wave incident upon a dipole at position $\mathbf{y}$ whose strength is specified by the force vector $\mathbf{f}$, then the power output can be written as

$$W = (1/2)\text{Re}\{[\mathbf{u}(\mathbf{y}) + \mathbf{u}_p(\mathbf{y})]^*.\mathbf{f}\}, \tag{9.12.5}$$

where $\mathbf{u}(\mathbf{y})$ is the particle velocity produced by the dipole source at the position of the dipole and $\mathbf{u}_p(\mathbf{y})$ is the particle velocity vector associated with the incident plane wave. Now, for simplicity, let us assume that the dipole has only a component in the x_1 direction, such that $\mathbf{f} = \mathbf{i}f$. Furthermore, assume that the particle velocity associated with the dipole in the x_1 direction can be written as $u_1(\mathbf{y}) = M(\mathbf{y}|\mathbf{y})f$, where $M(\mathbf{y}|\mathbf{y})$ is the "input mobility" of the acoustic dipole and that the incident plane wave velocity field is specified by $\mathbf{u}_p = u_{p1}\mathbf{i} + u_{p2}\mathbf{j} + u_{p3}\mathbf{k}$. Substitution into equation (9.12.5) then shows that

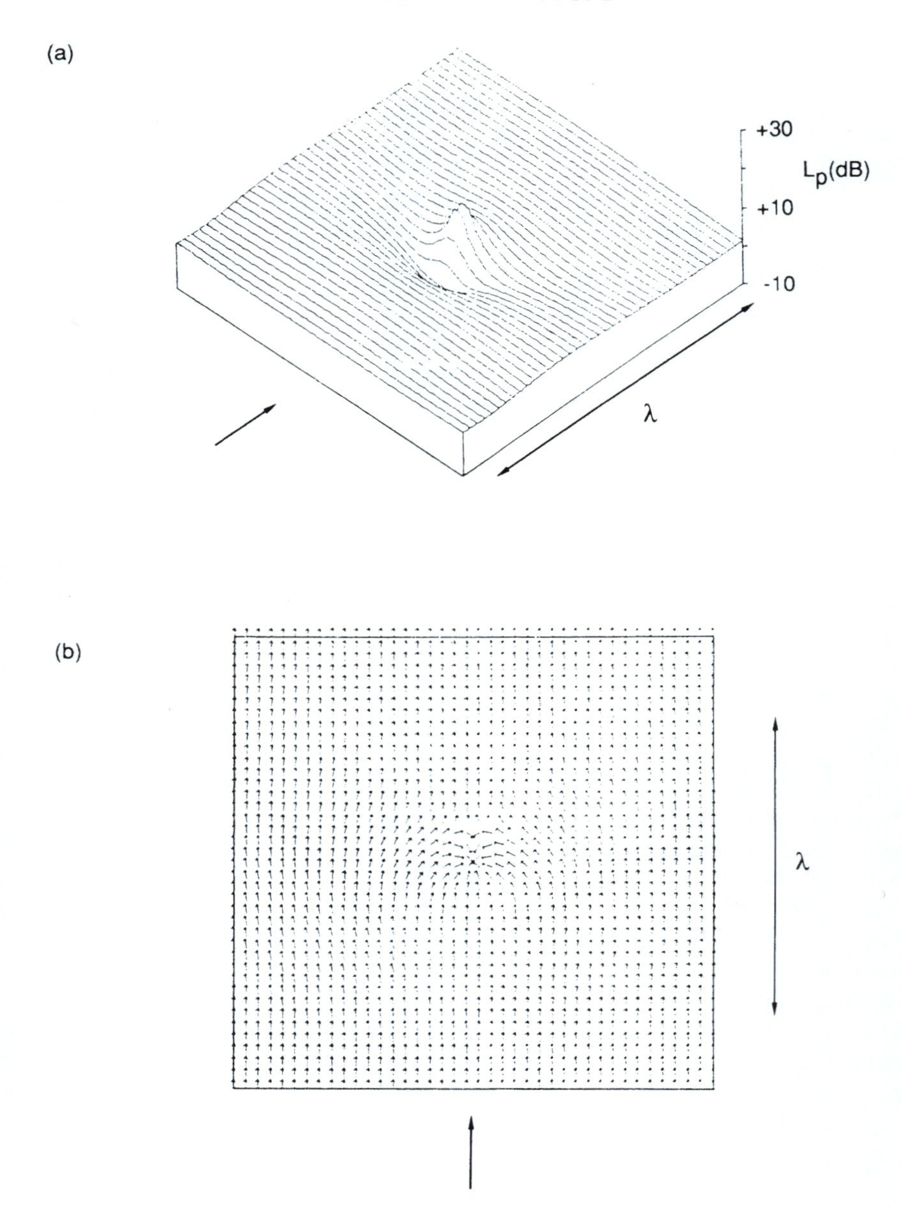

Fig. 9.6 The distribution of (a) sound pressure level and (b) intensity level in a plane through point monopole source whose strength is optimally chosen to maximise the energy absorbed from an incident plane wave. The arrow shows the direction of propagation of the incident plane wave of wavelength λ.

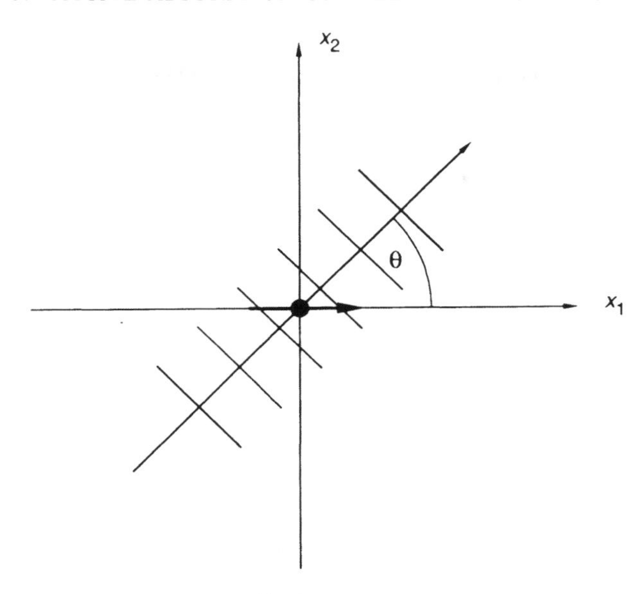

Fig. 9.7 A plane wave incident upon a dipole source.

$$W = (1/2)f^*\mathrm{Re}\{M(\mathbf{y}|\mathbf{y})\}f + (1/4)f^*u_{p1}(\mathbf{y}) + (1/4)u^*_{p1}(\mathbf{y})f. \qquad (9.12.6)$$

We can again identify this as a quadratic form of the type specified by equation (9.12.2) where $A = (1/2)\mathrm{Re}\{M(\mathbf{y}|\mathbf{y})\}$ and $b = (1/4)u_{p1}(\mathbf{y})$. It follows from equation (8.7.12) that the radial component of the particle velocity field associated with the dipole can be written as

$$u_r(r,\theta) = \frac{f\omega^2\cos\theta}{4\pi\rho_0 c_0^2}\left\{\frac{2kr + j[(kr)^2 - 2]}{(kr)^2}\right\}\frac{\mathrm{e}^{-jkr}}{kr}. \qquad (9.12.7)$$

Note that in deducing this equation use has been made of the relationship $f = j\omega\rho_0 qd$, where q is the strength of the constituent monopole sources forming the dipole described in Section 8.7. It also follows that in the limit of $kr \to 0$, $\mathrm{Re}\{M(\mathbf{y}|\mathbf{y})\} = \omega^2/12\pi\rho_0 c_0^3$. Now note that we can write the general form of the complex pressure associated with the plane wave travelling in some arbitrary direction in three dimensions as

$$p_p(\mathbf{x}) = p_p(\mathbf{y})\mathrm{e}^{-j\mathbf{k}.(\mathbf{x}-\mathbf{y})}, \qquad (9.12.8)$$

where $\mathbf{k}$ is the wavenumber vector which has modulus given by $|\mathbf{k}| = \omega/c_0 = k$. Thus, if we assume that the plane wave incident upon the dipole only has components in the x_1 and x_2 directions and travels at an angle θ to the x_1 axis, as illustrated in Fig. 9.7, then equation (9.12.8) can be written as

$$p_p(x_1, x_2) = p_p(\mathbf{y})\mathrm{e}^{-jk(x_1 - y_1)\cos\theta - jk(x_2 - y_2)\sin\theta}. \qquad (9.12.9)$$

The particle velocity in the x_1 component direction follows from the momentum equation and is given by $u_1 = -(1/j\omega\rho_0)\partial p/\partial x_1$. Differentiation of equation (9.12.9) therefore shows that

$$u_{p1}(x_1, x_2) = \frac{p_p(\mathbf{y})\cos\theta}{\rho_0 c_0} e^{-jk(x_1 - y_1)\cos\theta - jk(x_2 - y_2)\sin\theta}, \quad (9.12.10)$$

and therefore at the position of the dipole the particle velocity due to the plane wave in the x_1 direction is given by

$$u_{p1}(\mathbf{y}) = \frac{p_p(\mathbf{y})\cos\theta}{\rho_0 c_0}. \quad (9.12.11)$$

Using the optimal solutions $f_0 = -b/A$ and $W_0 = -|b|^2/A$ then shows that

$$f_0 = -\frac{6\pi c_0^2}{\omega^2}\cos\theta\, p_p(\mathbf{y}), \quad W_0 = -\frac{3\pi c_0}{2\omega^2\rho_0}\cos^2\theta|p_p(\mathbf{y})|^2. \quad (9.12.12a,b)$$

This shows first that the optimal dipole strength is such that the force applied to the fluid is 180° out of phase with the plane wave pressure produced at the position at which the force is applied. In addition, the minimum power output, which again corresponds to the maximum power absorbed, can be as much as *three times* that associated with the absorption that can be produced by a monopole only (see equation (9.12.4b)). Note also that, as one would expect, the optimal force and the minimum power output have a dependence on the direction of the incident plane wave. The absorption is maximised when the plane wave direction is along the axis of the dipole, whilst no absorption can be produced by a dipole from a plane wave which is incident at 90° to the axis of the dipole.

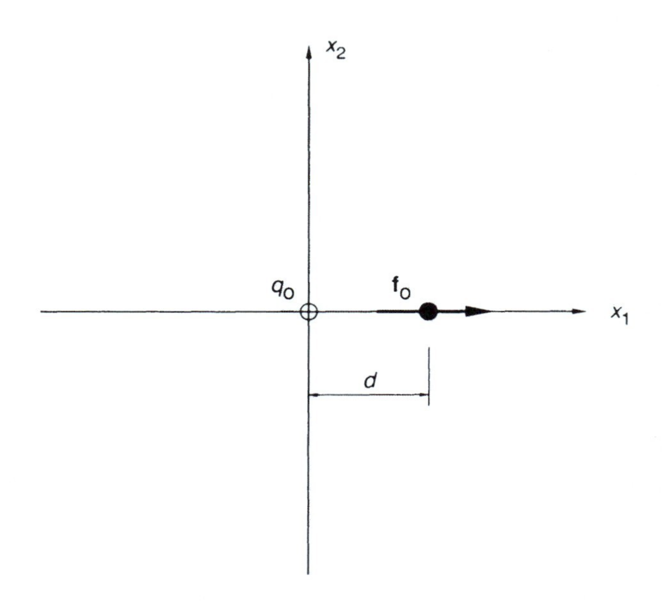

Fig. 9.8 A monopole and a dipole source separated by a small distance d.

9.13 The superposition of maximally absorbing monopole and dipole sources

We will now demonstrate that the superposition, at the same point, of a monopole and a dipole source whose strengths are chosen to maximise their power absorption from an incident plane wave, results in a net power absorption which is the sum of those powers absorbed by the monopole and dipole sources alone. This can be demonstrated by evaluating the power outputs of the monopole and dipole sources when separated by a small distance d as illustrated in Fig. 9.8. With a plane wave incident axially upon the two sources, we can express the sound power output of the monopole source as

$$W_m = (1/2)\mathrm{Re}\{[p_p(\mathbf{y}_m) + p_m(\mathbf{y}_m) + p_d(\mathbf{y}_m)]^* q_0\}, \tag{9.13.1}$$

where the monopole source strength chosen, q_0, is that specified by equation (9.12.3a). We can write equation (9.13.1) as

$$W_m = W_{m0} + W_1, \tag{9.13.2}$$

where W_{m0} is the sound power output of the source of strength q_0 in the absence of the dipole source, and $W_1 = (1/2)\ \mathrm{Re}\{p_d^*(\mathbf{y}_m)q_0\}$, which is the power output produced by the monopole source due to the influence of the dipole source. Similarly, the expression for the sound power output of the dipole source is given by

$$W_d = (1/2)\mathrm{Re}\{f_0^*[u_{p1}(\mathbf{y}_d) + u_{d1}(\mathbf{y}_d) + u_{m1}(\mathbf{y}_d)]\}, \tag{9.13.3}$$

which can be written as

$$W_d = W_{d0} + W_2\ , \tag{9.13.4}$$

where W_{d0} is the sound power output of the dipole of strength f_0 in the absence of the monopole source, and W_2 represents the additional power output of the dipole source due to the presence of the monopole source. This is given by $W_2 = (1/2)\mathrm{Re}\{f_0^* u_{m1}(\mathbf{y}_d)\}$. Note first that it follows from equation (8.7.10) with $qd = f/j\omega\rho_0$, that the pressure field at a radial distance r from a dipole source can be written as

$$p_d(r) = \frac{f e^{-jkr}\cos\theta}{4\pi r}\left(jk + \frac{1}{r}\right). \tag{9.13.5}$$

Thus the pressure produced at the position of the point monopole by the presence of the dipole of strength f_0 is given by

$$p_d(\mathbf{y}_m) = -\frac{f_0 e^{-jkd}}{4\pi d}\left(jk + \frac{1}{d}\right), \tag{9.13.6}$$

where the minus sign results from the fact that $\theta = 180°$. The velocity field due to the monopole can be deduced from the application of the linearised momentum equation, where its radial component is given by $u_r = -(1/j\omega\rho_0)\partial p/\partial r$. It therefore follows that

$$u_r(r) = \frac{q e^{-jkr}}{4\pi r}\left(jk + \frac{1}{r}\right). \tag{9.13.7}$$

Thus, the velocity produced in the direction of the dipole axis at the position of the dipole can be written as

$$u_{m1}(\mathbf{y}_d) = \frac{q_0 e^{-jkd}}{4\pi d}\left(jk + \frac{1}{d}\right). \tag{9.13.8}$$

We can now use equations (9.13.6) and (9.12.3a) to calculate the power output of the monopole produced by the presence of the dipole. Since, for an axial plane wave, the contributions of the plane wave pressures at $\mathbf{y}_d$ and $\mathbf{y}_m$ are given by $p_p(\mathbf{y}_d) = p_p(\mathbf{y}_m)e^{-jkd}$, it follows, after some algebra, that

$$W_1 = -\frac{3\pi c_0}{2\omega^2 \rho_0} |p_p(\mathbf{y}_m)|^2 \left(\frac{\cos 2kd}{(kd)^2} + \frac{\sin 2kd}{kd}\right). \tag{9.13.9}$$

In the limit of the separation between the monopole and dipole sources being small compared to the wavelength, series expansion of this relationship shows that

$$W_1 = -\frac{3\pi c_0}{2\omega^2 \rho_0} |p_p(\mathbf{y}_m)|^2 \left(\frac{1}{(kd)^2} + O(kd)^2\right), \quad \text{as } kd \to 0. \tag{9.13.10}$$

Similarly, we can use equations (9.13.8) and (9.12.2a) to calculate the power output of the dipole produced by the presence of the monopole. Again, after some algebra, it follows that

$$W_2 = \frac{3\pi c_0}{2\omega^2 \rho_0} |p_p(\mathbf{y}_m)|^2 \left(\frac{1}{(kd)^2}\right). \tag{9.13.11}$$

It is thus evident that as the monopole and dipole sources are brought increasingly close together relative to the acoustic wavelength, then the presence of the monopole produces an increasingly large power output from the dipole but this power output is exactly absorbed by the monopole itself. When the two sources are superposed at the same point, then, although there is an infinite flow of power between them, as far as the radiated field is concerned they act to absorb power to an amount which is given by the sum of the maximum values of the power that they absorb individually from the plane wave. Thus, we can write the total power output as

$$W_0 = W_{m0} + W_{d0} + W_1 + W_2, \tag{9.13.12}$$

and, since $W_1 = -W_2$, then using equations (9.12.4a) and (9.12.12b) we have

$$W_0 = -\frac{2\pi c_0}{\omega^2 \rho_0} |p_p(\mathbf{y})|^2. \tag{9.13.13}$$

This corresponds to an equivalent absorption cross section of *four* times that of a monopole acting alone, i.e. an equivalent absorbing area of λ^2/π. An exactly equivalent effect can be produced by using two monopoles of appropriate strength spaced a distance apart that is small compared to the wavelength. In this case the monopoles have strengths which effectively synthesise the requisite sum of monopole and dipole strengths (see Nelson and Elliott, 1987). The resulting sound pressure and intensity fields produced by using this arrangement are shown in Fig. 9.9.

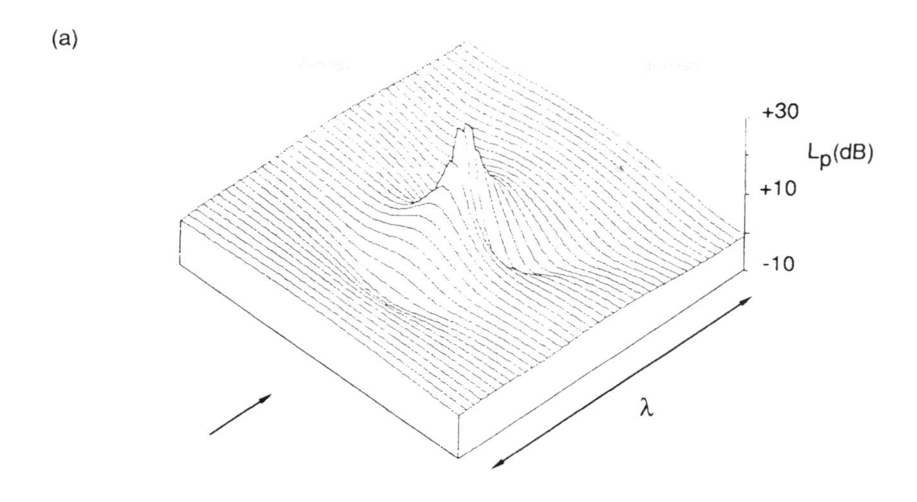

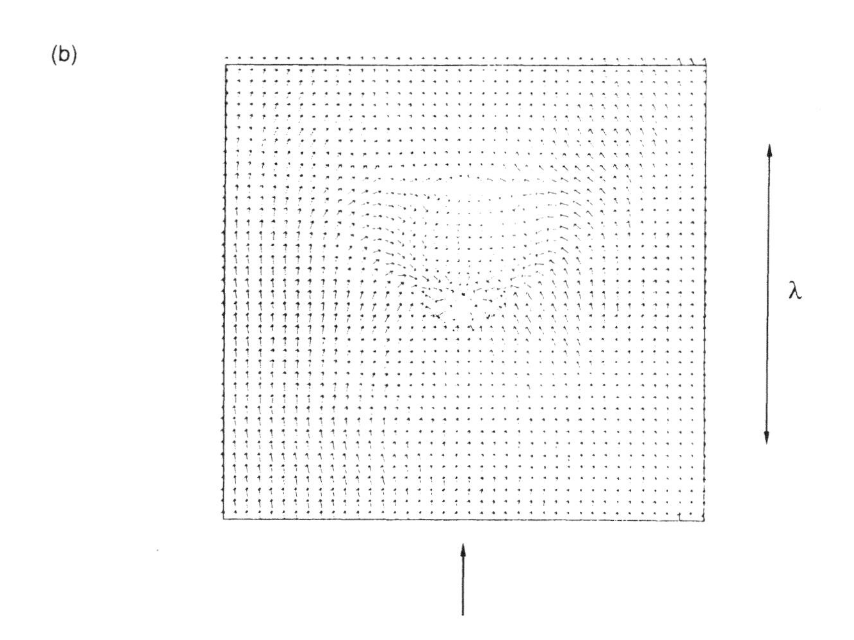

Fig. 9.9 The distribution of (a) sound pressure level and (b) intensity level in a plane through two monopole sources whose strengths are optimally chosen to maximise the net energy absorbed from an incident plane wave. The direction of propagation of the plane wave (shown by the arrow) was parallel with the axis of the two sources. The sources were separated by a distance of approximately 0.016λ in the case shown.

9.14 The discretisation of continuous absorbing surfaces

The analysis of the previous section has demonstrated that the requisite combination of a point monopole source and a point dipole source can act to produce an absorption cross section of λ^2/π. One therefore is tempted to speculate that an infinite two-dimensional array of such source combinations, spaced on a square grid at intervals of $\lambda/\sqrt{\pi}$, would act to produce perfect absorption of a normally incident plane wave (see Fig. 9.10). However, one cannot with certainty make such an assertion since there is no guarantee that the same absorption cross-section associated with the sources individually would be produced when the sources were arranged in an array. In order to calculate the net total absorption, one would have to evaluate the influence that all the sources have on the power outputs of all the others. However, if the sources are spaced at intervals of approximately $\lambda/2$, one would expect that their mutual interaction would be relatively small (see Section 8.4) and that a grid spacing of $\lambda/\sqrt{\pi}$ (or approximately $\lambda/2$) would, at least to a first approximation, result in a substantial absorption of a normally incident wave.

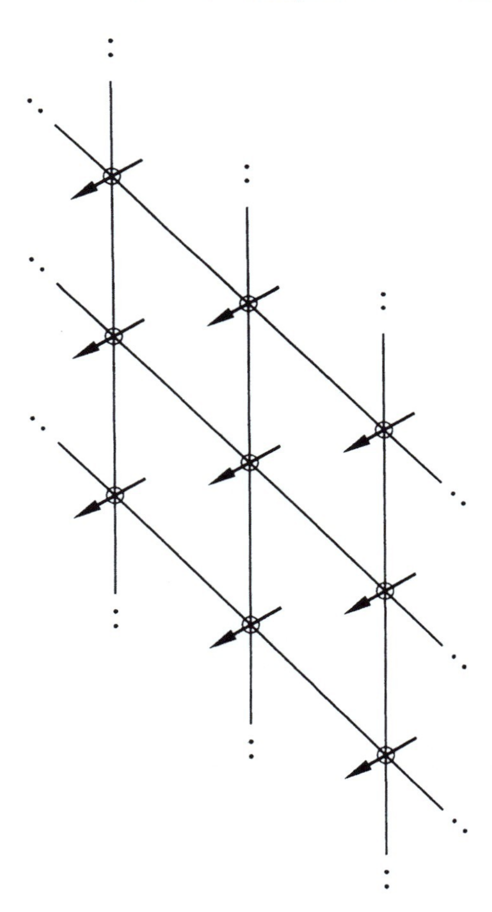

Fig. 9.10 An infinite two-dimensional array of point monopole/dipole sources.

Some support is given to this view by the various numerical studies undertaken to date on the discrete approximation of continuous absorbing source layers. Most of this work has been conducted in the Soviet Union. For example, Fedoryuk (1975) reported that, in numerical studies, the suppression of the field of a cylindrical radiator is satisfactorily realised if the necessary continuous source distribution is replaced by a discrete distribution with sources spaced at "two or three per wavelength". A more detailed numerical study has been presented by Konyaev *et al.* (1977). They considered a distribution of discrete monopole/dipole sources over a spherical surface and calculated the deviation (both inside and outside the sphere) in the field produced from that produced by the equivalent continuous surface source distribution used to synthesise the field of a plane wave. Although the degree of approximation achieved depended on the distance from the spherical surface, it was concluded that a good approximation is produced (with errors in the pressure amplitude of less than 10^{-3}) provided the parameter $(\lambda/2R)\sqrt{N/\pi} \geq 2.5$, where N is the number of sources and R is the radius of the sphere. This result implies that each discrete monopole/dipole source contributes an effective absorbing area to the spherical surface of $\lambda^2/(2.5)^2$ (i.e. somewhat less than the value of λ^2/π suggested for the plane array considered above).

A similar study was reported by Zavadskaya *et al.* (1976). In this case the equivalent two-dimensional problem was considered in which an attempt was made to suppress the field of a plane wave within a circular region, a discrete approximation again being made to the necessary continuous monopole/dipole source distribution on the boundary of the circle. The conclusion from this study was that a good approximation (again to within a factor of 10^{-3}) could be made by using discrete sources spaced apart around the circumference of the circle at intervals of no greater than $\lambda/2$. A more recent study by Konyaev and Fedoryuk (1988) deals in more detail with not only the problem of discretising the continuous radiating surface, but also that of using an array of discrete sensors to detect the radiation that is to be suppressed by the sources. In addition, each discrete monopole/dipole radiator was further synthesised by using two discrete monopoles. The "discrete approximation error" again appears to behave in a manner that is consistent with the earlier study reported by Konyaev *et al.* (1977) and in this case was found to increase rapidly for values of $(\lambda/2R)\sqrt{N/\pi} < 2$. This again requires an approximate linear separation distance between the sources that is not greater than $\lambda/2$.

Finally, the work of Mangiante (1977) and Mangiante and Vian (1977) addressed the problem of discretising a continuous spherical secondary source layer which provided perfect absorption of the radiation from a monopole placed at the centre of the sphere. They concluded that the minimum number of discrete secondary monopole–dipole sources to give a good suppression of the radiated field was given by $4R/\lambda$ where R again denotes the radius of the sphere. This implies an effective absorption cross-section of the individual sources of $\pi R\lambda$, a result which appears inconsistent with the above findings. However, Mangiante (1977) presented some impressive experimental results for the cancellation of the radiation from a primary loudspeaker source by using arrays of both 12 and 20 single monopole secondary sources to surround the primary source. The results presented show convincingly uniform reductions in the far field radiation for both pure tone and broad-band sound.

9.15 Sound power absorption by quadrupole sources

The techniques used to evaluate the potential for monopole and dipole sources in absorbing sound can be extended to deal with quadrupole type sources. This can be achieved relatively easily by assuming quadrupoles to consist of two dipole sources and using the relationship specified by equation (9.11.9) to calculate the sound power output of the constituent dipoles. Thus, given two dipoles at positions $\mathbf{y}_1$ and $\mathbf{y}_2$, whose associated forces are specified by $\mathbf{f}_1$ and $\mathbf{f}_2$, the total sound power output of the combination in the presence of an incident plane wave is given by

$$W = (1/2)\mathrm{Re}\{[\mathbf{u}_1(\mathbf{y}_1) + \mathbf{u}_2(\mathbf{y}_1) + \mathbf{u}_p(\mathbf{y}_1)]^*.\mathbf{f}_1 + [\mathbf{u}_1(\mathbf{y}_2) + \mathbf{u}_2(\mathbf{y}_2) + \mathbf{u}_p(\mathbf{y}_2)]^*.\mathbf{f}_2\}. \tag{9.15.1}$$

where $\mathbf{u}_1$ and $\mathbf{u}_2$ are the particle velocities produced by the two dipoles and $\mathbf{u}_p$ is that produced by the plane wave. First, we will examine the power absorbed by a longitudinal quadrupole such that the constituent dipoles are arranged as shown in Fig. 9.11 and the forces associated with the dipoles are given by $\mathbf{f}_1 = \mathbf{i}f$, $\mathbf{f}_2 = -\mathbf{i}f$. With the dipoles positioned at $\mathbf{y}_1$ and $\mathbf{y}_2$ respectively, and generalising the definition of the "mobility" of an acoustic dipole introduced in Section 9.11, it also follows that we can express the particle velocity at $\mathbf{y}_2$ due to the dipole at $\mathbf{y}_1$ as $u_1(\mathbf{y}_2) = M(\mathbf{y}_2|\mathbf{y}_1)f$. Similarly, the particle velocity at $\mathbf{y}_1$ due to the dipole at $\mathbf{y}_2$ can be written as $u_2(\mathbf{y}_1) = M(\mathbf{y}_1|\mathbf{y}_2)f$, where it is evident for this configuration that $M(\mathbf{y}_2|\mathbf{y}_1) = M(\mathbf{y}_1|\mathbf{y}_2)$. Both u_1 and u_2 are the particle velocities in the x_1 direction. It now follows that the expression for the sound power output is given by

$$W = f^*\mathrm{Re}\{M(\mathbf{y}_1|\mathbf{y}_1) - M(\mathbf{y}_1|\mathbf{y}_2)\}f + (1/4)f^*[u_p(\mathbf{y}_1) - u_p(\mathbf{y}_2)] + (1/4)[u_p(\mathbf{y}_1) - u_p(\mathbf{y}_2)]^*f. \tag{9.15.2}$$

where u_p is the particle velocity due to the plane wave in the x_1 direction.

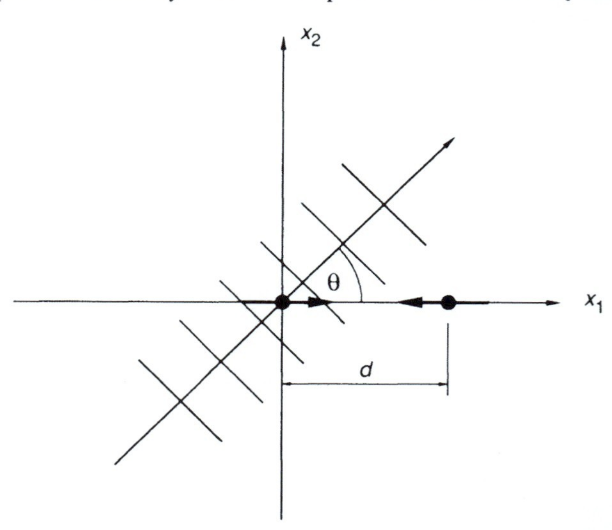

Fig. 9.11 Absorption of sound from an incident plane wave by a longitudinal quadrupole represented by two equal and opposite dipoles.

This is again a quadratic function of the complex force f of the type specified by equation (9.12.2) and we can therefore determine the value of f that minimises W. It also follows from the expression for the radial component of the particle velocity associated with the dipole given in Section 8.7 that

$$M(\mathbf{y}_1|\mathbf{y}_2) = \frac{\omega^2}{4\pi\rho_0 c_0^3}\left[\frac{2}{kr} + j\left(1 - \frac{2}{(kr)^2}\right)\right]\frac{e^{-jkr}}{kr}. \tag{9.15.3}$$

In the limit $kr \to 0$ then a series expansion of this expression shows that $\text{Re}\{M(\mathbf{y}_1|\mathbf{y}_1)\} = \omega^2/12\pi\rho_0 c_0^3$. Thus in terms of the parameters of the quadratic function defined by equation (9.12.2) we have

$$\begin{aligned} A &= \text{Re}\{M(\mathbf{y}_1|\mathbf{y}_1) - M(\mathbf{y}_1|\mathbf{y}_2)\} \\ &= \frac{\omega^2}{4\pi\rho_0 c_0^3}\left\{\frac{(kd)^3 - 6kd\cos kd - 3[(kd)^2 - 2]\sin kd}{3(kd)^3}\right\}, \end{aligned} \tag{9.15.4}$$

$$b = (1/4)[u_p(\mathbf{y}_1) - u_p(\mathbf{y}_2)] = \frac{p_p(\mathbf{y}_1)\cos\theta}{4\rho_0 c_0}(1 - e^{-jkd\cos\theta}). \tag{9.15.5}$$

Again we have assumed that the incident plane wave impinges on the two dipoles at an angle θ. The optimal force and minimum power output associated with this quadratic function are given by $f_0 = -b/A$ and $W_0 = -|b|^2/A$, and thus

$$f_0 = -\frac{\pi c_0}{\omega^2}\cos\theta\, p_p(\mathbf{y}_1)\left\{\frac{3(kd)^3(1 - e^{-jkd\cos\theta})}{(kd)^3 - 6kd\cos kd - 3[(kd)^2 - 2]\sin kd}\right\}, \tag{9.15.6}$$

$$W_0 = -\frac{\pi c_0}{2\omega^2\rho_0}\cos^2\theta\,|p_p(\mathbf{y}_1)|^2\left\{\frac{3(kd)^3\,(1 - \cos(kd\cos\theta))}{(kd)^3 - 6kd\cos kd - 3[(kd)^2 - 2]\sin kd}\right\}. \tag{9.15.7}$$

In the limit of $kd \to 0$, the term in the curly brackets of equation (9.15.6) can be shown by a series expansion to reduce to $10j\cos\theta/(kd)$. Thus in the low frequency limit the optimal force value which maximises the power absorption of the two constituent dipoles is given by

$$f_0 = -\frac{j10\pi c_0^2\cos^2\theta}{\omega^2(kd)}p_p(\mathbf{y}_1), \tag{9.15.8}$$

where we should note that $f_0 \to \infty$ as $kd \to 0$. Similarly, the term in the curly brackets of (9.15.7) can be shown in the limit of $kd \to 0$ to be given by $5\cos^2\theta$, and thus

$$W_0 = -\frac{5\pi c_0}{2\omega^2\rho_0}\cos^4\theta|p_p(\mathbf{y}_1)|^2. \tag{9.15.9}$$

It is therefore evident that in the low frequency limit the maximum power absorbed occurs when the plane wave is incident axially on the longitudinal quadrupole (i.e. $\cos\theta = 1$). Under these circumstances the power absorbed is *five times* that of a single point monopole (see Nelson *et al.*, 1986b). It is also evident that a plane wave impinging on the longitudinal quadrupole at 90° to the quadrupole axis can produce no absorption of power.

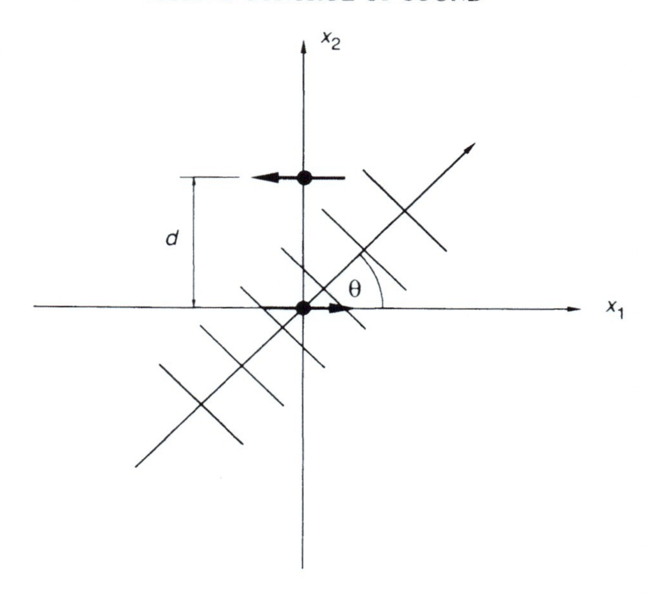

Fig. 9.12 Absorption of sound from an incident plane wave by a lateral quadrupole represented by two equal and opposite dipoles.

Similarly, we can evaluate the sound power absorbed by a lateral quadrupole. In this case we require a knowledge of the tangential component of the particle velocity field associated with the constituent dipoles. It follows from Section 8.7 that

$$u_\theta = \frac{f\omega^2 \sin\theta}{4\pi\rho_0 c_0^3}\left[\frac{j}{(kr)^2} - \frac{1}{kr}\right]\frac{e^{-jkr}}{kr}. \tag{9.15.10}$$

With the arrangement of dipoles depicted in Fig. 9.12, we can therefore write the "transfer mobility" as

$$M(\mathbf{y}_1|\mathbf{y}_2) = \frac{\omega^2}{4\pi\rho_0 c_0^3}\left[\frac{j}{(kd)^2} - \frac{1}{kd}\right]\frac{e^{-jkd}}{kd}. \tag{9.15.11}$$

The terms associated with the quadratic function specified by equation (9.15.2) when written in the form of equation (9.12.2) are thus given by

$$\begin{aligned} A &= \mathrm{Re}\{M(\mathbf{y}_1|\mathbf{y}_1) - M(\mathbf{y}_1|\mathbf{y}_2)\} \\ &= \frac{\omega^2}{4\pi\rho_0 c_0^3}\left[\frac{(kd)^3 - 3kd\sin kd + 3kd\cos kd}{3(kd)^3}\right], \end{aligned} \tag{9.15.12}$$

$$b = (1/4)[u_{p1}(\mathbf{y}_1) - u_{p1}(\mathbf{y}_2)] = \frac{p(\mathbf{y}_1)\cos\theta}{4\rho_0 c_0}(1 - e^{-jkd\sin\theta}). \tag{9.15.13}$$

The expressions for the optimal force and minimum power output are then given by

$$f_0 = -\frac{\pi c_0}{\omega^2}\cos\theta\, p_p(\mathbf{y}_1)\left[\frac{3(kd)^3(1 - e^{-jkd\sin\theta})}{(kd)^3 - 3\sin kd + 3\,kd\cos kd}\right], \tag{9.15.14}$$

$$W_0 = -\frac{\pi c_0}{2\omega^2 \rho_0} \cos^2\theta \, |p_p(\mathbf{y}_1)|^2 \left\{ \frac{3(kd)^3 \, [1 - \cos(kd \sin\theta)]}{(kd)^3 - 3 \sin kd + 3kd \cos kd} \right\}. \quad (9.15.15)$$

In the low frequency limit $kd \to 0$, the term in the square brackets of equation (9.15.14) can be shown by a series expansion to reduce to $30j \sin\theta/(kd)$. The optimal force then becomes

$$f_0 = -\frac{j30\pi c_0^2 \cos\theta \sin\theta}{\omega^2(kd)} p_p(\mathbf{y}_1) \quad (9.15.16)$$

where again $f_0 \to \infty$ as $kd \to 0$. Similarly, series expansion in the limit of $kd \to 0$ of the term in the curly brackets in equation (9.15.15) shows that the term becomes $15 \sin^2\theta$. Under these circumstances the expression for the power output becomes

$$W_0 = -\frac{15\pi c_0 \cos^2\theta \sin^2\theta}{2\omega^2 \rho_0} |p_p(\mathbf{y}_1)|^2. \quad (9.15.17)$$

In the low frequency limit, equation (9.15.16) shows that the maximum power absorption associated with this lateral quadrupole arrangement occurs when the angle of incidence of the plane wave is at 45°. Under this condition the maximum absorption is slightly less than that which can be produced by a longitudinal quadrupole but still amounts to a factor which is (15/4) times that associated with a single point monopole type source.

9.16 Sound power absorption by compact monopole arrays

It is also possible to calculate the maximum sound power absorbed by arbitrary arrays of point monopole sources. For example, assume that we have a cluster of monopole point sources in the (x_1, x_2) plane whose positions are specified by the vectors $\mathbf{y}_m$. We can calculate the complex pressure produced by a plane wave impinging on the source cluster from equation (9.12.8). The total power output of the sources can be written as

$$W = (1/2)\mathrm{Re}\{(\mathbf{p}_p + \mathbf{p})^{\mathrm{H}}\mathbf{q}\}, \quad (9.16.1)$$

where the vectors $\mathbf{p}_p$ and $\mathbf{p}$ are the vectors of complex pressures produced at the positions of the monopole sources by the plane wave and the monopoles themselves respectively. The latter vector is in turn related to the vector $\mathbf{q}$ of complex source strengths by $\mathbf{p} = \mathbf{Z}\mathbf{q}$ where $\mathbf{Z}$ is a matrix of complex acoustic input and transfer impedances. If we calculate the real part of the term in curly brackets by halving the sum of the term plus its complex conjugate, and further use the principle of reciprocity which ensures that $\mathbf{Z} = \mathbf{Z}^{\mathrm{T}}$, then equation (9.16.1) can be reduced to

$$W = (1/2)\mathbf{q}^{\mathrm{H}}\mathrm{Re}\{\mathbf{Z}\}\mathbf{q} + (1/4)\mathbf{p}_p^{\mathrm{H}}\mathbf{q} + (1/4)\mathbf{q}^{\mathrm{H}}\mathbf{p}_p. \quad (9.16.2)$$

This is a quadratic function of the form described by equation (8.11.14) and discussed in detail in the appendix. In this case the matrix $\mathbf{A}$ of equation (8.11.4) can be identified as $(1/2)\mathrm{Re}\{\mathbf{Z}\}$ whilst the vector $\mathbf{b}$ is given by $(1/4)\mathbf{p}_p$. Note again that the scalar constant c is equal to zero as is the case in equation (9.12.2) which is the scalar equivalent of equation (9.16.2). The optimal vector of secondary source

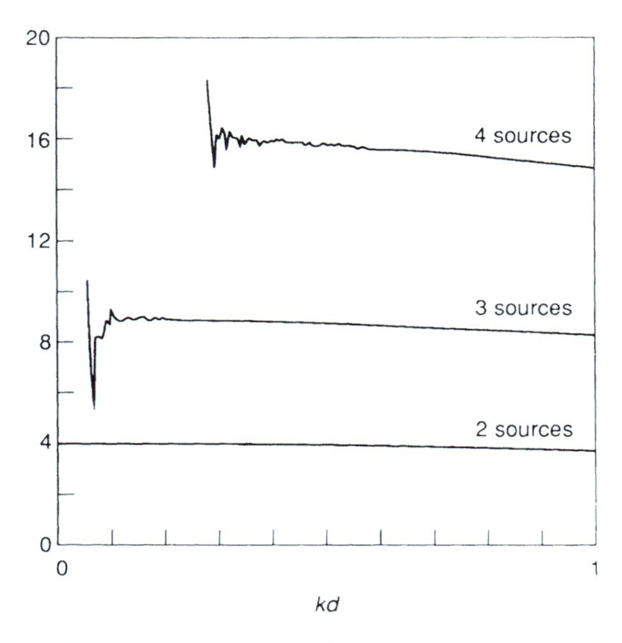

Fig. 9.13 Numerical results for the maximum sound power absorbed by linear arrays of 2, 3 and 4 point monopole sources from a plane wave that is axially incident on the sources. Note that d is the separation distance between the sources. The vertical axis gives the ratio of the maximum sound power absorbed to the maximum power absorbed by a single point monopole.

strengths is thus calculated from $\mathbf{q}_0 = -\mathbf{A}^{-1}\mathbf{b}$ and the corresponding minimum power output is given by $W_0 = -\mathbf{b}^{\mathrm{H}}\mathbf{A}^{-1}\mathbf{b}$. The latter is the maximum power that can be absorbed by the monopole array. It is therefore possible to calculate numerically the maximum power absorbed by an array by inversion of the real part of the acoustic transfer impedance matrix $\mathbf{Z}$. Some results are shown in Fig. 9.13 of undertaking this numerical evaluation for linear arrays of two, three and four point monopoles, respectively (see Nelson and Elliott, 1987). In each case, the monopoles were separated from each other by a distance d and the plane wave was incident axially along the line joining the sources. The results are plotted as a function of the parameter kd. It can be seen that as $kd \to 0$, the numerical results suggest that the maximum sound power that can be absorbed is respectively four, nine and sixteen times the maximum power that can be absorbed by a single point monopole. Note that the results for the arrays of three and four point monopoles exhibit numerical instabilities at small values of kd. This occurs as a result of the matrix $\mathbf{A}$ becoming increasingly ill-conditioned as the value of kd is reduced. Nevertheless, the results suggest that the source arrays effectively synthesise increasingly high orders of multipole sources which are capable of absorbing progressively increasing amounts of power. Thus it appears, for example, that the optimal use of just three monopole point sources is capable of absorbing *nine* times the maximum power that can be absorbed by a single point monopole. As we have seen in the previous section, then three such sources when constrained to form a longitudinal quadrupole could only absorb *five* times that absorbed by a single monopole. In the case of just two sources, it can be shown analytically (see Nelson and Elliott, 1987) that in the limit $kd \to 0$, the two sources combine to give the superposition of an optimally absorbing monopole plus an optimally absorbing dipole. Although the results suggest that very high levels of absorption can be produced by such source arrays it should be borne in mind that the individual strengths of the constituent monopoles will become increasingly large as $kd \to 0$ and that the physical implementation of such source arrays is unlikely to be a practical proposition.

10
Global Control of Enclosed Sound Fields

10.1 Introduction

In this chapter we will consider the active control of enclosed sound fields. In particular we will be concerned with the possibility of using secondary sources to reduce the sound pressure level at all positions in the enclosure; in other words, to achieve global control. We soon find that such a bold ambition is not necessary in practice; there is little point in trying to achieve further reductions in the pressure level at points in the enclosure where natural destructive interference takes place, at a pressure node of an acoustic mode for example. A more useful criterion turns out to be to reduce the total acoustic potential energy in the enclosure. This strategy may well increase the pressure level at some points in the enclosure, where the level was originally low, but it will always attempt to reduce the level where it was originally higher, and more offensive. In fact, it turns out that such a strategy has the effect of levelling out the spatial variation of the pressure level in the enclosure, since dominant acoustic modes in the enclosure are suppressed.

The majority of the published work on actively controlling enclosed sound fields has concentrated on creating "zones of quiet" within the enclosure rather than attempting global control. Such strategies will be discussed in the next chapter. Olson and May (1953) describe an "electronic sound absorber" using a single microphone and single loudspeaker placed close together, and coupled via a feedback system which can, in principle, be configured to either drive the pressure at the microphone to zero, or to absorb acoustic energy (see Chapter 7). These authors noted that active control will be most effective at low frequencies and suggested using several individual devices in their energy absorbing mode, in the corners of a room, to provide additional acoustic damping in the enclosure and thus effect global control.

Since the Olson and May paper, little work appears to have been published on the active control of enclosed sound fields until the early 1980s, since when several authors have considered the problem. Short (1980) presented experimental work in a reverberation chamber, showing that global reductions in level were possible if a secondary source were placed within half a wavelength of a compact primary source. Ross (1980, 1981) described experiments on active control in which an internal secondary source was used in enclosures excited by an external source, and reported reductions of about 10 dB in the sound pressure level at several points in a room, at frequencies close to the natural frequency of one of its acoustic modes. Warnaka *et al.* (1983) stressed the need for global control and reported experiments in a reverberation chamber using closely spaced sources. Nadim and Smith (1983) and Oswald (1984) implemented single channel feedforward active control systems in a tractor cab and car interior, respectively. Oswald presented some measurements of the extent of cancellation and noted that the reductions in noise were large, and occurred over the whole cabin volume, at frequencies close to cabin resonances. Chaplin (1983) has also discussed the suppression of dominant acoustic modes by secondary sources.

Piraux and Nayroles (1980) presented a theoretical formulation for the minimisation of the total acoustic potential energy within a volume of space, which was used as the basis of some experiments in an anechoic chamber (Mazzanti and Piraux, 1983). The theoretical approach suggested by Piraux and Nayroles was quadratic optimisation, which was applied to a modal model of enclosed sound fields by Nelson *et al.* (1985, 1987b). It is this approach which we will follow below. There are many possible theoretical models of enclosed sound fields, using, for example, modes, images or rays, or numerical descriptions using, for example, finite element models. It is, however, the modal model which is the most appropriate here. This is partly because, as we have seen, active control is most effective at low frequencies, and in this frequency range the modal model of an enclosed sound field requires relatively few parameters for an adequate description of the spatial and frequency dependence of the sound field. We shall also see that a modal model provides a convenient intuitive description of the behaviour of an active control system in an enclosure with lightly damped acoustic modes, and these insights can be useful in the more practical case of heavier acoustic damping. Other formulations suitable for numerical calculation have been presented by various authors; Molo and Bernhard (1987), for example, describe simulations of active control in enclosures in which the boundary element method is used.

The first task in this chapter is to introduce a modal expansion for the pressure in a *damped* enclosure. A physically reasonable form for this damping is then considered. The application of this model to active control in simple one- and two-dimensional enclosures is then discussed and this forms the main body of the chapter. A discussion is also presented of the effects of modal density and damping on the applicability of active control. The chapter concludes with the presentation of some results obtained by the authors in the practical application of active techniques in controlling the sound inside aircraft and automobiles.

10.2 The eigenfunctions and boundary conditions of a rectangular enclosure

In Chapter 1 we saw that the pressure at a point $\mathbf{x}$ in a *rigid walled rectangular* room could be expressed as the sum of a series of acoustic modes. These modes correspond to the eigenfunctions (or characteristic functions) of the homogeneous Helmholtz equation, so that

$$\nabla^2\psi_n(\mathbf{x}) + k_n^2\psi_n(\mathbf{x}) = 0, \tag{10.2.1}$$

within the enclosure, where $k_n^{\;2}$ is the eigenvalue corresponding to the eigenfunction

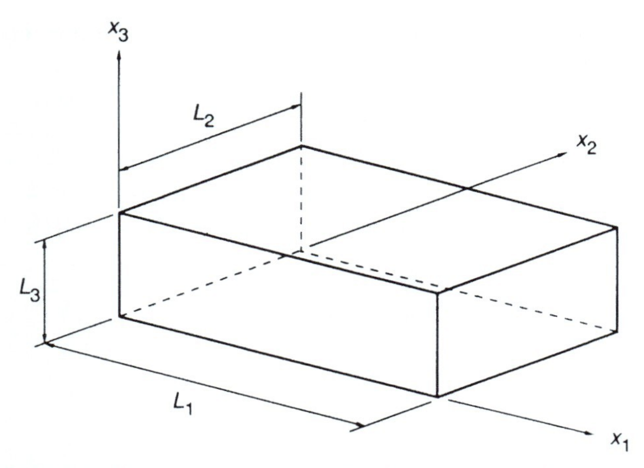

Fig. 10.1 The co-ordinate system and definitions of the enclosure dimensions.

$\psi_n(\mathbf{x})$, and may be expressed in terms of the room dimensions as

$$k_n^2 = \frac{\omega_n^2}{c_0^2} = \left[\frac{n_1\pi}{L_1}\right]^2 + \left[\frac{n_2\pi}{L_2}\right]^2 + \left[\frac{n_3\pi}{L_3}\right]^2. \tag{10.2.2}$$

These eigenfunctions also satisfy the rigid walled boundary condition

$$\nabla\psi_n(\mathbf{x}).\mathbf{n} = 0, \tag{10.2.3}$$

on the surface of the volume, where $\mathbf{n}$ is the unit vector pointing out of the volume on its surface. We can express these real mode shapes as

$$\psi_n(\mathbf{x}) = \sqrt{\varepsilon_{n_1}\varepsilon_{n_2}\varepsilon_{n_3}}\,\cos\frac{n_1\pi x_1}{L_1}\cos\frac{n_2\pi x_2}{L_2}\cos\frac{n_3\pi x_3}{L_3}, \tag{10.2.4}$$

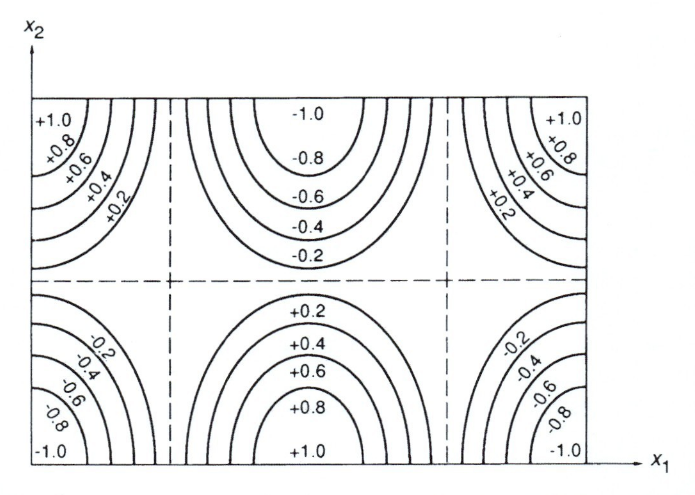

Fig. 10.2 Contours of constant pressure in the x_3 plane for the (2,1,0) acoustic mode. The dashed lines are contours of zero pressure or *nodal lines*.

where L_1, L_2 and L_3 are the dimensions of the room in the x_1, x_2 and x_3 co-ordinate directions, as indicated in Fig. 10.1, and n denotes the trio of modal integers n_1, n_2, n_3. The normalisation factors, ε_{n_1}, ε_{n_2}, ε_{n_3} will be defined below.

If $n_1 = n_2 = n_3 = 0$ the mode is referred to as the (0,0,0) mode and corresponds to the purely compliant behaviour of the air in the enclosure due to its bulk modulus. We will assume that $L_1 > L_2 > L_3$ so the mode with the next lowest natural frequency is the (1,0,0) mode. The spatial distribution of the (2,1,0) mode in the x_3 plane is shown in Fig. 10.2 for illustration. The modes have the property that the integral of the product of two dissimilar modes over the enclosure volume is zero, i.e. they are *orthogonal* and satisfy the relationship

$$\int_V \psi_n(\mathbf{x})\psi_m(\mathbf{x})\mathrm{d}V = 0, \; n \neq m. \tag{10.2.5}$$

We now choose the normalisation factors, ε_{n_1}, ε_{n_2} and ε_{n_3} such that the integral of the square of each mode shape function over the enclosure volume is equal to the volume of the enclosure. Thus we ensure that

$$\int_V \psi_n^2(\mathbf{x})\mathrm{d}V = V, \tag{10.2.6}$$

and therefore we require $\varepsilon_{n_i} = 1$ if $n_i = 0$ and $\varepsilon_{n_i} = 2$ if $n_i > 0$. The modes thus form an *orthonormal* set of functions and the two properties above can be summarised by the equation

$$\int_V \psi_n(\mathbf{x})\psi_m(\mathbf{x})\mathrm{d}V = V\,\delta_{nm}, \tag{10.2.7}$$

where δ_{nm} is the Kronecker delta function, which has the property $\delta_{nm} = 1$ if $n = m$ and $\delta_{nm} = 0$ if $n \neq m$.

Now we turn to the boundary conditions on the surface of an enclosure. In particular we are interested in the pressure distribution in an enclosure with walls that are not rigid, but which have *locally reacting* surfaces. A locally reacting surface is one from which no wave motion is possible *within* the walls of the enclosure and so the boundary condition at each point on the surface S can be described, for a harmonic sound field, in terms of its specific acoustic impedance at that point. Thus we write

$$\frac{p(\mathbf{y})}{u(\mathbf{y})} = z(\mathbf{y}) \text{ on } S, \tag{10.2.8}$$

where $u(\mathbf{y})$ is the complex acoustic particle velocity normal to the surface, at position $\mathbf{y}$ on the surface of the enclosure. We are assuming a harmonic excitation of the enclosure at frequency ω and so the particle velocity can be expressed in terms of the derivative of the complex pressure with respect to the direction normal to and into the wall. This direction is specified by the unit vector $\mathbf{n}$, and therefore using the momentum equation (see Section 1.10) shows that

$$\nabla p(\mathbf{y}).\mathbf{n} = -j\omega\rho_0\, u(\mathbf{y}). \tag{10.2.9}$$

The *normalised specific acoustic admittance* of the wall at position $\mathbf{y}$ is defined as

$$\beta(\mathbf{y}) = \frac{\rho_0 c_0}{z(\mathbf{y})} = \chi(\mathbf{y}) + j\sigma(\mathbf{y}), \tag{10.2.10}$$

where the real and imaginary parts of $\beta(\mathbf{y})$ are the normalised specific acoustic conductance, $\chi(\mathbf{y})$, and normalised specific acoustic susceptance, $\sigma(\mathbf{y})$. The homogeneous boundary condition on the surface can now be expressed as

$$\nabla p(\mathbf{y}).\mathbf{n} = -jk\beta(\mathbf{y})p(\mathbf{y}) \text{ on } S. \qquad (10.2.11)$$

Thus although we can readily determine the form of the sound field in a rigid walled enclosure in terms of the eigenfunctions satisfying the *rigid walled* boundary conditions, we are in general faced with the problem of determining the sound field in an enclosure having boundary conditions described by equation (10.2.11). We will show that by making some approximations we can accurately describe the sound field in enclosures whose surfaces have a *small* value of $\beta(\mathbf{y})$ (i.e. that are "lightly damped") by using the rigid walled eigenfunctions $\psi_n(\mathbf{x})$.

10.3 The sound field in a lightly damped enclosure

We will now proceed to determine the sound field in the enclosure produced by an imposed vibration of the enclosure walls. The analysis that follows is based on that presented by Pierce (1981, Ch.6). Other (earlier) treatments are given by Morse and Bolt (1944), Morse (1948) and Morse and Ingard (1961). First we determine the Green function $G(\mathbf{x}|\mathbf{y})$ for the enclosure. This satisfies

$$(\nabla^2 + k^2)G(\mathbf{x}|\mathbf{y}) = -\delta(\mathbf{x} - \mathbf{y}), \qquad (10.3.1)$$

and therefore quantifies the spatial dependence of the complex pressure field produced in the enclosure by a point source at a position $\mathbf{y}$ (see Ch. 9, Section 9.3). We will determine this Green function in terms of the rigid walled eigenfunctions $\psi_n(\mathbf{x})$ of the enclosure such that the Green function also satisfies rigid walled boundary conditions, i.e. $\nabla G(\mathbf{x}|\mathbf{y}).\mathbf{n} = 0$. Thus we assume that

$$G(\mathbf{x}|\mathbf{y}) = \sum_{m=0}^{\infty} b_m \psi_m(\mathbf{x}), \qquad (10.3.2)$$

where b_m are complex coefficients that have to be determined and which will depend on the source position $\mathbf{y}$. Note that the eigenfunctions $\psi_m(\mathbf{x})$ form a *complete set* of functions (see, for example, the discussion presented by Pierce, 1981, Ch. 6) and so any well-behaved function of $\mathbf{x}$ can be approximated by a linear combination of the eigenfunctions. The expansion is analogous to a half period Fourier series expansion of a waveform (see Papoulis, 1981, p.73 for example). Substitution of equation (10.3.2) into equation (10.3.1), together with the fact that $\nabla^2\psi_m(\mathbf{x}) = -k_m^2\psi_m(\mathbf{x})$ shows that

$$\sum_{m=0}^{\infty} b_m(k^2 - k_m^2)\psi_m(\mathbf{x}) = -\delta(\mathbf{x} - \mathbf{y}). \qquad (10.3.3)$$

We now multiply both sides of this equation by $\psi_n(\mathbf{x})$ and then integrate over the volume V of the enclosure. Using the orthonormality property of the eigenfunctions (equation (10.2.7)) and the sifting property of the three-dimensional Dirac delta function (equation (9.3.3)) then shows that $b_n = \psi_n(\mathbf{y})/(V(k_n^2 - k^2))$ and therefore the expression for the Green function can be written as

$$G(\mathbf{x}|\mathbf{y}) = \sum_{n=0}^{\infty} \frac{\psi_n(\mathbf{x})\psi_n(\mathbf{y})}{V(k_n^2 - k^2)} . \tag{10.3.4}$$

We can now use the general solution to the Helmholtz equation that we described in Chapter 9 (see Section 9.4, equation (9.4.5)) in order to derive an expression for the pressure field in the enclosure. In the absence of any "volume" sources within the enclosure (i.e. $Q_{\text{vol}} = 0$ in equation (9.4.5)) and since the Green function $G(\mathbf{x}|\mathbf{y})$ satisfies rigid walled boundary conditions ($\nabla_y G(\mathbf{x}|\mathbf{y}).\mathbf{n} = 0$ in equation (9.4.5)), the pressure field in the enclosure can be written as

$$p(\mathbf{x}) = \int_S G(\mathbf{x}|\mathbf{y})\nabla_y p(\mathbf{y}).\mathbf{n}\, dS. \tag{10.3.5}$$

Now note that $\nabla_y p(\mathbf{y})$ (the pressure gradient on the enclosure walls) can arise from two mechanisms. The first is through some imposed surface normal vibration $\mathbf{u}_i(\mathbf{y}).\mathbf{n}$ (say) and the second is through the surface normal particle velocity resulting from the finite impedance of the enclosure walls. Using equation (10.2.11) to describe the latter and also using the momentum equation (10.2.9) enables equation (10.3.5) to be written as

$$p(\mathbf{x}) = \int_S G(\mathbf{x}|\mathbf{y})[j\omega\rho_0\mathbf{u}_i(\mathbf{y}).\mathbf{n} - jk\beta(\mathbf{y})p(\mathbf{y})]dS. \tag{10.3.6}$$

We now have an integral equation for the pressure field in the enclosure; the pressure $p(\mathbf{x})$ is dependent not only on the prescribed surface velocity $\mathbf{u}_i(\mathbf{y})$ but also on the pressure $p(\mathbf{y})$ on the enclosure walls. We shall now show that we can make an approximation that enables the pressure field to be expressed in terms of a series of *damped* modal responses, where the second term in the square brackets of equation (10.3.6) is seen to be directly responsible for this damping mechanism. First, using the series expansion for the Green function given by equation (10.3.4) shows that equation (10.3.6) can be written as

$$p(\mathbf{x}) = \sum_{n=0}^{\infty} a_n\psi_n(\mathbf{x}), \tag{10.3.7}$$

where the coefficients a_n are given by

$$a_n = \frac{1}{V(k_n^2 - k^2)}\left[\int_S j\omega\rho_0\psi_n(\mathbf{y})\mathbf{u}_i(\mathbf{y}).\mathbf{n}\;\mathrm{d}S - \int_S jk\psi_n(\mathbf{y})\beta(\mathbf{y})p(\mathbf{y})\mathrm{d}S\right]. \tag{10.3.8}$$

We can also substitute a series expansion for $p(\mathbf{y})$ into this expression (where $p(\mathbf{y})$ is expressed as a sum over $a_m\psi_m(\mathbf{y})$) to show that

$$a_n = \frac{1}{V(k_n^2 - k^2)}\left[\int_S j\omega\rho_0\psi_n(\mathbf{y})\mathbf{u}_i(\mathbf{y}).\mathbf{n}\;\mathrm{d}S - \int_S jk\psi_n(\mathbf{y})\beta(\mathbf{y})\sum_{m=0}^{\infty} a_m\psi_m(\mathbf{y})\mathrm{d}S\right]. \tag{10.3.9}$$

We now have a set of coupled algebraic equations that determine the coefficients a_n. These can be written as

$$a_n(k_n^2 - k^2) = \frac{1}{V}\int_S j\omega\rho_0\psi_n(\mathbf{y})\mathbf{u}_i(\mathbf{y}).\mathbf{n}\,\mathrm{d}S - jk\sum_{m=0}^{\infty} a_m D_{nm}, \qquad (10.3.10)$$

where the coefficients D_{nm} are given by

$$D_{nm} = \frac{1}{V}\int_S \beta(\mathbf{y})\psi_n(\mathbf{y})\psi_m(\mathbf{y})\mathrm{d}S. \qquad (10.3.11)$$

These coupled equations can be solved by iteration (see, for example, Morse and Ingard, 1961). An approximate solution, valid in the limit of small surface admittance $\beta(\mathbf{y})$ and sufficient for our purposes, can be obtained here by noting that the cross coupling terms (D_{nm}, $n \neq m$) are relatively small. Indeed, the coupling terms are zero if $\beta(\mathbf{y})$ is uniform over each wall and any two corresponding modal integers are dissimilar in the trios n and m. Thus if we neglect D_{nm} if $n \neq m$, then equation (10.3.10) becomes much simpler and can be rearranged to show that

$$a_n(k_n^2 - k^2 + jkD_{nn}) = \frac{1}{V}\int_S j\omega\rho_0\psi_n(\mathbf{y})\mathbf{u}_i(\mathbf{y}).\mathbf{n}\,\mathrm{d}S. \qquad (10.3.12)$$

This equation is then readily solved for a_n and the series form of the solution for the acoustic pressure (equation (10.3.7)) can be expressed in the form

$$p(\mathbf{x}) = \sum_{n=0}^{\infty} \frac{\omega\rho_0\, c_0^2\, \psi_n(\mathbf{x})}{V\,[\omega c_0 D_{nn} + j(\omega^2 - \omega_n^2)]}\int_S \psi_n(\mathbf{y})\mathbf{u}_i(\mathbf{y}).\mathbf{n}\,\mathrm{d}S, \qquad (10.3.13)$$

where we have used $k = \omega/c_0$ and $k_n = \omega_n/c_0$. Note that an entirely equivalent argument can be used in order to derive an expression for the sound field excited by a distribution of *volume* sources within the enclosure. Thus one again works from the general solution (equation (9.4.5) in Chapter 9) of the inhomogeneous Helmholtz equation but in this case retaining the term involving $q_{\mathrm{vol}}(\mathbf{y})$. This leads to an expression for the pressure field that is given by

$$p(\mathbf{x}) = \sum_{n=0}^{\infty} \frac{\omega\rho_0\, c_0^2\, \psi_n(\mathbf{x})}{V\,[\omega c_0 D_{nn} + j(\omega^2 - \omega_n^2)]}\int_V \psi_n(\mathbf{y})q_{\mathrm{vol}}(\mathbf{y})\mathrm{d}V. \qquad (10.3.14)$$

Now note that the term D_{nn} appears explicitly as a *damping* term in the response of the nth mode. The frequency-dependent term outside the surface integral in equation (10.3.14) is directly comparable with the frequency response function of a second order resonant system (see, for example, equation (8.6.15)).

10.4 A model of the acoustic damping at the enclosure walls

We have seen that the term D_{nn} in the above solution for the sound field arises directly as a result of the finite impedance of the enclosure walls (equation (10.3.11)) and it is therefore not surprising that this is manifested as a damping of the response of each mode. This also explains how equation (10.3.13) describes the pressure field in an enclosure with *non-rigid* walls in terms of the eigenfunctions of the equivalent *rigid* walled enclosure. Further insight is given into the mechanism of modal damping by considering the form of the terms D_{nn} in more detail. The complex term

$$D_{nn} = \frac{1}{V} \int_S \psi_n^2(\mathbf{y})(\chi(\mathbf{y}) + j\sigma(\mathbf{y}))\mathrm{d}S \tag{10.4.1}$$

has an imaginary part due to the susceptance of the enclosure walls, which slightly shifts the modal resonance frequencies from those of the hard-walled enclosure, ω_n. This is not an important effect here, and will be ignored (see the discussion presented by Pierce, 1981, Ch.6). The real part of D_{nn}, however, provides the damping in each mode and this term is of considerable interest. This is given by

$$\mathrm{Re}\{D_{nn}\} = \frac{1}{V} \int_S \psi_n^2(\mathbf{y})\chi(\mathbf{y})\mathrm{d}S, \tag{10.4.2}$$

and is related to the "average wall coefficient" used by Morse (1948) which is a measure of the power loss from each mode at the walls. It can be evaluated for any given distribution of specific acoustic conductance on the walls. If we assume, however, that $\chi(\mathbf{y})$ is constant over each wall, and that on the wall where $x_1 = 0$ (of area S_1) it is χ_{10}, on the wall where $x_1 = L_1$ (also of area S_1) it is χ_{1L}, etc., the total surface integral can be split up into integrals over each of the six walls of the enclosure, and by making use of the simple form of $\psi_n(\mathbf{y})$ these integrals can be evaluated to show that

$$\mathrm{Re}\{D_{nn}\} = \frac{1}{V} [\varepsilon_{n1}S_1(\chi_{10} + \chi_{1L}) + \varepsilon_{n2}S_2(\chi_{20} + \chi_{2L}) + \varepsilon_{n3}S_3(\chi_{30} + \chi_{3L})]. \tag{10.4.3}$$

This is precisely the damping term derived by Morse and Bolt (1944) from an analysis of the complex, frequency-dependent eigenfunctions of the damped enclosure, in the limit of small χ (see also Bullmore, 1988).

Modes in which none of the three modal integers are zero are called *oblique*; if one modal integer is zero the mode is called *tangential*; if two modal integers are zero the mode is called *axial* (Morse, 1948); and if all three modal integers are zero, the mode may be called *compliant*. It is clear from the expression for D_{nn} that if the conductance of all the walls is equal, the most heavily damped modes will be the oblique modes, followed, in order, by the tangential and axial modes and the compliant mode. For the compliant (0,0,0) mode, assuming $\sigma(\mathbf{x}) = 0$, the damping term $\omega c_0 D_{nn}$ in the denominator of equation (10.3.13) becomes

$$\omega c_0 D_{00} = \frac{\omega \rho_0 c_0^2}{R_a V}, \tag{10.4.4}$$

where $\rho_0 c_0/R_a$ is the sum of the acoustic conductances of all six walls and R_a may therefore be regarded as the net acoustic resistance of the enclosure walls. When the enclosure is driven at a sufficiently low frequency that only this mode is significantly excited, equation (10.3.13) reduces to

$$\frac{p(\mathbf{x})}{q} = \frac{\omega \rho_0 c_0^2 /V}{\omega \rho_0 c_0^2 /R_a V + j\omega^2} = \frac{R_a}{1 + j\omega C_a R_a}, \tag{10.4.5}$$

where $C_a = V/\rho_0 c_0^2$ is the acoustic compliance of the enclosure volume (see Section 3.2 and Kinsler *et al.*, 1982) and q is the complex volume velocity injected into the enclosure evaluated by integrating the imposed normal velocity $\mathbf{u}_p.\mathbf{n}$ over the surface of the enclosure. The input impedance of the cavity in its lowest mode is thus the parallel combination of its compliant acoustic reactance, $1/j\omega C_a$, and its net acoustic resistance, R_a. The modal model is thus seen to be consistent with the lumped parameter model introduced in Chapter 3 and used in Chapter 7. The damping term also has a simple form for oblique modes, for which if $\sigma(\mathbf{y}) = 0$ again,

$$\omega c_0 D_{nn} = \frac{2\omega\rho_0 c_0^2}{R_a V} = \frac{2\omega}{R_a C_a}, \tag{10.4.6}$$

These simple expressions imply a frequency-independent damping mechanism. In fact the acoustic admittance of the walls $\beta(\mathbf{y})$, and hence the acoustic resistance R_a, will typically be due to materials on the walls of the enclosure whose absorptive behaviour is frequency dependent. The acoustic conductance of the surfaces of a practical enclosure is not generally well documented or easily measured. Morse (1948), however, has derived a relationship between the random incidence absorption coefficient, $\bar{\alpha}$, of a locally reacting material and its specific acoustic conductance. Provided $\bar{\alpha}$ is not too large this expression reduces to

$$\bar{\alpha} = 8\chi. \tag{10.4.7}$$

If $\bar{\alpha}$ could be estimated or measured for each of the surfaces in the enclosure the modal damping terms above could thus be calculated. In general, $\bar{\alpha}$ will be different for each surface in the enclosure and will depend strongly on the frequency of excitation. Here we seek to develop a physically reasonable model for the damping in the enclosure, which can be used in computer simulations of active control in rectangular enclosures. We will typically be dealing with enclosures whose dimensions are of the order of a few metres, and looking at the effect of active control up to a few hundred Hertz. In this frequency range the random incidence absorption coefficient of a variety of materials tends to increase nearly linearly with frequency (see, for example, Beranek, 1954, Fig. 10.13). Bullmore (1988) also found that measurements of acoustic transfer impedance in the interior of an aircraft cabin were best fitted by assuming a theoretical model with such a frequency dependence. In general then, within this low frequency range we will assume for the purposes of our discussions that

$$\bar{\alpha}(\omega) = a\omega, \tag{10.4.8}$$

where a is a real constant of proportionality with units s^{-1}, and so for small $\bar{\alpha}$

$$\chi(\omega) = \frac{a}{8}\omega. \tag{10.4.9}$$

Since the resonant response of the nth mode is significantly affected by the damping term only at frequencies near its natural frequency, ω_n, it is reasonable to assume that the specific acoustic wall conductance χ_n associated with that mode is frequency independent with a value equal to that at ω_n. Hence we make the further approximation

$$\chi_n = \frac{a}{8}\,\omega_n\,. \tag{10.4.10}$$

If the value of χ over all surfaces is the same, the damping term for the nth mode, equation (10.4.3) becomes

$$\omega c_0 \mathrm{Re}\{D_{nn}\} = \frac{2\omega c_0 \chi_n}{V}[\varepsilon_{n1}S_1 + \varepsilon_{n2}S_2 + \varepsilon_{n3}S_3], \tag{10.4.11}$$

which, on using equation (10.4.10) can be written as

$$\omega c_0 \mathrm{Re}\{D_{nn}\} = \frac{\omega c_0\,\omega_n a}{4V}[\varepsilon_{n1}S_1 + \varepsilon_{n2}S_2 + \varepsilon_{n3}S_3]. \tag{10.4.12}$$

The modal damping term is often written in terms of a viscous damping ratio ζ_n (e.g. Nelson *et al.*, 1987b) as

$$\omega c_0 \mathrm{Re}\{D_{nn}\} = 2\zeta_n\omega_n\omega. \tag{10.4.13}$$

When the expressions (10.4.12) and (10.4.13) above are compared, then the modal damping ratio ζ_n is seen to be given by

$$\zeta_n = \frac{c_0\,a}{8V}[\varepsilon_{n1}S_1 + \varepsilon_{n2}S_2 + \varepsilon_{n3}S_3]. \tag{10.4.14}$$

For oblique modes, for which $\varepsilon_{n1} = \varepsilon_{n2} = \varepsilon_{n3} = 2$, this expression reduces to

$$\zeta_n = \frac{c_0\,a\,S}{8V}, \tag{10.4.15}$$

where $S = 2(S_1 + S_2 + S_3)$ is the total surface area of the enclosure. Over the frequency range of interest here, the assumption of a linearly increasing random incidence absorption coefficient with frequency is thus seen to lead to a self-consistent and physically reasonable viscous damping model for the damping of each mode.

10.5 Global active control with a single secondary source

The solution for the complex acoustic pressure in a harmonically excited lightly damped enclosed sound field can be written using the series expansion introduced in Section 10.3 such that

$$p(\mathbf{x}) = \sum_{n=0}^{\infty} a_n\psi_n(\mathbf{x}). \tag{10.5.1}$$

The coefficients a_n are the mode amplitudes and are frequency-dependent complex numbers that quantify the extent to which a given mode is excited. For example, in the case of a source distribution inside the volume of the enclosure, it follows from equation (10.3.14) that

$$a_n = \frac{\omega\rho_0\,c_0^2}{V\,[2\zeta_n\omega_n\omega + j(\omega^2 - \omega_n^2)]}\int_V \psi_n(\mathbf{y})q_{\mathrm{vol}}(\mathbf{y})\mathrm{d}V, \tag{10.5.2}$$

where we have expressed the modal damping in terms of the viscous damping coefficient ζ_n introduced in equation (10.4.13) of the last section.

Now assume that the enclosure is excited by a single point monopole primary source of complex strength q_p at position $\mathbf{y}_p$ and that we introduce at position $\mathbf{y}_s$ a single point monopole secondary source of complex strength q_s in order to control the field. Thus, following the discussion of Section 9.4 we write

$$q_{\text{vol}} = q_p\delta(\mathbf{y} - \mathbf{y}_p) + q_s\delta(\mathbf{y} - \mathbf{y}_s), \qquad (10.5.3)$$

and when we undertake the volume integral in equation (10.5.2), the expression for a_n becomes

$$a_n = \frac{\omega\rho_0\, c_0^{\,2}}{V\,[2\zeta_n\omega_n\omega + j(\omega^2 - \omega_n^{\,2})]}\,[q_p\psi_n(\mathbf{y}_p) + q_s\psi_n(\mathbf{y}_s)]. \qquad (10.5.4)$$

This expression clearly shows that the extent to which a particular mode can be driven by a point source depends on the value of the mode shape function at the position of the source, i.e. through the terms $\psi_n(\mathbf{y}_p)$ and $\psi_n(\mathbf{y}_s)$. Thus if the primary source for example is situated at a pressure node of the nth mode, where $\psi_n(\mathbf{y}_p) = 0$, then it will produce no excitation of the nth mode. Also note that in general the extent to which a given mode is excited by a *source distribution* will be determined by the value of the volume integral in equation (10.5.2). It also follows from equation (10.5.4) that the response in any given mode can be made equal to *zero* by judicious choice of the strength and position of the secondary source. Thus, provided we choose the complex strength q_s such that

$$q_s = -q_p\psi_n(\mathbf{y}_p)/\psi_n(\mathbf{y}_s), \qquad (10.5.5)$$

then the value of a_n can be made equal to zero. In principle it is possible to choose $\mathbf{y}_s = \mathbf{y}_p$ thereby collocating the sources and cancelling the excitation of *all the modes* excited by the primary source (since then $\psi_n(\mathbf{y}_p) = \psi_n(\mathbf{y}_s)$ for all the modes in the series). However we can always ensure that the response of any *particular mode* (corresponding to a given value of n) can be made equal to zero by using a secondary source placed remotely from the primary source and adjusting its source strength according to equation (10.5.5). This is possible provided $\mathbf{y}_s$ is chosen in order to ensure that $\psi_n(\mathbf{y}_s)$ is non-zero for the mode shape in question.

One is therefore at first tempted to conclude that *global control* of the sound field is a very straightforward matter, since if a_n can be made equal to zero for a given mode, then the pressure field associated with that mode can be made *everywhere equal to zero*. As we shall see if the sound field at a given frequency is dominated by the response of a single mode (i.e. the excitation frequency ω is at or close to a given natural frequency ω_n and the value of ζ_n is small) then global control is sometimes possible. The important factor however is that in choosing q_s to suppress the response of a particular mode, we do not produce an increased excitation of *all the other modes* whose responses also contribute to the total pressure field at a given frequency.

We can establish the extent to which global control of the sound field can be achieved by finding the minimum of a suitable quadratic cost function (Nelson *et al.*, 1987b). In this case an appropriate single measure of the global response of the sound field is given by the *total time-averaged acoustic potential energy* in the enclosure. This is given by

$$E_p = \frac{1}{4\rho_0 c_0^2} \int_V |p(\mathbf{x})|^2 \mathrm{d}V. \tag{10.5.6}$$

This function evaluates the sum of the squared pressures at *all points* within the enclosure, and therefore in establishing the minimum value of the function, we determine the extent to which the sound field can be reduced at all points in the enclosure. We can substitute the series expansion for the pressure given by equation (10.5.1) into this expression to give

$$\begin{aligned} E_p &= \frac{1}{4\rho_0 c_0^2} \int_V p^*(\mathbf{x}) p(\mathbf{x}) \mathrm{d}V \\ &= \frac{1}{4\rho_0 c_0^2} \int_V \left(\sum_{n=0}^{\infty} a_n \psi_n(\mathbf{x}) \right)^* \left(\sum_{m=0}^{\infty} a_m \psi_m(\mathbf{x}) \right) \mathrm{d}V. \end{aligned} \tag{10.5.7}$$

Evaluation of the product of the series expansions and subsequent use of the modal orthogonality property described by equation (10.2.7) shows that only cross-product terms for which $n = m$ yield a non-zero value of the volume integral. It therefore follows that

$$E_p = \frac{V}{4\rho_0 c_0^2} \sum_{n=0}^{\infty} |a_n|^2, \tag{10.5.8}$$

and thus the total acoustic potential energy can be evaluated from the sum of the squares of the complex mode amplitudes (see, for example, Pierce, 1981, Ch.6). Now note that equation (10.5.4) can be written in the form

$$a_n = a_{pn} + B_n q_s, \tag{10.5.9}$$

where the values of a_{pn} and B_n are defined by

$$a_{pn} = \frac{\omega \rho_0 c_0^2 \psi_n(\mathbf{y}_p)}{V[2\zeta_n \omega_n \omega + j(\omega^2 - \omega_n^2)]} q_p, \tag{10.5.10}$$

$$B_n = \frac{\omega \rho_0 c_0^2 \psi_n(\mathbf{y}_s)}{V[2\zeta_n \omega_n \omega + j(\omega^2 - \omega_n^2)]}. \tag{10.5.11}$$

The expression for the total time-averaged acoustic potential energy given by equation (10.5.8) can therefore be written as

$$E_p = \frac{V}{4\rho_0 c_0^2} \sum_{n=0}^{\infty} |a_{pn} + B_n q_s|^2. \tag{10.5.12}$$

Expansion of the modulus squared term shows that E_p can be written as the quadratic function of q_s given by

$$E_p = q_s^* A q_s + q_s^* b + b^* q_s + c, \tag{10.5.13}$$

where the terms in this quadratic form are defined by

$$A = \frac{V}{4\rho_0 c_0^2} \sum_{n=0}^{\infty} |B_n|^2, \qquad b = \frac{V}{4\rho_0 c_0^2} \sum_{n=0}^{\infty} B_n^* a_{pn},$$

$$c = \frac{V}{4\rho_0 c_0^2} \sum_{n=0}^{\infty} |a_{pn}|^2. \tag{10.5.14a,b,c}$$

Thus as we have seen in, for example, Sections 5.5 and 5.6 of Chapter 5, the minimum value of this function is given by $c - |b|^2/A$ when the source strength q_s is adjusted to its optimal value given by $-b/A$. As usual, a unique minimum is assured since A must be positive. The optimal secondary source strength is thus given by

$$q_{s0} = \frac{-\sum_{n=0}^{\infty} B_n^* a_{pn}}{\sum_{n=0}^{\infty} |B_n|^2} = \frac{-\sum_{n=0}^{\infty} |A_n|^2 \psi_n(\mathbf{y}_p)\psi_n(\mathbf{y}_s)}{\sum_{n=0}^{\infty} |A_n|^2 \psi_n{}^2(\mathbf{y}_s)} q_p, \tag{10.5.15}$$

where the term A_n is defined by

$$A_n = \frac{\omega}{[2\zeta_n \omega_n \omega + j(\omega^2 - \omega_n^2)]}, \tag{10.5.16}$$

and thus quantifies the resonant response of the nth mode.

10.6 Global active control in one dimension

As an example of the application of the theory developed in the last section we consider the minimisation of the total acoustic potential energy in a one-dimensional acoustic system, a long thin duct. We assume that a primary harmonic volume velocity source of strength q_p is present at one end of the duct and a single secondary source (q_{s1}) is placed at the opposite end (Fig. 10.3). These sources are of the plane monopole type introduced in Section 5.2 and therefore are represented as sources whose source strength density can be described by for example $q_p\delta(y_1 - y_{1p})$ where y_{1p} is the co-ordinate position of the primary source in the y_1 direction. The theory developed in the last section is thus directly applicable although the source strength is concentrated in a plane rather than at a point. This general problem has already been discussed using a wave approach with no damping by Curtis *et al.* (1987), and in Chapter 5. It is instructive to use the modal formulation presented in this chapter to observe the effect of acoustic damping on the results, and to interpret the results in terms of the modal excitation. The duct was modelled as an enclosure measuring 3.4 m × 0.1 m × 0.1 m and 16 acoustic modes (all axial) were considered in the summation for the energy given by equation (10.5.8). The effect of doubling the number of modes considered in the summation was negligible for these simulations. The original acoustic potential energy due to the primary source, E_{pp}, and the minimum value, E_{p0}, with the secondary source strength optimally adjusted at each frequency according to equation (10.5.15), are shown in Fig. 10.4 for a modal damping ratio of $\zeta = 0.01$, and in Fig. 10.5 for a modal damping ratio of $\zeta = 0.1$.

At very low frequencies, large reductions in E_p can be observed for both values of damping ratio. In modal terms this is due to only the (0,0,0) mode being significantly excited by either source (the duct acts as an acoustic compliance). The excitation of all the other modes due to both primary and secondary sources is small, and very efficient control can be applied by arranging for $q_s = -q_p$. Similarly the sound field, when driven at frequencies close to the natural frequencies of the higher modes in the lightly damped case of Fig. 10.4, is dominated by a single modal contribution

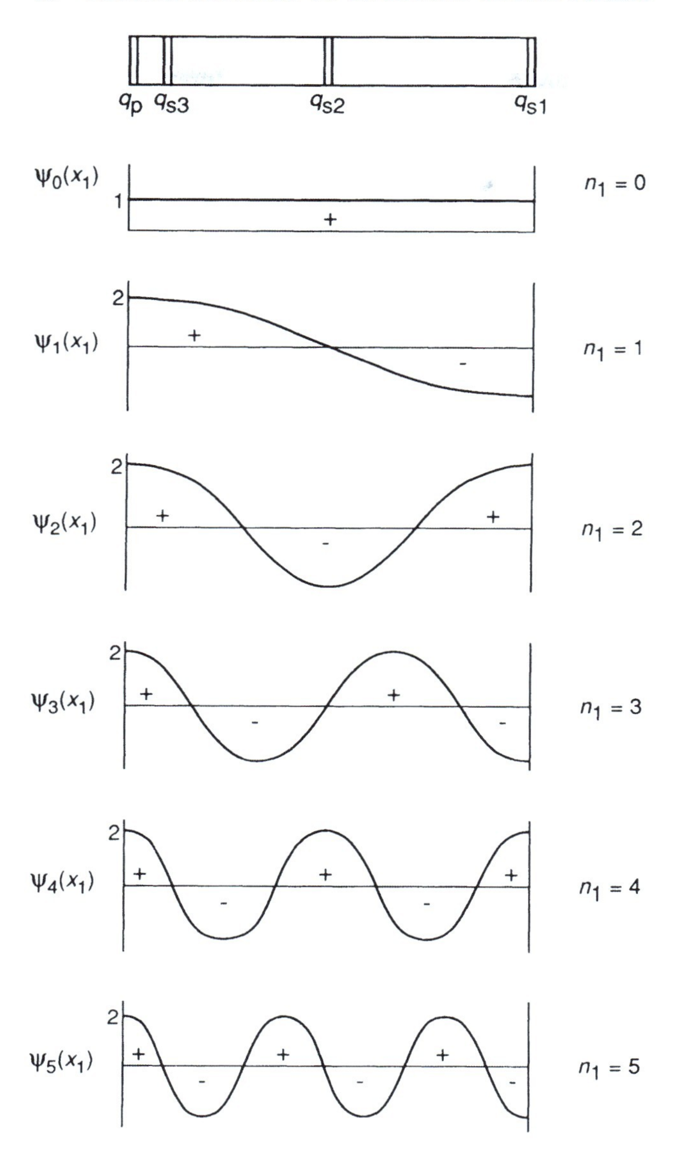

Fig. 10.3 The positions of the primary and three secondary sources in the duct and the mode shapes associated with the first six modes.

which can be significantly attenuated by the secondary source. Between the resonances in the duct, however, there are many modes contributing to E_p at a given frequency and the secondary source cannot control any one of these without significantly exciting the others, and active control with this secondary source has little effect.

This is illustrated by sketching the individual contributions to E_p from each of the modes when the enclosure is excited by the primary source only. This is shown in

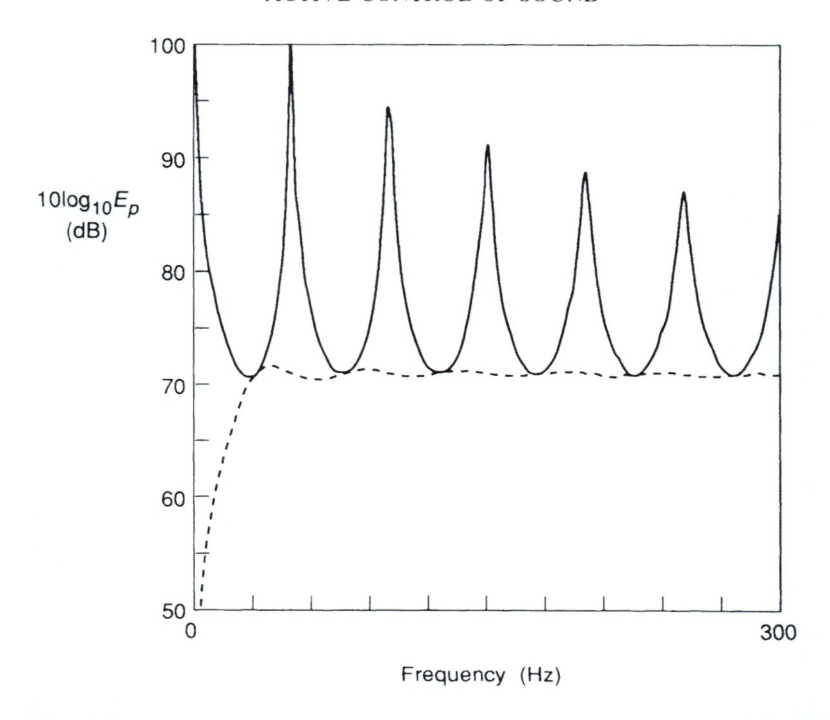

Fig. 10.4 The total acoustic potential energy in the $3.4 \times 0.1 \times 0.1$ m enclosure when excited by the primary source alone (——) and by the harmonic primary source and an optimally adjusted secondary source (q_{s1}) at the opposite end of the enclosure (-----). The damping ratio of the acoustic modes is 0.01.

Fig. 10.6. At a given value of frequency, the contributions to E_p may be considered to consist of a *dominant* modal contribution plus a contribution from a number of *residual* modes. It should also be noted that the phase of the modal excitation due to the primary and secondary source is the same for modes $n_1 = 0,2,4$, etc., and of opposite phase for the modes $n_1 = 1,3,5$, etc. (as illustrated in Fig. 10.3). When a large number of residually excited modes are present, as occurs between resonances, the secondary source cannot cancel the contribution from the primary source to the even numbered modes, for example, without increasing the excitation of the odd numbered modes. It is noteworthy that the total acoustic potential energy after control, E_{p0}, near the resonances is similar to that *between* the resonances before control. This demonstrates that the field after control is largely made up of the contribution from all the *residual* modes driven by the primary source which cannot be controlled, but are at least not unduly enhanced in amplitude by the secondary source. As the contribution to E_p from the dominant mode compared to the residual modes is decreased, with increasing frequency or damping, the reductions in E_p using active control are also seen to be reduced. The main effect of active control can be seen to be to suppress the resonances in the duct. These resonances, however, can be suppressed to only the level at which the residual modes become significant.

Figure 10.7 shows the effect of optimally controlling E_p with a secondary source placed halfway along the duct (q_{s2} in Fig. 10.3). Significant reductions are achieved

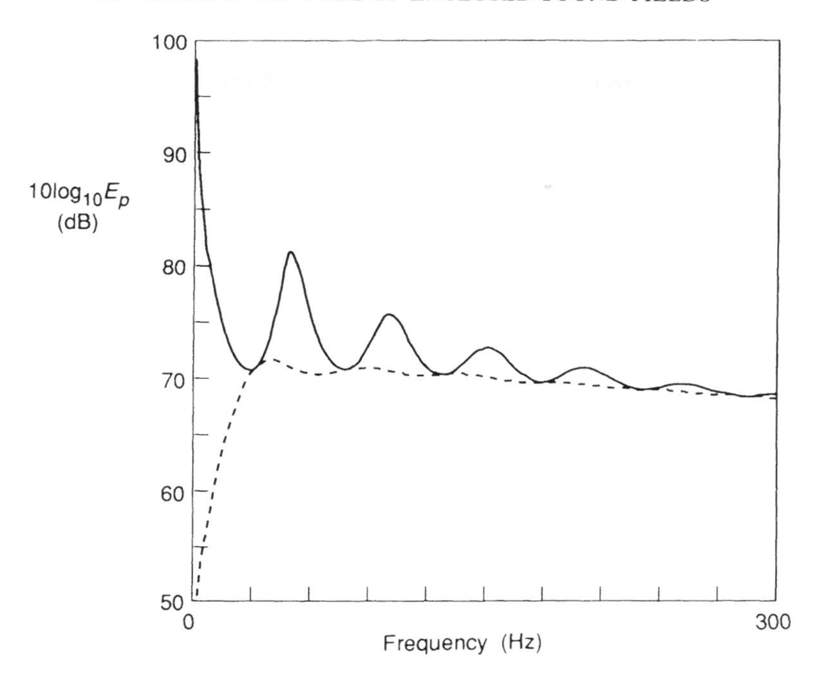

Fig. 10.5 As Fig. 10.4 but with a damping ratio of the acoustic modes of 0.1.

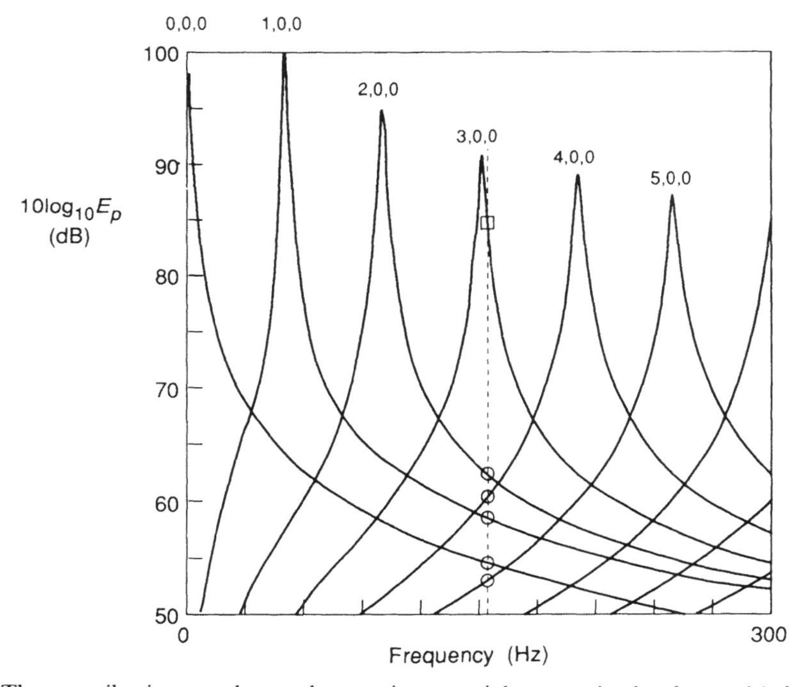

Fig. 10.6 The contributions to the total acoustic potential energy in the duct, with $\zeta = 0.01$, due to the individual acoustic modes, when driven by the primary source only. The contributions from the "dominant" (□) and residual (○) modes at a particular value of frequency are also illustrated.

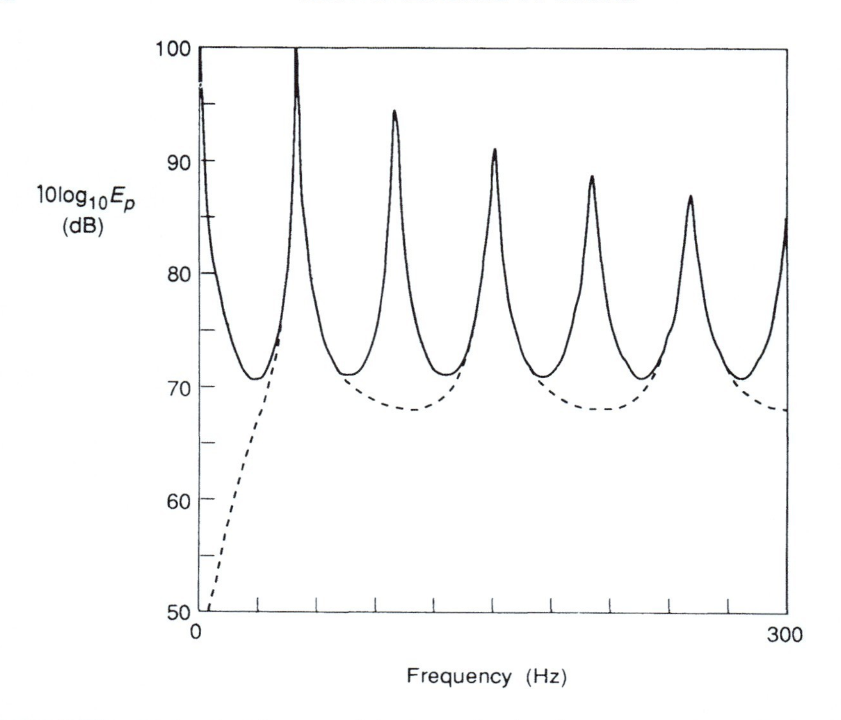

Fig. 10.7 The total acoustic potential energy in the duct due to the primary source alone (——) and with the harmonic primary source and an optimally adjusted secondary source (q_{s2}), placed halfway along the duct, in operation (-----). The damping ratio of the acoustic modes is 0.01.

at resonances corresponding to $n_1 = 0$, 2 and 4 but not those corresponding to $n_1 = 1$, 3 and 5. This is because the secondary source is placed at the nodes of the odd numbered modes (as sketched in Fig. 10.3) and so cannot couple into these modes. In general the degree of coupling is proportional to the mode shape evaluated at the source position (equation (10.5.4)), so if the secondary source is close to a nodal line, the coupling will be weak. The secondary source will thus have to work hard to drive this mode, which it cannot do without exciting the residual modes, and so little active control can be achieved.

If the secondary source were placed very close to the primary source, both sources could couple into each mode in a similar way. In particular, they would excite each mode with the same phase. Large reductions in E_p can be achieved under these conditions as illustrated in Fig. 10.8, for example, in which the secondary source q_{s3} in Fig. 10.3, placed one-tenth of the duct length from the primary source, has been optimally adjusted. Control of each mode up to $n_1 = 5$ can be obtained, using q_{s3} driving out of phase with q_p. Significant reductions can thus be achieved *between* the lowest four resonances in this case. No control over the $n_1 = 5$ mode can be achieved since q_{s3} lies on a nodal line of this mode (as seen from Fig. 10.3). In general, if the secondary source is placed a distance l from the primary source, the secondary source will couple into each mode with the same phase as the primary source until a nodal line occurs between the primary and secondary source. The

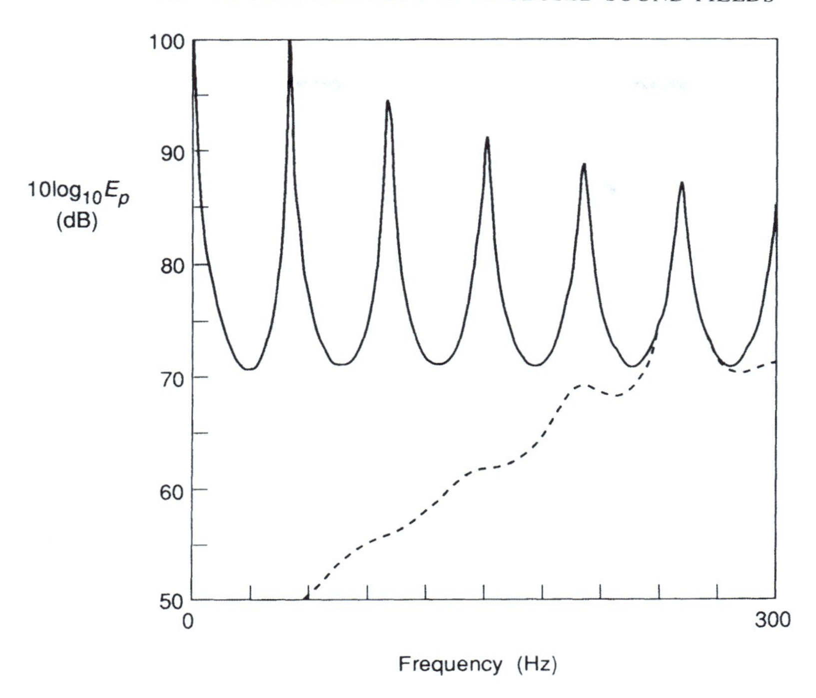

Fig. 10.8 As Fig. 10.7 but the dashed curve is for the harmonic primary and optimally adjusted secondary source q_{s3}, placed one-tenth of the way along the duct from the primary source.

secondary source will thus be able to control all modes with a natural frequency below that at which $l \approx \lambda/4$ in this case, since the first nodal line occurs $\lambda/4$ from the rigid end of the duct. It is interesting to compare this with the case of two monopoles in free space, separated by a distance l, for which active reductions in the power radiated occur up to a frequency corresponding to $l \approx \lambda/2$ (see Section 8.4). In contrast to the free field case, however, global control in the enclosed sound field can also be achieved with $l > \lambda/2$, provided the secondary sources can still couple with the same phase to the acoustic modes strongly excited by the primary source.

10.7 Global active control with multiple secondary sources

The theory outlined in Section 10.5 can readily be extended to deal with multiple secondary sources. Here we follow the analysis of Nelson *et al.* (1987b). First we write the complex amplitude of the nth mode as

$$a_n = a_{pn} + \sum_{m=1}^{M} B_{nm} q_{sm}, \tag{10.7.1}$$

where we have introduced M secondary sources of complex strengths q_{sm} in order to control the field. Note that this formulation allows the specification of the values of

a_{pn} and B_{nm} in a very general way. Thus for example if the sound field in the enclosure is excited by a "primary" distribution of surface velocity $\mathbf{u}_p(\mathbf{y})$ on the enclosure walls then it follows from equation (10.3.13) that

$$a_{pn} = \frac{\omega \rho_0 c_0^2}{V\,[2\zeta\,\omega_n \omega + j(\omega^2 - \omega_n^2)]} \int_S \psi_n(\mathbf{y})\mathbf{u}_p(\mathbf{y}).\mathbf{n}\,\mathrm{d}S, \tag{10.7.2}$$

where again we have described the modal damping by using equation (10.4.13). Similarly we can evaluate the coefficients B_{nm} for secondary sources which could consist of localised regions of "piston-like" wall vibrations (see Bullmore *et al.*, 1987 for full details). Both the terms a_{pn} and B_{nm} can also be evaluated (in principle) by using equation (10.3.14) for either extensive or localised distributions of primary and secondary source strength that exist within the volume of the enclosure.

We now consider the sound field to be represented by a *finite* number of modal contributions by truncating the series representation of equation (10.5.1) after N terms. The complex mode amplitudes a_n can then be expressed in the form of a vector $\mathbf{a}$ defined by

$$\mathbf{a}^{\mathrm{T}} = [a_0, a_1, \ldots, a_{N-1}]. \tag{10.7.3}$$

We can then write equation (10.7.1) in the vector form

$$\mathbf{a} = \mathbf{a}_p + \mathbf{B}\mathbf{q}_s, \tag{10.7.4}$$

where the vectors $\mathbf{a}_p$ and $\mathbf{q}_s$ are respectively defined as

$$\mathbf{a}_p^{\mathrm{T}} = [a_{p1}, a_{p2}, \ldots, a_{p(N-1)}],\ \mathbf{q}_s^{\mathrm{T}} = [q_{s1}, q_{s2}, \ldots, q_{sM}], \tag{10.7.5}$$

and the matrix $\mathbf{B}$ is defined by

$$\mathbf{B} = \begin{bmatrix} B_{01} & B_{02} & \ldots & B_{0M} \\ B_{11} & B_{12} & \ldots & B_{1M} \\ \cdot & \cdot & & \cdot \\ \cdot & \cdot & & \cdot \\ B_{(N-1)1} & B_{(N-1)2} & & B_{(N-1)M} \end{bmatrix}. \tag{10.7.6}$$

With the truncated series representation of the sound field it also follows from equation (10.5.8) that the total time-averaged acoustic potential energy can be written as

$$E_p = \frac{V}{4\rho_0 c_0^2} \sum_{n=0}^{N-1} |a_n|^2 = \frac{V}{4\rho_0 c_0^2} \mathbf{a}^{\mathrm{H}}\mathbf{a}, \tag{10.7.8}$$

where we have adopted the vector notation for the sum of the squared mode amplitudes (see Section 8.11). If we now substitute equation (10.7.4) into this expression then it follows that E_p can be expressed in the Hermitian quadratic form

$$E_p = \frac{V}{4\rho_0 c_0^2} \quad [\mathbf{q}_s^{\mathrm{H}}\mathbf{B}^{\mathrm{H}}\mathbf{B}\mathbf{q}_s + \mathbf{q}_s^{\mathrm{H}}\mathbf{B}^{\mathrm{H}}\mathbf{a}_p + \mathbf{a}_p^{\mathrm{H}}\mathbf{B}\mathbf{q}_s + \mathbf{a}_p^{\mathrm{H}}\mathbf{a}_p]. \tag{10.7.9}$$

The total acoustic potential energy due to the secondary sources acting alone is proportional to $\mathbf{q}_s^{\mathrm{H}}\mathbf{B}^{\mathrm{H}}\mathbf{B}\mathbf{q}_s$. Since this real quantity is clearly greater than zero for any $\mathbf{q}_s$, the matrix $\mathbf{B}^{\mathrm{H}}\mathbf{B}$ must be positive definite, and equation (10.7.9) describes a bowl-shaped surface in terms of the real and imaginary components of $\mathbf{q}_s$, with a unique global minimum. (See the discussion of this type of quadratic function presented in the appendix and introduced in Section 8.11.)

It is tempting at this stage to assume that if we have as many acoustic modes significantly excited by the primary source as there are secondary sources, only these modes need be considered in the modal summation, i.e. $N = M$; in which case the matrix $\mathbf{B}$ is square, and assuming it is not singular, the amplitude of each of the N modes can be independently controlled using the N secondary sources. In particular, if we set

$$\mathbf{q}_s = \mathbf{q}_{s0} = -\mathbf{B}^{-1}\mathbf{a}_p, \tag{10.7.10}$$

then the amplitude of each of the N modes, and hence the estimate of E_p given by equation (10.7.8), is driven to zero. This in turn implies that the sound pressure at all points in the enclosure can be driven to zero by the action of the secondary sources.

This approach may be valid at very low frequencies where only the (0,0,0) mode is significantly excited by both the primary and a single secondary source. In general, however, such an approach runs the danger of driving the M modes originally strongly excited by the primary source to zero at the expense of the secondary sources significantly exciting *other* modes not originally dominant and not considered in the analysis above. This effect is often called *control spillover* in the context of active control (see, for example, Balas, 1978). What is required is for the action of the secondary sources to strike a balance between controlling the originally dominant modes, and not significantly exciting the *residual* modes. Such a balance is automatically performed if $\mathbf{q}_s$ is adjusted to minimise E_p, with the pressure described as the sum of many *more* modes than there are secondary sources, so that the effect of the residual modes is accounted for.

Thus, when $N > M$, the optimal set of secondary source strengths which minimises the quadratic function given by equation (10.7.9) for E_p is then given by

$$\mathbf{q}_{s0} = -[\mathbf{B}^{\mathrm{H}}\mathbf{B}]^{-1}\mathbf{B}^{\mathrm{H}}\mathbf{a}_p, \tag{10.7.11}$$

and the corresponding minimum value of E_p, the residual acoustic potential energy, is given by

$$E_{p0} = \frac{V}{4\rho_0\, c_0^2}\, \mathbf{a}_p^{\mathrm{H}}[\mathbf{I} - \mathbf{B}[\mathbf{B}^{\mathrm{H}}\mathbf{B}]^{-1}\mathbf{B}^{\mathrm{H}}]\mathbf{a}_p. \tag{10.7.12}$$

This formulation has been used by Bullmore *et al.* (1987, 1988, 1990) as the basis for evaluating the potential for global active control in a number of cases of practical interest. In the next section we describe briefly one of the simplest examples examined.

10.8 Minimisation of the total acoustic potential energy in a harmonically excited two-dimensional enclosure

Consideration of active control in an essentially two-dimensional enclosure introduces some of the spatial complexity found in practical, three-dimensional,

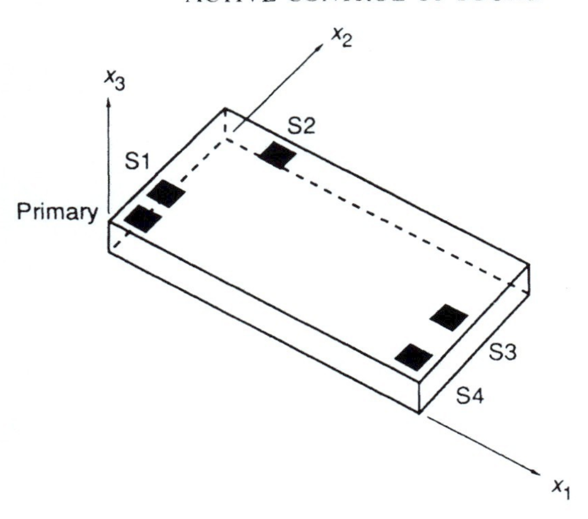

Fig. 10.9 Schematic diagram of the two-dimensional enclosure modelled in the computer simulations described in Section 10.8.

enclosures, whilst still enabling the pressure field to be easily visualised. The enclosure considered by Bullmore *et al.* (1987) is shown in Fig. 10.9. The depth of the enclosure (L_3 = 0.186 m) is small compared with the length (L_1 = 2.264 m) and the width (L_2 = 1.132 m), so that the natural frequency of the first mode having a variation in pressure along the x_3 axis (the (0,0,1) mode with natural frequency 860 Hz) is well above the frequency range considered (up to 300 Hz). The sound field thus consists mainly of contributions from axial and tangential modes in the x_2 and x_3 directions and the pressure can be visualised as being essentially two-dimensional. The rather curious dimensions were chosen to be consistent with some early experimental work by Lewers (1984). The primary and four secondary sources (S1 to S4 in Fig. 10.9) were modelled as 0.15 m by 0.15 m square pistons having a uniform surface velocity. Thus the pressure field produced in the enclosure was calculated in accordance with equation (10.3.14) as described in the previous section. Bullmore investigated the convergence of the modal summation for the pressure in the enclosure and found that the pressure in front of one of the piston sources had converged to within 1 dB of its "true" value (calculated by summing 20000 modes) at all frequencies below 300 Hz if 7000 modes were taken into account in the modal summation. The use of a finite size source is important in the model, since if point sources are considered an infinite number of modes would be needed to reproduce their extreme near fields (see the discussion presented by Morse and Ingard, 1968, Ch.9).

Figure 10.10 shows the total acoustic potential energy (E_p) in the enclosure when excited by only the harmonic primary source, at various discrete frequencies up to 300 Hz. This result is calculated from the sum of 7000 modes each with ζ_n = 0.01. The modal integers associated with the modes corresponding to each resonance are indicated. The natural frequency of both the (2,0,0) and (0,1,0) modes occur at the same frequency, about 150 Hz, since the length of the enclosure is exactly twice its width. The theory outlined in the previous section was used to calculate the optimal

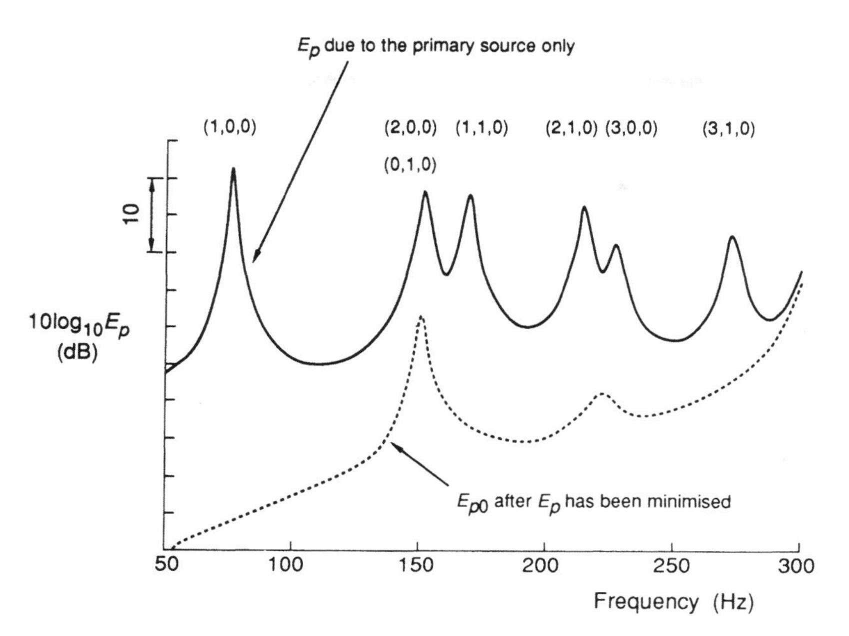

Fig. 10.10 The total acoustic potential energy, E_p, in the enclosure of Fig. 10.9 before and after E_p has been minimised by using the single secondary source S1.

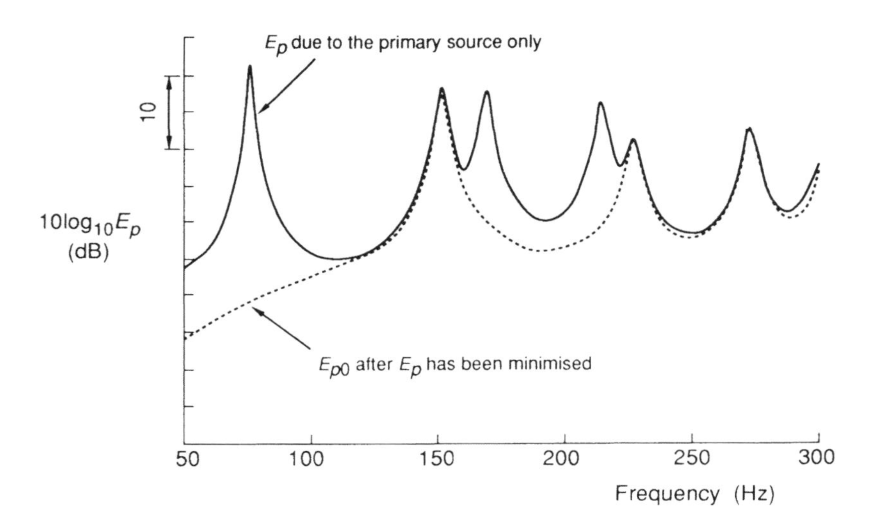

Fig. 10.11 The total acoustic potential energy, E_p, in the enclosure of Fig. 10.9 before and after E_p has been minimised by using the single secondary source S2.

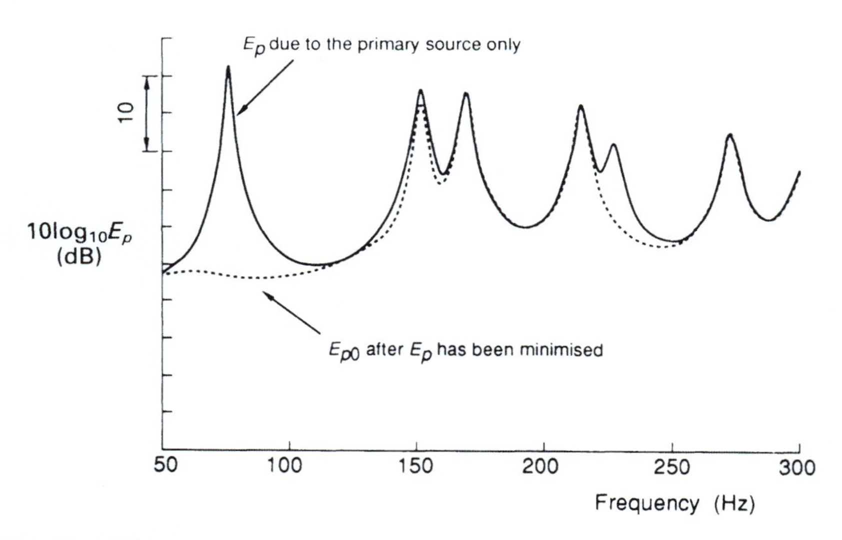

Fig. 10.12 The total acoustic potential energy, E_p, in the enclosure of Fig. 10.9 before and after E_p has been minimised by using the single secondary source S3.

source strength of each of the secondary sources, S1 to S4, such that when each source was operating individually with the primary source it minimised the total acoustic potential energy. These source strengths were then used to calculate E_p after control (E_{p0}) with each of the secondary sources individually, and the results are shown in Figs 10.10–10.13. The action of secondary source S1 when optimally adjusted at all frequencies, Fig. 10.10, is clearly to reduce the total acoustic potential energy by more than 20 dB at all frequencies, except near the resonance at 150 Hz, and near 300 Hz. This global control of the sound field is expected since the source S1 is positioned immediately adjacent to the primary source, and can thus couple into all modes in a similar way to the primary source. The optimal reductions due to the other sources acting alone are more variable and the results are most easily explained with reference to Fig. 10.14. This shows the *nodal lines* of each of the modes having a natural frequency in the range of interest. Broadly speaking, a given source can suppress a dominant mode provided it can effectively drive that mode. In terms used by conventional control theory, acoustic modes whose nodal lines exactly bisected a source would be said to not be "controllable" by that source (see, for example, Owens, 1981); for example, modes (0,1,0), (1,1,0), (2,1,0) and (3,1,0) are not controllable using source S3. Even if a mode is controllable by a source, however, there is no guarantee that in controlling this mode a number of other modes will not also be inadvertently excited.

Figure 10.15 illustrates the use of sources S2, S3 and S4, in combination, whose source strengths have been optimally adjusted to minimise E_p. The minimum value of E_p must be below that of any source acting alone, and since every mode can be controlled by at least one of the sources, the net result is control of all the resonances. The minimum value is also considerably below that achieved with sources S2, S3 or S4 alone, since in addition to controlling the dominant mode the source combination can independently couple into and control some of the less

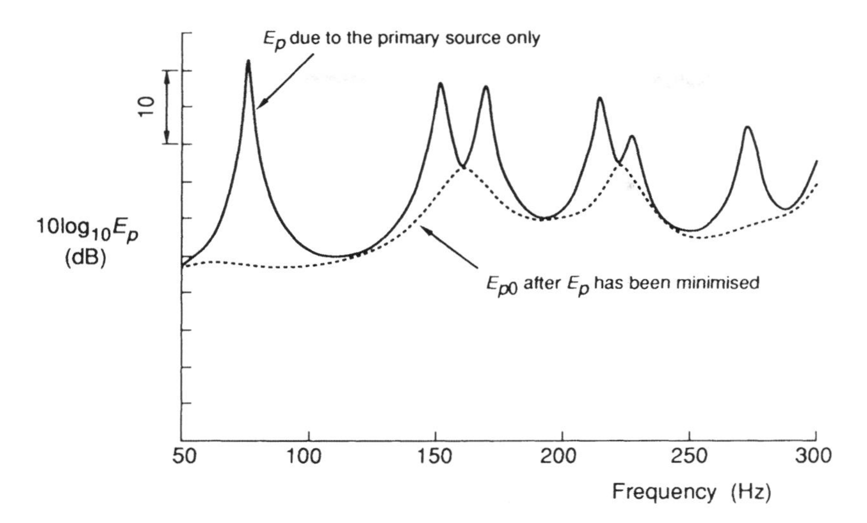

Fig. 10.13 The total acoustic potential energy, E_p, in the enclosure of Fig. 10.9 before and after E_p has been minimised by using the single secondary source S4.

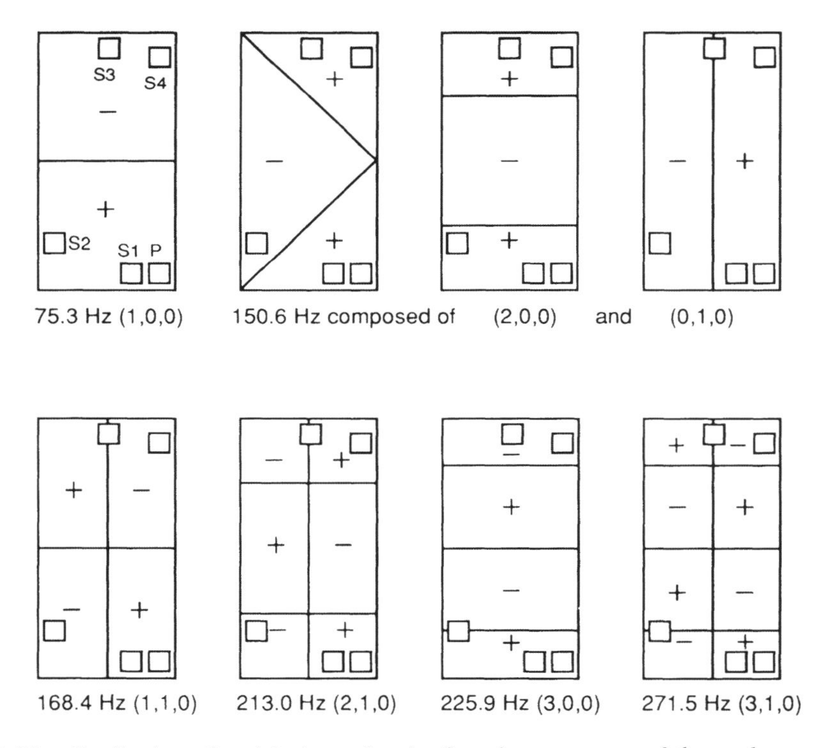

Fig. 10.14 The distribution of nodal planes for the first six resonances of the enclosure, when driven by the primary source only.

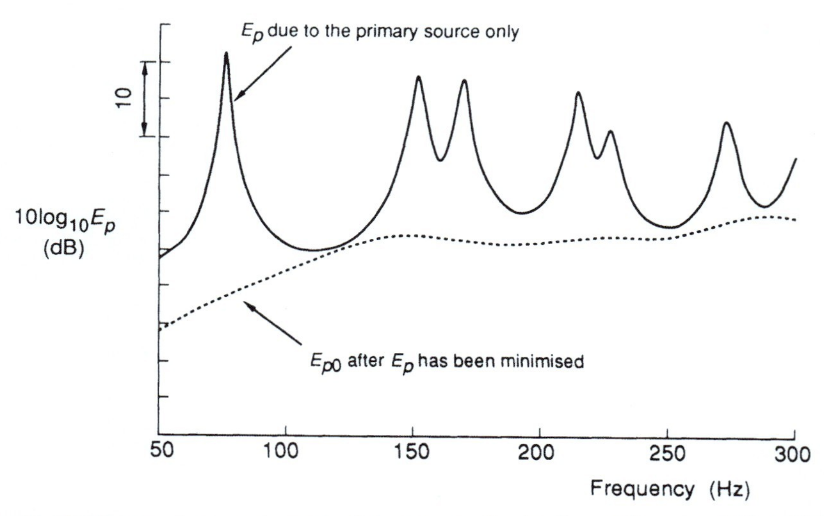

Fig. 10.15 The total acoustic potential energy, E_p, in the enclosure of Fig. 10.9 before and after E_p has been minimised by using the three secondary sources S2, S3 and S4.

dominant modes excited by the primary source at some frequencies. A more detailed discussion of these results has been presented by Bullmore *et al.* (1987).

10.9 Global active control in a randomly excited enclosure

In the examples described in the last section and in Section 10.6, the sound fields dealt with have been assumed to be harmonic and the optimal values of source strength have been determined at each frequency in a given range. These results can thus be applied to the control of primary sources whose waveform is deterministic, with no regard needing to be taken of the causal relationship between the output of the primary source and that of the secondary source. If the waveform of the primary source strength were random, however, and one conceived of implementing an active control system in which the primary source strength was measured, and used to drive the secondary source via some practical electronic controller, the issue of causality becomes extremely important.

A formulation for predicting the effect of active control of low frequency *random* sound in enclosures has been presented by Nelson *et al.* (1988c, 1990a). The optimal causal controller for some model problems has been deduced by Joplin and Nelson (1990) by using a numerical solution to the Wiener–Hopf integral equation which can be formulated for this problem. The result of applying this theory to an enclosure similar to that discussed in the last section is presented in Fig. 10.16. The frequency range in this case is taken from 0 to 300 Hz, with a heavily damped (0,0,0) mode, and only seven other modes considered in the summation for the pressure at any point. The severe modal truncation was necessary because of the intensive computation needed in this case. The primary source is driven by white noise, low pass filtered at 190 Hz, which results in the power spectral density of E_p marked

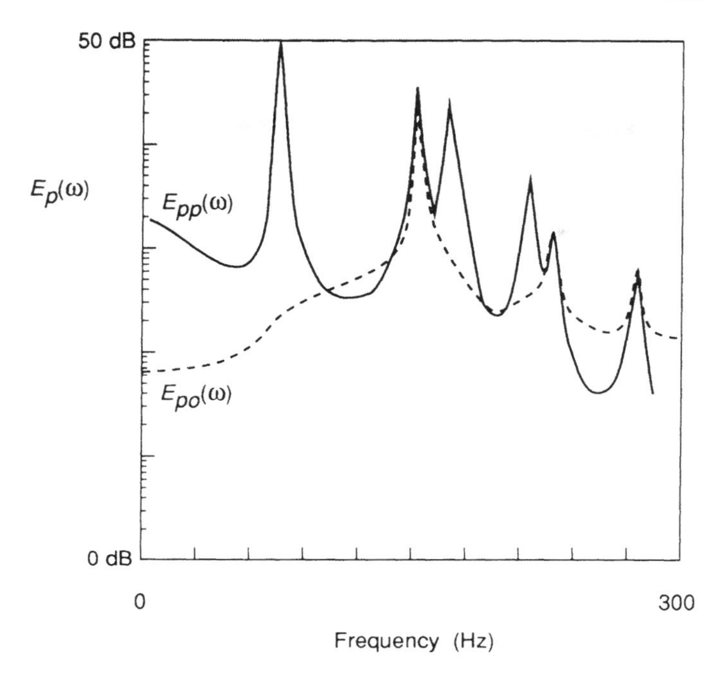

Fig. 10.16 The residual energy with broadband active control by using source S2 and with a damping ratio of 0.01. The spectral density of the total acoustic potential energy when control is applied is shown dashed. (After Joplin and Nelson, 1990.)

$E_{pp}(\omega)$ in this figure. The graph of $E_{p0}(\omega)$ is the power spectral density of E_p in the enclosure with source S2 optimally driven from a causally filtered version of this signal. When this figure is compared with Fig. 10.11, which shows the comparable unconstrained (frequency domain) minimisation, the resultant reductions are seen to be similar, with significant suppression of resonances corresponding to the modes into which source S2 can effectively couple. Increases in the level of the spectrum of E_p are, however, observed at some frequencies corresponding to a low initial excitation by the primary source. The optimisation process minimises the mean square residual E_p across the whole frequency band, and so will produce large reductions at resonance even if this means small increases in E_p at other frequencies.

10.10 Minimisation of the sum of the squared pressures at a discrete number of locations

In the previous sections the minimisation of the total acoustic potential energy E_p in an enclosure was used as a criterion for active control. This allowed us to calculate the "best possible" global performance of an active control system, with specified primary and secondary source distributions in the enclosure. Calculations such as these could thus be used to estimate the number and positions of the secondary sources required to achieve a given level of global attenuation in an enclosure with a specified primary source distribution. In order to achieve these best possible reductions by using a practical control system, an infinite number of microphones would, in principle, be necessary to estimate E_p. In this section we investigate the approximation of E_p by the sum of the squares of the pressures at a discrete number of sensors in the enclosure.

We begin by again considering harmonic sound fields and determine the complex secondary source strengths necessary to minimise a cost function similar to that introduced in Section 8.11. Here we define the cost function as

$$J_p = \frac{V}{4\rho_0 c_0^2 L} \sum_{l=1}^{L} |p(\mathbf{x}_l)|^2, \tag{10.10.1}$$

where $p(\mathbf{x}_l)$ is the complex pressure at the lth sensor location $(\mathbf{x}_l)$, and the summation is taken over a total of L sensors. The constant of proportionality in the equation for J_p is chosen such that if the sensors are uniformly distributed in the enclosure J_p tends to E_p in the limit of a large number of sensors, i.e. that

$$\lim_{L\to\infty} J_p \to E_p. \tag{10.10.2}$$

The analysis and discussion of Section 8.11 is therefore directly applicable to the problem considered here and the inclusion of the factor $V/4\rho_0 c_0^2$ will make no difference to the solution for the optimal secondary source strengths given by equation (8.11.15a).

As pointed out in the discussion of Section 8.11 it is always possible to produce *zero* pressure at (say) L sensor locations by introducing the same number L of secondary sources of appropriate complex strength. Unfortunately, in driving the pressures at these positions completely to zero, it is often found that the pressures at positions between the sensors may be increased by the action of the secondary sources. An approach which tends to avoid this problem is to ensure that there are more sensors than sources (i.e. $L > M$), in which case complete cancellation of the sound field at the sensors is not required of the secondary sources, and the least squares solution is better conditioned. This will be discussed in more detail in Chapter 12.

Some examples of applying this approach to the enclosure discussed in Section 10.8 are given by Bullmore *et al.* (1987). It is clear that J_p must be reduced by the action of the secondary sources, but we have to establish how this affects the sound field away from the microphones. This can be readily achieved by using computer simulations, because once the optimal vector of secondary source strengths necessary to minimise J_p has been calculated, this set of source strengths can then be used to calculate the resulting total acoustic potential energy in the enclosure. The global effect of applying control at discrete points can hence be evaluated.

Figure 10.17 shows a sketch of the enclosure considered in Section 10.8 with the position of secondary source S4 illustrated. Also shown are the positions of the three sensors, M1, M2 and M3, placed exactly halfway along the length of the enclosure. Figure 10.18 compares the total acoustic potential energy in the enclosure (E_p) due to the primary source only, and its value when source S4 is also operating, its complex source strength being adjusted at each frequency to minimise the sum of the squares of the pressures at the sensors M1, M2 and M3. Minimising J_p in this

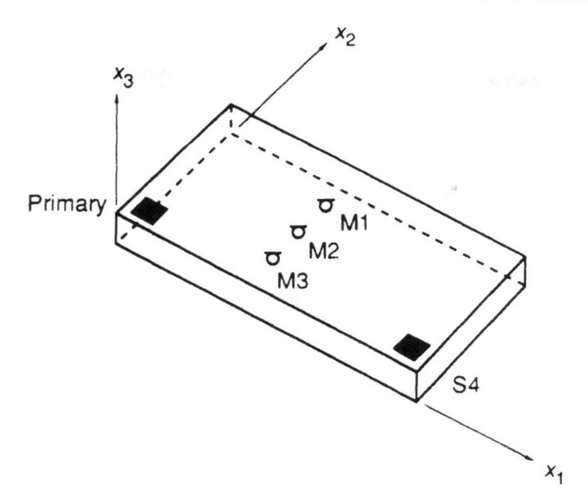

Fig. 10.17 Schematic diagram showing the position of the three microphones controlled by source S4 in Fig. 10.18.

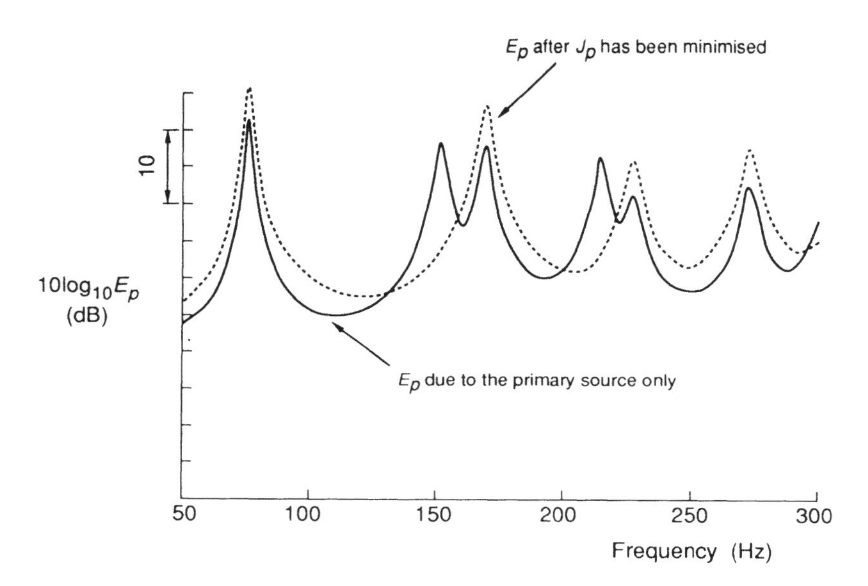

Fig. 10.18 The total acoustic potential energy, E_p, before and after the sum of the squared pressures, J_p, at the three error sensor locations M1, M2 and M3 (see Fig. 10.17) has been minimised by using the single secondary source S4.

case clearly has a detrimental effect on the energy in the enclosure over most of this frequency range. The reasons for this are not difficult to discover, however, if the positions of the sensors are compared with the positions of the nodal planes in this enclosure, as shown in Fig. 10.14. The sensors all fall on the central nodal line present in the (1,0,0) and (1,1,0), (3,0,0) and (3,1,0) modes, and so J_p is not affected by the amplitudes of any of these modes. By adjusting S4 to minimise J_p, the

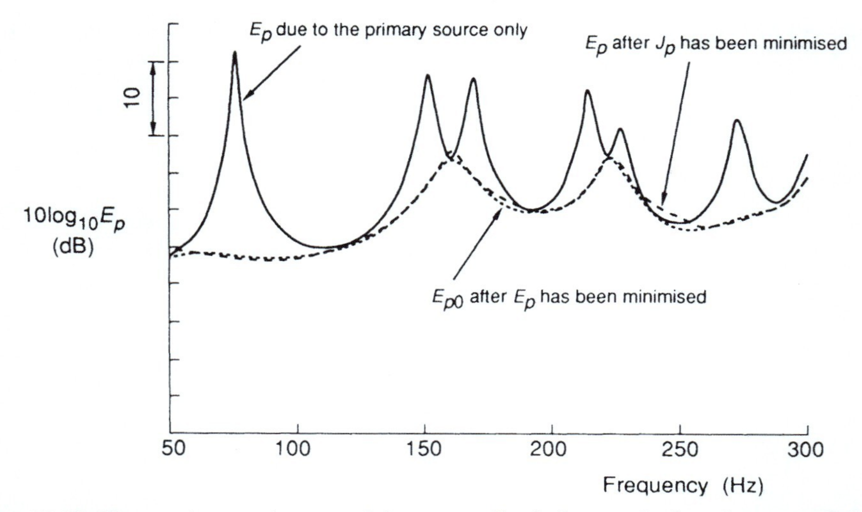

Fig. 10.19 The total acoustic potential energy, E_p, before and after the sum of the squared pressures, J_p, at the corners of the enclosure has been minimised by using the single secondary source S4. Also shown for comparison is the minimum value of E_p which can be achieved by using this secondary source.

acoustic modes in the enclosure which do contribute to J_p will be reduced, but in so doing the modes listed above will be inadvertently amplified, and so cause uncontrolled increases in E_p. The modes listed above are said not to be "observable" by the sensors M1, M2 and M3. (See, for example, the discussion of observability presented by Owens, 1981.)

Figure 10.19 shows a similar comparison of the total acoustic potential energy in the enclosure before and after minimising J_p with the same primary and secondary source, but now J_p is the sum of the squared outputs from four sensors in the *corners* of the enclosure. The energy in the enclosure after control is now below that before control at nearly all frequencies. Indeed the reductions in E_p achieved by minimising J_p in this case are almost as great as could optimally be achieved with this source by minimising E_p, as derived in the previous section, and also plotted (as E_{p0}) in Fig. 10.19 for comparison.

Experimental confirmation of these results, with excellent agreement between theory and experiment, has been presented by Elliott *et al.* (1987b). Further extensive computer simulation work aimed at evaluating the potential for active control in the cabins of propeller driven aircraft has also been undertaken by Bullmore (1988) and Bullmore *et al.* (1990). Some work has also been undertaken by Hough (1988) on the equivalent problem with random excitation.

10.11 The influence of sensor locations

In order to shed some light on the observations of the previous section, the theoretical relationship between J_p and E_p can be investigated (following Bullmore *et al.*, 1987) by expressing the vector of pressures at the L sensors as

$$\mathbf{p} = \mathbf{\Psi}_L^{\mathrm{T}}\mathbf{a}, \tag{10.11.1}$$

where $\mathbf{\Psi}_L$ is the $N \times L$ matrix of N mode shapes at the L sensor positions, and $\mathbf{a}$ is the vector of N total mode amplitudes, defined in Section 10.7. The cost function J_p can now be written as

$$J_p = \frac{V}{4\rho_0\, c_0^2 L}\ \mathbf{a}^{\mathrm{H}}\mathbf{\Psi}_L\mathbf{\Psi}_L^{\mathrm{T}}\mathbf{a}. \tag{10.11.2}$$

It is clear that if $\mathbf{\Psi}_L\mathbf{\Psi}_L^{\mathrm{T}}/L$ were to equal the identity matrix, then J_p would equal E_p as defined by equation (10.7.8), and a control system which minimised J_p would automatically achieve the *optimal* reductions in the total acoustic potential energy in the enclosure. The matrix $\mathbf{\Psi}_L\mathbf{\Psi}_L^{\mathrm{T}}$ may be expanded as

$$\sum_{l=1}^{L}\begin{bmatrix} \psi_1^2(\mathbf{x}_l) & \psi_1(\mathbf{x}_l)\psi_2(\mathbf{x}_l) & \cdots & \psi_1(\mathbf{x}_l)\psi_N(\mathbf{x}_l) \\ \psi_2(\mathbf{x}_l)\psi_1(\mathbf{x}_l) & \psi_2^2(\mathbf{x}_l) & \cdots & \psi_2(\mathbf{x}_l)\psi_N(\mathbf{x}_l) \\ \cdot & \cdot & & \cdot \\ \cdot & \cdot & & \cdot \\ \psi_N(\mathbf{x}_l)\psi_1(\mathbf{x}_l) & \psi_N(\mathbf{x}_l)\psi_2(\mathbf{x}_l) & \cdots & \psi_N^2(\mathbf{x}_l) \end{bmatrix}. \tag{10.11.3}$$

If the number of sensors were very large, then the orthogonality property of the mode shapes shows that

$$\lim_{L\to\infty}\frac{1}{L}\sum_{l=1}^{L}\psi_n(\mathbf{x}_l)\psi_m(\mathbf{x}_l) \to \begin{Bmatrix} 1, \text{ for } n = m \\ 0, \text{ for } n \neq m \end{Bmatrix} \tag{10.11.4}$$

where m is used again here as an alternative to the index n denoting a trio of modal integers.

For the two-dimensional rectangular enclosure considered here, an examination of the mode shapes of Fig. 10.14 shows that for the four corner microphones

$$\sum_{l=1}^{L}\psi_n(\mathbf{x}_l)\psi_m(\mathbf{x}_l) = 0, \tag{10.11.5}$$

for all *adjacent* pairs of modes (n,m) with natural frequencies below 300 Hz. J_p is thus generally a good indicator of E_p over this frequency range. In general, the outputs of L error microphones could be decomposed into estimates of the amplitudes of L specific modes, provided the mode shapes were known and at least one sensor could detect each mode. The total acoustic potential energy could then be estimated from the sum of the squares of these L mode amplitudes. This would be valid over the frequency range where only *these* L modes were significantly excited.

The approach used here, however, is *not* modal decomposition. We use the sum of the squares of the microphone outputs as a *direct* estimate of the total acoustic potential energy in the enclosure, which will be valid in the frequency range over which *any* set of only L modes, which can be independently detected by the sensors, is significantly excited. In the example above the four microphones could only be used up to about 160 Hz if decomposition into the four lowest modes were used, but the sum of the squares of their outputs has been shown to provide a good direct estimate of the energy right up to 300 Hz. In a three-dimensional rectangular enclosure, the sum of the squares of eight corner microphones would thus provide a reasonable estimate of the total acoustic potential energy in the enclosure at all frequencies at which almost any set of only eight modes were significantly excited.

The corner microphones will always avoid the nodal lines of any mode in a rectangular enclosure. Unfortunately, in enclosures which do not have a simple rectangular shape (a car interior for example) it is generally not possible to identify microphone positions which lie at the maximum of the mode shape for all significantly contributing modes.

10.12 Mechanisms of active control at low modal densities

Further insight into the mechanism of active control can be obtained by considering the acoustic power output of the sources when active control is implemented. Bullmore (1988) has computed the power output of the primary source in an enclosure similar to that considered in Section 10.8, using the expression

$$W_p = \tfrac{1}{2}\int_{S_p} \mathrm{Re}\{p^*(\mathbf{y})u_p\}\mathrm{d}S, \tag{10.12.1}$$

where u_p is the uniform normal surface velocity of the primary source, which has an area of S_p. The enclosure used by Bullmore for these studies is illustrated in Fig. 10.20. When secondary source S5 has been optimally adjusted to minimise the total acoustic potential energy in the enclosure, as described in Section 10.7, the resulting total acoustic potential energy is shown in Fig. 10.21(a), over a restricted range of frequencies. The acoustic power output of the primary source when operating alone is shown in Fig. 10.21(b). Near to the natural frequencies of either of the two modes excited within this frequency range, the power output from the primary source is seen to increase significantly. This can be explained by considering the input impedance of an enclosure, driven by a compact piston source, whose response is

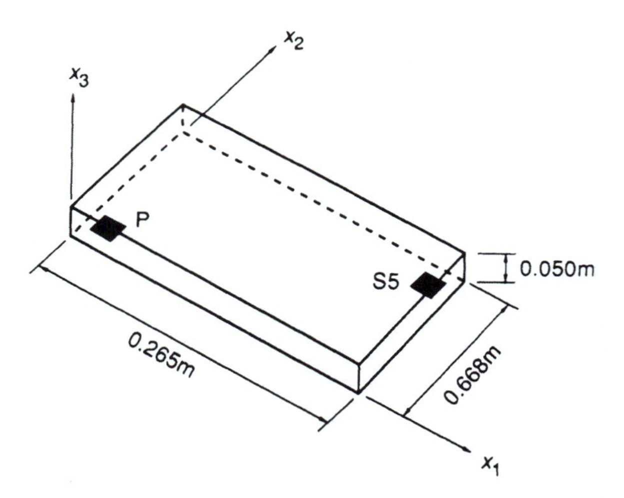

Fig. 10.20 The enclosure used by Bullmore (1988) for a study of energy flow in the active control of enclosed sound fields. Both sources were modelled as $0.02\,\mathrm{m}^2$ pistons with centres located at $x_3 = 0$ with (x_1,x_2) co-ordinates of (0.025, 0.025) and (0.643, 0.24) for the primary source and secondary source S5, respectively.

dominated by a single mode. This can be written as

$$\frac{p(\mathbf{y})}{q(\mathbf{y})} = Z(\mathbf{y}|\mathbf{y}) \approx \frac{\omega\rho_0 c_0^2}{V\,[2\zeta_n\omega_n\omega + j(\omega^2 - \omega_n^2)]}\psi_n^2(\mathbf{y}), \qquad (10.12.2)$$

where the source strength $q(\mathbf{y})$ is calculated from the integral of the surface velocity over the area of the piston source. The power output of such a compact source can be written as

$$W = \tfrac{1}{2}\,\mathrm{Re}\{p^*(\mathbf{y})q(\mathbf{y})\} = \tfrac{1}{2}\,|q(\mathbf{y})|^2\mathrm{Re}\{Z(\mathbf{y}|\mathbf{y})\}. \qquad (10.12.3)$$

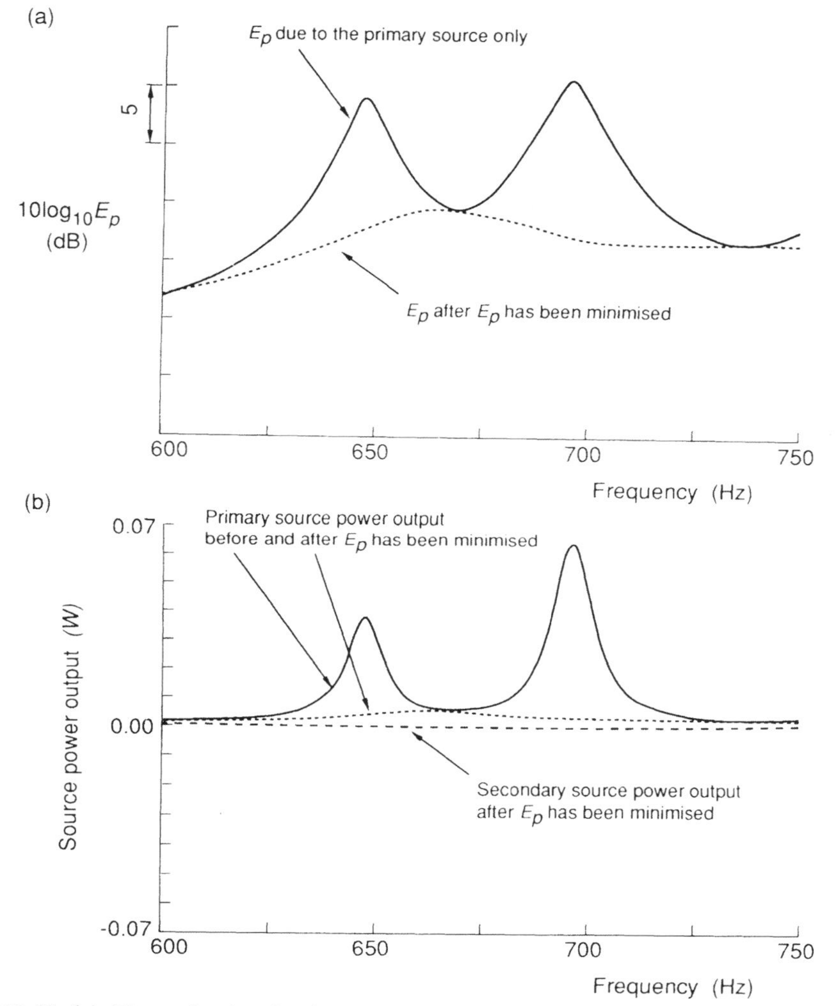

Fig. 10.21 (a) The reduction in the total acoustic potential energy, E_p, when E_p has been minimised by using secondary source S5. (b) The effect on the primary and secondary source power outputs when the total acoustic potential energy, E_p, has been minimised using secondary source S5.

The real part of the input impedance increases considerably as ω approaches ω_n, and at $\omega = \omega_n$ the impedance is entirely real and is equal to

$$Z(\mathbf{y}|\mathbf{y}) = \frac{\rho_0 c_0^2}{2\zeta_n \omega_n V} \psi_n^2(\mathbf{y}). \tag{10.12.4}$$

The primary source is seen to be "loaded" by a large, real, acoustic impedance at resonance.

The power outputs from both the primary and the secondary source (S5) after the complex source strength of S5 has been adjusted in order to minimise the total acoustic potential energy in the enclosure, are also shown in the lower part of Fig. 10.21(b). The secondary source has now substantially removed the contribution of the dominant mode from the pressure in front of the primary source. The pressure in front of the primary source and hence the impedance this source experiences in the enclosure are thus considerably reduced and although the primary source has the same velocity as when operating alone, the power it can supply to the sound field in the enclosure is much diminished. The mechanism of control can be said to be an "unloading" of the primary source by the action of the secondary source. The secondary source itself is seen to radiate very little power in Fig. 10.21(b), largely because its action is to cancel the contribution to the pressure from the dominant mode, and so the pressure in front of the secondary source is also small and it is also not significantly loaded.

It is clear from the results presented in Fig. 10.21(b) that energy *absorption* by the secondary source is not a major mechanism of active control in resonant enclosures at low modal densities. To reinforce this point, Bullmore (1988) has calculated the effect of maximising the power absorption of the secondary source. This is equivalent to *minimising* the acoustic power *output* of this source, W_s, which in this case will be negative, as discussed in Chapter 9. The resultant total acoustic potential energy (E_p) is shown in Fig. 10.22, and although E_p is somewhat reduced at the

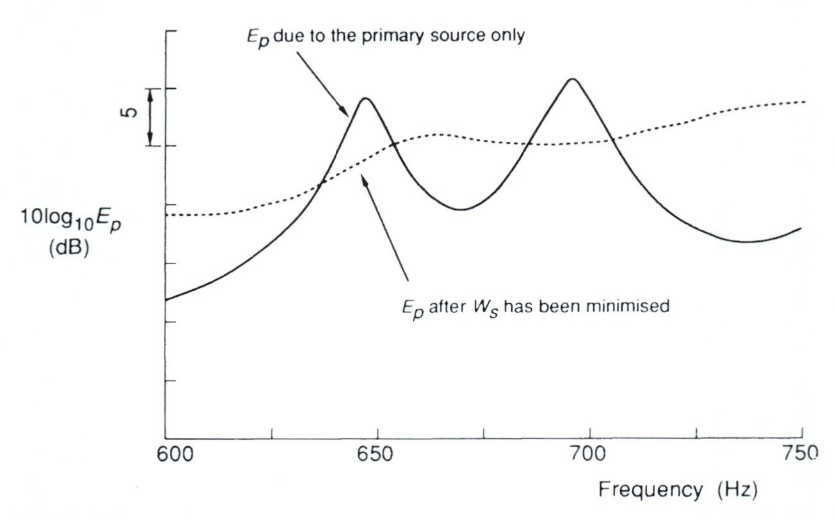

Fig. 10.22 The change in the total acoustic potential energy, E_p, when the power output, W_s, of the secondary source S5 has been minimised.

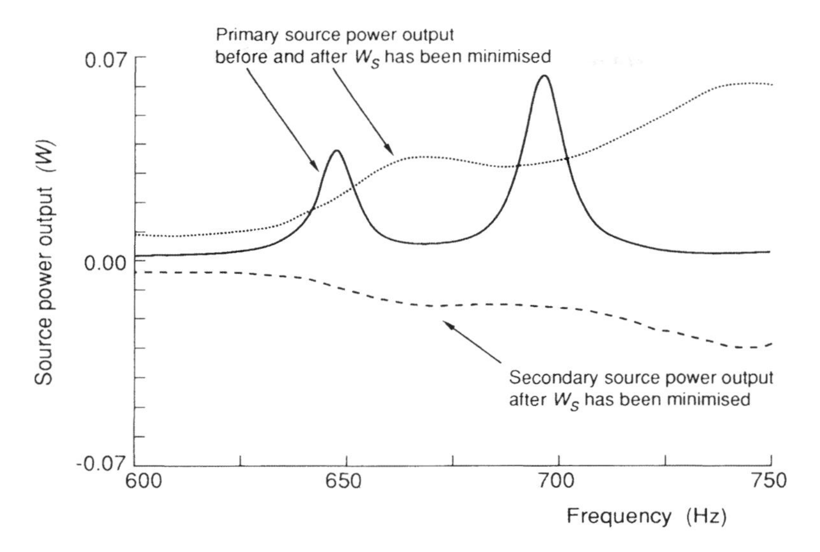

Fig. 10.23 The effect on the power outputs of the primary and secondary sources when the power output, W_s, of secondary source S5 has been minimised.

resonance frequencies, it has been increased, by up to 10 dB, away from these frequencies. The acoustic power output of the primary and secondary sources under these conditions is shown in Fig. 10.23. The power absorbed by the secondary source at non-resonant frequencies is greater than that originally radiated by the primary source. At these frequencies, however, the power output of the primary source is *increased* by the action of the secondary source, and this results in a rise in the total acoustic potential energy in the enclosure. Whereas the primary source originally experienced a small, mostly reactive, input impedance when operating alone off resonance, the action of the secondary source is now to *load* the primary source with a larger and more resistive impedance so its power output increases and the power absorbed by the secondary source is maximised. When active control does reduce the energy in the enclosure, at the resonance frequencies in Fig. 10.22, the mechanism is similar to the absorbing termination in the one-dimensional duct, discussed in Chapter 5. Sound power absorption in enclosures, together with total power output minimisation, are further discussed by Elliott *et al.* (1991b).

10.13 The influence of modal density and damping

We have seen that global active control in enclosures using a single secondary source is most effective at excitation frequencies close to the natural frequencies of isolated, lightly damped, acoustic resonances of the enclosure. In general the extent to which global active control can be achieved in a harmonically excited enclosure depends on the relative contribution to the total energy of any "dominant" modal response relative to the contribution of the "residual" modes (see the discussion of Section

10.5). The frequency separation between modes and their damping (quantified for example by the 3 dB bandwidth of any one mode) are thus the features of the enclosure which, generally speaking, determine the success of global active control. A parameter which quantifies the probable number of modes which exist within the 3 dB bandwidth of any one mode is the *modal overlap* $M(\omega)$ (Lyon, 1969, 1975), a parameter widely used in statistical energy analysis, which is defined by the expression

$$M(\omega) = \Delta\omega n(\omega), \tag{10.13.1}$$

where $\Delta\omega$ is the 3 dB bandwidth of any one mode at a frequency ω, and $n(\omega)$ is the corresponding modal density (average number of modes per unit angular frequency).

Assuming a viscous damping model, the 3 dB bandwidth of the nth mode is given by

$$\Delta\omega_n = 2\zeta_n\omega_n, \tag{10.13.2}$$

where ζ_n is the damping ratio of the nth mode. The density of various types of acoustic mode in a rectangular enclosure is discussed by Morse (1948). We can classify the mode types as "axial", "tangential" and "oblique" as described in Section 10.4. Morse (1948, Ch.8) shows that for one-dimensional enclosures of length l, two-dimensional enclosures of cross-section S and three-dimensional enclosures of volume V, the modal density of these three types can be approximated by, respectively,

$$n_{\text{ax}}(\omega) = \frac{l}{\pi c_0},\ n_{\text{ta}}(\omega) = \frac{\omega S}{2\pi c_0^2},\ n_{\text{ob}}(\omega) = \frac{\omega^2 V}{2\pi^2 c_0^3}. \tag{10.13.3a,b,c}$$

Now if we consider modes with natural frequencies $\omega_n \approx \omega$, and assume that the damping ratio $\zeta_n = \zeta$ is the same for each mode, then the corresponding values of the modal overlap are given by

$$M_{\text{ax}}(\omega) = \frac{2\zeta\omega l}{\pi c_0},\ M_{\text{ta}}(\omega) = \frac{\zeta\omega^2 S}{\pi c_0^2},\ M_{\text{ob}}(\omega) = \frac{\zeta\omega^3 V}{\pi^2 c_0^3}. \tag{10.13.4a,b,c}$$

Tohyama and Suzuki (1987) have derived an expression in terms of the modal overlap for the average reduction of the total acoustic potential energy in a three-dimensional acoustic field when a dominant resonance is suppressed by the action of a secondary source placed remotely from the primary excitation. The argument proceeds by using equations (10.5.8) and (10.5.4) to derive an expression for E_p in terms of the strengths of single point primary and secondary sources in the enclosure. It follows that

$$E_p = \frac{\rho_0 c_0^2 \omega^2}{4V} \sum_{n=0}^{\infty} \left| \frac{q_p\psi_n(\mathbf{y}_p) + q_s\psi_n(\mathbf{y}_s)}{2\zeta_n\omega_n\omega + j(\omega^2 - \omega_n^2)} \right|^2. \tag{10.13.5}$$

It is now assumed that there is one "dominant" mode excited by the primary and secondary sources, for which the trio of modal integers n is denoted d. The effect of all the other modes is gathered into "residual" contributions. This enables the expression for the total acoustic potential energy to be written as

$$E_p \approx D|\psi_d(\mathbf{y}_p)q_p + \psi_d(\mathbf{y}_s)q_s|^2 + \varepsilon_p|q_p|^2 + \varepsilon_s|q_s|^2, \tag{10.13.6}$$

where the terms D and ε_p are given by

$$D = \frac{\rho_0 c_0^2}{4V} \frac{\omega^2}{4\,\zeta_d^2\,\omega_d^2\,\omega^2 + (\omega^2 - \omega_d^2)^2}, \tag{10.13.7}$$

$$\varepsilon_p = \frac{\rho_0 c_0^2}{4V} \sum_{n \neq d} \frac{\omega^2}{4\,\zeta_n^2\,\omega_n^2\,\omega^2 + (\omega^2 - \omega_n^2)^2} \psi_n^2(\mathbf{y}_p), \tag{10.13.8}$$

and ε_s is similarly defined. It is assumed here that cross terms in the expansion of the residual modes in equation (10.3.5), of the form $\psi_n(\mathbf{y}_p)\psi_m(\mathbf{y}_s)$, have random signs and so their sum over all residual modes is small and may be ignored. The acoustic potential energy expressed in terms of the dominant and residual modal contributions can be written as a quadratic function of the type defined by equation (10.5.13) where the terms in the equation are defined by $A = [D\psi_d^2(\mathbf{y}_s) + \varepsilon_s]$, $b = [D\psi_d(\mathbf{y}_s)\psi_d(\mathbf{y}_p)q_p]$ and $c = [(D\psi_d^2(\mathbf{y}_p) + \varepsilon_p)|q_p|^2]$. The optimal secondary source strength which minimises E_p is given by $q_{s0} = -b/A$ and thus

$$q_{s0} = -\frac{D\,\psi_d(\mathbf{y}_s)\,\psi_d(\mathbf{y}_p)}{D\,\psi_d^2(\mathbf{y}_s) + \varepsilon_s} q_p. \tag{10.13.9}$$

If ε_s were zero, the secondary source could control the dominant mode without exciting any residual modes, and q_{s0} would be $-\psi_d(\mathbf{y}_p)q_p/\psi_d(\mathbf{y}_s)$ as in equation (10.5.5). In general, the minimum value of E_p with $q_s = q_{s0}$ is given by $E_{p0} = c - |b|^2/A$, and since the acoustic potential energy due to the primary source alone, E_{pp}, is equal to c, we can write $E_{p0}/E_{pp} = 1 - |b|^2/cA$, and so

$$\frac{E_{p0}}{E_{pp}} = 1 - \frac{D^2\psi_d^2\,(\mathbf{y}_s)\;\psi_d^2\,(\mathbf{y}_p)}{(D\psi_d^2\,(\mathbf{y}_s) + \varepsilon_s)(D\psi_d^2\,(\mathbf{y}_p) + \varepsilon_p)}. \tag{10.13.10}$$

Assuming $\psi_d^2(\mathbf{y}_s) \approx \psi_d^2(\mathbf{y}_p) \approx 1$ and $\varepsilon_s \approx \varepsilon_p = \varepsilon$ then

$$\frac{E_{p0}}{E_{pp}} = 1 - \frac{D^2}{(D + \varepsilon)^2} = 1 - \left[1 + \frac{\varepsilon}{D}\right]^{-2}. \tag{10.13.11}$$

If it is now assumed that the contribution from the residual modes is much less than the contribution of the dominant mode, such that $\varepsilon/D << 1$, then this expression can be approximated by

$$\frac{E_{p0}}{E_{pp}} \approx \frac{2\varepsilon}{D}. \tag{10.13.12}$$

This therefore quantifies the dependence of the reductions in energy on the relative excitation of the dominant and residual modal terms.

Tohyama and Suzuki (1987) have estimated the magnitude of the terms in this expression for an enclosure dominated by oblique modes. Assuming the frequency of excitation is equal to the natural frequency of the dominant mode, it follows from equation (10.13.7) that the contribution to the energy from the dominant mode is given by

$$D = \frac{\rho_0\,c_0^2}{16\zeta_d^2\omega^2 V}. \tag{10.13.13}$$

The level of the residual excitation of the enclosure due to the primary or secondary source can be assumed to be the diffuse field value, which, assuming $\zeta_n = \zeta$ for each mode with $\omega_n \approx \omega$, is given by Nelson *et al.* (1987b) as

$$\varepsilon = \frac{\omega \rho_0}{32\,\pi\, c_0 \zeta}. \tag{10.13.14}$$

The ratio of the residual to the dominant component of the energy in the enclosure assuming $\zeta_d \approx \zeta$ is thus given by

$$\frac{\varepsilon}{D} \approx \frac{\zeta \omega^3 V}{2\pi c_0^3} = \frac{\pi}{2} M_{\mathrm{ob}}(\omega), \tag{10.13.15}$$

where $M_{\mathrm{ob}}(\omega)$ is the modal overlap for an acoustic field dominated by oblique modes (equation (10.13.4c)). The reduction in the total acoustic potential energy due to the suppression of a dominant mode in a three-dimensional field, with modal overlap $M_{\mathrm{ob}}(\omega)$ is thus, on average, given by

$$\frac{E_{p0}}{E_{pp}} = 1 - \left[1 + \frac{\pi}{2} M_{\mathrm{ob}}(\omega)\right]^{-2}. \tag{10.13.16}$$

This result is the same as that given by Tohyama and Suzuki (1987), although they used the energy bandwidth in defining modal overlap, so that the factor of $\pi/2$ does not appear in their expression. For the case of a modal overlap of very much less than unity (i.e. $M_{\mathrm{ob}}(\omega) << 1$) then we can make the further approximation

$$\frac{E_{p0}}{E_{pp}} \approx \pi M_{\mathrm{ob}}(\omega). \tag{10.13.17}$$

Elliott (1989) has undertaken a calculation similar to that above for an enclosure dominated by axial acoustic modes. In this case it can be shown that

$$\frac{\varepsilon}{D} = 2M_{\mathrm{ax}}^2(\omega), \tag{10.13.18}$$

and therefore substitution into equation (10.13.11) shows that in this case

$$\frac{E_{p0}}{E_{pp}} = 1 - (1 + 2M_{\mathrm{ax}}^2(\omega))^{-2}. \tag{10.13.19}$$

Again, in the case of $M_{\mathrm{ax}}(\omega) << 1$, we can make the further approximation

$$\frac{E_{p0}}{E_{pp}} \approx 4M_{\mathrm{ax}}^2(\omega). \tag{10.13.20}$$

The predictions of equations (10.13.16) and (10.13.19) are plotted in Fig. 10.24 and show the fractional reductions in E_p that can be produced in enclosures dominated by axial and oblique modes as a function of the modal overlap. In the case of axial modes, the predictions of equation (10.13.19) are generally found to be in good agreement with the results of the computer simulations presented in Sections 10.6 and 10.8 (see Elliott, 1989).

It is clear from these results that very little global reduction can be produced in enclosure energy when a secondary source is placed remotely from a primary source if the modal bandwidth approaches the frequency spacing between modes since

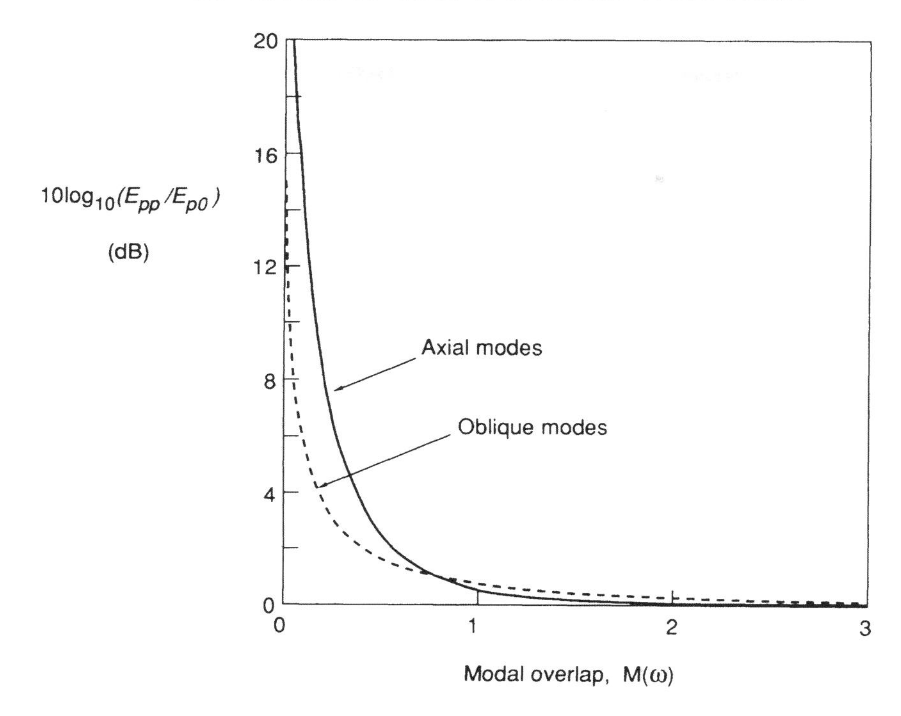

Fig. 10.24 The average reduction in the total acoustic potential energy achieved by suppressing a resonant mode with a remote secondary source, plotted as a function of modal overlap for enclosures dominated by axial and oblique modes, respectively.

$$\frac{E_{p0}}{E_{pp}} \to 1 \text{ for } M(\omega) \geq 1. \tag{10.13.21}$$

As we have seen from equation (10.13.4c), $M(\omega)$ is roughly proportional to ω^3 for oblique modes in enclosed sound fields and therefore increases very rapidly with frequency, thus restricting the opportunities for global control very much to the low frequency range. In structural vibration however the situation is very different; for flexural waves in plates for example $M(\omega)$ is generally proportional to ω (see, for example, Fahy, 1982). This difference is further emphasised by the very small modal damping ratios of structures which are typically several orders of magnitude less than for an enclosed sound field. The response of structures is thus more often dominated by lightly damped isolated resonances and control formulations relying purely on a modal approach are appropriate to these problems (Meirovitch, 1987).

10.14 Global control at high modal densities

Figure 10.25 shows the acoustic potential energy in an enclosure of dimensions 3.14 m × 2.72 m × 1 m with a constant damping ratio ζ of 0.01 up to a frequency of 1 kHz when excited by a monopole source in one corner. The transition is reasonably clear between isolated resonances below a few hundred Hertz to a more

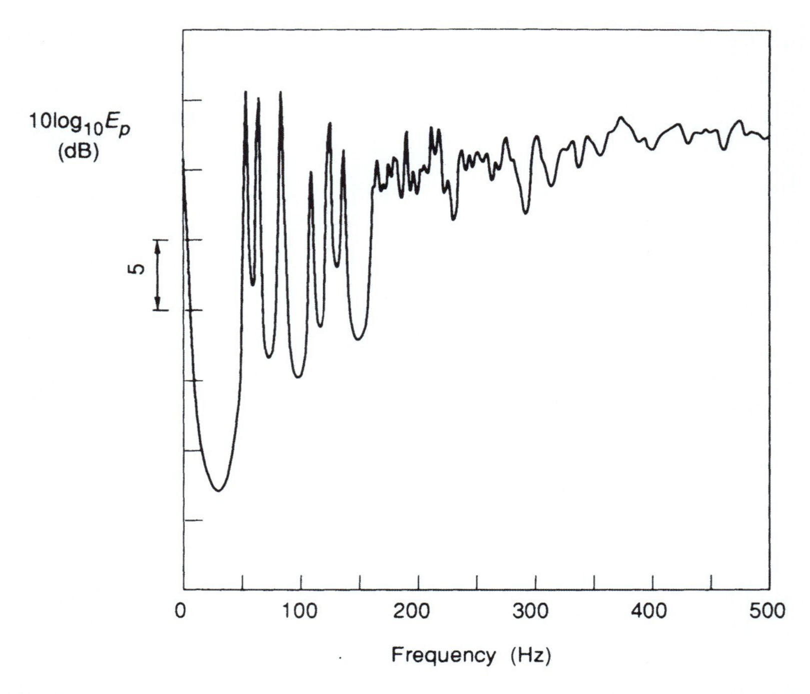

Fig. 10.25 The acoustic potential energy E_p in an enclosure of dimensions 3.14 m × 2.72 m × 1 m and with a modal damping ratio (ζ) = 0.01, excited by a source in one corner.

uniform response, with occasional, broader, peaks due to clustering of the modes. In terms of the discussion above, this corresponds to $M(\omega) < 1$ below about 260 Hz and $M(\omega) > 1$ above this frequency. In acoustics the transition between the low modal density frequency region and the high modal density region (or, more properly, between the low and high modal overlap frequency regions) is typified by the Schroeder frequency (Schroeder, 1954; Schroeder and Kuttruff, 1962; see also the discussion presented by Pierce, 1981). The modal overlap is usually defined to be three at the Schroeder frequency, f_{Sch}, so $M(\omega) = 3$ at $\omega = 2\pi f_{Sch}$. Thus for a field dominated by oblique waves, using equation (10.13.4c) for the modal overlap gives $M(\omega) = 3(f/f_{Sch})^3$. The Schroeder frequency for the enclosure whose response is shown in Fig. 10.25 is about 380 Hz. We shall return to the significance of the Schroeder frequency in the next chapter.

As the frequency of excitation increases, and hence the modal density and modal overlap in the enclosure also increase, we have seen that the reductions in total acoustic potential energy achievable by using a secondary source *remote* from the primary source become very small. What can always be achieved are local reductions in the pressure, close to individual control microphones, and such "zones of quiet" will be discussed in the next chapter. In this section we consider a special case of global control at high modal densities where the primary source is *compact* and a compact secondary source can be positioned *close* to the primary source. Specifically we consider the model problem where the primary and secondary sources are point monopoles, placed a distance r apart in an enclosure whose dimensions are much greater than r.

The analysis of Section 10.5 is therefore directly applicable to this problem and it follows that the optimal secondary source strength and minimum value of total time-averaged acoustic potential energy can be expressed as

$$q_{s0} = -\frac{\sum_{n=0}^{\infty} B_n^* a_{pn}}{\sum_{n=0}^{\infty} |B_n|^2}, \quad \frac{E_{p0}}{E_{pp}} = 1 - \frac{\left|\sum_{n=0}^{\infty} B_n^* a_{pn}\right|_n^2}{\sum_{n=0}^{\infty} |a_{pn}|^2 \sum_{n=0}^{\infty} |B_n|^2}, \quad (10.14.1a,b)$$

where the terms a_{pn} and B_n are defined by equations (10.5.10) and (10.5.11). Using these definitions, it follows that the expression for the optimal secondary source strength can be written in the form

$$q_{s0} = -q_p \frac{\sum_{n=0}^{\infty} |A_n(\omega)|^2 \psi_n(\mathbf{y}_p)\psi_n(\mathbf{y}_s)}{\sum_{n=0}^{\infty} |A_n(\omega)|^2 \psi_n^2(\mathbf{y}_s)}, \quad (10.14.2)$$

where we have used the definition

$$|A_n(\omega)|^2 = \omega^2/[(2\zeta_n\omega_n\,\omega)^2 + (\omega_n^2 - \omega^2)^2]. \quad (10.14.3)$$

We now follow the analysis presented by Nelson *et al.* (1987b) and derive an approximate value of this solution that is valid in the limit of high modal overlap. A modal summation of the form appearing in the numerator of equation (10.14.2) can be approximated by an integral provided the frequency of interest is above the Schroeder frequency for the enclosure considered. Pierce (1981), for example, has pointed out that we can write

$$\sum_{n=0}^{\infty} |A_n(\omega)|^2 \psi_n(\mathbf{y}_p)\psi_n(\mathbf{y}_s)$$
$$\approx \int_0^{\infty} <|A_n(\omega)|^2>_u <\psi_n(\mathbf{y}_p)\psi_n(\mathbf{y}_s)>_u n(\omega)|_{\omega=u}\, du, \quad (10.14.4)$$

where u is the continuous variable which replaces the discrete variable ω_n, $<|A_n(\omega)|^2>_u$ and $<\psi_n(\mathbf{y}_p)\psi_n(\mathbf{y}_s)>_u$ are the values of $|A_n(\omega)|^2$ and $\psi_n(\mathbf{y}_p)\psi_n(\mathbf{y}_s)$ averaged over a small bandwidth centred on u and $n(\omega)$ is the modal density. The average value $<|A_n(\omega)|^2>_u$ can be expressed as

$$<|A_n(\omega)|^2>_u = \omega^2/[(2\zeta\omega u)^2 + (u^2 - \omega^2)^2], \quad (10.14.5)$$

where ζ is the damping ratio which is assumed to be constant. The average value $<\psi_n(\mathbf{y}_p)\psi_n(\mathbf{y}_s)>_u$ has been evaluated by Nelson *et al.* (1987b) and is given by

$$<\psi_n(\mathbf{y}_p)\psi_n(\mathbf{y}_s)>_u = \frac{\sin(ur/c_0)}{ur/c_0}, \quad (10.14.6)$$

where $r = |\mathbf{y}_s - \mathbf{y}_p|$ is the distance between the two sources. If it is assumed that only oblique modes need be considered then using equation (10.13.3c) we can write the modal density as $(u^2 V/2\pi^2 c_0^3)$. Then equation (10.14.4) can be written as

$$\sum_{n=0}^{\infty} |A_n(\omega)|^2 \psi_n(\mathbf{y}_p)\psi_n(\mathbf{y}_s)$$

$$\approx \frac{\omega^2 V}{2\pi^2 c_0^3} \int_0^{\infty} \frac{u^2}{(2\zeta\omega u)^2 + (u^2 - \omega^2)^2} \cdot \frac{\sin(ur/c_0)}{(ur/c_0)} du. \quad (10.14.7)$$

Nelson *et al.* (1987b) have evaluated this integral exactly and provided it is assumed that $\zeta << 1$, then equation (10.14.7) reduces to

$$\sum_{n=0}^{\infty} |A_n(\omega)|^2 \psi_n(\mathbf{y}_p)\psi_n(\mathbf{y}_s) \approx \frac{\omega V}{8\pi c_0^3 \zeta} \cdot \frac{\sin kr}{kr}. \quad (10.14.8)$$

It follows directly from this approximation that equation (10.14.2) for the optimal secondary source strength can be written as

$$q_{s0} = -q_p \frac{\sin kr}{kr}. \quad (10.14.9)$$

It also follows that the fractional reduction in the total time-averaged potential energy can be written as

$$\frac{E_p}{E_{p0}} = 1 - \left(\frac{\sin kr}{kr}\right)^2. \quad (10.14.10)$$

Note that these are *exactly* the results for the optimal secondary source strength and fractional reduction in total source power output that are computed when we minimise the total power output of a pair of free-field point monopole sources (see Section 8.4 of Chapter 8). Thus we again see that, in an enclosed sound field excited at a frequency above the Schroeder frequency, global reductions in energy are only possible provided the point secondary source is placed closer than one-half wavelength from the primary source.

This problem has also been examined by Joseph (1990) who uses the statistical properties of high modal density sound fields that will be described in the next chapter. Joseph has computed the minimum *power output* of a primary/secondary source pair and shown that, when averaged over an ensemble of positions of the source pair, the *mean* value of the minimum power output is consistent with the above results, but there are also considerable variations about this mean value as the position of the source pair is varied within an enclosure.

10.15 Some examples of the application of active control in enclosed sound fields

Active control techniques in enclosures are potentially most effective at low frequencies, specifically for frequencies at which the modal overlap is small. It is also technologically easier to implement an active control system for periodic sound fields than for random sound fields. Economically, the installation cost and maintenance of an active control system, compared with conventional, passive, noise control solutions, can generally be justified only where space or weight are at a premium. All of these considerations limit the range of useful application of active control methods in practical enclosures. However, one field in which low frequency periodic noise is a significant problem, and where conventional solutions cannot generally be applied without a weight penalty whose operating cost is significant, is that of interior noise in various transportation vehicles. In this section we briefly review two such areas of application: the control of propeller noise in the passenger cabins of aircraft, and the control of low frequency engine-induced noise inside cars.

Propeller propulsion is common for small aircraft (up to 50 seats), and the development of prop-fan engines promises significant increases in fuel efficiency over conventional jet engines for larger aircraft. The spectrum of the sound pressure in the passenger compartments of such aircraft contains strong tonal components at harmonics of the blade passage frequency (BPF) of the propellers, which are difficult to attenuate using passive absorption (Metzger, 1981; Wilby *et al.*, 1980). Bullmore *et al.* (1987, 1988, 1990) have used a computer model to predict the effect of using loudspeakers in the cabin in order to minimise the sum of the squares of the pressures at a number of microphones at head height. We will briefly outline here the results of some flight trials of a practical active control system operating in the passenger cabin of a B.Ae. 748 twin turboprop aircraft, undertaken with the co-operation of British Aerospace in early 1988 (Elliott *et al.*, 1989a, b, 1990; Dorling *et al.*, 1989).

The aircraft used for the trials had a fully trimmed passenger cabin about 9 m long by 2.6 m diameter and for these trials was generally flown at an altitude of 10 000 feet under straight and level cruise with the engines running at a nominal 14 200 rpm. As a result of the gearing in the engine, this produces a blade passage frequency (BPF) of 88 Hz. The control system used in the experiments discussed here used a tachometer on one of the engines to generate a reference signal at the BPF and its second and third harmonics, i.e. 88 Hz, 176 Hz and 264 Hz. These reference signals were passed through an array of adaptive digital filters driving 16 loudspeakers. The coefficients of the digital filters were adjusted using the Multiple Error LMS algorithm (Elliott and Nelson, 1985a; see also Chapter 12) in order to minimise the sum of the square of 32 microphone signals. Many different configurations of loudspeaker and microphone position were investigated. The results presented here are for the distribution illustrated in Fig. 10.26, with eight of the 200 mm diameter loudspeakers distributed at the front of the passenger cabin near the plane of the propellers and eight distributed more evenly further down the cabin.

The reductions achieved in the sum of the mean squared pressures at all the control microphones are listed in Table 10.1 at the three frequencies. These were achieved by using the control system with all the loudspeakers operating on the pressure field from either the port or starboard propeller alone. It is clear that significant reductions are achieved at all frequencies. The normalised acoustic pressures at 88 Hz, before and after control, are plotted in terms of their physical position in the cabin in Fig. 10.27 for a similar distribution of loudspeakers and microphones. Also shown in Table 10.1 are the reductions achieved at the three harmonics if the sum of the squares of the 32 microphone outputs is minimised using either only the eight loudspeakers in the propeller plane, or only the eight

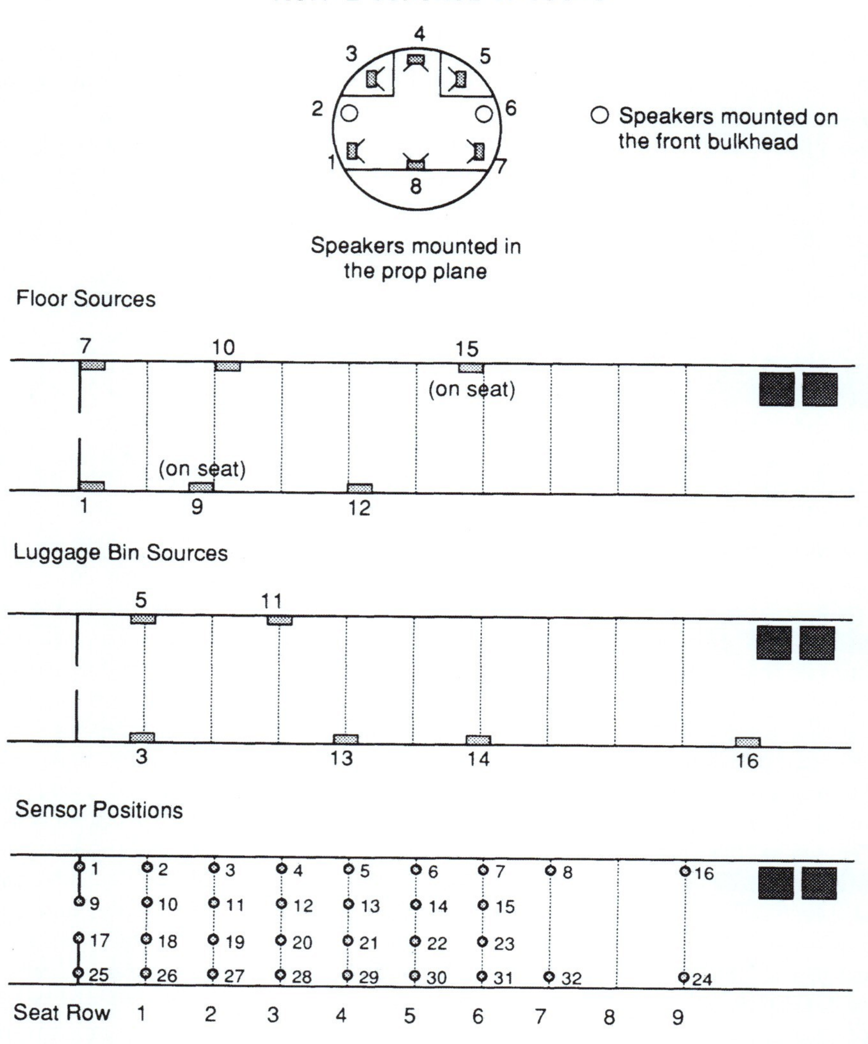

Fig. 10.26 Location of loudspeakers in the aircraft cabin, on the floor (upper) and in the luggage bins (middle), and of the microphones (lower) which are all placed at seated head height. The shaded blocks at the rear of the cabin (right hand side of the figure) show the position of the control system.

loudspeakers distributed in the cabin. With only the eight loudspeakers distributed in the cabin some control is still obtained at 88 Hz, but virtually no reductions are achieved at 176 Hz and 264 Hz. These experiments indicate that at 88 Hz the mechanism of control is largely through suppression of dominant modes, with similar reductions being obtained with the eight loudspeakers in either position. However, at 176 Hz and 264 Hz, where the acoustic modal overlap in the cabin is much larger, significant reductions can only be achieved by matching the distribution of secondary sources to the primary source distribution, which is mostly concentrated in the propeller plane.

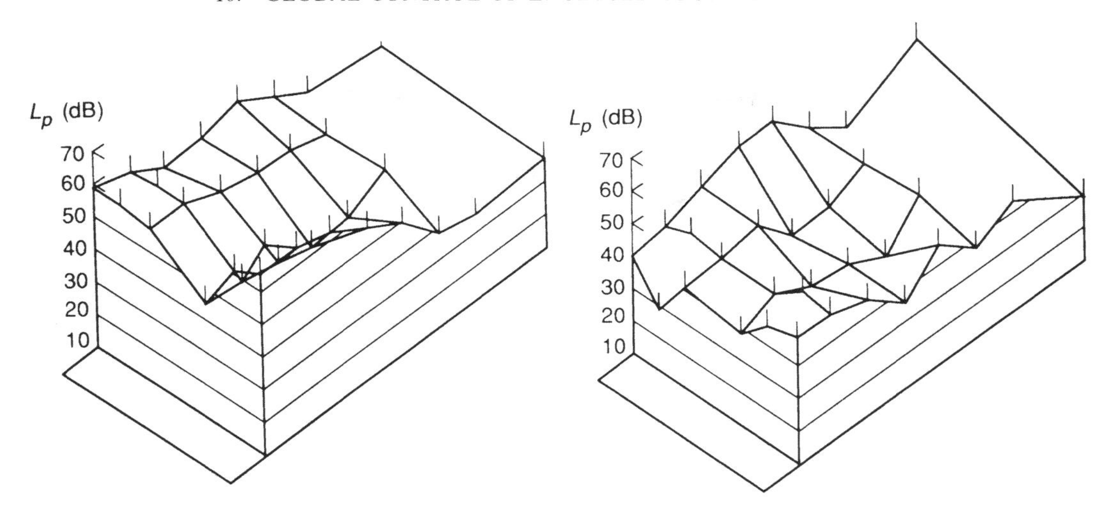

Fig. 10.27 Pressure distribution at 88 Hz measured at the 32 control microphones with the control system off (left hand plots) and on (right hand plots) (after Elliott *et al.*, 1990).

Table 10.1 Changes in the level of the sum of the squared output of 32 control microphones measured during flight trials of a 16 loudspeaker 32 microphone control system in a B.Ae 748 aircraft (Elliott *et al.*, 1990).

Configuration	**Propeller**	**88 Hz** ΔJ_{p32} (dB)	**176 Hz** ΔJ_{p32} (dB)	**264 Hz** ΔJ_{p32} (dB)
8 loudspeakers in prop plane + 8 distributed	Port	−13.0	−7.9	−9.5
	Starboard	−7.9	−10.0	−10.8
8 loudspeakers in prop plane only	Port	−6.0	−7.6	−7.9
	Starboard	−3.8	−7.8	−9.8
8 distributed loudspeakers only	Port	−6.4	−0.6	−3.5
	Starboard	−4.4	−0.7	−1.3

Although the control system would not have to respond very quickly under straight and level cruise conditions if the speeds of the two propellers were locked together, this was not possible on the aircraft used for the flight trials. The results presented in Table 10.1 were obtained with only one engine operating at 14 200 rpm and the other detuned to 12 700 rpm, so that its contribution to the sound field was ignored by the control system. As the speeds of the two engines are brought closer together, the sound fields due to the two propellers beat together and the amplitude tracking properties of the adaptive algorithm allow the control system to follow these beats provided they are not quicker than about two beats per second. Reductions in overall A-weighted sound pressure level of up to 7 dB(A) were measured with the active control system operating at all three harmonics (Elliott *et al.*, 1990).

An application where the frequency of the primary excitation, as well as its amplitude, constantly changes, is the control of engine-induced noise inside cars. A convenient reference signal is still available, however, from the ignition circuit of the engine for example, which tracks these excitation frequency changes. It is the engine firing frequency which dominates the internal noise level in a number of cars, particularly at higher engine speeds. For a four-cylinder engine the firing frequency varies from 20 to 200 Hz as the engine speed changes from 600 to 6000 rpm. The continuous variation of excitation frequency with engine speed makes it more difficult, in some ways, to understand the acoustic properties of the car than those of the aircraft. This complexity in the acoustics is compounded by the variability observed in the level of the engine-induced noise in the car from sample to sample off a production line, making a detailed prediction of the excitation in a single car not only difficult, but also of dubious value. This variability makes noise control using conventional passive methods very difficult, since the exact mechanism by which noise is transmitted into the vehicle from the engine changes from car to car, and on one car from time to time. The likelihood of low frequency, engine noise "boom" problems is increased by the current trend towards lighter car bodies and more powerful engines.

Apart from having to track the excitation frequency, however, the control task in the car is less intensive than in the aircraft case since the volume of the enclosure is smaller, and fewer loudspeakers and microphones need to be used. At the very lowest frequency, only a single loudspeaker and microphone would be needed, which could control a single dominant acoustic mode, as was pointed out by Oswald (1984). He experimented with an analogue control system with an adaptation time of about 2 seconds, which was considered too slow for general application. The use of adaptive digital filters gives a much improved convergence time and allows the use of multiple loudspeakers and microphones. An experimental version of such a system using two loudspeakers and four microphones with a convergence time of about a tenth of a second was reported by Elliott and Stothers (1986), and this system has

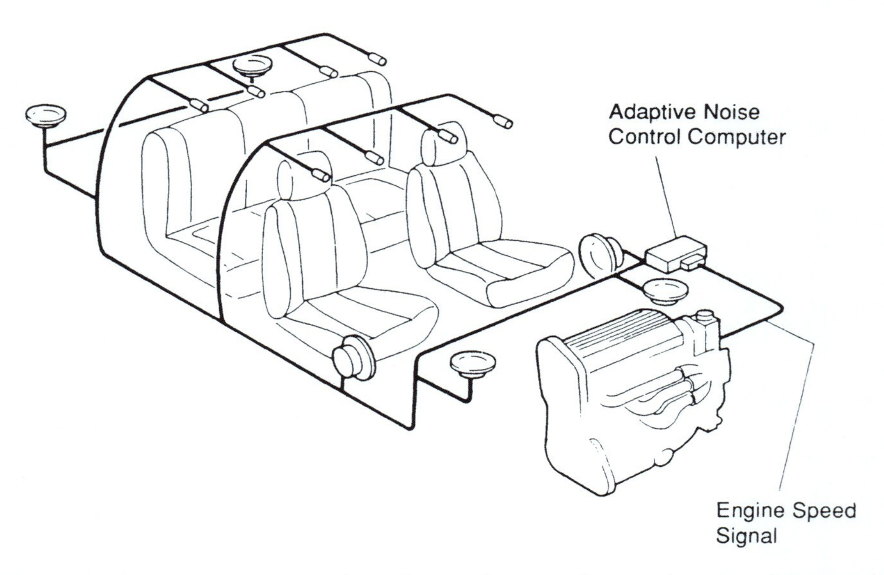

Fig. 10.28 A schematic representation of an active sound control system, integrated into a car, to control the low frequency engine noise "boom". The control system uses engine speed reference signal and drives up to six loudspeakers (which can be shared with the in-car entertainment system) to minimise the sum of the squared outputs from eight microphones.

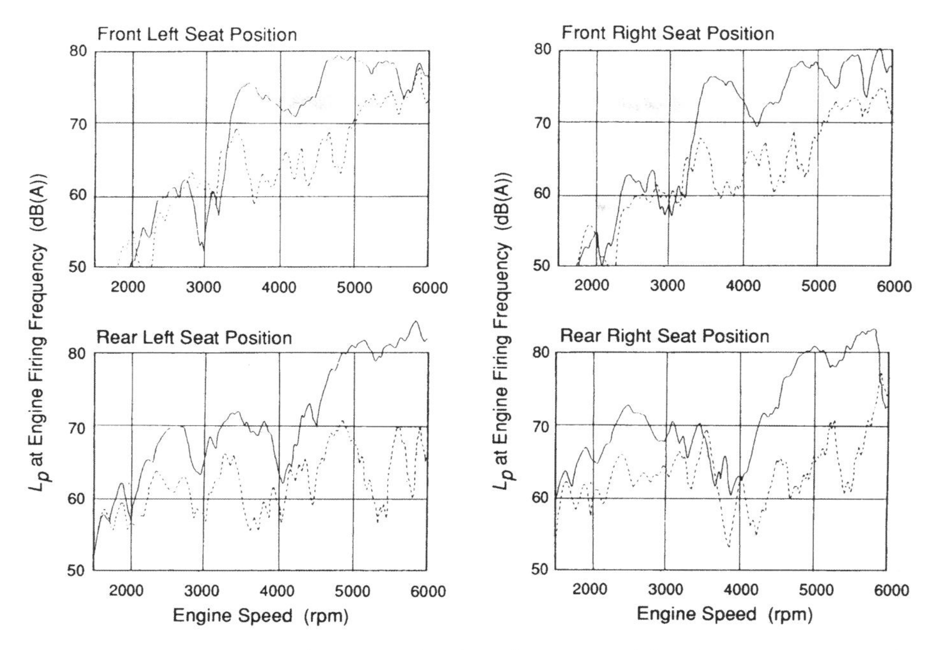

Fig. 10.29 The A-weighted sound pressure level at the engine firing frequency measured at head height in the four seat positions of a 1.1 litre small saloon car as it was accelerated hard in second gear with (——) and without (-----) a four loudspeaker, eight microphone active noise control system in operation.

subsequently been developed and expanded based on experience in over 30 different vehicles (McDonald *et al.*, 1988; Elliott *et al.*, 1988c). The current system is able to control six loudspeakers, which are shared with the in-car entertainment system, so as to minimise two harmonics of the engine firing frequency at eight microphone positions (Perry *et al.*, 1989). Figure 10.28 shows a diagrammatic representation of such an active control system in a car, and Fig. 10.29 shows the results obtained with a four loudspeaker, eight microphone system installed in a small saloon car, with a 1.1 litre, 4 cylinder engine. The four graphs show the A-weighted pressures, due to the engine firing frequency only, at head height in the four seat positions in the car; the horizontal axis is the engine speed. The car was accelerated under full load in second gear to produce the results shown, the acceleration from an engine speed of 2000 to 6000 rpm taking about 9 seconds. The reductions in sound pressure level at the firing frequency are about 10 dB in the front and rear seats above 3000–4000 rpm, with a significant reduction in the rear seats at lower engine speeds. These reductions in the sound at the firing frequency give improvements in *overall* A-weighted sound pressure level of 4–5 dB(A) at higher engine speeds in front and rear seats and 2–3 dB(A) at lower engine speeds in the rear. Similar reductions in the noise levels inside the car would be very difficult to achieve using conventional, passive, damping methods without a considerable increase in the body weight.

11
Local Control of Enclosed Sound Fields

11.1 Cancellation of a pure tone sound field at a point in an enclosure

In the previous chapter we examined the possibilities of using active control to reduce the total acoustic potential energy in the enclosure, and so achieve global control over the whole of the enclosed sound field. It was found that large global reductions can be achieved, even with secondary sources remotely placed from the primary source, provided the enclosure is excited at a frequency close to a lightly damped, isolated resonance. In this chapter we will consider the consequences of using a secondary source to control the sound at only one point in the enclosure, and so achieve local control.

If the sound field in the room is produced by a pure tone primary source of known characteristics, we can calculate the effects of active control using a frequency domain analysis. The problem we will initially consider is illustrated in Fig. 11.1. Point monopole primary and secondary sources of complex source strengths q_p and q_s are placed at positions $\mathbf{y}_p$ and $\mathbf{y}_s$ in the enclosure. A single error microphone is positioned at $\mathbf{x}_0$ in the enclosure. With linearity assumed, the complex pressure at this location is given by

$$p(\mathbf{x}_0) = Z(\mathbf{x}_0|\mathbf{y}_p)q_p + Z(\mathbf{x}_0|\mathbf{y}_s)q_s, \qquad (11.1.1)$$

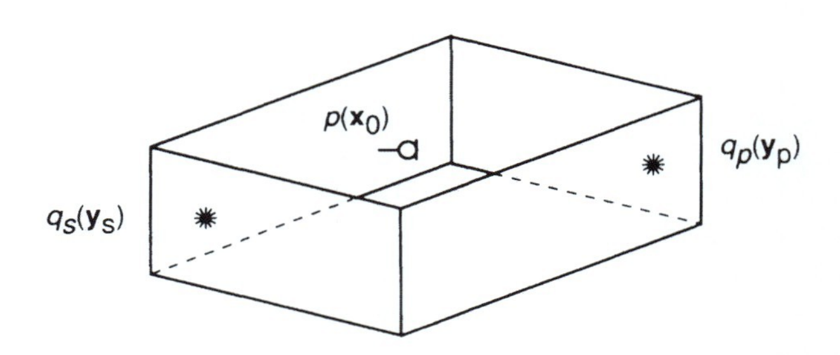

Fig. 11.1 Monopole primary and secondary sources at positions $\mathbf{y}_p$ and $\mathbf{y}_s$ in an enclosure. The secondary source is driven so that the pressure at $\mathbf{x}_0$ is zero in order to achieve local active control in the region of $\mathbf{x}_0$.

where we have again used $Z(\mathbf{x}|\mathbf{y})$ to denote the complex acoustic transfer impedance between points $\mathbf{x}$ and $\mathbf{y}$ in the enclosure. Since we are dealing with a single frequency, the secondary source can be driven with any desired amplitude and phase with respect to the primary source, without any causality restriction (see Section 8.6), and thus the pressure at $\mathbf{x}_0$ can be driven to zero by setting the secondary source strength to the value given by

$$q_{s0} = -\frac{Z(\mathbf{x}_0|\mathbf{y}_p)}{Z(\mathbf{x}_0|\mathbf{y}_s)} q_p. \tag{11.1.2}$$

It is the objective of the first part of this chapter to explore the consequences of this apparently simple expression.

By using the modal formulation introduced in the previous chapter, the optimal secondary source strength can alternatively be expressed as

$$q_{s0} = -\frac{\sum_{n=0}^{\infty} A_n \psi_n(\mathbf{x}_0)\psi_n(\mathbf{y}_p)}{\sum_{n=0}^{\infty} A_n \psi_n(\mathbf{x}_0)\psi_n(\mathbf{y}_s)} q_p, \tag{11.1.3}$$

where $\psi_n(\mathbf{x})$ and A_n are respectively the mode shape function and complex mode resonance terms defined in the last chapter (see equations (10.2.4) and (10.5.16)). If the frequency of excitation is close to the natural frequency of a lightly damped, isolated acoustic mode (the modal overlap is very small), then the modal summations in both the numerator and denominator of equation (11.1.3) will be dominated by the response of a single mode, with $n = d$, say, so that

$$q_{s0} \approx -\frac{\psi_d(\mathbf{y}_p)}{\psi_d(\mathbf{y}_s)} q_p. \tag{11.1.4}$$

Notice that this expression is not a function of the microphone position $\mathbf{x}_0$. Provided the value of the mode shape at the secondary source position, $\psi_d(\mathbf{y}_s)$, is not too small, this secondary source strength will achieve not only cancellation at the microphone location, $\mathbf{x}_0$, but also global control over the whole sound field. In fact, in the limit of the contribution from the "residual modes" excited by the secondary source tending to zero, equation (11.1.4) reduces to the source strength necessary to minimise the acoustic potential energy in the field in accordance with equation (10.5.5). A particular case of such control is when the excitation frequency is very low, so that only the (0,0,0) mode is significantly excited, in which case $q_{s0} = -q_p$.

We are more interested here in the case in which many modes are contributing to the pressure at $\mathbf{x}_0$ due to sources at $\mathbf{y}_p$ or $\mathbf{y}_s$ (i.e. the modal overlap is large and the sound field can be considered diffuse). We will assume for the time being that the secondary source is not only remote from the primary source but also some distance away from the error microphone. The fact that the secondary source is remote from the primary source, i.e. $|\mathbf{y}_p - \mathbf{y}_s| > \lambda$, implies that global control is not possible at frequencies where the modal overlap is high, as we have discussed in the previous chapter. The condition that the secondary source is remote from the error microphone, in particular it is assumed that the spacing of the error microphone from the secondary source is several times the "radius of reverberation" (Pierce,

1981), means that the microphone picks up very little of the *direct* field of the secondary source. The statistical behaviour of the pressure close to the cancellation microphone (the "zone of quiet") is determined by the statistical properties of the diffuse sound field, and a brief review of these properties will be given in the next section. In Sections 11.6 and 11.7 we look at the consequences of moving the secondary source close to the point of cancellation and in Section 11.8 we consider the consequences of adjusting a secondary source to optimally absorb acoustic power in a diffuse sound field.

11.2 Statistical properties of diffuse sound fields

In Chapter 1 we introduced the notion of a diffuse sound field (see Section 1.15). The model that we used consisted of an infinite number of randomly phased plane waves whose contribution to the local space-averaged time average energy density was equally shared among all possible directions of propagation. We also pointed out that this model had a certain similarity to the modal model of a sound field at high frequencies, where the many modes excited could each be thought of as contributing a number of plane waves to the total field. The model used in Chapter 1 enables us to derivation of the usual relationships between local space-averaged, time-averaged field variables such as squared pressure, energy density and ("one-sided") intensity. A further important property of a diffuse sound field that we will find useful is that the real and imaginary parts of the complex pressure in the sound field may each be considered to behave like Gaussian random variables for variations in spatial position. This follows from the assumption that the complex pressure at a position $\mathbf{x}$ (see equation (1.15.1)) is described by

$$p(\mathbf{x}) = \lim_{N \to \infty} \frac{1}{\sqrt{N}} \sum_{n=1}^{N} |p_n| e^{j\phi_n(\mathbf{x})}, \qquad (11.2.1)$$

where $|p_n|$ and $\phi_n(\mathbf{x})$ are the modulus and phase of the nth plane wave which are assumed to be statistically independent of one another. It is also assumed that these values for each n are statistically independent. Upon assuming that the values of $\phi_n(\mathbf{x})$ are uniformly distributed over the range $0 \to 2\pi$, it can be argued that the real and imaginary parts of the total pressure each have a Gaussian probability density function (see Chapter 2, Section 2.4). This follows from the central limit theorem in statistics (see, for example, Mood *et al.*, 1976) and is a result which applies when a random variable is itself the sum of a large number of independent random variables. The real and imaginary parts of the complex pressure have zero mean and are statistically independent, i.e.

$$<\mathrm{Re}\{p(\mathbf{x})\}> = <\mathrm{Im}\{p(\mathbf{x})\}> = 0, \; <\mathrm{Re}\{p(\mathbf{x})\}\mathrm{Im}\{p(\mathbf{x})\}> = 0, \qquad (11.2.2\mathrm{a,b})$$

where we have again used $< >$ to denote the operation of spatial averaging. It also follows that the real and imaginary parts of the pressure have equal spatial variances. (See Ebeling, 1984, for a comprehensive discussion on the statistical properties of pure tone diffuse fields.) Schroeder (1954) first demonstrated that such a treatment of reverberant sound fields becomes a statistically valid treatment of the physics above a frequency at which the modal overlap $M(\omega)$ (introduced in the last chapter) exceeds a given value. Schroeder and Kuttruff (1962) give this value as 3. It can then be argued (see the discussion presented by Pierce, 1981, Ch. 6) that this corresponds to a frequency that is predicted from the approximate formula

$$f_{\mathrm{Sch}} = 2000(T_{60}/V)^{1/2}, \tag{11.2.3}$$

where a sound speed of $340\,\mathrm{m\,s^{-1}}$ is assumed, T_{60} is the reverberation time (in seconds) and V is the volume (in m^3) of the enclosure considered. This frequency is thus often used to determine the onset of "diffuse" behaviour in an enclosed sound field.

Another important statistical property of pure tone diffuse sound fields is the spatial cross-correlation function. This is defined by

$$R(\Delta x) = \langle p^*(\mathbf{x}_1)p(\mathbf{x}_2)\rangle, \tag{11.2.4}$$

and since the statistical properties of the field are assumed to be stationary and ergodic with respect to position (see the discussion presented by Jacobsen, 1979), this correlation function depends only on the modulus of the separation distance between the two positions defined by $\Delta x = |\mathbf{x}_2 - \mathbf{x}_1|$. We can derive an expression for this function using the plane wave model introduced in Chapter 1. First note that for the nth plane wave, whose direction of propagation is at an angle θ_n to the line joining the points $\mathbf{x}_1$ and $\mathbf{x}_2$, the complex pressures at the points $\mathbf{x}_1$ and $\mathbf{x}_2$ are related by

$$p_n(\mathbf{x}_2) = p_n(\mathbf{x}_1)\mathrm{e}^{-jk\Delta x\cos\theta_n}. \tag{11.2.5}$$

The expression for the correlation function can thus be written as

$$R(\Delta x) = \left\langle \left[\lim_{N\to\infty} \frac{1}{\sqrt{N}} \sum_{n=1}^{N} p_n^* \right]\left[\lim_{N\to\infty} \frac{1}{\sqrt{N}} \sum_{n=1}^{N} p_n \mathrm{e}^{-jk\Delta x\cos\theta_n} \right]\right\rangle. \tag{11.2.6}$$

where $p_n = (p_n)\mathrm{e}^{j\phi_n(x)}$. We now undertake the space averaging operation and following exactly the reasoning presented in Chapter 1 (see Section 1.15) we assume that contributions to the products of terms in equation (11.2.6) average to zero for dissimilar propagation directions. This leads to the expression

$$R(\Delta x) = \lim_{N\to\infty} \frac{1}{N} \sum_{n=1}^{N} |p_n|^2 \mathrm{e}^{-jk\Delta x\cos\theta_n}. \tag{11.2.7}$$

We also again use the assumption that contributions to the summation of the terms $|p_n|^2$ are the same for all directions of propagation and write this expression as

$$R(\Delta x) = \langle|p(\mathbf{x})|^2\rangle \lim_{N\to\infty} \frac{1}{N} \sum_{n=1}^{N} \mathrm{e}^{-jk\Delta x\cos\theta_n}, \tag{11.2.8}$$

where $\langle|p(\mathbf{x})|^2\rangle$ is the local space-averaged modulus squared pressure. A more rigorous discussion of this step is given by Pierce (1981, Ch.6). In exactly the same way that we derived an expression for the acoustic intensity in a diffuse sound field (see Section 1.16, equations (1.16.5)–(1.16.8)) we can reduce the summation remaining in equation (11.2.8) to an integral. Thus we write, again using the notation of Chapter 1,

$$\lim_{N \to \infty} \frac{1}{N} \sum_{n=1}^{N} \mathrm{e}^{-jk\Delta x \cos \theta_n} = \frac{1}{4\pi} \int_0^{2\pi} \int_0^{\pi} \mathrm{e}^{-jk\Delta x \cos \theta} \sin \theta \mathrm{d}\theta \mathrm{d}\psi. \quad (11.2.9)$$

Evaluating the real part of the integral produces a result which is simply $[\sin(k\Delta x)]/k\Delta x$ whilst the imaginary part integrates to zero. The final expression for the spatial correlation function thus becomes

$$R(\Delta x) = <|p(\mathbf{x})|^2> \operatorname{sinc} k\Delta x. \quad (11.2.10)$$

This result for the diffuse field spatial correlation function was first derived by Cook *et al.* (1955) and an equivalent result may be derived by using the modal model of an enclosed sound field (see, for example, Morrow, 1971). Note that the function is real and thus

$$R(\Delta x) = <p^*(\mathbf{x}_1)p(\mathbf{x}_2)> = <p(\mathbf{x}_1)p^*(\mathbf{x}_2)>. \quad (11.2.11)$$

We will also find it useful to define a *normalised* spatial correlation function given by

$$\rho(\Delta x) = \frac{R(\Delta x)}{<|p(\mathbf{x})|^2>} = \operatorname{sinc} k\Delta x. \quad (11.2.12)$$

11.3 The space-averaged diffuse field transfer impedance

Now we will consider the relationship between the space-averaged squared pressure in a diffuse field and the strength of a point monopole source responsible for exciting the field. We will use as our starting point the modal model of an enclosed sound field that was introduced in the last chapter. It follows from equations (10.3.14) and (10.5.16) that the sound field due to a point source of strength q at a position $\mathbf{y}$ can be written as

$$p(\mathbf{x}) = \sum_{n=0}^{\infty} \frac{\rho_0 c_0^2}{V} A_n(\omega) \psi_n(\mathbf{x}) \psi_n(\mathbf{y}) q(\mathbf{y}). \quad (11.3.1)$$

Note that the damping of each mode is associated with a viscous damping ratio ζ_n as described in the last chapter (see Section 10.4). We can now calculate the space-averaged modulus squared pressure by multiplying equation (11.3.1) by its complex conjugate and noting that the orthonormality property of the mode shape functions (equation (10.2.7)) can be expressed as

$$<\psi_n(\mathbf{x})\psi_m(\mathbf{x})> = \begin{Bmatrix} 1, & n = m \\ 0, & n \neq m \end{Bmatrix}, \quad (11.3.2)$$

where the space-averaging operation has been interpreted as integration over (and division by) the entire volume V of the enclosure. Thus, if we average over both source *and* receiver positions, the expression for the space-averaged modulus squared pressure thus reduces simply to

$$<|p(\mathbf{x})|^2> = \frac{\rho_0^2\, c_0^4}{V^2}|q(\mathbf{y})|^2 \sum_{n=0}^{\infty} |A_n(\omega)|^2. \tag{11.3.3}$$

Another principle identified by Schroeder is that for excitation frequencies above the cut-off frequency defined by equation (11.2.3), the density of modal contributions is sufficient for a modal summation such as that appearing in equation (11.3.3) to be replaced by an integral. Thus, following the argument presented in Section 10.14, it can be shown that the summation of $|A_n(\omega)|^2$ over n takes the value $(\omega V/8\pi c_0^3 \zeta)$ where the damping ratio ζ has been assumed to be constant and to take a value much less than unity. The expression for the space-averaged modulus squared pressure (equation (11.3.4)) now reduces to

$$<|p(\mathbf{x})|^2> = \frac{\omega\, \rho_0^2\, c_0}{8\pi V\zeta}|q(\mathbf{y})|^2. \tag{11.3.4}$$

It is interesting to note that if we assume an average absorption coefficient in the enclosure that depends linearly upon frequency in accordance with equation (10.4.8) and if we use equation (10.4.15) for the modal damping ratio for oblique modes, this expression can be written as

$$<|p(\mathbf{x})|^2> = \frac{\omega^2 \rho_0^2}{\pi S\bar{\alpha}}|q(\mathbf{y})|^2. \tag{11.3.5}$$

We now use the result derived by Lyon (1969) that the average power output of a monopole source in a diffuse sound field is the same as that in the free field (equation (1.13.11)) so that

$$<W> = \frac{\omega^2 \rho_0}{8\pi c_0}|q(\mathbf{y})|^2 . \tag{11.3.6}$$

We can now express the space-averaged squared pressure as a function of the space-averaged power output of the source:

$$<|p(\mathbf{x})|^2> = \frac{8\rho_0\, c_0}{\bar{\alpha} S} <W>, \tag{11.3.7}$$

which is consistent with equation (1.6.13) which was derived using somewhat different assumptions.

Returning to equation (11.3.4), it follows that the space-averaged squared value of the transfer impedance relating the pressure at a remote point **x** to the strength of a source at **y** is given by

$$<|Z_r(\mathbf{x}|\mathbf{y})|^2> = \frac{\omega \rho_0^2 c_0}{8\pi V\zeta}, \tag{11.3.8}$$

where we have used $Z_r(\mathbf{x}|\mathbf{y})$ to denote the transfer impedance due to only the "reverberant field" at **x** produced by the source at **y**. Note that this does not therefore include the influence of the "direct field" radiated by the source that is not reflected by the enclosure walls. We can compare this value with the real part of the radiation impedance seen by a point source in a free field. We saw in Chapter 1 that the latter is given by $Z_0 = \omega^2 \rho_0 / 4\pi c_0$ and therefore we can write

$$\frac{<|Z_r(\mathbf{x}|\mathbf{y})|^2>}{Z_0^2} = \frac{2\,\pi\,c_0^3}{\omega^2 V\,\zeta}. \tag{11.3.9}$$

This ratio can be simply represented in terms of the modal overlap in an enclosure whose response is dominated by oblique modes. Thus using equation (10.13.4c) shows that

$$\frac{<|Z_r(\mathbf{x}|\mathbf{y})|^2>}{Z_0^2} = \frac{2}{\pi M_{\text{ob}}(\omega)} = \frac{2}{3\pi}\left(\frac{f_{\text{Sch}}}{f}\right)^3, \tag{11.3.10}$$

where we have also expressed the ratio in terms of the ratio of the excitation frequency to the Schroeder frequency (Section 10.14). The impedance ratio therefore becomes progressively smaller as the modal overlap $M_{\text{ob}}(\omega)$ in the enclosure increases as the excitation frequency is increased above the Schroeder frequency.

11.4 The diffuse field zone of quiet

In this section we will evaluate the pressure close to an error microphone when the output of the microphone has been driven to zero by a secondary source placed remotely from the microphone in a pure tone diffuse sound field. We first consider the complex pressure at position $\mathbf{x}_0 + \Delta\mathbf{x}$ (where $\mathbf{x}_0$ is again the error microphone position) to be decomposed into two spatially "orthogonal" components (Elliott *et al.*, 1988a) such that

$$p(\mathbf{x}_0 + \Delta\mathbf{x}) = p_c(\mathbf{x}_0 + \Delta\mathbf{x}) + p_u(\mathbf{x}_0 + \Delta\mathbf{x}). \tag{11.4.1}$$

The first of these components is perfectly correlated *spatially* with the pressure at $\mathbf{x}_0$, in other words regardless of the actual position of the microphone in the enclosure a linear relationship exists between $p_c(\mathbf{x}_0 + \Delta\mathbf{x})$ and $p(\mathbf{x}_0)$, with a complex constant of proportionality which depends only on $\Delta\mathbf{x}$. (We again assume that the statistical properties of the sound field are spatially stationary.) This complex constant can be written as $H(\Delta\mathbf{x})$ so that

$$p_c(\mathbf{x}_0 + \Delta\mathbf{x}) = H(\Delta\mathbf{x})\,p(\mathbf{x}_0). \tag{11.4.2}$$

The second component is perfectly uncorrelated with the pressure at $\mathbf{x}_0$, so that *on average spatially* the product of $p_u^*(\mathbf{x}_0 + \Delta\mathbf{x})$ and $p(\mathbf{x}_0)$ is zero. Thus we write

$$<p_u^*(\mathbf{x}_0 + \Delta\mathbf{x})p(\mathbf{x}_0)> = 0, \tag{11.4.3}$$

where $< >$ has again been used to denote the operation of space averaging. This decomposition into orthogonal components is standard practice for temporally correlated signals in signal processing (see, for example, Chapter 3, Section 3.8 or Papoulis, 1981), but its use does not appear to be widespread for spatially correlated components in acoustics. This approach to signal decomposition is usually applied to functions of time with the time-averaging operation used in place of the space-averaging operation used here. Thus we are replacing the notion of an ensemble average over various time histories (see Section 2.4) by an ensemble average over spatial position. We saw in Section 3.8 that a function of time $f(t + T)$ can be decomposed into a component that is uncorrelated with $f(t)$ and a component that is perfectly correlated with $f(t)$. The latter was shown to be related to $f(t)$ via the (temporal) autocorrelation function of the signal. In the case of spatially dependent functions used here, the spatial correlation function for the sound field plays a similar role. This is readily demonstrated by using the properties described by equations (11.4.1), (11.4.2) and (11.4.3) to show that

$$<p^*(\mathbf{x}_0)p(\mathbf{x}_0 + \Delta\mathbf{x})> = <|p(\mathbf{x})|^2> H(\Delta\mathbf{x}). \quad (11.4.4)$$

For a diffuse sound field, as demonstrated by equations (11.2.4) and (11.2.12), $H(\Delta\mathbf{x}) = \rho(\Delta x) = \text{sinc}\ k\Delta x$ is exactly the normalised spatial correlation function derived in Section 11.2.

We can now use this decomposition of the pressure field to establish the form of the sound field at $(\mathbf{x}_0 + \Delta\mathbf{x})$ when the pressure at $\mathbf{x}_0$ is driven to zero. Since $<p_c^*(\mathbf{x}_0 + \Delta\mathbf{x})p_u(\mathbf{x}_0 + \Delta\mathbf{x})> = 0$, the space average square pressure at $(\mathbf{x}_0 + \Delta\mathbf{x})$ can be expressed as

$$<|p(\mathbf{x}_0 + \Delta\mathbf{x})|^2> = <|p_c(\mathbf{x}_0 + \Delta\mathbf{x})|^2> + <|p_u(\mathbf{x}_0 + \Delta\mathbf{x})|^2>. \quad (11.4.5)$$

The space average squared modulus of the correlated and uncorrelated components of $p(\mathbf{x}_0 + \Delta\mathbf{x})$ can now be expressed using equation (11.4.2), the fact that $<|p(\mathbf{x}_0 + \Delta\mathbf{x})|^2> = <|p(\mathbf{x})|^2>$ and equation (11.4.5) as

$$<|p_c(\mathbf{x}_0 + \Delta\mathbf{x})|^2> = <|p(\mathbf{x})|^2>\ \text{sinc}^2 k\Delta x, \quad (11.4.6)$$

$$<|p_u(\mathbf{x}_0 + \Delta\mathbf{x})|^2> = <|p(\mathbf{x})|^2>\ [1 - \text{sinc}^2 k\Delta x]. \quad (11.4.7)$$

If $p(\mathbf{x}_0)$ is driven to zero through the action of the secondary source, then clearly $p_c(\mathbf{x}_0 + \Delta\mathbf{x})$ is also set to zero, according to equation (11.4.2), and the space-averaged squared pressure at $\mathbf{x}_0 + \Delta\mathbf{x}$ is due solely to the component $p_u(\mathbf{x}_0 + \Delta\mathbf{x})$. Thus, if $p(\mathbf{x}_0) = 0$, then

$$<|p(\mathbf{x}_0 + \Delta\mathbf{x})|^2> = <|p(\mathbf{x})|^2>\ [1 - \text{sinc}^2 k\Delta x], \quad (11.4.8)$$

in which $<|p(\mathbf{x})|^2>$ is the space-averaged squared modulus of the pressure in the enclosure *after control* (Elliott *et al.*, 1988a).

In general, if the spatial correlation properties of the sound field are such that

$$<p^*(\mathbf{x}_0)p(\mathbf{x}_0 + \Delta\mathbf{x})> = <|p(\mathbf{x})|^2>\ \rho(\Delta x), \quad (11.4.9)$$

where $\rho(\Delta x)$ is the normalised spatial correlation coefficient and if $p(\mathbf{x}_0)$ is actively driven to zero, then

$$<|p(\mathbf{x}_0 + \Delta\mathbf{x})|^2> = <|p(\mathbf{x})|^2>\ [1 - \rho^2(\Delta x)]. \quad (11.4.10)$$

These expressions define the *average* zone of quiet which will be found close to the error microphone in a sound field with high modal overlap, which is purely a function of the space-averaged square pressure in the enclosure after control, and the spatial correlation properties of the sound field.

Confirmation of this result from computer simulations has been presented by Elliott *et al.* (1988a). Experimental verification has been performed by Joseph (1990) in a $13.3\,\text{m}^3$ reverberant chamber with a reverberation time of about 0.4 s, giving a Schroeder frequency of about 330 Hz. The experiments were performed using a loudspeaker driven at 572 Hz ($\lambda \approx 0.6\,\text{m}$) with the primary source placed remotely, i.e. several wavelengths away, from both the control microphone and secondary source. Fifty sets of measurements were made with different primary source

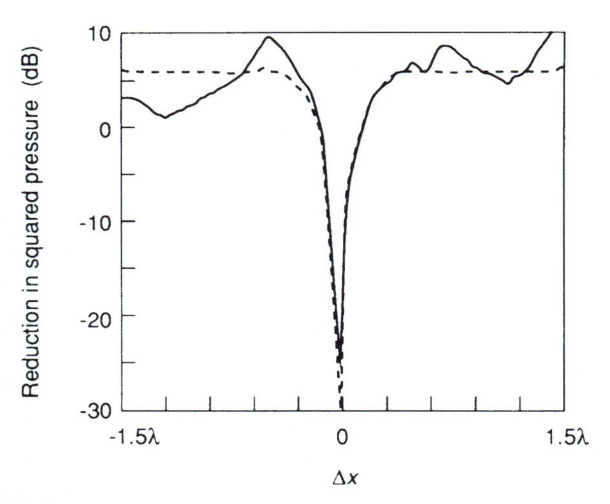

Fig. 11.2 The average squared pressure, compared to that in the primary field, evaluated from 50 pure tone measurements in a reverberation chamber at various distances, Δx, from a control microphone whose output is driven to zero by a remote secondary source (——). Also shown (-----) is the theoretical prediction for the zone of quiet which in this case is given by $3.8\ (1 - \text{sinc}^2 k\Delta x)$.

locations, and for these measurements it was found that the space-averaged squared pressure in the room with active control was some 3.8 times its value with only the primary field in operation. This increase in space-averaged squared pressure is worrying, and furthermore was found to change with a different set of source and microphone positions. A detailed discussion of this aspect of control is, however, postponed until the next section. The average results from the experiments on the zone of quiet, and the theoretical result in this case,

$$<|p(\mathbf{x}_0 + \Delta\mathbf{x})|^2> \approx 3.8 <|p_p(\mathbf{x})|^2>(1 - \text{sinc}^2\, k\Delta x), \qquad (11.4.11)$$

are plotted in Fig. 11.2. Note that $<|p_p(\mathbf{x})|^2>$ is the mean square pressure in the enclosure due to the primary source alone, which is plotted as 0 dB in this figure. The theoretical prediction of the zone of quiet and the experimental results agree well near to the cancellation point. These results imply that the zone of quiet, within which the pressure is at least 10 dB below that due to the primary source, is a sphere with a diameter of about one-tenth of a wavelength.

In an attempt to increase the extent of the zone of quiet, and to limit the increase in the space-averaged square pressure away from the point of control, some computer simulations were performed in which the secondary source was again randomly positioned, but now adjusted to *minimise* the sum of the squared outputs of *two* closely spaced microphones (Elliott *et al.*, 1988b; Joseph, 1990). The results are shown in Fig. 11.3 for various separations between the microphones (Δr), and distances from their centre point (Δx) along the axis of the two microphones. Although the zone of quiet defined by a 3 dB reduction over the primary field gets slightly broader as the microphone separation, Δr, is increased, the zone of quiet defined by a 10 dB reduction over the primary field gradually decreases in size and disappears at a microphone separation of about $\Delta r \approx 0.2\lambda$. The advantage of

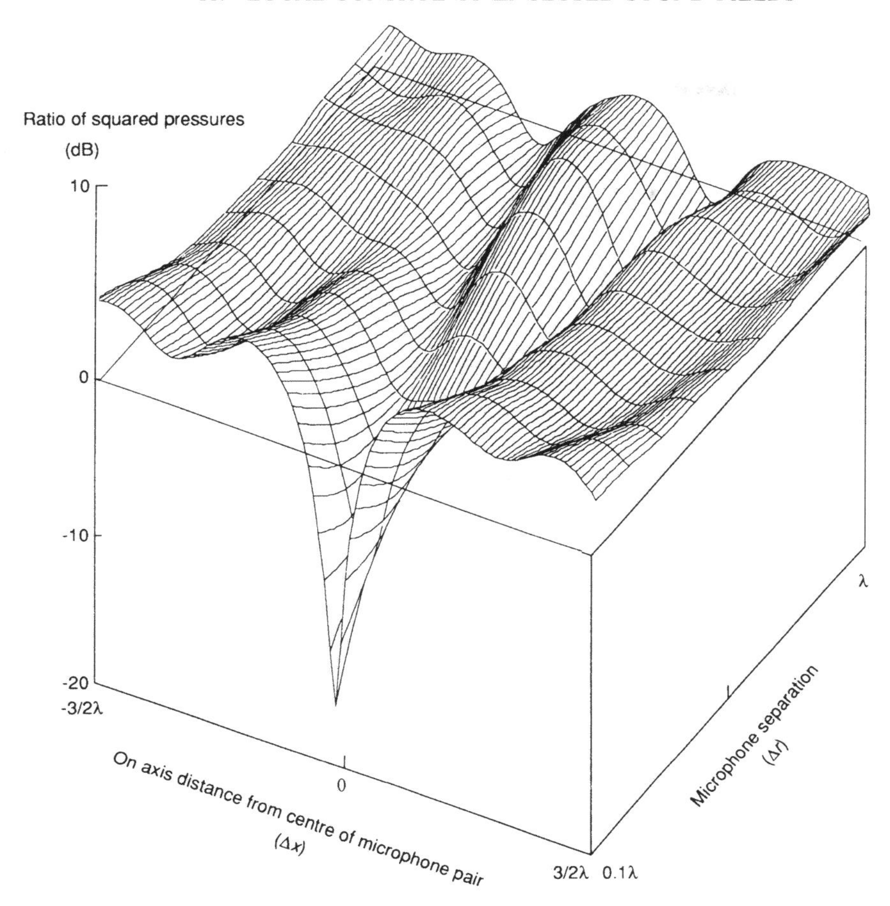

Fig. 11.3 The average squared pressure, divided by that of the primary field, at various distances (Δx) from the centre of a pair of control microphones, a distance Δr apart. The sum of the squared outputs of the microphones are minimised by a remote secondary source.

minimising the pressures at two microphones is the smaller increase in the space-averaged square pressure away from the cancellation point, compared to cancellation at one microphone, as we shall see in the next section.

It is interesting that the single zone of quiet splits into two zones, around the individual control microphones, for $\Delta r \gtrsim 0.5\lambda$. Each of these zones around the individual microphones shows a reduction of about 3 dB in pressure compared to the primary field, and the space-averaged square pressure well away from the control microphones is now increased by about 3 dB. A more detailed discussion of this behaviour is presented by Joseph (1990), who likens the zones of quiet around the two microphones as they are separated to the pulling apart of a soap bubble. The "surface tension" in the bubble can be thought of as being due to the spatial correlation in the sound field which tends to hold the quiet zones around the two microphones together. The forces due to this "surface tension" are just overcome when the microphones are about half a wavelength apart, so that two separate soap bubbles are created, and the zones of quiet around each of the two microphones become distinct.

11.5 The statistical behaviour of the secondary source strength

The volume of the zone of quiet near the control microphone is very small in comparison to the volume of the rest of the enclosure. The total acoustic potential energy in the enclosure is thus hardly affected by the behaviour near the control microphone, and, because the primary and secondary sources are well separated, its average value depends almost entirely on the square of the strength of these two sources. Thus, on average, we can write

$$<E_p> \approx \frac{\omega \rho_0}{32\pi c_0 \zeta} (|q_p|^2 + |q_s|^2), \tag{11.5.1}$$

where $< >$ indicates that E_p is evaluated from an average over a number of source positions and where the constant of proportionality has been deduced from the definition of E_p (equation (10.5.6)) and equation (11.3.5). If the total acoustic potential energy due to the primary source alone is denoted by E_{pp} the relative increase in E_p due to the action of the secondary source is therefore given by

$$\frac{<E_p>}{<E_{pp}>} \approx 1 + \left| \frac{q_s}{q_p} \right|^2. \tag{11.5.2}$$

The values of q_s and E_p will vary depending on the exact positioning of the primary and secondary source, and it is principally the spatial ensemble average of E_p/E_{pp} which concerns us, this again being denoted by the symbol $< >$. This leads to an examination of the statistical properties of $|q_s/q_p|^2$. If the secondary source at $\mathbf{y}_s$ is optimally adjusted to cancel the pressure at $\mathbf{x}_0$ due to the primary source at $\mathbf{y}_p$, then from equation (11.1.2) it follows that

$$\frac{q_{s0}}{q_p} = - \frac{Z(\mathbf{x}_0|\mathbf{y}_p)}{Z(\mathbf{x}_0|\mathbf{y}_s)}. \tag{11.5.3}$$

As pointed out in Section 11.2, in a fully diffuse field the real and imaginary components of the pressure are due to the random contributions from a large number of modes. The probability density functions of the real and imaginary components of the pressure, and hence of the transfer impedances above (assuming sources with an infinite internal impedance), then tend towards a Gaussian distribution, as a result of the central limit theorem (Schroeder, 1954). Each of these components is independent, so the modulus squared value of each of the impedances is due to the sum of the square of two independent Gaussian processes, and will therefore have a chi-squared distribution with two degrees of freedom (see, for example, Mood *et al.*, 1974 and Chapter 2 Section 2.2). The two impedances themselves will be due to independent processes since the separation between the primary and secondary sources is assumed to be much larger than a wavelength. The probability distribution function of the ratio given by

$$\left|\frac{q_{s0}}{q_p}\right|^2 = \frac{|Z(\mathbf{x}_0|\mathbf{y}_p)|^2}{|Z(\mathbf{x}_0|\mathbf{y}_s)|^2}, \tag{11.5.4}$$

where the probability density functions of $|Z(\mathbf{x}_0|\mathbf{y}_p)|^2$ and $|Z(\mathbf{x}_0|\mathbf{y}_s)|^2$ are independent chi-squared distributions, is defined by the $F(2,2)$ distribution (Mood *et al.*, 1974). This is described by the probability density function

$$p(v) = \frac{1}{(1+v)^2}, \tag{11.5.5}$$

where the variable $v = |q_{s0}/q_p|^2$ in this case.

Figure 11.4 shows the probability density of $|q_{s0}/q_p|^2$ observed in the computer simulations described by Elliott *et al.* (1988a), compared to the $F(2,2)$ distribution. The two distributions are seen to be in good agreement, implying that we have made valid assumptions of independent Gaussian distributions of the various pressure components. One unfortunate property of the $F(2,2)$ distribution is that its moments do not converge: i.e., it does not, for example, have a finite mean or variance, so that $<|q_{s0}|^2>$ is infinite. This explains the lack of repeatability in the computer simulations and experiments described above, when changes were made to the ensemble of source positions over which an average was taken. Physically, this ill-conditioning is due to the possibility that the secondary source may be very poorly coupled to the error microphone at some frequencies. This results in a very small

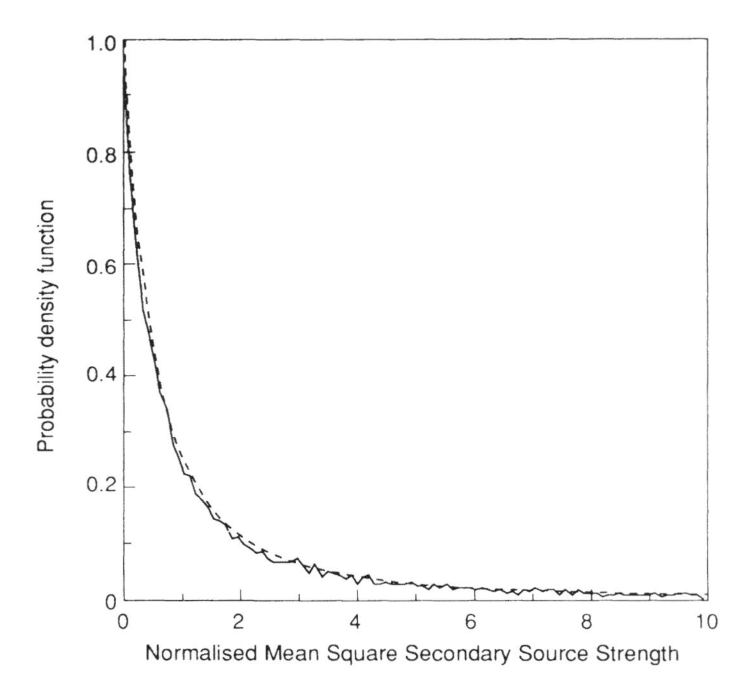

Fig. 11.4 Probability density function for the mean square secondary source strength divided by the mean square primary source strength for the computer simulations of active cancellation at a point in a diffuse sound field (——). The results are compared with the theoretically predicted $F(2,2)$ distribution (-----).

value of the transfer impedance $Z(\mathbf{x}_0|\mathbf{y}_s)$ and the required value of q_{s0} is very large.

Any practical secondary source will always, however, have an upper limit to its output amplitude. This "hard limiting" of the secondary source strength will truncate the probability density function given by equation (11.5.5) at some upper limiting amplitude, and the mean square source strength over the ensemble of source positions then becomes finite (Joseph, 1990). A number of other practical strategies for limiting the secondary source strength when cancelling at a remote microphone have been discussed by Elliott *et al.* (1988b) and Joseph (1990). These include "soft limiting" (where a cost function of the form $J = |p(x_0)|^2 + \beta|q_s|^2$ is minimised) and the minimisation of the sum of the squares of the outputs of two closely spaced error microphones (as mentioned in the previous section). Any of these strategies can limit the rise in $|q_s|^2$. The space-averaged value of $|q_{s0}/q_p|^2$, and hence $E_p \,/\, E_{pp}$, is still, however, considerably greater than unity if any significant zone of quiet is to be preserved using a secondary source placed remotely from the primary source.

One strategy which can give good cancellation at a point, with moderate secondary source strengths compared to the primary source, is to use *two* randomly positioned secondary sources to drive the pressure at *one* point in the enclosure to zero. It is interesting to note that Miyoshi and Kaneda (1988b) have used a similar system, with more secondary sources than error sensors, for the active control of broad-band noise in an enclosure to create zones of quiet around microphone arrays. The motivation for their approach, however, is the use of multiple channels in order to produce an exact solution to the digital filtering problem when dealing with broad-band noise. In general, such an arrangement would be under-determined at a single frequency (as discussed in the next chapter) with an infinite number of combinations of the two source strengths which give perfect control at a point. If, however, the additional constraint is introduced that the pressure at the cancellation point should be driven to zero with the sum of the two secondary source strengths being minimised, a well-conditioned solution for the optimal vector of secondary source strengths is found. This is discussed in detail in the appendix where it is shown that the optimal vector of the secondary source strengths is given by

$$\mathbf{q}_s = -\mathbf{Z}(\mathbf{Z}\mathbf{Z}^{\mathrm{H}})^{-1}p_p \qquad (11.5.6)$$

where in the case of two secondary sources and one microphone, $\mathbf{q}_s = (q_{s1}, q_{s2})^{\mathrm{T}}$ and $\mathbf{Z} = [Z(\mathbf{x}_0|\mathbf{y}_{s1}), Z(\mathbf{x}_0|\mathbf{y}_{s2})]$. Even this strategy, however, potentially increases the averaged squared pressure in the enclosure by a factor of three on average (Joseph, 1990), and a 5 dB increase in pressure level away from the cancellation point is still a high price to pay for a zone of quiet of diameter $\lambda/10$.

11.6 Cancellation in the near field of a secondary source

When a secondary source is used to cancel the pressure at a microphone in a diffuse field, and the secondary source is remotely located from the microphone ($|\mathbf{x}_0 - \mathbf{y}_s| > \lambda$), we have shown in the previous two sections that the zone of quiet around $\mathbf{x}_0$ is predictable only in a statistical sense, and depends on the spatial correlation properties of the sound field near $\mathbf{x}_0$. In addition, the secondary source strength required to cancel $p(\mathbf{x}_0)$ is predictable only in a statistical sense and it may

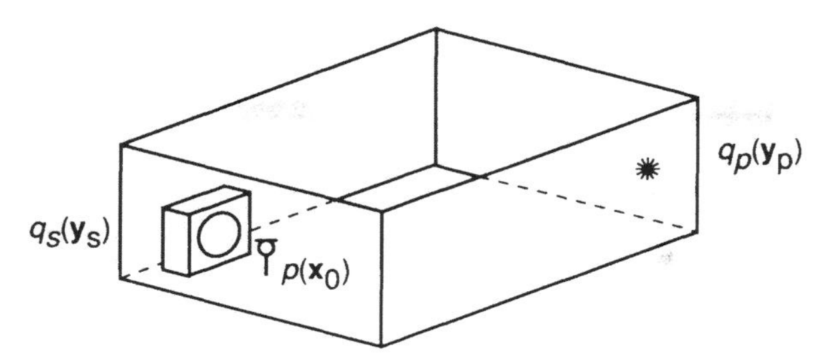

Fig. 11.5 Active control of the pressure, $p(\mathbf{x}_0)$, at the error microphone which is in the near field of a secondary source centred on position $\mathbf{y}_s$. The secondary source is of finite size and has a total source strength q_s.

take very large values, which can enormously increase the total acoustic potential energy in the enclosure. In this section we try to overcome some of these problems by locating the error microphone close to the secondary source, as illustrated in Fig. 11.5. We will find that the zone of quiet around $\mathbf{x}_0$ is now almost deterministically controlled by the near-field characteristics of the secondary source. Also the secondary source strength required to achieve cancellation is very much reduced, and largely independent of the statistical properties of the sound field.

It is interesting to note that minimising the pressure close to a finite size loudspeaker is one mode of operation of the "electronic sound reducer" described by Olson and May in 1953 (see also Chapter 7), although the control system in their case was feedback rather than feedforward. The other mode of operation of Olson and May's device as an "electronic sound absorber", i.e. controlling the loudspeaker to absorb power from the incident sound field, has been discussed for free field propagating waves in Chapter 9, and for enclosed sound fields with low modal densities in Chapter 10, Section 10.7. We will return to a discussion of this strategy for high frequency enclosed sound fields later in this chapter.

The low secondary source strength requirements of the "sound reducer" are reasonable intuitively if one considers the transfer impedance between the secondary source and error microphone to have two components: one due to the direct field of the secondary source $Z_d(\mathbf{x}_0|\mathbf{y}_s)$ and one due to reflected sound waves, i.e. the reverberant component $Z_r(\mathbf{x}_0|\mathbf{y}_s)$. This enables equation (11.5.3) to be written as

$$q_{s0} = -\frac{Z(\mathbf{x}_0|\mathbf{y}_p)q_p}{Z_d(\mathbf{x}_0|\mathbf{y}_s) + Z_r(\mathbf{x}_0|\mathbf{y}_s)}. \tag{11.6.1}$$

Large secondary source strengths could arise with a secondary source placed remotely from the microphone, since only the reverberant path would be present and $Z_r(\mathbf{x}_0|\mathbf{y}_s)$ could be very small at some frequencies and source positions. This lack of coupling between the secondary source and error sensor is overcome if the control microphone is close to the secondary source. The dominant, direct field contribution $Z_d(\mathbf{x}_0|\mathbf{y}_s)$ then ensures that the denominator in the expression for q_{s0} is large and so q_{s0} is, in general, relatively small compared with q_p.

In fact $Z_r(\mathbf{x}_0|\mathbf{y}_s)$ can be ignored compared to $Z_d(\mathbf{x}_0|\mathbf{y}_s)$, to a first approximation if the error microphone is sufficiently close to the secondary source, and the resulting expression for q_{s0} can be used to calculate the extent of the near-field zone of quiet.

If the primary pressure field at some small distance from $\mathbf{x}_0$ is $p_p(\mathbf{x}_0 + \Delta\mathbf{x})$, the *total* pressure at this position, with an optimally adjusted secondary source, will then be approximately

$$p(\mathbf{x}_0 + \Delta\mathbf{x}) \approx p_p(\mathbf{x}_0 + \Delta\mathbf{x}) + Z_d(\mathbf{x}_0 + \Delta\mathbf{x}|\mathbf{y}_s)q_{s0}, \tag{11.6.2}$$

so, using equation (11.6.1) with $Z_d(\mathbf{x}_0|\mathbf{y}_s) >> Z_r(\mathbf{x}_0|\mathbf{y}_s)$ together with the relationship $p_p(\mathbf{x}_0) = Z(\mathbf{x}_0|\mathbf{y}_p)q_p$ then shows that

$$p(\mathbf{x}_0 + \Delta\mathbf{x}) \approx p_p(\mathbf{x}_0 + \Delta\mathbf{x}) - \frac{Z_d(\mathbf{x}_0 + \Delta\mathbf{x}|\mathbf{y}_s)}{Z_d(\mathbf{x}_0|\mathbf{y}_s)} p_p(\mathbf{x}_0). \tag{11.6.3}$$

The ratio of the space-averaged squared value of this pressure to the space-averaged squared pressure in the primary field is thus given by

$$\frac{<|p(\mathbf{x}_0 + \Delta\mathbf{x})|^2>}{<|p_p(\mathbf{x}_0)|^2>} \approx 1 - 2\text{Re}\left\{\frac{Z_d(\mathbf{x}_0 + \Delta\mathbf{x}|\mathbf{y}_s)}{Z_d(\mathbf{x}_0|\mathbf{y}_s)}\right\}\rho(\Delta\mathbf{x}) + \left|\frac{Z_d(\mathbf{x}_0 + \Delta\mathbf{x}|\mathbf{y}_s)}{Z_d(\mathbf{x}_0|\mathbf{y}_s)}\right|^2, \tag{11.6.4}$$

where $\rho(\Delta\mathbf{x})$ is the normalised spatial correlation function of the primary field. If we consider only small distances from the point of control such that $k|\Delta\mathbf{x}| < 1$, then $\rho(\Delta\mathbf{x})$ can be taken as being unity, with only a few per cent error. The expression for the zone of quiet under these conditions then depends only on the direct field properties of the secondary source.

The direct field impedance at $\mathbf{x}_0 + \Delta\mathbf{x}$ is now approximated by the first two terms in its Taylor series expansion, about $\mathbf{x}_0$. Thus using the series expansion described in Section 8.7 shows that we can write

$$Z_d(\mathbf{x}_0 + \Delta\mathbf{x}|\mathbf{y}_s) \approx Z_d(\mathbf{x}_0|\mathbf{y}_s) + \nabla Z_d(\mathbf{x}_0|\mathbf{y}_s).\,\Delta\mathbf{x}, \tag{11.6.5}$$

where the second term in the series expansion is given by

$$\nabla Z_d(\mathbf{x}_0|\mathbf{y}_s).\Delta\mathbf{x} = \frac{\partial Z_d(\mathbf{x}_0|\mathbf{y}_s)}{\partial x_1}\Delta x_1 + \frac{\partial Z_d(\mathbf{x}_0|\mathbf{y}_s)}{\partial x_2}\Delta x_2 + \frac{\partial Z_d(\mathbf{x}_0|\mathbf{y}_s)}{\partial x_3}\Delta x_3. \tag{11.6.6}$$

This expression can be substituted into that for the space-averaged square pressure at $\mathbf{x}_0 + \Delta\mathbf{x}$, with $\rho(\Delta\mathbf{x}) = 1$, which gives (Joseph *et al.*, 1989; Joseph, 1990)

$$\frac{<|p(\mathbf{x}_0 + \Delta\mathbf{x})|^2>}{<|p_p(\mathbf{x}_0)|^2>} \approx \left|\frac{\nabla Z_d(\mathbf{x}_0|\mathbf{y}_s).\,\Delta\mathbf{x}}{Z_d(\mathbf{x}_0|\mathbf{y}_s)}\right|^2. \tag{11.6.7}$$

This remarkably simple result shows that the near field zone of quiet is, to first order, determined by the ratio of the gradient to the absolute value of the pressure in the near field of the secondary source. The important assumptions in the derivation are that, first, the pressure field due to the secondary source is dominated by the direct field contribution (the microphone is placed within the reverberation radius of the secondary source; Pierce, 1981), and second, that the primary field is reasonably uniform between the secondary source and control microphone. In fact, equation (11.6.7) can be deduced directly from equations (11.6.3) and (11.6.5) if it is assumed that $<p_p(\mathbf{x}_0 + \Delta\mathbf{x})> = <p_p(\mathbf{x}_0)>$. Note also that

$$\left| \frac{\nabla Z_d(\mathbf{x}_0|\mathbf{y}_s).\, \Delta\mathbf{x}}{Z_d(\mathbf{x}_0|\mathbf{y}_s)} \right|^2 = \left| \frac{\nabla p_d(\mathbf{y}_s).\Delta\mathbf{x}}{p_d(\mathbf{y}_s)} \right|^2, \tag{11.6.8}$$

where p_d is the pressure at $\mathbf{y}_s$ due to the secondary source at $\mathbf{x}_0$. Using the momentum equation for this harmonic pressure we can express ∇p_d as $-j\omega\rho_0\, \mathbf{u}_d(\mathbf{y}_s)$ where $\mathbf{u}_d(\mathbf{y}_s)$ is the acoustic particle velocity due to the secondary source at $\mathbf{y}_s$. Thus we can express the ratio of the pressures as

$$\frac{<|p(\mathbf{x}_0 + \Delta x)|^2>}{<|p_p(\mathbf{x}_0)|^2>} = \omega^2\rho_0^2 \left| \frac{\mathbf{u}_d(\mathbf{y}_s).\Delta\mathbf{x}}{p_d(\mathbf{y}_s)} \right|^2. \tag{11.6.9}$$

Assuming that the particle velocity is evaluated in the same direction as $\Delta\mathbf{x}$, this can be expressed as (Joseph *et al.*, 1991)

$$\frac{<|p(\mathbf{x}_0 + \Delta\mathbf{x})|^2>}{<|p_p(\mathbf{x}_0)|^2>} = \left| \frac{\rho_0\, c_0}{z_d(\mathbf{y}_s)} \right|^2 k^2 \Delta x^2, \tag{11.6.10}$$

where $z_d(\mathbf{y}_s)$ is the specific acoustic impedance of the pressure field due to the secondary source (equation (1.7.7)).

As a simple example of the use of this formula, consider a point monopole secondary source placed at the origin of the co-ordinate system, and driven in order to drive the pressure to zero at a radial distance r_0 away. The specific acoustic impedance at the cancellation point for such a source is given from equation (1.11.11) as

$$z(r_0) = \rho_0\, c_0 \left(\frac{jkr_0}{1 + jkr_0} \right), \tag{11.6.11}$$

where $r_0 = |\mathbf{x}_0 - \mathbf{y}_s|$. Hence equation (11.6.10) reduces to

$$\frac{<|p(r_0 + \Delta r)|^2>}{<|p_p(r_0)|^2>} \approx (1 + k^2 r_0^2) \left(\frac{\Delta r}{r_0} \right)^2. \tag{11.6.12}$$

If we assume that $(kr_0)^2 << 1$, the zone of quiet within which the space averaged square pressure is reduced by a factor ε is defined by

$$\frac{<|p(r_0 + \Delta r)|^2>}{<|p_p(r_0)|^2>} \approx \left(\frac{\Delta r}{r_0} \right)^2 < \varepsilon. \tag{11.6.13}$$

This zone of quiet is thus bounded by a radial distance from the monopole source given by twice the value resulting from equation (11.6.13) and thus

$$\Delta r_\varepsilon = 2r_0\sqrt{\varepsilon}. \tag{11.6.14}$$

If a 10 dB reduction is required, the extent of the zone of quiet is thus $\Delta r_{0.1} \approx 0.6\, r_0$. In particular, if the pressure very close to the monopole secondary source is driven to zero, i.e. $r_0 \to 0$, there will be a vanishingly small zone of quiet and the effect of the point monopole will be to generate an infinitesimally small "pin-prick" in the sound field, with zero pressure very close to the secondary source but with the primary pressure being unaffected by the secondary source at very small distances away from it.

If the point of cancellation is some way from the monopole secondary source, specifically if $r_0 >> 0.2\lambda$, so that $k^2 r_0^2 >> 1$, then the zone within which the space-averaged square pressure is reduced by a factor ε is defined by

$$\frac{<|p(r_0 + \Delta r)|^2>}{<|p_p(r_0)|^2>} \approx k^2 \Delta r^2 < \varepsilon. \quad (11.6.15)$$

This zone of quiet is bounded by a radial distance given by

$$\Delta r_\varepsilon = \frac{2\sqrt{\varepsilon}}{k} = \frac{\sqrt{\varepsilon}}{\pi}\lambda. \quad (11.6.16)$$

If a 10 dB reduction is required, the zone of quiet is thus $\Delta r_{0.1} \approx \lambda/10$ which is similar to that deduced in Section 11.4 under somewhat different assumptions.

11.7 Experiments on the zone of quiet in the near field of a secondary source

Joseph (1990) has investigated the near field zone of quiet experimentally, and compared it to the theoretical prediction derived above. The error microphone was placed at various positions on the axis of a series of loudspeakers used as secondary sources, of effective radii 10 mm, 55 mm and 110 mm. One set of results, when the error microphone was placed one-tenth of a wavelength from the centre of the loudspeaker at a frequency of 572 Hz, is shown in Fig. 11.6. The dashed lines were computed from equation (11.6.7) using the well-known expressions for the pressure on the axis of a circular piston in an infinite baffle (see, for example, Kinsler *et al.*, 1982). The agreement between the experiments and theory in the zone of quiet is generally good. The same theoretical model has been used to investigate the optimal

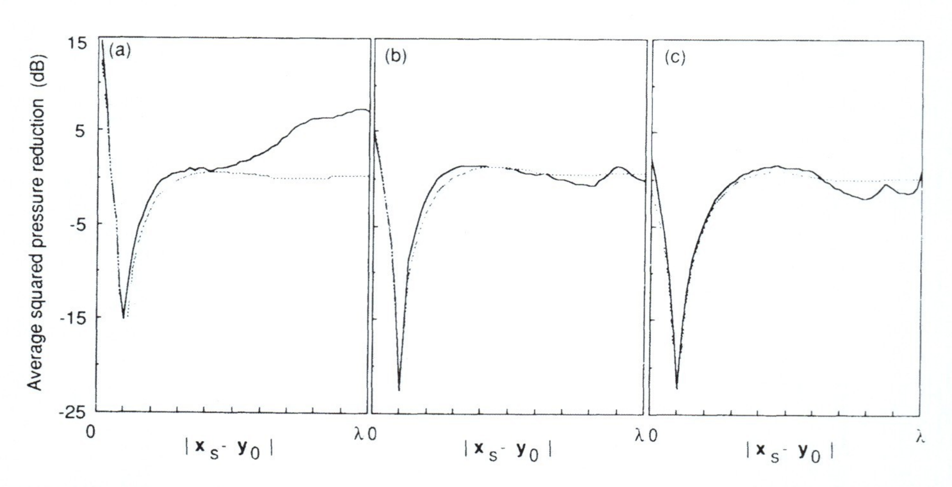

Fig. 11.6 The squared pressure reductions at various distances from the secondary source, from the average of 10 measurements, around the point of cancellation at an "on axis" point one-tenth of a wavelength from the cone of a secondary loudspeaker of radius (a) 10 mm, (b) 55 mm, (c) 110 mm (——). Also shown (-----) are the theoretical predictions of the pressure distributions in this case.

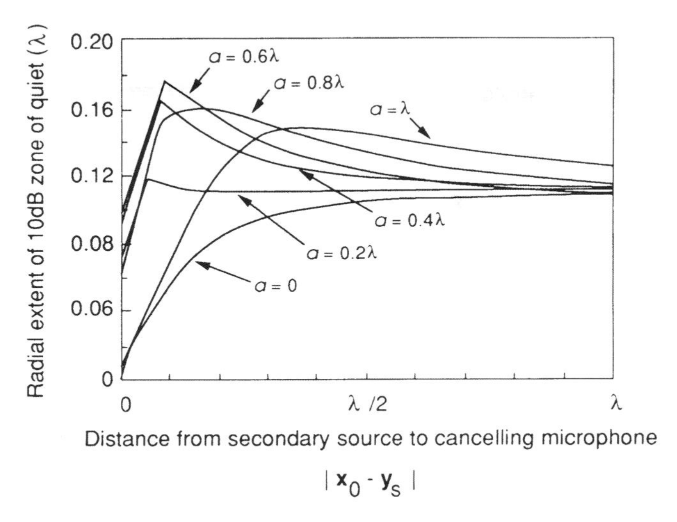

Fig. 11.7 The theoretical expectation for the near field zone of quiet in the region of a piston-like secondary source. Results are shown as a function of the distance of the point of cancellation from the source and the piston radius a, normalised by the acoustic wavelength.

size of a piston-like secondary source, and the optimal position of the error microphone along the axis from this source, in order to give the greatest zone of quiet. The size of the quiet zone is again defined when the pressure is reduced by 10 dB compared to the original sound field. One set of such theoretical predictions (Joseph, 1990) is illustrated in Fig. 11.7 for various secondary source radii (a) from $a = 0$ (a point monopole source) to $a = \lambda$, and various secondary source, error microphone separations of up to $|\mathbf{x}_0 - \mathbf{y}_s| = \lambda$. It is interesting to note that for all secondary source sizes the zone of quiet becomes about $\lambda/10$ if the error microphone is further than about $\lambda/2$ from the source, as predicted above. If the microphone is placed more closely, however, a zone of quiet of diameter of about 0.18λ can be obtained by using a piston source with diameter of about a wavelength ($a \approx 0.5\lambda$) and an error microphone placed about 0.1λ away, on axis. Such a secondary source is probably inconveniently large in practice (the diameter would be approximately 0.7 m at 500 Hz), but the chart shown in Fig. 11.7 can provide a useful guide to optimal error microphone placement and expected zones of quiet in cases of more practical interest.

In Fig. 11.7 all dimensions are normalised by the acoustic wavelength. In the practical case, the piston radius and cancellation position will be fixed and the zone of quiet is difficult to predict from Fig. 11.7 over a *range* of frequencies, because ka and kr_0 vary. An example of the 10 dB zone of quiet when the pressure is cancelled close to a 0.1 m diameter piston over a range of frequencies is shown in Fig. 11.8 (David, 1991). When the error microphone is placed close to the secondary source ($r_0 = 10$ mm) the size of the zone of quiet is almost independent of frequency up to 1 kHz, although it is impractically small. As the error microphone is moved away, the zone of quiet becomes larger at low frequencies, but the size decreases as the frequency rises, until at higher frequencies it approaches the curve defined by the zone being one-tenth of a wavelength.

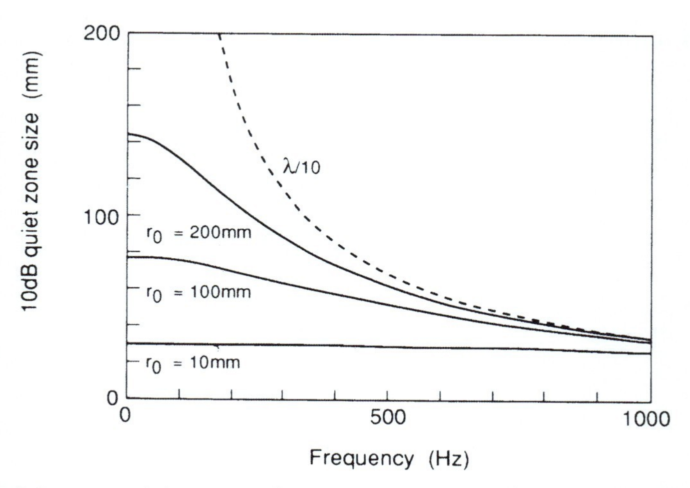

Fig. 11.8 The axial extent of the zone of quiet, within which the pressure has been reduced by 10 dB, when a secondary piston source of diameter 0.1 m is used to cancel the primary field at an axial distance r_0 from the piston centre.

11.8 The power output of a monopole source used to cancel the pressure in its near field

In this section we calculate the power output of a monopole source which is excited in order to drive the total pressure to zero at some point in its near field. The monopole secondary source is again assumed to be at the origin of the co-ordinate system and to generate a negligible reverberant field compared with the primary field (as above), so that from equation 11.6.2 and using the results of Section 1.12 we can write

$$p(\mathbf{x}) = p_p(\mathbf{x}) + jZ_0\, q_s \frac{e^{-jkr}}{kr}, \tag{11.8.1}$$

where $Z_0 = \omega^2\rho_0/4\pi c_0$ and $r = |\mathbf{x}|$. If $p(\mathbf{x}_0)$ is driven to zero, then the optimal secondary source strength q_{s0} is given by $[j\,kr_0\,e^{jkr_0}p_p(\mathbf{x}_0)/Z_0]$, where $r_0 = |\mathbf{x}_0|$. Thus the expression for the pressure at $\mathbf{x}$ becomes

$$p(\mathbf{x}) = p_p(\mathbf{x}) - \frac{r_0}{r}\, e^{-jk(r - r_0)} p_p(\mathbf{x}_0). \tag{11.8.2}$$

The power output of the monopole source can now be calculated by using the technique described in Section 9.11. Thus

$$W_s = \lim_{r \to 0} \frac{1}{2}\, \mathrm{Re}\{p(\mathbf{x})q_{s0}^*\}. \tag{11.8.3}$$

Substituting the expressions for $p(\mathbf{x})$ and q_{s0} gives

$$W_s = |p_p(\mathbf{x}_0)|^2\, \frac{k^2r_0^2}{2Z_0} - \frac{kr_0}{2Z_0}\mathrm{Re}\{\,j p_p^*(\mathbf{x}_0)p_p(\mathbf{x})e^{-jkr_0}\}. \tag{11.8.4}$$

Taking the spatial average of this equation and using the spatial correlation properties of a three-dimensional diffuse field (see Section 11.2) then shows that the space-averaged power output from the secondary source is given by

$$<W_s> = \frac{<|p_p(\mathbf{x})|^2>}{2Z_0}(k^2r_0^2 - \sin^2 kr_0). \qquad (11.8.5)$$

If the primary source were a monopole of source strength q_p, its space-averaged power output in the enclosure would be equal to the free field power output (Maling, 1967) and therefore we can write

$$<W_p> = \frac{1}{2} Z_0|q_p|^2. \qquad (11.8.6)$$

The spaced-averaged squared pressure in the enclosure due to this monopole primary source which is placed in a position remote from the secondary can also be written as

$$<|p_p(\mathbf{x})|^2> = <|Z_r(\mathbf{x}|\mathbf{y}_p)|^2>|q_p|^2, \qquad (11.8.7)$$

and thus the ratio of the average power output of the secondary source to that of the primary source is given by

$$\frac{<W_s>}{<W_p>} = \frac{<|Z_r(\mathbf{x}|\mathbf{y}_p)|^2>}{Z_0^2}(k^2r_0{}^2 - \sin^2 kr_0). \qquad (11.8.8)$$

If $k^2r_0^2 < 1$ and the series expansion for $\sin kr_0$ is substituted in the above expression, it becomes clear that the power output from the secondary source, compared to that from the primary source, varies as $k^4r_0^4$. The constant of proportionality is one-third of the ratio of the mean square diffuse field impedance to the square of the free field impedance. This quantity was shown in Section 11.3 to be inversely proportional to the modal overlap of the enclosure, $M(\omega)$ (see equation (11.3.10)). Thus for $(kr_0)^2 < 1$, we can write (Joseph, 1990)

$$\frac{<W_s>}{<W_p>} \approx \frac{2}{\pi M(\omega)}\left(\frac{k^4r_0^4}{3}\right) \approx 110\left(\frac{f_{\mathrm{Sch}}}{f}\right)^3\left(\frac{r_0}{\lambda}\right)^4. \qquad (11.8.9)$$

Provided the field is diffuse ($f > f_{\mathrm{Sch}}$) and the distance from the point of cancellation to the secondary source is small compared with a wavelength, $r_0 << \lambda$, then the power output of a secondary source adjusted to cancel the pressure in the near field will be small compared to the power output of the primary source in the enclosure: that is

$$<W_s> << <W_p>. \qquad (11.8.10)$$

For example, if the pressure is cancelled at $r_0 = 0.1\lambda$, at an excitation frequency equal to the Schroeder frequency, then $<W_s> \approx 0.01 <W_p>$. The total acoustic potential energy in the enclosure is proportional to the sum of W_p and W_s. If W_s is small, the act of creating a near field zone of quiet is seen not to unduly affect the pressure in the rest of the enclosure. For the example given above, the increase in E_p is only about 0.05 dB.

11.9 The Olson "sound absorber" in a diffuse sound field

One mode of operation of the "electronic sound absorber" described by Olson and May (1953) was to absorb sound power, and thus provide additional acoustic damping, when placed in an enclosure. The theory for a monopole acoustic source used as an optimal active absorber has already been developed and applied to the free field case in Chapter 9. Here we will outline briefly its effect in an enclosed sound field with high modal overlap, both in terms of the effective absorption and the local changes to the sound field. In Chapter 9 it was shown that the power absorption of a monopole secondary source at $\mathbf{y}_s$ is maximised if

$$q_s = q_{s0} = \frac{-p_p(\mathbf{y}_s)}{2\mathrm{Re}\{Z(\mathbf{y}_s|\mathbf{y}_s)\}}, \tag{11.9.1}$$

where $\mathrm{Re}\{Z(\mathbf{y}_s|\mathbf{y}_s)\}$ is the real part of the acoustic radiation impedance experienced by the secondary source. For a monopole source in a reverberant enclosure we can assume that

$$\mathrm{Re}\{Z(\mathbf{y}_s|\mathbf{y}_s)\} = Z_0 + \mathrm{Re}\{Z_r(\mathbf{y}_s|\mathbf{y}_s)\}, \tag{11.9.2}$$

where Z_0 is the real part of the impedance due to the direct field and $Z_r(\mathbf{y}_s|\mathbf{y}_s)$ is the impedance due to the reverberant part of the field. It was also shown in Section 11.3 that this reverberant impedance is small compared to Z_0 if the frequency is larger than the Schroeder frequency, i.e. the modal overlap is reasonably high. Under these conditions the reverberant component can be ignored in equation (11.9.1) and the result reduces to the equivalent free field result given by equation (9.9.3a). Thus

$$q_{s0} \approx \frac{-p_p(\mathbf{y}_s)}{2Z_0}. \tag{11.9.3}$$

By using the results of Section 9.9, the power absorbed by the secondary source under these conditions can also be written as

$$<W_0> = \frac{-<|p_p(\mathbf{y}_s)|^2>}{8\mathrm{Re}\{Z(\mathbf{y}_s|\mathbf{y}_s)\}}. \tag{11.9.4}$$

If it is again assumed that the reverberant component of the impedance can be assumed small, the expectation of the power absorption over the secondary source position is given by

$$<W_0> \approx \frac{-<|p_p(\mathbf{x})|^2>}{8Z_0}, \tag{11.9.5}$$

an average result similar to equation (9.9.3b), in which $<|p_p(\mathbf{x})|^2>$ is now the space-averaged squared modulus of the pressure in the enclosure due to the primary source.

The steady-state power balance in a diffuse sound field (equation (1.16.12)) is given by

$$W = IS\bar{\alpha}, \tag{11.9.6}$$

where W is the total power injected into the field, $S\bar{\alpha}$ is the net absorption area, and $I = <|p_p(\mathbf{x})|^2>/8\rho_0c_0$ is the "one-sided" acoustic intensity. The total power injected into the field is the sum of that due to the primary source W_p and that due to the remotely placed but optimally adjusted secondary source, W_0. The latter is negative in this case. The power output of the primary source is on average unaffected by the presence of the secondary source since they are assumed to be well separated. If the average value of the secondary source power output is taken over to the right side of equation (11.9.6), then we can write

$$<W_p> = \frac{<|p_p(\mathbf{x})|^2>}{8\rho_0 c_0} S\bar{\alpha} - <W_0>, \tag{11.9.7}$$

and using equation (11.9.5) then shows that

$$<W_p> = \frac{<|p_p(\mathbf{x})|^2>}{8\rho_0 c_0}\left(S\bar{\alpha} + \frac{\rho_0 c_0}{Z_0} \right). \tag{11.9.8}$$

Since $Z_0 = \omega^2 \rho_0 / 4\pi c_0$ the extra absorption term due to the optimally adjusted secondary source has an equivalent absorbing area of

$$\frac{\rho_0 c_0}{Z_0} = \frac{4\pi}{k^2} = \frac{\lambda^2}{\pi}, \tag{11.9.9}$$

which can be thought of as the surface of a sphere of radius equal to the reciprocal of the wave number (Joseph, 1990). Although the equivalent absorbing area of an optimally adjusted monopole source is rather small at high frequencies, being about $0.04\,\text{m}^2$ at 1 kHz, it gets larger at low frequencies, where conventional, passive, absorptive materials do not generally work well. At 50 Hz, for example, the equivalent absorbing area calculated using equation (11.9.9) is about $15\,\text{m}^2$. It should be emphasised, however, that this is an average result for a diffuse sound field, which may well not be valid in smaller enclosures at 50 Hz. A practical arrangement for measuring the acoustic power output of a loudspeaker, and its use in preliminary experiments on active power absorption in an enclosure, have been presented by Anthony and Elliott (1991).

The average mean square pressure close to such an optimally absorbing monopole may be calculated by considering the total pressure at some position $\Delta\mathbf{y}$ away from the position of the absorbing source ($\mathbf{y}_s$), and assuming again that the direct field contribution to the pressure at this point is much larger than the reverberant contribution, so that equation (11.7.1) becomes

$$p(\mathbf{y}_s + \Delta\mathbf{y}) = p_p(\mathbf{y}_s + \Delta\mathbf{y}) + jZ_0\, q_s \frac{\mathrm{e}^{-jkr}}{kr}, \tag{11.9.10}$$

where $r = |\Delta\mathbf{y}|$. If q_s is given by the optimal secondary source strength specified by equation (11.9.3), then this expression reduces to

$$p(\mathbf{y}_s + \Delta\mathbf{y}) = p_p(\mathbf{y}_s + \Delta\mathbf{y}) - j\,\frac{1}{2kr}\, p_p(\mathbf{y}_s)\, \mathrm{e}^{-jkr}. \tag{11.9.11}$$

We can now evaluate the squared modulus of the pressure. This is given by

$$|p(\mathbf{y}_s + \Delta\mathbf{y})|^2 = |p_p(\mathbf{y}_s + \Delta\mathbf{y})|^2 + \frac{1}{4k^2r^2}|p_p(\mathbf{y}_s)|^2$$
$$- \frac{j}{2kr}\left[\, p_p(\mathbf{y}_s)p_p^*(\mathbf{y}_s + \Delta\mathbf{y})\mathrm{e}^{-jkr} - p_p^*(\mathbf{y}_s)p_p(\mathbf{y}_s + \Delta\mathbf{y})\mathrm{e}^{jkr}\,\right]. \tag{11.9.12}$$

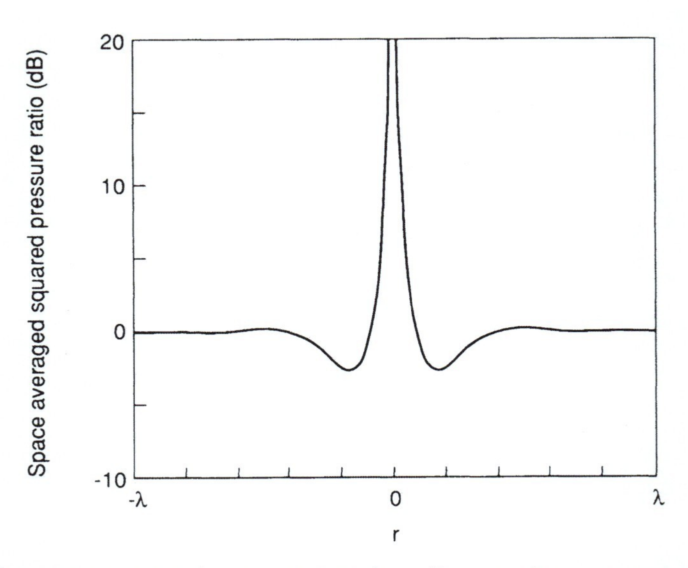

Fig. 11.9 The average squared pressure at various distances from a monopole source whose strength has been adjusted to be optimally absorbing in a diffuse sound field. The resulting mean squared pressure is plotted relative to the diffuse field mean squared pressure in the absence of the absorbing source.

If we now take expectations over the spatial variable $\mathbf{y}_s$, and note that

$$<|p_p(\mathbf{y}_s)|^2> = <|p_p(\mathbf{y}_s + \Delta\mathbf{y})|^2> = <|p_p(\mathbf{x})|^2>, \qquad (11.9.13)$$

which is the space-averaged squared modulus of the pressure due to the primary source, and that from Section 11.2

$$<p_p^*(\mathbf{y}_s)p_p(\mathbf{y}_s + \Delta\mathbf{y})> = <p_p(\mathbf{y}_s)p_p^*(\mathbf{y}_s + \Delta\mathbf{y})> = <|p_p(\mathbf{x})|^2> \text{ sinc } kr, \qquad (11.9.14)$$

for a three-dimensional diffuse field, then the space-averaged modulus squared pressure at a distance r from the absorbing monopole becomes

$$<|p(\mathbf{y}_s + \Delta\mathbf{y})|^2> = <|p_p(\mathbf{x})|^2> \left(1 + \frac{1}{4k^2r^2} - \text{sinc}^2\, kr\right). \qquad (11.9.15)$$

This function is plotted in Fig. 11.9, which shows that the space-averaged square pressure is reduced, by up to about 3 dB, in a "shell" at distances from about 0.1λ to 0.3λ from the secondary source. The near field of the monopole begins to dominate when the distance from the source is less than about 0.1λ, and the average squared pressure is thus increased very close to the source. It is interesting to compare this space-averaged change in the local pressure for a diffuse sound field with Fig. 9.6a which shows the equivalent result for the pressure close to an absorbing monopole in a plane wave sound field.

12
Multi-channel Feedforward Control

12.1 Multi-channel control of periodic sound

The profusion of practically important noise sources which are *periodic* has already been remarked upon. We have also seen how active techniques may be used to control such sources at low frequencies, and how *multiple* secondary sources are often required to prevent radiation (see Chapter 8) or control an enclosed sound field (see Chapter 10). A practical control strategy which is often used to adjust the output of these sources is to minimise the sum of the mean square outputs of a number of microphones. The purpose of this chapter is to discuss the various methods by which this may be achieved in practice. We begin by adopting the strategy discussed in Chapter 6 and consider the control problem at each harmonic of the periodic sound. With the acoustic and electroacoustic parts of the control system assumed to be linear, the control of each harmonic will be independent. The principles of operation of a control system can thus be demonstrated by describing the behaviour at one discrete frequency only.

Also following the approach of Chapter 6, we will formulate the control problem purely in terms of observable electrical signals. A simple example is shown in Fig. 12.1, in which the harmonic signals from three microphones are indicated, each of which is influenced by the action of a harmonic primary source, and two secondary sources, each fed by harmonic reference signals at the same frequency (ω) as the primary source. In general we assume that there are L sensors (microphones) and M sources (loudspeakers). If linearity is assumed, and additionally the important assumption is made that the signals are in their steady state, the complex output from the lth sensor, e_l, will be the superposition of the complex output due to the

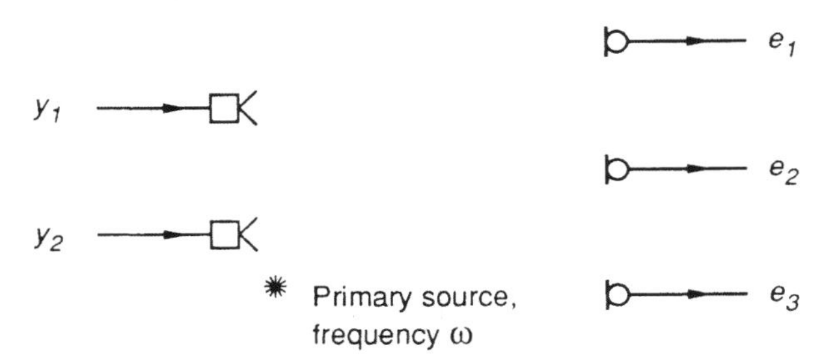

Fig. 12.1 A simple multiple channel active control system operating at a single frequency.

primary source operating alone, d_l say, and the outputs due to each of the M secondary sources operating alone. By writing the signal feeding the mth secondary source as y_m, the total output from the lth error sensor can be expressed in its complex form as

$$e_l = d_l + C_{l1}(j\omega)y_1 + C_{l2}(j\omega)y_2 + \dots C_{lM}(j\omega)y_M, \tag{12.1.1}$$

where $C_{lm}(j\omega)$ is the electrical transfer response between the signal driving the mth secondary source and the output from the lth sensor, at the frequency ω. (Note that as in Chapter 6, we will have no requirement to use the symbols x and y to denote spatial co-ordinates and we will therefore use them to denote complex values of the input and output signals to the physical system.) It has been implicitly assumed that a reference signal at frequency ω is available to generate driving signals, y_m, of this form. No "detection sensors" are needed in this case, and in terms of the discussion in Chapter 6, all the microphones in Fig. 12.1 are "error sensors". The great advantage of working in terms of electrical variables is that no assumptions need to be made about the physical system under control, except that it must be linear. The secondary sources could, for example, be loudspeakers with finite internal acoustic impedance, whose volume velocities depended on the pressure in front of each loudspeaker due to each of the other sources, as a result of acoustic loading effects. The physical interpretation of the electrical transfer responses would be difficult under such conditions: for example, they would not just be proportional to the *acoustic* transfer impedance. The superposition of the electrical variables accounts for all such effects in the definition of the electrical transfer responses and so provides a very general framework in which to discuss the practical control problem.

We can express the complex response at each microphone by using the matrix representation given by

$$\begin{bmatrix} e_1 \\ e_2 \\ \cdot \\ \cdot \\ \cdot \\ e_L \end{bmatrix} = \begin{bmatrix} d_1 \\ d_2 \\ \cdot \\ \cdot \\ \cdot \\ d_L \end{bmatrix} + \begin{bmatrix} C_{11}(j\omega) & C_{12}(j\omega) & \dots & C_{1M}(j\omega) \\ C_{21}(j\omega) & C_{22}(j\omega) & & \\ \cdot & & & \\ \cdot & & & \\ \cdot & & & \\ C_{L1}(j\omega) & C_{L2}(j\omega) & \dots & C_{LM}(j\omega) \end{bmatrix} \begin{bmatrix} y_1 \\ y_2 \\ \cdot \\ \cdot \\ \cdot \\ y_M \end{bmatrix}. \tag{12.1.2}$$

This leads to a convenient matrix form for the response of the system which can be written as

$$\mathbf{e} = \mathbf{d} + \mathbf{C}\,\mathbf{y}, \tag{12.1.3}$$

where the definitions of the complex vectors $\mathbf{e}$, $\mathbf{d}$ and $\mathbf{y}$ and the complex matrix $\mathbf{C}$ of electrical transfer responses follow from equation (12.1.2), the explicit dependence of $\mathbf{C}$ on $j\omega$ having been dropped for convenience. A block diagram representation of this equation is shown in Fig. 12.2, and this forms the basis for the discussion of solutions to this control problem as set out below. It is worth reiterating that the equation above can describe only the response of the system in the *steady state*, when all transients (due to, for example, changes in the secondary source amplitudes or phases) have died away. In Section 12.2, exact least squares solutions to this equation are discussed, for which the steady-state assumption is not a restriction. We will also, however, use this expression as the basis for discussing iterative methods of

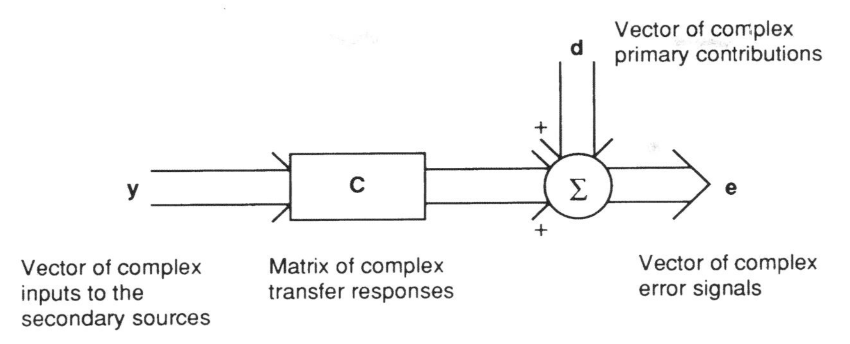

Fig. 12.2 Block diagram representation of the electrical response of a multi-channel control system in the steady state.

adjusting the signals driving the secondary sources, and some care will be needed in the use of the equation at that stage.

12.2 Exact least squares solutions

We choose, for the time being, to adjust the complex inputs to the M secondary sources to minimise the sum of the squared outputs from each of the L sensors. In this section we will present the exact solutions to this least squares problem for various relative numbers of sources and sensors. This provides a background for the discussion of iterative algorithms and introduces some of the geometric properties of the *error surface* which are important in understanding the behaviour of these iterative algorithms.

The error surface is the $2M$ dimensional surface generated by plotting the error criterion against the real and imaginary parts of the inputs to each of the M secondary sources. The error criterion (also called the *cost function*, or *performance index*) can be defined here as the sum of the squared moduli of the outputs from all the error microphones. This is given by

$$J = \sum_{l=1}^{L} |e_l|^2 = \mathbf{e}^{\mathrm{H}}\mathbf{e}, \tag{12.2.1}$$

which, on substitution of equation (12.1.3), results in the Hermitian quadratic form

$$J = \mathbf{d}^{\mathrm{H}}\mathbf{d} + \mathbf{d}^{\mathrm{H}}\mathbf{C}\ \mathbf{y} + \mathbf{y}^{\mathrm{H}}\mathbf{C}^{\mathrm{H}}\mathbf{d} + \mathbf{y}^{\mathrm{H}}\mathbf{C}^{\mathrm{H}}\mathbf{C}\ \mathbf{y}. \tag{12.2.2}$$

If the complex secondary source signals are split into their real and imaginary parts, one three-dimensional part of the error surface can be plotted either as an isometric projection or as a contour diagram, as illustrated in Fig. 12.3. This shows the variation of J with the real part of one secondary source signal ($y_{1\mathrm{R}}$) and the imaginary part of another secondary source signal ($y_{2\mathrm{I}}$). Understanding the shape of the multi-dimensional error surface is considerably aided by simple three-dimensional examples. Although the extension to more than three dimensions is impossible to visualise, the geometric properties of position, distance and slope, for example, can

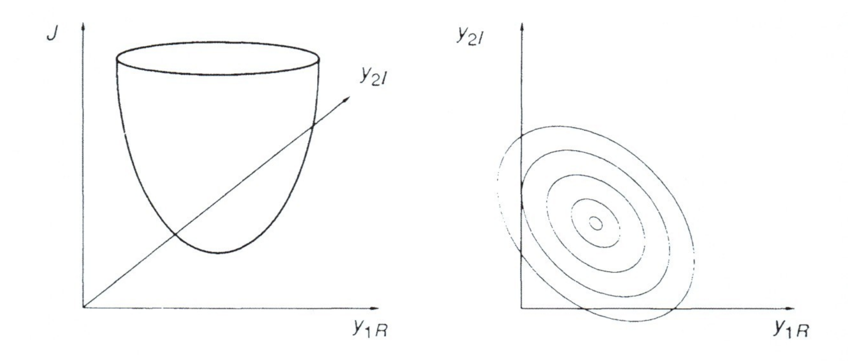

Fig. 12.3 The variation of error criterion J as a function of the two variables y_{1R}, y_{2I}. The function is plotted as the isometric projection of a three-dimensional surface and as a two-dimensional map in the y_{1R}, y_{2I} plane, showing contours of constant J.

all be generalised and still serve as useful concepts in the multi-dimensional case.

The exact least squares solution we are seeking is the bottom point of this bowl-shaped error surface. We begin with the case which is by far the most important in practice: when the number of sensors L is greater than the number of sources M. Mathematically this is known as the *overdetermined* case since there are more equations to solve of the form

$$e_l = d_l + \sum_{m=1}^{M} C_{lm}(j\omega)y_m, \tag{12.2.3}$$

than there are unknowns (the y_ms). Provided that $\mathbf{C}^H\mathbf{C}$ is positive definite, the error surface has a unique global minimum, as illustrated in Fig. 12.3.

The mathematical details of the minimisation are discussed in the appendix. The resulting optimal set of input signals to the secondary sources, $\mathbf{y}_0$, which minimises $\mathbf{J}$ is given by

$$\mathbf{y}_0 = -(\mathbf{C}^H\mathbf{C})^{-1}\mathbf{C}^H\mathbf{d}. \tag{12.2.4}$$

If this expression for $\mathbf{y}_0$ is substituted into the expression for J (equation (12.2.2)), the bottom of the bowl is seen to have a minimum value of J given by

$$J_0 = \mathbf{d}^H[\mathbf{I} - \mathbf{C}(\mathbf{C}^H\mathbf{C})^{-1}\mathbf{C}^H]\mathbf{d}. \tag{12.2.5}$$

The mathematically fully determined case, when $L = M$, is easier to solve algebraically, but less useful in practice. With the matrix $\mathbf{C}$ assumed to be non-singular, the optimal set of signals feeding the secondary sources which drives the vector of error signals to zero, $\mathbf{e} = 0$, is

$$\mathbf{y}_0 = -\mathbf{C}^{-1}\mathbf{d}. \tag{12.2.6}$$

In this case the bottom of the error bowl just touches the $J = 0$ plane, and $J_0 = 0$. Each of the individual error signals is driven to zero by this strategy and while this solution may be mathematically neat, it is often found in practice to have detrimental effects on the sound field being controlled. This is a result of the requirement that the pressure in the sound field is exactly zero at the microphones, which can cause unrestricted increases in the pressure away from these points.

If the "square" case (with $L = M$) seems to be physically unwise, to implement a harmonic active control system with more secondary sources than error sensors would appear to be positively perverse! This mathematically *underdetermined* case is worthy of a brief mention, however, partly because it was touched on in Chapter 11 and partly because it is the extreme limit of an *ill-conditioning* in the overdetermined set of equations. This ill-conditioning can sometimes occur in practice and will be discussed in Section 12.3. Consider the situation in which two loudspeakers are being used to control the sound at a single microphone as illustrated in Fig. 12.4. We assume for simplicity that the transfer responses from both loudspeakers to the microphone are real and both equal to C. We also assume that the primary contribution to the error signal, d, is also real, so that

$$e = d + C(y_1 + y_2). \tag{12.2.7}$$

Clearly any combination of y_1 and y_2 which satisfies the expression

$$y_1 + y_2 = -\frac{d}{C}, \tag{12.2.8}$$

will drive the error signal to zero. The contours of the error surface for this problem are illustrated in Fig. 12.5. Since there are an infinite number of combinations of y_1 and y_2 which satisfy the above expression, some other constraint must be introduced into the problem to obtain a unique solution.

A physically reasonable constraint is that the cost function should be driven to zero with least "effort". So in the problem above $y_1^2 + y_2^2$ should be minimised, while keeping $y_1 + y_2 = -d/C$. The "effort" can be geometrically interpreted as the distance of a point on the y_1, y_2 plane away from the origin. The geometric interpretation of the least effort constraint is that the solution nearest to the origin is chosen. The general multi-channel solution to this constrained optimisation problem is discussed in the appendix, and the result is

$$\mathbf{y}_0 = -\mathbf{C}^{\mathrm{H}}(\mathbf{C}\mathbf{C}^{\mathrm{H}})^{-1}\mathbf{d}. \tag{12.2.9}$$

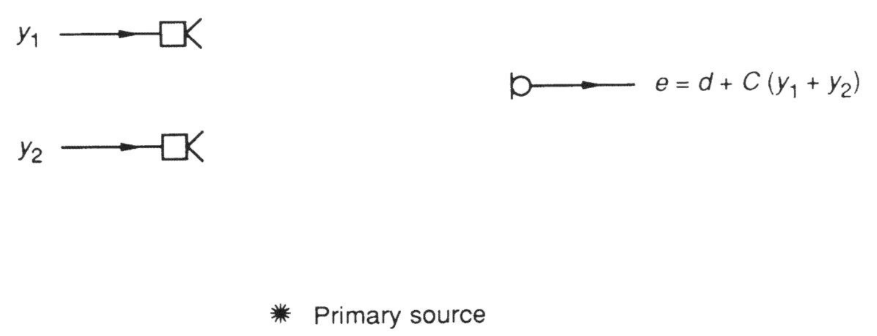

y_1

y_2

$e = d + C(y_1 + y_2)$

✱ Primary source

Fig. 12.4 An example of an underdetermined control system.

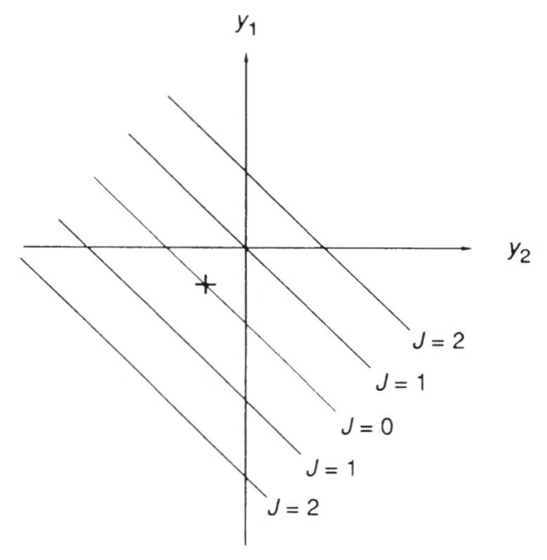

Fig. 12.5 Contours of constant mean squared error for an underdetermined control system. The cross indicates the solution which minimises the error criterion *and* minimises the "effort" $y_1^2 + y_2^2$.

For the particular example above the least effort solution is

$$y_1 = y_2 = -\frac{d}{2C}. \tag{12.2.10}$$

It is interesting to note that if the signals y_1 and y_2 were voltages feeding the loudspeakers, and the electrical impedances of each of the two loudspeakers were equal and independent of the action of the other loudspeaker, then $y_1^2 + y_2^2$ would be proportional to the total electrical power supplied to the two loudspeakers. This analysis suggests that if both loudspeakers are driven equally according to equation (12.2.10) then $y_1^2 + y_2^2 = d^2/2C^2$, which corresponds to *half* the electrical power required to cancel the error signal with either one loudspeaker acting on its own (Ford, 1984). This has generally been found to be the case in practice (see Eatwell and Hughes, 1988).

12.3 Matrix conditioning and the shape of the error surface

Even though the underdetermined example (with more sources than sensors) is an extreme case, similar effects on the shape of the error surface can be caused by unfavourable acoustical conditions in the overdetermined case most often encountered in practice. If, for example, two secondary sources are placed very close together, their effect on each of many microphones will be very similar, and linear combinations of input signals to the two sources will produce very similar values of the sum of the squared errors, J. This will cause the elliptic contours of constant error to be stretched out, until in the limit when the sources are in the same position, the error surface has an identical form to that of the underdetermined case of Fig. 12.5.

(a)

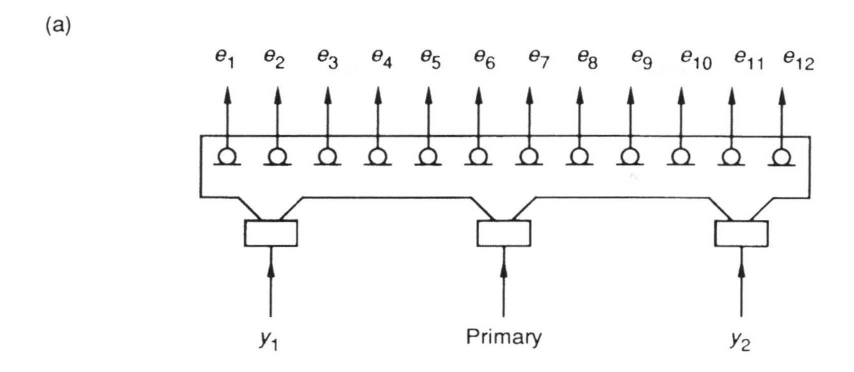

(b)

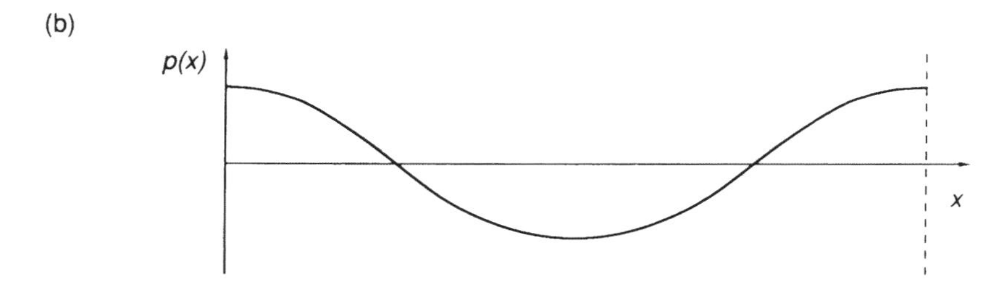

Fig. 12.6 (a) Active control of the sound in a long duct by using two secondary sources near the ends of the duct. (b) The primary source is driven by a pure tone at a frequency close to the second longitudinal "organ pipe" resonance of the duct, whose pressure distribution is also shown.

It is perhaps not surprising that placing two sources physically close together can give rise to the sort of ambiguity described above. Such ambiguity may also arise in less obvious ways, however, and one simple example of this is illustrated in Fig. 12.6. The control objective here is to minimise the sum of the squared pressures at 12 points in a long thin duct, which is excited by a primary source somewhere in the middle of the duct. The sound field is controlled by two secondary sources near the two ends. If the primary source is driven with a pure tone, at a frequency close to the natural frequency of the lightly damped (2,0,0) mode in the duct, the pressure distribution in the duct will be as illustrated in the lower part of Fig. 12.6. The two secondary sources can couple into this mode with equal efficiency and the mode could be suppressed with either source acting independently. Even though the two secondary sources are now physically remote from one another, they give rise to exactly the same type of ambiguity as if they were physically close. The contours of the error surface for this control problem are shown in Fig. 12.7, in which the elliptical contours of equal error have been elongated into a long "valley" in the error surface. The exact least squares solution to this problem is still mathematically defined, but is now rather sensitive to the exact positioning of the secondary sources or the contributions from the residual acoustic modes in the duct. The algebraic reason for the sensitivity is that two columns in the matrix

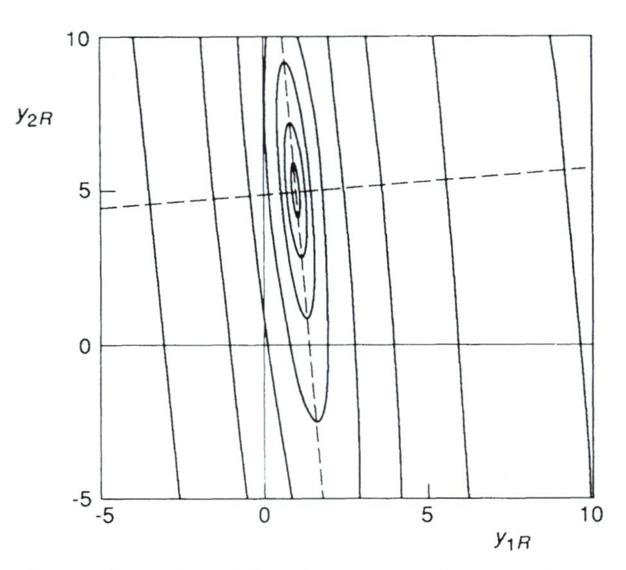

Fig. 12.7 Contours of equal cost function J in 5 dB steps, for an ill-conditioned control problem. The dashed lines indicate the principal axes of the error surface.

$$\mathbf{C}^{\mathrm{H}}\mathbf{C} = \sum_{l=1}^{L} \begin{bmatrix} |C_{l1}(j\omega)|^2 & C_{l1}^*(j\omega)C_{l2}(j\omega) & \ldots & C_{l1}^*(j\omega)C_{lM}(j\omega) \\ C_{l2}^*(j\omega)C_{l1}(j\omega) & |C_{l2}(j\omega)|^2 & \ldots & C_{l2}^*(j\omega)C_{lM}(j\omega) \\ . & . & \ldots & . \\ . & . & \ldots & . \\ C_{lM}^*(j\omega)\,C_{l1}(j\omega) & C_{lM}^*(j\omega)C_{l2}(j\omega) & \ldots & |C_{lM}(j\omega)|^2 \end{bmatrix}. \tag{12.3.1}$$

are very similar. Such a matrix is termed "ill-conditioned". The implication of this is that the determinant of the matrix is small and the inverse of this matrix, which is used to calculate $\mathbf{y}_0$, is very sensitive to small changes in any of the elements of $\mathbf{C}$. Table 12.1 summarises the conditions under which the exact least squares solution is ill-conditioned for the overdetermined, fully determined and underdetermined cases.

Table 12.1 Conditioning for various exact least squares solutions.

Size of control system	Matrix to invert for solutions	Conditions for which determinant of matrix ≈ 0 (i.e. ill-conditioning)
$L > M$ (overdetermined)	$\mathbf{C}^{\mathrm{H}}\mathbf{C}$	Two sources close together (two columns of $\mathbf{C}^{\mathrm{H}}\mathbf{C}$ nearly equal)
$L = M$ (fully determined)	$\mathbf{C}$	Either two sources or two sensors close together
$L < M$ (underdetermined)	$\mathbf{C}\mathbf{C}^{\mathrm{H}}$	Two sensors close together (two columns of $\mathbf{C}\mathbf{C}^{\mathrm{H}}$ nearly equal)

Another view of conditioning can be obtained by manipulation of the equation for the cost function. In the appendix it is shown that the cost function can be written as

$$J = J_0 + (\mathbf{y} - \mathbf{y}_0)^{\mathrm{H}}(\mathbf{C}^{\mathrm{H}}\mathbf{C})(\mathbf{y} - \mathbf{y}_0), \tag{12.3.2}$$

where we assume an overdetermined system, for which J_0 and $\mathbf{y}_0$ are defined by equations (12.2.4) and (12.2.5). We now expand the Hermitian matrix $\mathbf{C}^{\mathrm{H}}\mathbf{C}$ in terms of its matrix of normalised eigenvectors, $\mathbf{Q}$, and the diagonal matrix of its eigenvalues, $\boldsymbol{\Lambda} = \mathrm{diag}(\lambda_1,\lambda_2,\ldots,\lambda_M)$ so that

$$\mathbf{C}^{\mathrm{H}}\mathbf{C} = \mathbf{Q}\boldsymbol{\Lambda}\,\mathbf{Q}^{\mathrm{H}}. \tag{12.3.3}$$

If we define a new set of co-ordinates for the error surface as

$$\mathbf{v} = \mathbf{Q}^{\mathrm{H}}(\mathbf{y} - \mathbf{y}_0) = (v_1 v_2 \ldots v_M)^{\mathrm{T}}, \tag{12.3.4}$$

we can express the cost function as

$$J = J_0 + \mathbf{v}^{\mathrm{H}}\boldsymbol{\Lambda}\,\mathbf{v} = J_0 + \sum_{m=1}^{M} \lambda_m |v_m|^2. \tag{12.3.5}$$

The variables $v_1,v_2,\ldots$ represent a co-ordinate system for the error surface which has been translated (by subtracting $\mathbf{y}_0$) and rotated (by multiplication with $\mathbf{Q}$). It has the very important property that the variation of J with any one element of $\mathbf{v}$ is *independent* of the values of the other elements. For this reason they are known as the *principal co-ordinates* of the error surface. The principal axes defined by these co-ordinates are indicated as dotted lines on Fig. 12.7 by way of example. The principal axes are orthogonal because of the independence of the variation of J with each element of $\mathbf{v}$.

The steepness of the error surface in the direction of the principal axis v_m is proportional to the associated eigenvalue, λ_m. For the ill-conditioned case illustrated in Fig. 12.7 the eigenvalues associated with the two principal co-ordinates are clearly very different. The ratio of the maximum eigenvalue to the minimum eigenvalue of a matrix is sometimes known as the *condition number* and used as a measure of the extent to which the matrix is ill-conditioned. For the error surface of Fig. 12.7 the ratio of the maximum to the minimum eigenvalue of $\mathbf{C}^{\mathrm{H}}\mathbf{C}$ was 100. The rather formal indication of conditioning provided by the eigenvector/eigenvalue expansion of $\mathbf{C}^{\mathrm{H}}\mathbf{C}$, and the definition of the principle co-ordinates, is important for the *analysis* of iterative gradient descent algorithms presented below. A clear *physical* understanding of these algorithms can be obtained, however, by considering their action geometrically, as iterating down the slopes of the "valleys" in the error surface.

12.4 Iterative gradient descent methods

Although the exact least squares solution discussed above is an important tool for theoretical studies and computer simulations of active control, there are a number of practical problems associated with the direct computation of the least squares solution in a practical control system: (1) the primary field ($\mathbf{d}$) may change with time; (2) the computing facilities to calculate the inverse of a matrix may not be available; (3) the matrix may be difficult to invert because of ill-conditioning; (4) the system under control may be weakly non-linear. Each of these problems motivates the development of iterative methods of adjusting the signals driving the secondary sources, which do not use a matrix inversion, and which are able to track changes in the amplitude and phase of the primary field and compensate for slight non-linearities. Gradient descent algorithms have been found to provide many of these advantages.

Gradient descent algorithms can be applied with some confidence to the overdetermined control problem since the error surface is guaranteed to be quadratic, assuming only that the system under control is reasonably linear. In practice, the matrix $\mathbf{C}^{\mathrm{H}}\mathbf{C}$ is almost always positive definite and the surface must have a unique global minimum. Thus if we consider the error surface with no primary source present, so that

$$J = \mathbf{y}^{\mathrm{H}}\mathbf{C}^{\mathrm{H}}\mathbf{C}\;\mathbf{y}, \tag{12.4.1}$$

the sum of the squared errors must be positive for any value of $\mathbf{y}$, and in practice will generally be greater than zero for all non-zero values of $\mathbf{y}$. The matrix $\mathbf{C}^{\mathrm{H}}\mathbf{C}$ must then be positive definite, and this ensures a unique global minimum in the more general case with the primary source present (equation (12.3.2)). Since the shape of the error surface is determined by such general properties of the system under control, gradient descent algorithms are guaranteed to converge to the unique global minimum under a wide variety of conditions.

The simplest such algorithm would involve making an adjustment to one of the control variables, observing the effect on the cost function (J), retaining the adjustment if J were reduced and backtracking if J were increased. This type of "trial and error" adjustment is a generalisation of that described by Smith and Chaplin (1983) as the "power sensing algorithm". In a multi-channel control system, however, it is not just sufficient to individually adjust each of the control variables once to minimise J, the process must be repeated many times for each control variable to ensure that the global minimum is reached. This is because the matrix $\mathbf{C}^{\mathrm{H}}\mathbf{C}$ is not diagonal, so the value of one control variable which minimises J will depend on the values of all the other control variables. The procedure is illustrated for a three-dimensional error surface in Fig. 12.8. For each iteration of each control variable, time must be allowed for the transients in the system to settle, and for J to be measured accurately. This control strategy can clearly take some time since each iteration is repeated a number of times for one control variable, there are many control variables, and the whole procedure must be repeated many times to ensure the global minimum is reached. One great advantage that such trial and error algorithms do have, however, is that no estimates of the transfer responses, $C_{lm}(j\omega)$, of the system under control are needed in their implementation. Provided that the sound field is reasonably stationary with not too many control channels, such algorithms can provide a very simple and robust control strategy.

Once the number of control variables becomes significant, it is far quicker to adjust *all* of these variables at each iteration in an attempt to reduce the error, rather than to adjust them one at a time. One well-known algorithm which does this is the method of steepest descent, in which the set of control variables at the kth iteration is set equal to that at the previous iteration minus an amount proportional to the gradient of the cost function with respect to that control variable. In this case we have $2M$ control variables, which are the real and imaginary parts of the M elements of the vector of complex inputs to the secondary sources given by $\mathbf{y} = \mathbf{y}_{\mathrm{R}} + j\mathbf{y}_{\mathrm{I}}$. The method of steepest descent can be written as the pair of equations

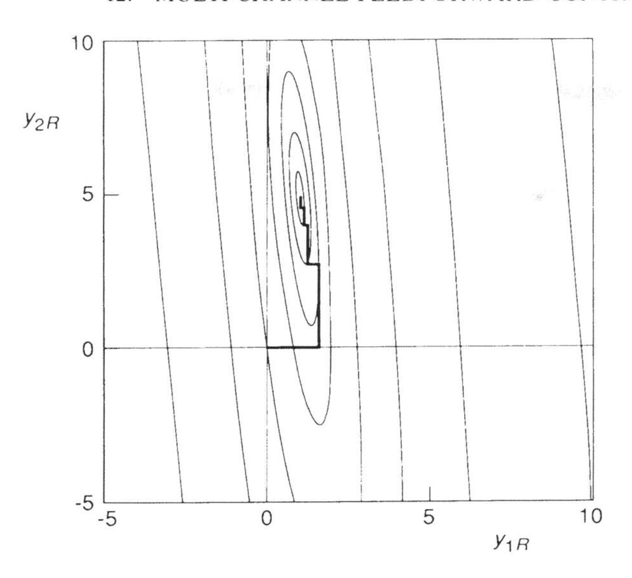

Fig. 12.8 The convergence of a trial and error gradient descent algorithm, in which the cost function is first minimised by adjusting y_{1R}, secondly by adjusting y_{2R}, and the process is continuously repeated.

$$\mathbf{y}_R(k) = \mathbf{y}_R(k-1) - \mu \frac{\partial J}{\partial \mathbf{y}_R(k-1)}, \quad \mathbf{y}_I(k) = \mathbf{y}_I(k-1) - \mu \frac{\partial J}{\partial \mathbf{y}_I(k-1)}, \tag{12.4.2a,b}$$

in which μ is a convergence coefficient and the $M \times 1$ gradient vectors are defined by

$$\frac{\partial J}{\partial \mathbf{y}_R(k-1)} = \left[\frac{\partial J}{\partial y_{1R}(k-1)}, \frac{\partial J}{\partial y_{2R}(k-1)}, \ldots, \frac{\partial J}{\partial y_{MR}(k-1)} \right]^T \tag{12.4.3a}$$

$$\frac{\partial J}{\partial \mathbf{y}_I(k-1)} = \left[\frac{\partial J}{\partial y_{1I}(k-1)}, \frac{\partial J}{\partial y_{2I}(k-1)}, \ldots, \frac{\partial J}{\partial y_{MI}(k-1)} \right]^T \tag{12.4.3b}$$

In the appendix we evaluate the derivative of a Hermitian quadratic function of a complex variable (such as the error criterion J here) with respect to the real or imaginary parts of the complex control variables. Both of these derivatives are found to be entirely real quantities. We can now combine the two individual expressions describing the method of steepest descent for the real and imaginary parts of the control variables, each of which contains entirely real variables, into a single complex form such that

$$\mathbf{y}(k) = \mathbf{y}(k-1) - \mu\mathbf{g}(k-1), \tag{12.4.4}$$

where $\mathbf{g}$, which we call the *complex gradient vector*, is defined to be

$$\mathbf{g} = \frac{\partial J}{\partial \mathbf{y}_R} + j\,\frac{\partial J}{\partial \mathbf{y}_I}. \tag{12.4.5}$$

It is also shown in the appendix that $\mathbf{g}$ can be written as

$$\mathbf{g} = 2(\mathbf{C}^H\mathbf{C}\ \mathbf{y} + \mathbf{C}^H\mathbf{d}). \tag{12.4.6}$$

Since $\mathbf{e} = \mathbf{d} + \mathbf{C}\mathbf{y}$, this can be written as $\mathbf{g} = 2(\mathbf{C}^H\mathbf{e})$, and the algorithm which implements the method of steepest descent can be written as

$$\mathbf{y}(k) = \mathbf{y}(k-1) - \alpha\mathbf{C}^H\mathbf{e}(k-1), \tag{12.4.7}$$

where α is another convergence coefficient which has a value equal to 2μ.

The properties of the method of steepest descent will now be considered in some detail and related to the convergence in the error surface. As well as illustrating important points about the steepest descent algorithm, such a discussion also throws light on the inherent compromises which have to be made in any iterative algorithm. We begin by considering the convergence of the control variables, following the approach of Widrow and Stearns (1985). From the discussion above we know that the method of steepest descent can be expressed as

$$\mathbf{y}(k) = \mathbf{y}(k-1) - \alpha[\mathbf{C}^H\mathbf{C}\mathbf{y}(k-1) + \mathbf{C}^H\mathbf{d}], \tag{12.4.8}$$

and since $\mathbf{y}_0 = -(\mathbf{C}^H\mathbf{C})^{-1}\mathbf{C}^H\mathbf{d}$, then $\mathbf{C}^H\mathbf{d}$ can be written as $-(\mathbf{C}^H\mathbf{C})\mathbf{y}_0$, so one iteration of the steepest descent method is described by

$$[\mathbf{y}(k) - \mathbf{y}_0] = (\mathbf{I} - \alpha\mathbf{C}^H\mathbf{C})[\mathbf{y}(k-1) - \mathbf{y}_0]. \tag{12.4.9}$$

If the iterative procedure begins with $\mathbf{y}(0) = \mathbf{0}$, and the equation above is applied recursively, then it can be seen that

$$[\mathbf{y}(k) - \mathbf{y}_0] = (\mathbf{I} - \alpha\mathbf{C}^H\mathbf{C})^k[-\mathbf{y}_0]. \tag{12.4.10}$$

We now use the eigenvector/eigenvalue expansion of $\mathbf{C}^H\mathbf{C} = \mathbf{Q}\mathbf{\Lambda}\mathbf{Q}^H$ as defined in Section 12.3, where $\mathbf{Q}$ is the unitary matrix of normalised eigenvectors ($\mathbf{Q}\,\mathbf{Q}^H = \mathbf{Q}^H\mathbf{Q} = \mathbf{I}$), to show that we can write

$$(\mathbf{I} - \alpha\mathbf{C}^H\mathbf{C})^k = \mathbf{Q}(\mathbf{I} - \alpha\mathbf{\Lambda})^k\mathbf{Q}^H. \tag{12.4.11}$$

Equation (12.4.10) may now be written in terms of the principal co-ordinates $\mathbf{v}(k)$ introduced in equation (12.3.4), where $\mathbf{v}(k) = \mathbf{Q}^H[\mathbf{y}(k) - \mathbf{y}_0]$. Use of this relationship, together with equation (12.4.11), then leads to

$$\mathbf{v}(k) = (\mathbf{I} - \alpha\mathbf{\Lambda})^k\mathbf{v}(0). \tag{12.4.12}$$

The term in brackets is a diagonal matrix, so each component of the vector $\mathbf{v}(k)$ can be written in scalar form such that

$$v_m(k) = (1 - \alpha\lambda_m)^k v_m(0). \tag{12.4.13}$$

This equation shows that the principle co-ordinates of the control variables will converge *independently* in the method of steepest descent. The condition for the principal co-ordinates to converge to zero, i.e. for $\mathbf{y}(k) \to \mathbf{y}_0$, is that $|1 - \alpha\lambda_m| < 1$, i.e.

$$0 < \alpha < \frac{2}{\lambda_m} \text{ for all } m. \tag{12.4.14}$$

The maximum value of the convergence coefficient α which can be used in the algorithm is thus governed by the *largest* eigenvalue (λ_{max}) of the matrix $\mathbf{C}^H\mathbf{C}$. Thus

$$0 < \alpha < \frac{2}{\lambda_{max}}. \tag{12.4.15}$$

In general, however, $\alpha\lambda_m$ will be small, and under these conditions $(1 - \alpha\lambda_m) \approx e^{-\alpha\lambda_m}$ and we may therefore write

$$(1 - \alpha\lambda_m)^k \approx e^{-\alpha\lambda_m k}, \tag{12.4.16}$$

and thus the time constant of convergence associated with the mth eigenvalue is given by $\tau_m \approx 1/\alpha\lambda_m$ iterations. The longest time constant will be associated with the smallest eigenvalue, λ_{min}. Thus $\tau_{max} \approx 1/\alpha\lambda_{min}$, and since $\alpha \lesssim 1/\lambda_{max}$, then we have

$$\tau_{max} \gtrsim \frac{\lambda_{max}}{\lambda_{min}} \quad \text{iterations.} \tag{12.4.17}$$

The longest time constant associated with the method of steepest descent is therefore due to the convergence of the control variables along the principal axis associated with the smallest eigenvalue, i.e. the most shallow "valley" in the error surface. The overall speed of convergence is set by the convergence coefficient, and the maximum value of this coefficient is fixed by the need to prevent the algorithm from "overshooting" during convergence of the control variables along the principal axis associated with the largest eigenvalue, i.e. the steepest "valley" in the error surface. The convergence of the method of steepest descent on a three-dimensional error surface is illustrated in Fig. 12.9, which demonstrates the rapid initial

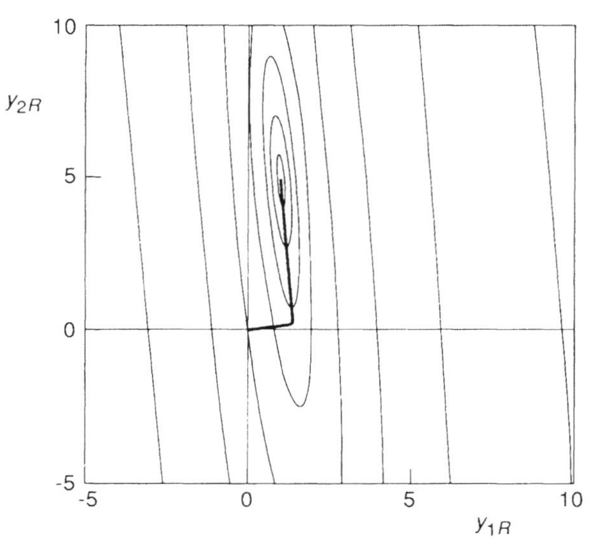

Fig. 12.9 The convergence of a steepest descent algorithm in which y_{1R} and y_{2R} are simultaneously adjusted.

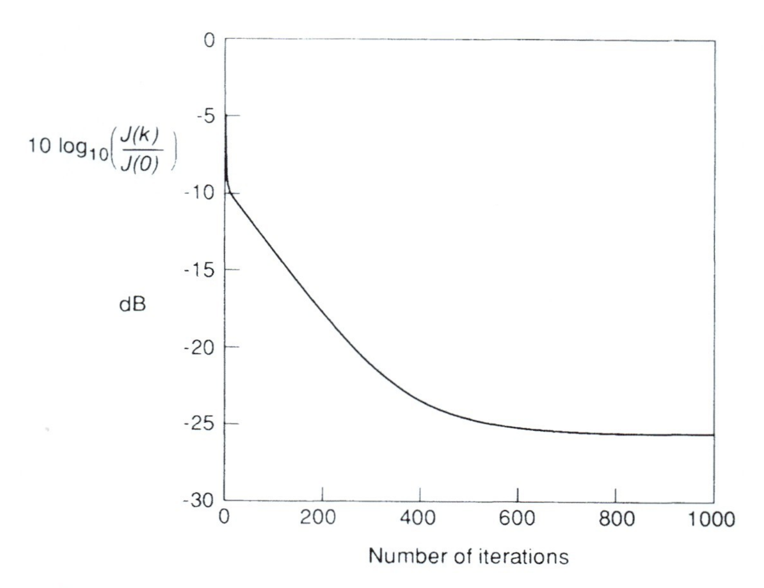

Fig. 12.10 The reduction in the cost function J as a result of the convergence, using the method of steepest descent, illustrated in Fig. 12.9.

convergence in the direction of the principal axis associated with the largest eigenvalue (steepest slope), and the slower final convergence in the direction of the other principal axis associated with the smaller eigenvalue (shallow valley). Figure 12.10 is a plot of the cost function J during the convergence shown in Fig. 12.9, and clearly shows the dual slope associated with the convergence along the two principal axes. In this example a reduction of some 10 dB in the sum of squares of errors is obtained within the first 10 iterations, but a further 500 iterations are needed to increase the reduction in the cost function to 25 dB, which is the reduction associated with the exact least squares solution for this example. The sum of the squares of the control variables, which is a measure of the power needed to drive the secondary sources, is proportional to the distance from the origin on the contour plot of Fig. 12.9. It can be seen that the secondary source power required to achieve a reduction of 25 dB in the cost function is about four times that required to achieve a reduction of 10 dB. In a practical active control system it is sometimes debatable whether achieving further reductions in the sound field at the control microphones at the expense of large secondary source strengths is a physically sensible strategy (Elliott *et al.*, 1991a).

12.5 Other iterative algorithms

In this section we discuss a class of algorithm which attempts to avoid the problems of slow convergence along shallow valleys in the error surface which are associated with gradient descent methods. One way of viewing these algorithms is that the direction of the changes in the control variables is adjusted so that the path of the iterative solution is more directly towards the optimal solution, rather than following the rather indirect path of the gradient descent algorithms. This is achieved by introducing a matrix $\mathbf{D}^{-1}$ which pre-multiplies the gradient term such that the algorithm is written as

$$\mathbf{y}(k) = \mathbf{y}(k-1) - \alpha \mathbf{D}^{-1}\mathbf{g}(k-1), \tag{12.5.1}$$

where the derivative of the cost function with respect to the real and imaginary parts of $\mathbf{y}(k)$ is contained in the real and imaginary parts of $\mathbf{g}(k)$, as described above.

The best known algorithm of this type is the Gauss–Newton method (discussed, for example, by Widrow and Stearns, 1985), in which $\mathbf{D} = \mathbf{C}^{\mathrm{H}}\mathbf{C}$ and the complete algorithm becomes

$$\mathbf{y}(k) = \mathbf{y}(k-1) - \alpha[\mathbf{C}^{\mathrm{H}}\mathbf{C}]^{-1}\mathbf{C}^{\mathrm{H}}\mathbf{e}(k-1). \tag{12.5.2}$$

The convergence properties of this algorithm in the linear case can be readily deduced by substituting $\mathbf{d} + \mathbf{C}\mathbf{y}(k-1)$ for $\mathbf{e}(k-1)$ in equation (12.5.2), and noting that $\mathbf{y}_0 = -(\mathbf{C}^{\mathrm{H}}\mathbf{C})^{-1}\mathbf{C}^{\mathrm{H}}\mathbf{d}$, so that the convergence equation becomes

$$[\mathbf{y}(k) - \mathbf{y}_0] = (1-\alpha)[\mathbf{y}(k-1) - \mathbf{y}_0]. \tag{12.5.3}$$

Repeated application of this equation with $\mathbf{y}(0) = \mathbf{0}$ gives the expression

$$[\mathbf{y}(k) - \mathbf{y}_0] = (1-\alpha)^k[-\mathbf{y}_0], \tag{12.5.4}$$

which can be written in the alternative form

$$\mathbf{y}(k) = [1 - (1-\alpha)^k]\mathbf{y}_0. \tag{12.5.5}$$

The fact that the term in square brackets is a scalar demonstrates that each of the physical control variables now converges *independently*, and with equal time constants. The path of convergence is illustrated in Fig. 12.11, and it is clear that the algorithm converges directly to the optimal value. Convergence of the algorithm is ensured if $|1-\alpha| < 1$ or

$$0 < \alpha < 2. \tag{12.5.6}$$

If the convergence coefficient is set to unity, then in the absence of any measurement noise or errors in the estimation of the elements of $\mathbf{C}$ the algorithm converges in one step, since the result of the first iteration is given by

$$\mathbf{y}(1) = -(\mathbf{C}^{\mathrm{H}}\mathbf{C})^{-1}\mathbf{C}^{\mathrm{H}}\mathbf{e}(0) \tag{12.5.7}$$

where $\mathbf{e}(0) = \mathbf{d}$. This is therefore exactly the optimal result. Continued iteration allows the algorithm to track changes in the primary field. A convergence coefficient of less than unity does allow some element of "caution" to be introduced into the algorithm, in case of errors in the estimation of $\mathbf{C}$ or $\mathbf{e}(k)$, for example, especially if the control problem is not well conditioned. Newton's method with $\alpha < 1$ is sometimes known as the "damped least squares" algorithm (Adby and Dempster, 1974). One disadvantage of this algorithm is that the inverse of the matrix $\mathbf{C}^{\mathrm{H}}\mathbf{C}$ must be calculated. This can require considerable computational power in a control system with many channels, but it does only need to be performed once, rather than at every iteration, provided the matrix $\mathbf{C}$ does not change with time.

Finally note that in the case where there are as many secondary sources as error

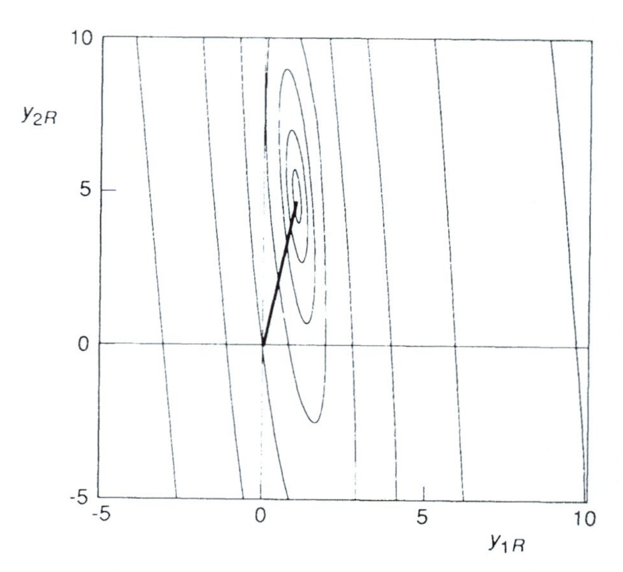

Fig. 12.11 The convergence of a Newton's method algorithm, in which y_{1R} and y_{2R} converge independently and with equal time constants towards the optimal solution.

sensors, the matrix **C** is square, and assuming it is invertible, Newton's method reduces to

$$\mathbf{y}(k) = \mathbf{y}(k-1) - \alpha \mathbf{C}^{-1}\mathbf{e}(k-1), \tag{12.5.8}$$

as discussed by Pierce (1985). If $\alpha = 1$ in the above expression, the algorithm is identical to that proposed by White and Cooper (1984). Newton's method provides a very direct convergence to the exact least squares solution. Indeed if $\alpha = 1$ it could be regarded as a continuous recalculation of this solution. It is perhaps worth reiterating, however, that if the problem is ill-conditioned the exact least squares solution can, in some active control applications, involve excessively large signals driving the secondary sources.

12.6 Alternative cost functions

Although the principal purpose of this book is to discuss the active control of sound, there is a large body of literature on the "higher harmonic control" of helicopter vibration, in which the control problem is so similar to the one being considered here that a brief review of these control algorithms will be worthwhile. The basic idea of higher harmonic control is to introduce a sinusoidal modulation, at a higher harmonic of the blade rotation frequency, into the signals driving the hydraulic actuators on the swash plate of the helicopter. These signals are adjusted in amplitude and phase to reduce the vibration measured by a number of accelerometers on various parts of the helicopter. It is a modulation of the aerodynamic lift on the helicopter blades which provides the secondary inputs in this case to reduce the vibration (Wernicke and Drees, 1963; Daughaday, 1968). The transfer response between the input signals to the hydraulic actuators and the output signals from the accelerometers thus depends on the aerodynamic characteristics of the helicopter blades. The aerodynamic component makes these transfer responses time dependent (due to varying flight conditions) and non-linear (due to the nature of the aerodynamic effects). So the problems which are encountered in higher harmonic control are in some ways considerably more severe than encountered in the active control of sound, and the strategies which have been developed to cope with these problems may be useful under difficult acoustical conditions.

The transfer response in higher harmonic control is sometimes modelled as being "locally linear". In other words small perturbations in the outputs are assumed to be linearly related to small perturbations in the control variables. To be consistent with the work discussed above, however, we choose what is described in the higher harmonic control literature as a "global system model" (Johnson, 1982; Davis, 1984) and write the complex response at a single frequency $\mathbf{e}$ due to the original excitation $\mathbf{d}$ and the control variables $\mathbf{y}$ as

$$\mathbf{e} = \mathbf{d} + \mathbf{C}\mathbf{y}. \tag{12.6.1}$$

One of the big differences between higher harmonic control algorithms and those discussed above is the use of a more complicated but more general cost function, which the control system attempts to minimise. For example, Davis (1984) discusses the use of a cost function given by

$$J_D(k) = [\mathbf{e}(k) - \mathbf{e}_{\text{des}}]^{\text{H}}\mathbf{W}_e[\mathbf{e}(k) - \mathbf{e}_{\text{des}}] + \mathbf{y}^{\text{H}}(k)\mathbf{W}_y\mathbf{y}(k) + \Delta\mathbf{y}^{\text{H}}(k)\mathbf{W}_\Delta\Delta\mathbf{y}(k). \tag{12.6.2}$$

The three terms in this expression are generally called the "error" term, the "effort" term and the "rate" term. The variables are defined as follows: $\mathbf{e}(k)$ is the vector of error signals at the kth iteration; $\mathbf{e}_{\text{des}}$ is a set of desired error outputs; $\mathbf{W}_e$ is a weighting matrix, generally diagonal, which allows some error signals to be given greater priority than others; $\mathbf{y}(k)$ is the vector of control signals; $\mathbf{W}_y$ is a weighting matrix, generally diagonal, for the control signals and has the effect of preventing small reductions at the error sensors at the expense of large control signals; $\Delta\mathbf{y}(k)$ is $\mathbf{y}(k) - \mathbf{y}(k-1)$; $\mathbf{W}_\Delta$ is a weighting matrix, generally diagonal, which penalises large changes in the control variables and can sometimes have a stabilising effect on iterative algorithms.

In the cost functions considered in the sections above, $\mathbf{e}_{\text{des}} = 0$, $\mathbf{W}_e = \mathbf{I}$, $\mathbf{W}_y = \mathbf{0}$ and $\mathbf{W}_\Delta = \mathbf{0}$. By substituting equation (12.6.1) into (12.6.2) and expanding out the equation for the cost function, it can again be put into a complex quadratic form. By differentiating this quadratic function with respect to the real and imaginary parts of $\mathbf{y}(k)$, with $\mathbf{y}(k-1)$ taken as being independent of $\mathbf{y}(k)$, the complex gradient vector, given by equation (12.4.5), in this case becomes

$$\mathbf{g} = 2\mathbf{C}^{\text{H}}\mathbf{W}_e[\mathbf{d} - \mathbf{e}_{\text{des}}] - 2\mathbf{W}_\Delta\mathbf{y}(k-1) + 2[\mathbf{C}^{\text{H}}\mathbf{W}_e\mathbf{C} + \mathbf{W}_y + \mathbf{W}_\Delta]\mathbf{y}(k). \tag{12.6.3}$$

Under steady-state conditions $\mathbf{y}(k) = \mathbf{y}(k-1)$, and the rate term in the cost function disappears. The optimal set of control variables can now be found by setting the vector $\mathbf{g}$ to zero. This gives

$$\mathbf{y}_0 = -(\mathbf{C}^{\mathrm{H}}\mathbf{W}_e\mathbf{C} + \mathbf{W}_y)^{-1}\mathbf{C}^{\mathrm{H}}\mathbf{W}_e(\mathbf{d} - \mathbf{e}_{\mathrm{des}}). \quad (12.6.4)$$

Note that the complex gradient vector can also be written as

$$\mathbf{g}(k) = 2\mathbf{C}^{\mathrm{H}}\mathbf{W}_e[\mathbf{e}(k) - \mathbf{e}_{\mathrm{des}}] + {}^2\mathbf{W}_y\mathbf{y}(k) + {}^2\mathbf{W}_\Delta[\mathbf{y}(k) - \mathbf{y}(k-1)]. \quad (12.6.5)$$

The steepest descent algorithm (equation 12.4.4) for this cost function thus becomes

$$\mathbf{y}(k) = (\mathbf{I} - \alpha\mathbf{W}_y)\mathbf{y}(k-1) - \alpha\mathbf{W}_\Delta[\mathbf{y}(k-1) - \mathbf{y}(k-2)] - \alpha\mathbf{C}^{\mathrm{H}}\mathbf{W}_e[\mathbf{e}(k^{-1}) - \mathbf{e}_{\mathrm{des}}]. \quad (12.6.6)$$

The equivalent algorithm using Newton's method is obtained by premultiplying $\mathbf{g}(k)$ by the inverse of the complex Hessian matrix, where the complex Hessian matrix in this case is given by

$$\mathbf{A} = (\mathbf{C}^{\mathrm{H}}\mathbf{W}_e\mathbf{C} + \mathbf{W}_y + \mathbf{W}_\Delta). \quad (12.6.7)$$

The relative advantages and disadvantages of effort weighting and rate weighting, together with a discussion of various algorithms for the on-line estimation of the matrix $\mathbf{C}$ and many other practical issues, are discussed with application to higher harmonic control in helicopters by Johnson (1982) and Davis (1984).

12.7 Multi-channel control of random sound

When the sound field to be controlled is not composed of a series of discrete frequencies, the additional complexity of using detection sensors to generate the reference signal has generally to be introduced. As in the single channel case discussed in Chapter 6, there is now the possibility of feedback from the secondary sources to the detection sensors which somewhat complicates the design of the controller. We assume that there are K detection sensors, in addition to the M secondary sources and L error sensors, so the controller is now a *matrix* of $K \times M$ electronic filters connecting the detection sensors to the secondary sources. This arrangement is illustrated in Fig. 12.12, in which $K = 2$, $M = 2$ and $L = 3$. Although the detection sensors are shown well separated from the primary sources of sound, if the location of the sources were known, clearly one would make an effort to locate individual sensors close to each source. The analysis that follows, however, does not make the assumption of separately identifiable primary sources. The multi-channel generalisation of the block diagram introduced in Chapter 6 under these conditions is illustrated in Fig. 12.13. Here we follow the analysis presented by Nelson *et al.* (1990b) which offers an alternative to an earlier approach to the problem presented by Elliott and Nelson (1985c).

First we will deal with the problem in the frequency domain. This enables one to derive the optimal reductions in level that can be achieved when using filters which are not constrained to be causal. It is useful both in this case and in the causally constrained case (dealt with in the next section) to work with a modified form of the basic block diagram illustrated in Fig. 12.13. The signals due to the primary sources are detected by L error sensors whose outputs prior to control are represented by the vector $\mathbf{d}$. Note that in addition to the sound due to the primary sources, the

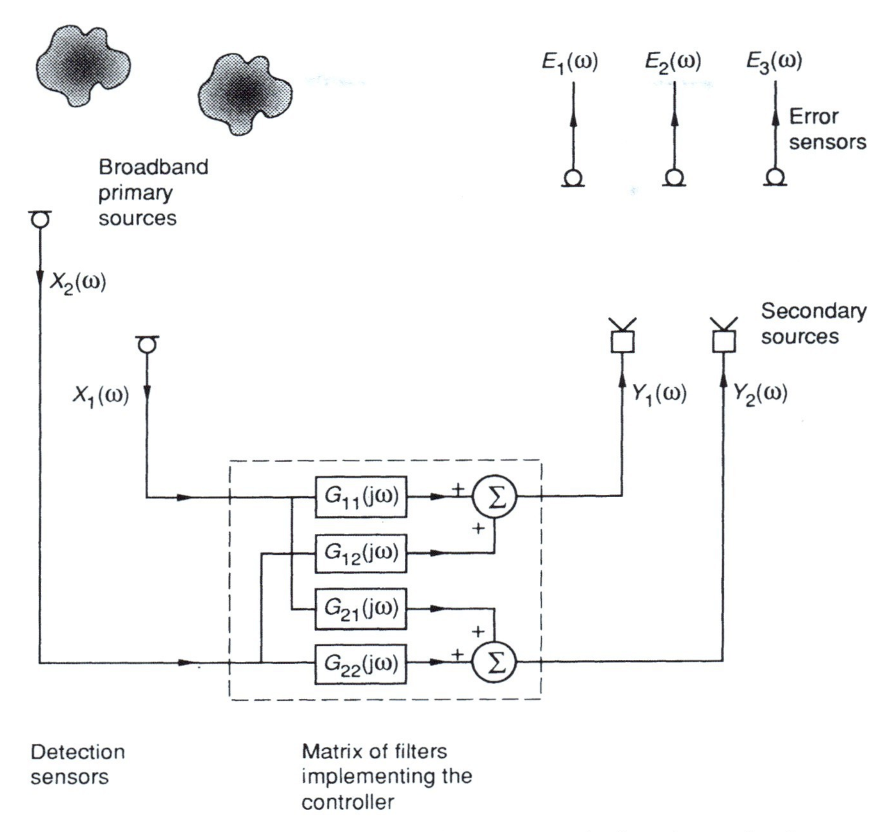

Fig. 12.12 A simple multi-channel system for the control of random noise from two primary sources.

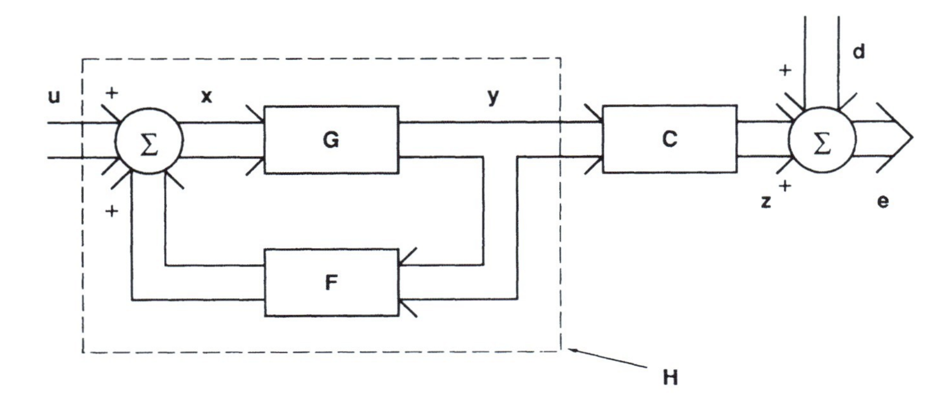

Fig. 12.13 Block diagram of the multi-channel system for the control of random noise.

sensors may be prone to measurement noise which is included in the signal vector **d**. The sound due to the primary sources is also detected by K detection sensors whose output signals *prior to control* have Fourier transforms represented by the vector **u**. Note that these signals may also contain measurement noise. These signals are passed through a matrix of electrical control filters **G** to produce the vector **y** whose elements are the Fourier transforms of the signals input to M secondary sources. These signals are also in general fed back via the transfer function matrix **F** and result in the corruption of the signals from the detection sensors. Thus we use **x** to represent the vector of signals input to the matrix **G** when control is applied. We will assume that **F** can be determined from measurements on the system and set out to determine the optimal form of the filter matrix **G** by first determining the optimal form of the dummy filter matrix **H**. The latter is defined as the filter matrix whose input is the signal vector **u** and whose output is the signal vector **y** (see Fig. 12.13). Once **H** is determined then, in principle **F** is known, and the frequency response of the physical controller **G** can be determined. This is readily demonstrated using manipulations in the frequency domain. We may write

$$\mathbf{y} = \mathbf{G}(\mathbf{u} + \mathbf{F}\mathbf{y}), \tag{12.7.1}$$

where **u** and **y** are vectors of Fourier transforms of the input and output signals and **F** and **G** are frequency response function matrices. It follows that

$$\mathbf{y} = (\mathbf{I} - \mathbf{G}\mathbf{F})^{-1}\mathbf{G}\mathbf{u}, \tag{12.7.2}$$

where **I** is the identity matrix and therefore the dummy filter matrix

$$\mathbf{H} = (\mathbf{I} - \mathbf{G}\mathbf{F})^{-1}\mathbf{G}. \tag{12.7.3}$$

This expression can be used to deduce **G** from **H**, given **F**, since it follows, after some rearrangement, that the expression for **G** may be written in the alternative forms

$$\mathbf{G} = \mathbf{H}(\mathbf{I} + \mathbf{F}\mathbf{H})^{-1} = (\mathbf{I} + \mathbf{H}\mathbf{F})^{-1}\mathbf{H}, \tag{12.7.4}$$

assuming that neither of the matrices being inverted are singular. (Proof of the second equality in equation (12.7.4) follows readily from successive pre-multiplication by $(\mathbf{I} + \mathbf{HF})$ and post-multiplication by $(\mathbf{I} + \mathbf{FH})$.) The results given by equation (12.7.4) suggest the two alternative filter structures depicted in Fig. 12.14. The first term on the right side of equation (12.7.4) suggests the "recursive" structure shown in Fig. 12.14(a) whilst the second term suggests a filter implementation which ensures the cancellation of any feedback signals illustrated in Fig. 12.14(b). Note that if **H** is constrained to be causal (as in the case dealt with in the next section) then since in any physical system **F** must be causal, it is evident from Fig. 12.14 that **G** will then also be causal. (The stability of **G** is, however, not guaranteed.) Henceforth we will proceed to determine the optimal filter **H** and assume that in practice the problem of feedback can be dealt with using this approach. There is of course a large class of problems in active noise control in which **F** is zero and the determination of the optimal value of **H** is all that is required.

The problem can thus be represented in the form illustrated in Fig. 12.15. The matrix **C** is the matrix of electroacoustic transfer functions relating the M secondary source input signals **y** to the L signals **z**. This vector represents the signals produced

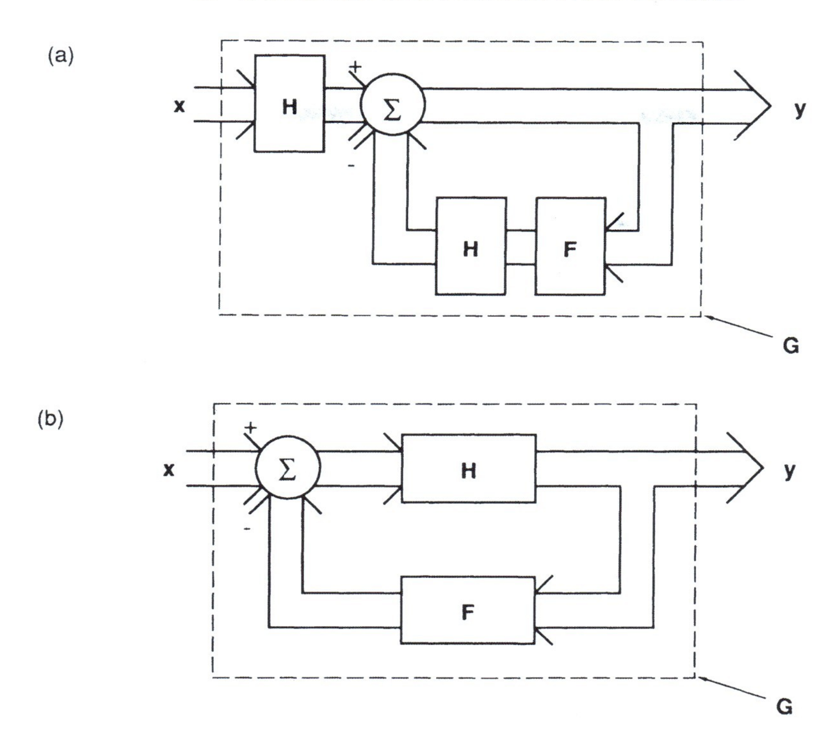

Fig. 12.14 The interpretation of the physically implemented controller, **G**, in terms of a "recursive" structure illustrated in (a) and the "feedback cancellation" structure illustrated in (b).

by the secondary sources at the L error sensors. The error signals are thus defined by $\mathbf{e} = \mathbf{d} + \mathbf{z}$. Before proceeding further we undertake another rearrangement of the block diagram that is illustrated in Fig. 12.15. Again, working in the frequency domain we first write the matrix equations

$$\mathbf{y} = \mathbf{Hu}, \qquad \mathbf{z} = \mathbf{Cy}, \tag{12.7.5a,b}$$

as

$$\mathbf{z} = \mathbf{CHu}, \tag{12.7.6}$$

or in terms of the double summation for the lth element of $\mathbf{z}$

$$Z_l(\omega) = \sum_{m=1}^{M} C_{lm}(j\omega) \sum_{k=1}^{K} H_{mk}(j\omega)U_k(\omega). \tag{12.7.7}$$

The summation can be arranged into the form

$$Z_l(\omega) = \sum_{k=1}^{K} \sum_{m=1}^{M} H_{mk}(j\omega)C_{lm}(j\omega)U_k(\omega). \tag{12.7.8}$$

This shows that the block diagram of Fig. 12.15 may be rewritten in the form depicted in Fig. 12.16. The latter shows that the lth contribution to the vector **d** may

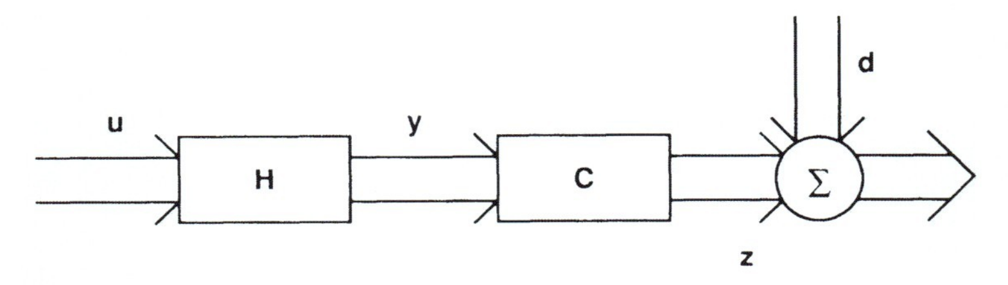

Fig. 12.15 The reduced block diagram of the multi-channel active control problem.

be written in terms of the "filtered reference signals" $R_{lmk}(\omega)$. These are the signals generated by passing the kth detected signal $U_k(\omega)$ through the transfer function $C_{lm}(j\omega)$ which comprises the l,mth element of the matrix **C**. Using these filtered reference signals, equation (12.7.8) can be rewritten as

$$Z_l(\omega) = \sum_{k=1}^{K} \sum_{m=1}^{M} H_{mk}(j\omega) R_{lmk}(\omega), \tag{12.7.9}$$

which may be written as an inner product by defining the vectors

$$\mathbf{r}_l = [R_{l11}(\omega) \ldots R_{lM1}(\omega) | R_{l12}(\omega) \ldots R_{lM2}(\omega)| \ldots |R_{l1K}(\omega) \ldots R_{lMK}(\omega)]^{\mathrm{T}}, \tag{12.7.10}$$

$$\mathbf{h} = [H_{11}(j\omega) \ldots H_{M1}(j\omega) | H_{12}(j\omega) \ldots H_{M2}(j\omega)| \ldots |H_{1K}(j\omega) \ldots H_{MK}(j\omega)]^{\mathrm{T}}, \tag{12.7.11}$$

where the vector **h** now contains all the elements of the matrix **H**. Thus we can write

$$Z_l(\omega) = \mathbf{h}^{\mathrm{T}} \mathbf{r}_l = \mathbf{r}_l^{\mathrm{T}} \mathbf{h}. \tag{12.7.12}$$

The vector of error sensor outputs due to the secondary sources may now be written as

$$\mathbf{z} = \mathbf{R}\mathbf{h}, \tag{12.7.14}$$

where the matrix **R** is defined by

$$\mathbf{R} = (\mathbf{r}_1^{\mathrm{T}}, \mathbf{r}_2^{\mathrm{T}}, \ldots, \mathbf{r}_L^{\mathrm{T}})^{\mathrm{T}}.$$

The vector of total error signals is then

$$\mathbf{e} = \mathbf{d} + \mathbf{R}\mathbf{h}. \tag{12.7.15}$$

We now seek to determine the optimal vector **h** of the frequency responses of the control filters which ensures the minimum value of the sum of the power spectral densities of L error signals at the frequency ω. Thus we define a cost function given by

$$J(\omega) = E[\mathbf{e}^{\mathrm{H}}\mathbf{e}], \tag{12.7.16}$$

where H denotes the Hermitian transpose and where the expectation operator E is used in the manner defined in Section 2.7. Thus we assume that all the signals dealt with are stationary random processes and the expectation operator refers to an ensemble average (see Chapter 2). Therefore the true power spectral densities of the

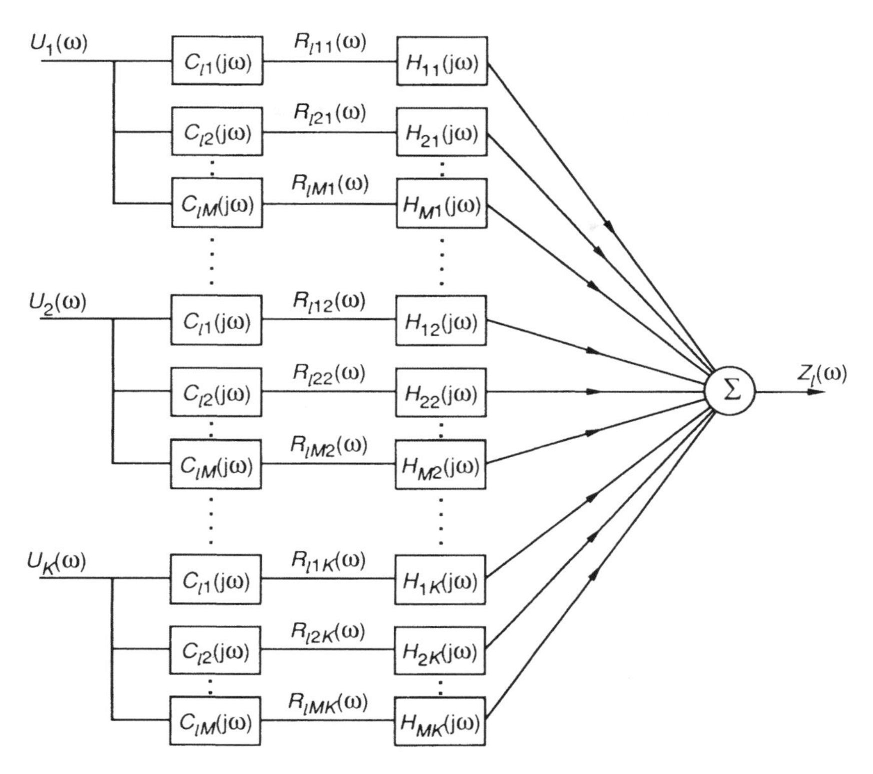

Fig. 12.16 The equivalent block diagram showing the generation of the *l*th secondary signal with the operation of the elements of the matrices **H** and **C** reversed.

signals are derived only from averaging over the ensemble of records of infinite duration. This should be borne in mind when using the results derived below in practice, since practical estimates of the true power spectrum are necessarily based on time averaging over a number of records of finite duration.

Using equation (12.7.15), the cost function can be written in the Hermitian quadratic form

$$J(\omega) = \mathbf{h}^{\mathrm{H}}\mathbf{A}\mathbf{h} + \mathbf{h}^{\mathrm{H}}\mathbf{b} + \mathbf{b}^{\mathrm{H}}\mathbf{h} + c, \tag{12.7.17}$$

where the matrix **A**, vector **b** and scalar c are defined by

$$\mathbf{A} = E[\mathbf{R}^{\mathrm{H}}\mathbf{R}], \qquad \mathbf{b} = E[\mathbf{R}^{\mathrm{H}}\mathbf{d}], \qquad c = E[\mathbf{d}^{\mathrm{H}}\mathbf{d}]. \qquad (12.7.18\text{a,b,c})$$

The latter can be recognised as the sum of the power spectra of the signals produced at the error sensors by the primary field alone. The optimal value of the filter vector **h** and the corresponding minimum value of the sum of the error sensor power spectra are thus given by $-\mathbf{A}^{-1}\mathbf{b}$ and $c - \mathbf{b}^{\mathrm{H}}\mathbf{A}^{-1}\mathbf{b}$ respectively and thus

$$\mathbf{h}_0 = -E[\mathbf{R}^{\mathrm{H}}\mathbf{R}]^{-1}E[\mathbf{R}^{\mathrm{H}}\mathbf{d}], \tag{12.7.19}$$

$$J_0(\omega) = E[\mathbf{d}^{\mathrm{H}}\mathbf{d}] - E[\mathbf{d}^{\mathrm{H}}\mathbf{R}]E[\mathbf{R}^{\mathrm{H}}\mathbf{R}]^{-1}E[\mathbf{R}^{\mathrm{H}}\mathbf{d}]. \tag{12.7.20}$$

The condition for the existence of this minimum is that $\mathbf{A} = E[\mathbf{R}^{\mathrm{H}}\mathbf{R}]$ be positive definite. It is clear that in the absence of the primary field (i.e. all $D_l(\omega)$ are zero) then the sum of the power spectra due to the secondary sources alone is given by $\mathbf{h}^{\mathrm{H}}\mathbf{A}\mathbf{h}$. This will clearly be greater than zero for all non-zero values of $\mathbf{h}$ provided all the elements of the vector $\mathbf{u}$ of detected signals are non-zero. This therefore defines a sufficient condition for the positive definiteness of $\mathbf{A}$ and the existence of a unique minimum of $J(\omega)$. Equations (12.7.19) and (12.7.20) thus define the optimal values of the filter frequency response vector $\mathbf{h}$ (and thus the optimal values of the elements of the filter matrix $\mathbf{H}$) and also the maximum reduction of the chosen cost function that can be achieved. To evaluate these optimal results, access is required to both the filtered reference signals $R_{lmk}(\omega)$ and the primary field signals $Z_l(\omega)$. In general, the former require a knowledge of both the detected signals $U_k(\omega)$ and the transfer functions $C_{lm}(j\omega)$, although in several special cases a knowledge of the transfer functions $C_{lm}(j\omega)$ is not required. The frequency response of the physically realised filter matrix $\mathbf{G}$ in Fig. 12.13 can now be calculated at each frequency by transforming the vector of filter responses $\mathbf{h}_0$ back into matrix form, $\mathbf{H}$, of equation (12.7.5a), and using the matrix of frequency responses of the feedback path, $\mathbf{F}$, so that $\mathbf{G}$ can be calculated using equation (12.7.4).

In the particular case of a random sound field generated by a number of primary sources which may or may not be correlated with one another, we may seek to minimise the power spectral density at *one* location by operating on K detection signals. One would clearly seek to ensure that the detected signals between them were able to characterise the output of the primary sources. In this case the matrix $\mathbf{A}$ can be written as

$$\mathbf{A} = E[\{C(j\omega)\mathbf{u}\}^*(C(j\omega)\mathbf{u}\}^{\mathrm{T}}] = |C(j\omega)|^2\mathbf{S}_{uu}, \tag{12.7.21}$$

where $\mathbf{S}_{uu}$ is the matrix of cross-spectra of the signals from the detection sensors. The vector $\mathbf{b}$ reduces to

$$\mathbf{b} = E[(C(j\omega)\mathbf{u})^*d] = C^*(j\omega)\mathbf{s}_{ud}, \tag{12.7.22}$$

where the vector $\mathbf{s}_{ud}$ is the vector of cross-spectra between the detected signals and the signal at the error sensor due to the primary field and measurement noise. The solution for the optimal filter vector thus reduces to

$$\mathbf{h}_0 = -\frac{1}{C(j\omega)}\mathbf{S}_{uu}^{-1}\,\mathbf{s}_{ud}, \tag{12.7.23}$$

and the minimum value of the error sensor spectral density is given by

$$J_0(\omega) = S_{dd}(\omega) - \mathbf{s}_{ud}^{\mathrm{H}}\,\mathbf{S}_{uu}^{-1}\,\mathbf{s}_{ud}, \tag{12.7.24}$$

where $S_{dd}(\omega)$ is the power spectral density of the signal at the error sensor due to the primary field and measurement noise. Note that this expression does not depend on the frequency response functions $C_{lm}(j\omega)$ and depends only on the properties of the reference and error signals measured before control. Putting $J_d(\omega) = S_{dd}(\omega)$, this expression can be written in non-dimensional form as

$$\frac{J_0(\omega)}{J_d(\omega)} = 1 - \frac{\mathbf{s}_{ud}^{\mathrm{H}}\,\mathbf{S}_{uu}^{-1}\,\mathbf{s}_{ud}}{S_{dd}(\omega)} = 1 - \eta^2{}_{ud}(\omega), \tag{12.7.25}$$

where $\eta_{ud}(\omega)$ is the *multiple coherence function* between the detection sensor inputs and the output from the single error sensor. It can be shown that this value varies between zero and unity and its estimation in a given practical situation therefore immediately evaluates the potential for active control. A concise description of the multiple coherence function and its use in the analysis of random signals is given by, for example, Newland (1984). The significance of the multiple coherence function within the context of active control was first pointed out by Ross (1980). In the single channel case, where we have only *one* detection sensor, this result reduces to that derived in Chapter 6 where we showed that the ultimate performance limit of a feedforward controller was determined by the *ordinary coherence function* relating the detection and error sensor outputs. We also pointed out in Chapter 6 that the coherence function should be used with some caution in determining the potential performance of an active controller. The major drawback is, of course, that since the above result has been established from manipulations in the frequency domain, there is no guarantee of the realisability of the optimal filters derived. However, we can conduct an analysis in the time domain (in discrete time) and establish the optimal form of a matrix of digital FIR filters for the multi-channel feedforward control of random sound. This is treated in the next section.

12.8 A digital filtering formulation

The frequency domain formulation for multi-channel active control, and the discussion of iterative algorithms in the frequency domain, is of considerable help in understanding the general behaviour of these control systems. However, such algorithms have still to be implemented in some electronic form to be of any practical use in controlling sound. The early implementations were purely analogue, and the control algorithms often restricted to manual adjustment of analogue circuits which altered the amplitude and phase of pure tone reference signals as, for example, in Fig. 12.17, which is taken from Conover (1956). As the number of

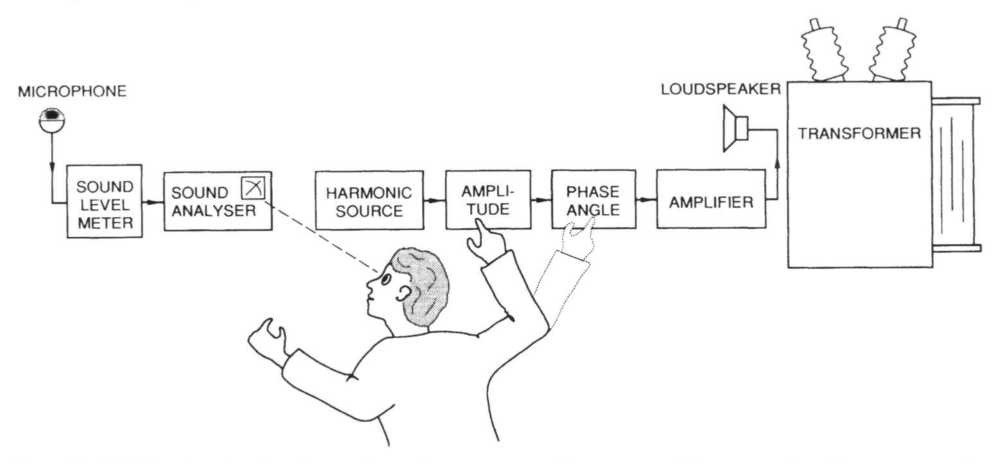

Fig. 12.17 Simple single channel analogue control system with manual adjustment, after Conover (1956).

channels increases and the need for automatic adjustment becomes more important, such manual systems become impractical. One or two successful analogue implementations of multiple channel, automatically adjusting, active noise control systems have been reported in the literature (see, for example, Kido, 1975). The more recent generation of multi-channel control systems all appear to use analogue-to-digital converters to transform the error signals into digital form, some form of digital computer as the controller, and digital-to-analogue converters to generate the control signals.

The most obvious way of using such digital controllers is to provide a direct implementation of the frequency domain algorithms discussed in the previous sections. In other words, some kind of transformation is used to estimate the in-phase and quadrature parts of the error signals at a number of discrete frequencies and the iterative algorithms described in Section 12.4 are programmed directly into the controller to operate independently at each frequency. Typically it is the harmonics of a set of periodic error signals which need to be obtained, and the fast Fourier transform can provide a convenient algorithm to generate the required frequency domain signals. Another way of using a digital control system is to work directly in the time domain by digitally filtering a reference sequence to generate the output signals driving each of the secondary sources. If the reference sequence were of sinusoidal form, only two coefficients would be required in the control filter to enable the amplitude and phase of the signal fed to the secondary source to be controlled.

Thus implementing multi-channel feedforward control by using digital filters may have clear advantages when dealing with periodic sound. It is also useful, however, to use a digital filtering formulation when dealing with the multi-channel control of random sound. As we have pointed out in the last section, we need to undertake an analysis in the time domain to enable the derivation of the matrix of optimal *causal* control filters. We can use the block diagram presented in the last section together with its rearranged form (see Figs 12.13, 12.15 and 12.16) to determine a causally constrained solution to the problem. The minimisation of the sum of the time-averaged squared error signals by using a matrix of analogue filters $\mathbf{H}$ whose elements are constrained to have causal impulse response functions has been dealt with by Nelson *et al.* (1988b). In that work it was shown that the vector of causal impulse response functions associated with the composite vector $\underset{\sim}{\mathbf{h}}$ defined above must satisfy a matrix Wiener–Hopf integral equation. This equation is difficult to solve analytically except in certain cases (see, for example, the problems addressed by Nelson *et al.*, 1990a). Here we constrain the optimal filter to have an impulse response which is of finite duration as well as being causal and analyse the problem in discrete time. This is therefore the case that is of most relevance in practice, where the controller is implemented as a matrix of digital FIR filters. The analysis presented here is a generalisation of that first presented by Elliott *et al.* (1987a,b) so that we may deal with multiple detected signals. Some further slight modifications are given to the analysis of Elliott *et al.* (1987a,b) in order to more easily facilitate numerical studies of multi-channel control of random sound. We work with the time domain version of the rearranged block diagram of Fig. 12.16 and follow the analysis of Nelson *et al.* (1990b). (See also Nelson and Elliott, 1989b.)

Thus, working in discrete time, the nth sample of the lth signal z_l can be written as the summation

$$z_l(n) = \sum_{k=1}^{K} \sum_{m=1}^{M} \sum_{i=0}^{I-1} h_{mk}(i) r_{lmk}(n-i), \tag{12.8.1}$$

where $h_{mk}(i)$ is the ith coefficient of the FIR control filter driving the mth secondary source from the kth reference signal, and each control filter is assumed to have an impulse response of I samples in duration. z_l can also be written as the convolution

$$z_l(n) = \mathbf{h}^{\mathrm{T}}(0)\mathbf{r}_l(n) + \mathbf{h}^{\mathrm{T}}(1)\mathbf{r}_l(n-1) + \ldots \mathbf{h}^{\mathrm{T}}(I-1)\mathbf{r}_l(n-I+1), \tag{12.8.2}$$

where we have defined a composite tap weight vector and a sampled reference signal vector respectively by

$$\mathbf{h}^{\mathrm{T}}(i) = [h_{11}(i)h_{21}(i) \ldots h_{M1}(i)|h_{12}(i)h_{22}(i) \ldots h_{M2}(i)| \ldots |h_{1K}(i)h_{2K}(i) \ldots h_{MK}(i)|], \tag{12.8.3}$$

$$\mathbf{r}_l^{\mathrm{T}}(n) = [r_{l11}(n)r_{l21}(n)..r_{lM1}(n)|r_{l12}(n)r_{l22}(n)..r_{lM2}(n)|..|r_{l1K}(n)r_{l2K}(n)..r_{lMK}(n)|]. \tag{12.8.4}$$

We now define a further composite tap weight vector $\mathbf{w}$ which consists of all the I tap weights of all the $L \times M$ filters such that

$$\mathbf{w}^{\mathrm{T}} = [\mathbf{h}^{\mathrm{T}}(0)\ \mathbf{h}^{\mathrm{T}}(1) \ldots \mathbf{h}^{\mathrm{T}}(I-1)]. \tag{12.8.5}$$

This enables us to write the Lth order vector of sampled error signals as

$$\mathbf{e}(n) = \mathbf{d}(n) + \mathbf{R}(n)\mathbf{w}, \tag{12.8.6}$$

where the matrix $\mathbf{R}(n)$ is defined by

$$\mathbf{R}(n) = \begin{bmatrix} \mathbf{r}_1^{\mathrm{T}}(n) & \mathbf{r}_1^{\mathrm{T}}(n-1) & \ldots & \mathbf{r}_1^{\mathrm{T}}(n-I+1) \\ \mathbf{r}_2^{\mathrm{T}}(n) & \mathbf{r}_2^{\mathrm{T}}(n-1) & \ldots & \mathbf{r}_2^{\mathrm{T}}(n-I+1) \\ \cdot & & & \\ \cdot & & & \\ \mathbf{r}_L^{\mathrm{T}}(n) & \mathbf{r}_L^{\mathrm{T}}(n-1) & \ldots & \mathbf{r}_L^{\mathrm{T}}(n-I+1) \end{bmatrix}. \tag{12.8.7}$$

We now define a cost function given by the sum of the L time-averaged error signals such that

$$J = E[\mathbf{e}^{\mathrm{T}}(n)\mathbf{e}(n)], \tag{12.8.8}$$

where E denotes the expectation operator, then substitution of equation (12.8.6) and subsequent expansion shows that

$$J = \mathbf{w}^{\mathrm{T}}E[\mathbf{R}^{\mathrm{T}}(n)\mathbf{R}(n)]\mathbf{w} + 2\mathbf{w}^{\mathrm{T}}E[\mathbf{R}^{\mathrm{T}}(n)\mathbf{d}(n)] + E[\mathbf{d}^{\mathrm{T}}(n)\mathbf{d}(n)]. \tag{12.8.9}$$

This is a quadratic function of the tap weight vector $\mathbf{w}$ which is minimised by the optimal vector

$$\mathbf{w}_0 = -\{E[\mathbf{R}^{\mathrm{T}}(n)\mathbf{R}(n)]\}^{-1}\{E[\mathbf{R}^{\mathrm{T}}(n)\mathbf{d}(n)]\}, \tag{12.8.10}$$

and has the corresponding minimum value given by

$$J_0 = E[\mathbf{d}^{\mathrm{T}}(n)\mathbf{d}(n)] - \{E[\mathbf{R}^{\mathrm{T}}(n)\mathbf{d}(n)]\}^{\mathrm{T}}\{E[\mathbf{R}^{\mathrm{T}}(n)\mathbf{R}(n)]\}^{-1}\{E[\mathbf{R}^{\mathrm{T}}(n)\mathbf{d}(n)]\}. \tag{12.8.11}$$

The matrix $E[\mathbf{R}^{\mathrm{T}}(n)\mathbf{R}(n)]$ which has to be inverted is clearly of high order, but by using the definition of the composite tap weight vector given by equation (12.8.5) this matrix can be shown to have a block Toeplitz structure and use can be made of recursive algorithms for its efficient inversion (see, for example, Robinson, 1978). Note that the definition of the composite tap weight vector given here is not the same as that adopted by Elliott *et al.* (1987a) but is a generalisation of that used by Hough (1988) to obtain a block Toeplitz structure in the case of a single reference signal.

It is interesting to note that if each of the control filters had single complex coefficients, and the signals are considered to be at a particular frequency in the frequency domain, rather than at a particular time in the time domain, equation (12.8.6) may be used as the basis of the frequency domain formulation expressed by equation (12.7.15) and the frequency domain optimal filter given by equation (12.7.19) can be deduced from the more general result of equation (12.8.10).

As an example of the difference beween the results of an unconstrained, frequency domain, optimisation and those obtained using a constrained time domain optimisation, Fig. 12.18 shows some computer simulations presented by Sutton *et al.* (1990) from a study of the active control of broad-band road noise in cars. The upper curve is the power spectral density of the internal pressure measured at one point inside the car, as it was driven over a road with a rough concrete surface. The outputs from six accelerometers attached to the car body were recorded at the same time as the pressure signal, and these signals used to calculate the multiple coherence function relating these six reference signals and the pressure signal. The multiple coherence function was then used in equation (12.7.25) to calculate the best possible reductions in the pressure spectrum, and the resulting residual power spectral density is plotted in Fig. 12.18. This simulation of the unconstrained

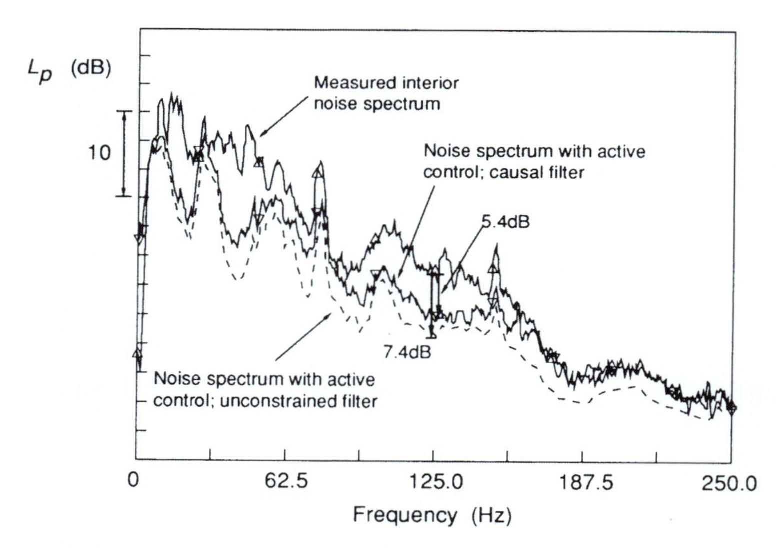

Fig. 12.18 Results of computer simulations of the effectiveness of active control in reducing low frequency random noise in a small hatchback car (after Sutton *et al.*, 1990). The predicted reductions are shown when the control filters are unconstrained and when they are constrained to have a causal impulse response. The filters operate on six reference signals derived from measurements of the vibrations on the underside of the car body.

optimisation predicts, for example, a reduction of about 7 dB at frequencies around 100 Hz. The sampled time histories from the six accelerometers and the microphone were then used to calculate the optimal digital FIR filter array to minimise the pressure using equation (12.8.10), with the constraints that the filters had a delay of 6 ms and were each composed of only 128 coefficients. The power spectral densities of the predicted residual pressure in this case are higher than for the unconstrained case, with only about 5 dB reduction at frequencies near 100 Hz.

Another approach to the design of realisable digital control filters is that presented by Miyoshi and Kaneda (1988b). Their approach is to use a number of secondary sources that is one *greater* than the number of microphones at which perfect cancellation of the sound field is required. The filter design technique relies on the multiple input/output inverse theorem described in detail by Miyoshi and Kaneda (1988a) which shows that for a general single input–single output linear transmission system an exact inverse of the system can be produced by using two transmission channels. A discussion of this principle, and its relationship to the formulation presented here, has been presented by Nelson *et al.* (1991a).

12.9 Adaptive digital filtering algorithms

In order to track changes in the primary field it is important to have ways of continuously readjusting the filter coefficients to give the best performance. One way of achieving this would be to implement algorithms in which use is made of the derivative of the cost function above with respect to the vector of filter coefficients. This gradient vector is given by using equation (12.8.9) and equation (A5.9) in the appendix:

$$\frac{\partial J}{\partial \mathbf{w}} = 2E[\mathbf{R}^{\mathrm{T}}(n)\mathbf{R}(n)\mathbf{w} + \mathbf{R}^{\mathrm{T}}(n)\mathbf{d}(n)], \tag{12.9.1}$$

and using equation (12.8.6) shows that this can be written as

$$\frac{\partial J}{\partial \mathbf{w}} = 2E[\mathbf{R}^{\mathrm{T}}(n)\mathbf{e}(n)]. \tag{12.9.2}$$

Either a steepest descent algorithm or a Newton's method algorithm, with use of the Hessian matrix $E[\mathbf{R}^{\mathrm{T}}(n)\mathbf{R}(n)]$, could be envisaged. In practice the statistical expectation associated with the derivative would have to be realised by performing an average over time after each iteration of the filter coefficients. This could be very time consuming. An alternative approach is to use the gradient of the sum of the squares of the *instantaneous* error signals to update *each* of the filter coefficients at *every* sample time. This is an approach pioneered by Widrow for electrical noise cancellation (Widrow and Hoff, 1960; Widrow and Stearns, 1985; see also Chapter 4), where such algorithms are very widely used in practical applications. In the context being considered here the gradient of the sum of the squares of the instantaneous error signals is given by

$$\frac{\partial[\mathbf{e}^{\mathrm{T}}(n)\mathbf{e}(n)]}{\partial \mathbf{w}} = 2\mathbf{R}^{\mathrm{T}}(n)\mathbf{e}(n). \tag{12.9.3}$$

The instantaneous equivalent of the method of steepest descent is now given by (Elliott and Nelson, 1985a)

$$\mathbf{w}(n) = \mathbf{w}(n-1) - \alpha\mathbf{R}^{\mathrm{T}}(n-1)\mathbf{e}(n-1). \tag{12.9.4}$$

This algorithm, which has become known as the *Multiple Error LMS algorithm*, is a multi-channel generalisation of the filtered-x LMS algorithm, discussed in Chapter 6. It should be noted that the definition of the composite tap weight vector, equation

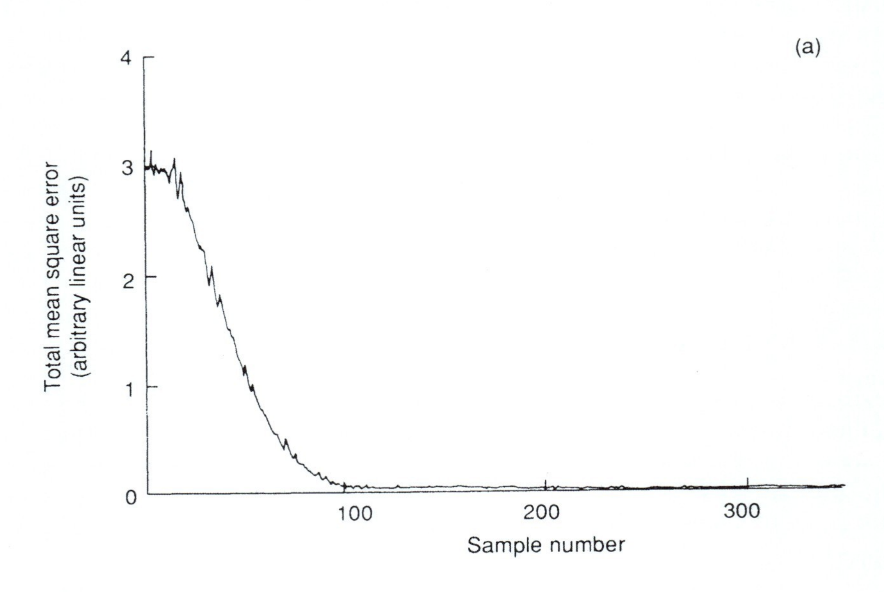

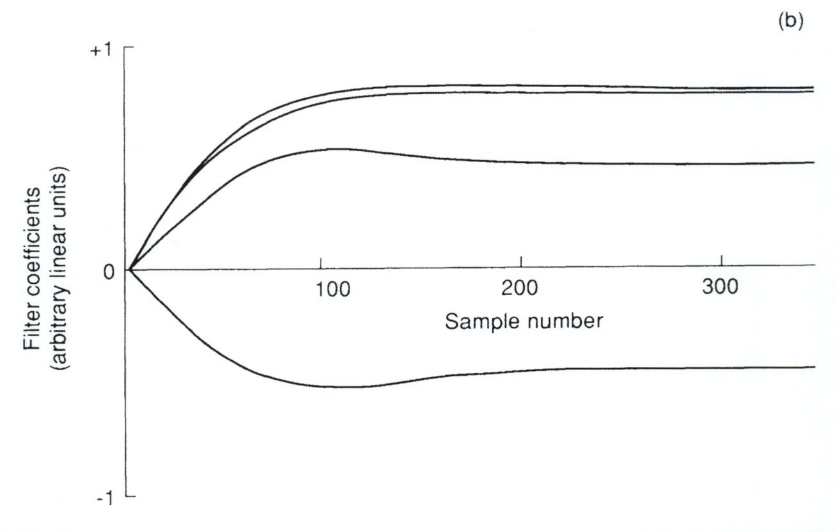

Fig. 12.19 Convergence of the sum of the squared microphone signals and filter coefficients in an experiment using the Multiple Error LMS algorithm (after Elliott *et al.*, 1987a).

(12.8.2), does not influence the implementation of this algorithm. An example of the convergence of the sum of the squares of the outputs from four error microphones in a simple laboratory experiment described by Elliott *et al.* (1987a) is shown in Fig. 12.19. A pure tone primary field was generated and two secondary sources were used to control the sound field, each driven by digital filters with two coefficients. The convergence of the four filter coefficients, adapted using the algorithm above, is also shown in Fig. 12.19. Algorithms of this form are often called *stochastic gradient* algorithms, since the update quantity is an instantaneous estimate of the true gradient, which is correct on average but at any one instant has a random or stochastic error associated with it. The algorithm has also been found very useful in the design of inverse filters in the multi-channel reproduction of sound fields (see Nelson *et al.*, 1991b, for a full description of this work).

The instantaneous equivalent of Newton's method could be written in a variety of ways, using, for example, an instantaneous estimate of the Hessian matrix, $E[\mathbf{R}^{\mathrm{T}}(n)\mathbf{R}(n)]$. The Hessian matrix does not, however, change during convergence unlike the vector of gradients, and so an estimate of the true, averaged, Hessian matrix could be pre-computed before the algorithm was implemented and used to normalise the gradient estimate in an algorithm:

$$\mathbf{w}(n) = \mathbf{w}(n-1) - \alpha E[\mathbf{R}^{\mathrm{T}}(n)\mathbf{R}(n)]^{-1}\mathbf{R}^{\mathrm{T}}(n-1)\mathbf{e}(n-1), \qquad (12.9.5)$$

which has been called the stochastic Newton algorithm (Elliott and Nelson, 1988). A similar algorithm is used in the adaptive cancellation of electrical noise, where it is called the recursive least squares or RLS algorithm (Haykin, 1986), although in that application the Hessian matrix cannot be estimated beforehand and must be iteratively estimated at the same time as the filter coefficients are being adjusted, which leads to considerable complications.

The simple justification for such stochastic algorithms presented above does not, however, bear close theoretical inspection, since the basis of the equation for the gradient of $\mathbf{e}^{\mathrm{T}}(n)\mathbf{e}(n)$ given by equation (12.9.2) is ultimately given by

$$\mathbf{e}(n) = \mathbf{d}(n) + \mathbf{R}(n)\mathbf{w}, \qquad (12.9.6)$$

which was derived in the previous section under the assumption of *constant* filter coefficients! As with the single channel filtered-x LMS algorithm, the approximation involved in using this equation for the error sequence is clearly a reasonable one if the filter coefficients are changing on a timescale which is long compared with the delays and dynamic response of the system under control. It is found, however, that the stochastic gradient algorithm still converges when the convergence coefficient, α, is large enough for the filter coefficients to change on a timescale which is *comparable* with the delays in the system (Elliott *et al.*, 1987b). Under such circumstances, equation (12.9.6) for the error sequence is no longer valid, and indeed the error surface itself is no longer quadratic under such conditions (Elliott *et al.*, 1988d).

Clearly no adaptive algorithm optimally control the error signals more quickly than the physical propagation delays in the system being controlled. The stochastic gradient algorithm has, however, been found to respond on a timescale which is comparable with this delay. A more detailed study of the way the error signal and filter coefficients change with time during adaption of the filtered-x LMS algorithm under these conditions is presented by Tsujino and Elliott (1991). These authors show that the *trajectories* of the filter coefficients when driven by the filtered-x LMS algorithm are similar to those required to minimise the sum of the squared values of the error sequence over the duration of the transient (the globally optimal solution). The transient response of even relatively simple electrical adaptive filters is, however, still not fully understood (Haykin, 1986). In the current application the transient behaviour of the adaptive filter combines with the transient behaviour of the acoustic system to make the complete behaviour even more difficult to analyse exactly. In the majority of applications, however, the simplified analysis presented above provides a reasonable model for the physical behaviour of iterative algorithms. In particular the geometric interpretation of the behaviour of these algorithms, in terms of the steady-state error surface, has been found to be very useful in understanding the behaviour of practical multi-channel control systems (Boucher and Elliott, 1989; Boucher *et al.*, 1991; Elliott *et al.*, 1991a).

Appendix
A Little Linear Algebra

In this appendix we briefly review the definition of vectors and matrices and the important properties of linear algebra. We have seen that the use of linear algebra considerably simplifies the formulation of multiple channel active control problems. Our purpose here is to provide a convenient reference for the important properties of matrix manipulation. We will concentrate on the case where all variables are complex, unless otherwise stated. Most of the results are taken from Noble (1969), Noble and Daniel (1977) and Bellman (1960), in which proofs of the various properties will be found.

A.1 Vectors

A *vector* is an ordered series of numbers, $x_1, x_2, \ldots, x_N$, which are generally complex. A vector is denoted here by a lower case, bold variable and, unless otherwise stated, corresponds to the numbers being listed as a single *column*:

$$\mathbf{x} = \begin{bmatrix} x_1 \\ x_2 \\ \cdot \\ \cdot \\ \cdot \\ x_N \end{bmatrix}. \tag{A1.1}$$

The *transpose* of a vector, denoted by the superscript T, is the corresponding *row* of ordered numbers:

$$\mathbf{x}^{\mathrm{T}} = [x_1, x_2, \ldots, x_N]. \tag{A1.2}$$

The *Hermitian* transpose of a vector, denoted by the superscript H, is the complex conjugate of the transpose

$$\mathbf{x}^{\mathrm{H}} = [x_1^*, x_2^*, \ldots, x_N^*], \tag{A1.3}$$

where the superscript * denotes complex conjugation. The *inner product* (or scalar product) of two vectors, $\mathbf{x}$ and $\mathbf{y}$, of equal size, is defined to be

$$\mathbf{x}^{\mathrm{H}}\,\mathbf{y} = x_1^* y_1 + x_2^* y_2 + x_3^* y_3 + \ldots + x_N^* y_N, \tag{A1.4}$$

and is often written as $(\mathbf{x},\mathbf{y})$. Note that $\mathbf{y}^{\mathrm{H}}\,\mathbf{x} = (\mathbf{x}^{\mathrm{H}}\,\mathbf{y})^*$. If $\mathbf{x}^{\mathrm{H}}\,\mathbf{y} = 0$, then the vectors $\mathbf{x}$ and $\mathbf{y}$ are said to be *orthogonal*.

A.2 Matrices

A *matrix* is a set of numbers, generally complex, arranged in a rectangular array. A matrix is denoted here by an upper case, bold, variable. For example

$$\mathbf{A} = \begin{bmatrix} a_{11} & a_{12} & a_{13} & \dots & a_{1N} \\ a_{21} & a_{22} & a_{23} & \dots & a_{2N} \\ \cdot & \cdot & \cdot & \cdot & \cdot \\ \cdot & \cdot & \cdot & \cdot & \cdot \\ a_{M1} & a_{M2} & a_{M3} & \dots & a_{MN} \end{bmatrix}, \tag{A2.1}$$

is an $M \times N$ matrix (it has M rows and N columns). The element in the ith row and jth column is denoted a_{ij}. If $N = M$, the matrix is said to be *square*. The matrix **0** has all zero elements and is called the *null* or *zero* matrix. The square matrix

$$\mathbf{I} = \begin{bmatrix} 1\,0\,0 & & \cdot\ \cdot \\ 0\,1\,0 & & \cdot\ \cdot \\ 0\,0\,1 & & \cdot\ \cdot \\ \cdot\ \cdot\ \cdot & & \\ \cdot\ \cdot\ \cdot & & 1\,0 \\ \cdot\ \cdot\ \cdot & & 0\,1 \end{bmatrix}, \tag{A2.2}$$

with all elements on the leading (or principal) diagonal being unity, and all off-diagonal elements being zero, is called the *identity* or *unit* matrix. The matrices **A** and **B** are said to be *equal* if, and only if, (a) both have the same number of rows and columns, and (b) all corresponding elements are the same in the two matrices, i.e.

$$a_{ij} = b_{ij} \quad \text{for all } i \text{ and } j. \tag{A2.3}$$

The *sum* of two matrices (**A** and **B**) is defined only if they both have the same number of rows and columns, and is itself a matrix of the same size (**C**) whose elements are the sum of the corresponding elements of **A** and **B**:

$$c_{ij} = a_{ij} + b_{ij}. \tag{A2.4}$$

The following laws apply to matrix addition:

$$(\mathbf{A} + \mathbf{B}) + \mathbf{C} = \mathbf{A} + (\mathbf{B} + \mathbf{C}) \quad \text{associative law} \tag{A2.5}$$

$$\mathbf{A} + \mathbf{B} = \mathbf{B} + \mathbf{A} \quad \text{commutative law.} \tag{A2.6}$$

The *multiplication* of a matrix **A** by a matrix **B** to form the matrix product **AB** is defined only if the number of columns of **A** is equal to the number of rows of **B**. If **A** is an $M \times N$ matrix with elements a_{ij} and **B** is an $N \times P$ matrix with elements b_{ij}, then $\mathbf{C} = \mathbf{AB}$ is an $M \times P$ matrix whosc elements are given by

$$c_{ik} = \sum_{j=1}^{N} a_{ij} b_{jk}, \tag{A2.7}$$

i.e. the elements are the sum of the products of the elements of the ith row of **A** and the elements of the kth column of **B**. In the product **AB**, the matrix **A** is said to pre-multiply **B**, or **B** post-multiplies **A**. The order of multiplication is important since in general

$$\mathbf{AB} \neq \mathbf{BA}, \tag{A2.8}$$

even if both operations are defined, i.e. **A** and **B** are square. Although the commutative law is not true for matrix multiplication, the associative and distributive laws do hold, i.e.

$$(\mathbf{AB})\mathbf{C} = \mathbf{A}(\mathbf{BC}), \tag{A2.9}$$

$$\mathbf{A}(\mathbf{B} + \mathbf{C}) = \mathbf{AB} + \mathbf{AC}. \tag{A2.10}$$

Note also that $\mathbf{AB} = \mathbf{0}$ does *not* imply that either **A** or **B** are zero, and if we have an equation of the form $\mathbf{AB} = \mathbf{AC}$, we *cannot*, in general, conclude that either $\mathbf{A} = \mathbf{0}$ or $\mathbf{B} = \mathbf{C}$.

The identity matrix has particularly simple properties under multiplication with a square matrix (**A**):

$$\mathbf{IA} = \mathbf{AI} = \mathbf{A}. \tag{A2.11}$$

The *transpose* of the $M \times N$ matrix **A** is an $N \times M$ matrix denoted $\mathbf{A}^{\mathrm{T}}$ obtained by interchanging the rows and columns of **A**. If $\mathbf{A} = \mathbf{A}^{\mathrm{T}}$ the square matrix **A** is said to be symmetric. The *Hermitian transpose* of a matrix denoted $\mathbf{A}^{\mathrm{H}}$ is the complex conjugate of the transpose. So if

$$\mathbf{A} = \begin{bmatrix} a_{11} & a_{12} & \cdots & a_{M1} \\ a_{21} & a_{22} & \cdots & a_{M2} \\ \cdot & \cdot & \cdot & \cdot \\ \cdot & \cdot & \cdot & \cdot \\ a_{1N} & a_{2N} & \cdots & a_{MN} \end{bmatrix}, \quad \mathbf{A}^{\mathrm{H}} = \begin{bmatrix} a_{11}^* & a_{21}^* & \cdots & a_{1N}^* \\ a_{12}^* & a_{22}^* & \cdots & a_{2N}^* \\ \cdot & \cdot & \cdot & \cdot \\ \cdot & \cdots & & \\ a_{M1}^* & a_{M2}^* & \cdots & a_{MN}^* \end{bmatrix}. \tag{A2.12}$$

The properties of the Hermitian transpose, from which the properties of the normal transpose are readily deduced, are:

$$(\mathbf{A} + \mathbf{B})^{\mathrm{H}} = \mathbf{A}^{\mathrm{H}} + \mathbf{B}^{\mathrm{H}}, \tag{A2.13}$$

$$(\mathbf{A}^{\mathrm{H}})^{\mathrm{H}} = \mathbf{A}, \tag{A2.14}$$

$$(\mathbf{A\,B})^{\mathrm{H}} = \mathbf{B}^{\mathrm{H}}\,\mathbf{A}^{\mathrm{H}}. \tag{A2.15}$$

If a square matrix **A** has the property that

$$\mathbf{A}^{\mathrm{H}}\mathbf{A} = \mathbf{A}\mathbf{A}^{\mathrm{H}} = \mathbf{I},$$

it is said to be *unitary*, and can be interpreted as having columns which are mutually orthogonal, but normalised such that their inner products with themselves are unity, i.e. they are *orthonormal*. Unitary matrices have the important property that their Hermitian transpose is equal to their inverse. If a square matrix **A** is equal to its own Hermitian transpose it is said to be a *Hermitian* matrix, i.e.

$$\text{if } \mathbf{A} = \mathbf{A}^{\mathrm{H}}, \quad \mathbf{A} \text{ is Hermitian.} \tag{A2.16}$$

A simple example of a Hermitian matrix, used in the main text, is $\mathbf{A} = \mathbf{C}^{\mathrm{H}}\mathbf{C}$. Note that a Hermitian matrix must have real elements along its principal diagonal. If **A** is Hermitian, then products of the form $\mathbf{x}^{\mathrm{H}}\mathbf{A}\mathbf{x}$ will be entirely real scalars. If **A** is also *positive definite* then also

$$\mathbf{x}^{\mathrm{H}}\mathbf{A}\mathbf{x} > 0 \text{ for all } \mathbf{x} \neq 0, \tag{A2.17}$$

and if **A** is *positive semi-definite* then

$$\mathbf{x}^{\mathrm{H}}\mathbf{A}\mathbf{x} \geq 0 \text{ for all } \mathbf{x} \neq 0. \tag{A2.18}$$

A.3 Determinants and the inverse matrix

The *determinant* of a 2 × 2 matrix, **A**, is defined as

$$|\mathbf{A}| = \begin{vmatrix} a_{11} & a_{12} \\ a_{21} & a_{22} \end{vmatrix} = a_{11}a_{22} - a_{12}a_{21}. \tag{A3.1}$$

The *minor*, M_{ij}, of the element a_{ij} of the general square matrix **A** is the determinant of the matrix formed by deleting the ith row and jth column from **A**. For example, the minor M_{21} of the 3 × 3 matrix

$$\mathbf{A} = \begin{vmatrix} a_{11} & a_{12} & a_{13} \\ a_{21} & a_{22} & a_{23} \\ a_{31} & a_{32} & a_{33} \end{vmatrix}, \tag{A3.2}$$

is given by striking out the second row and first column to give

$$M_{21} = \begin{vmatrix} a_{12} & a_{13} \\ a_{32} & a_{33} \end{vmatrix}. \tag{A3.3}$$

The *cofactor* C_{ij} of element a_{ij} of the matrix **A** is defined to be

$$C_{ij} = (-1)^{i+j}M_{ij}. \tag{A3.4}$$

The determinant of a square matrix of any size can be expanded as the sum of the products of elements and their cofactors along any row or column. So the determinant of the 3 × 3 matrix above is, for example, equal to

$$|\mathbf{A}| = a_{11}C_{11} + a_{12}C_{12} + a_{13}C_{13}. \tag{A3.5}$$

The determinant of a square matrix of arbitrary order can thus be broken up into smaller and smaller matrices, and so be readily calculated. Note that if **A** and **B** are square and of the same order

$$|\mathbf{AB}| = |\mathbf{A}|\,|\mathbf{B}|. \tag{A3.6}$$

If the determinant of a matrix is zero the matrix is said to be *singular*.

The *inverse* $\mathbf{A}^{-1}$ of a matrix **A** is defined by the equation

$$\mathbf{A}\mathbf{A}^{-1} = \mathbf{A}^{-1}\mathbf{A} = \mathbf{I}. \tag{A3.7}$$

This inverse matrix exists only if (a) the matrix **A** is square and (b) the determinant of the matrix is not zero.

The inverse of a matrix **A** can be derived by first defining the *adjoint*, $\hat{\mathbf{A}}$, of the matrix **A**, as the transpose matrix of cofactors of **A**. For example, the adjoint of an $N \times N$ matrix **A** is

$$\hat{\mathbf{A}} = \begin{bmatrix} C_{11} & C_{21} & \ldots & C_{N1} \\ C_{12} & C_{22} & & \\ \cdot & & & \\ \cdot & & & \\ C_{1N} & \cdot & \cdot & C_{NN} \end{bmatrix}. \tag{A3.8}$$

The inverse matrix $\mathbf{A}^{-1}$ is now equal to the reciprocal of the determinant of $\mathbf{A}$, multiplied by its adjoint

$$\mathbf{A}^{-1} = \frac{1}{|\mathbf{A}|}\,\hat{\mathbf{A}}. \tag{A3.9}$$

The inverse of the 2×2 matrix

$$\mathbf{A} = \begin{bmatrix} a_{11} & a_{12} \\ a_{21} & a_{22} \end{bmatrix} \tag{A3.10}$$

for example, is

$$\mathbf{A}^{-1} = \frac{1}{a_{11}a_{22} - a_{12}a_{21}} \begin{bmatrix} a_{22} & -a_{12} \\ -a_{21} & a_{11} \end{bmatrix}. \tag{A3.11}$$

Note that

$$[\mathbf{AB}]^{-1} = \mathbf{B}^{-1}\mathbf{A}^{-1}, \tag{A3.12}$$

if $\mathbf{A}$ and $\mathbf{B}$ are non-singular. It should be emphasised that although equation (A3.9) is a useful theoretical representation of a matrix inverse, it is not very efficient for numerical evaluation. For larger N, it requires of the order of $N!$ operations to compute the inverse, compared to the order of N^3 operations for other general methods (Noble and Daniel, 1977). Matrices with particular structures may have even more efficient algorithms for their inversion. If the matrix is *Toeplitz* for example, i.e. all the elements along any diagonal are equal and it is symmetric about the leading diagonal, the Levinson recursion method can be used, which requires of the order of N^2 operations (Markel and Gray, 1976).

A.4 Eigenvalue/eigenvector decomposition

The *eigenvalues* λ_i and corresponding *eigenvectors* $\mathbf{q}_i$ of a square matrix $\mathbf{A}$ are given by solutions of the equation

$$\mathbf{A}\mathbf{q}_i = \lambda_i\,\mathbf{q}_i, \tag{A4.1}$$

which will have non-trivial solutions only if the determinant below is zero:

$$|\mathbf{A} - \lambda_i\mathbf{I}| = 0. \tag{A4.2}$$

The resulting equation for λ_i is called the *characteristic equation*.

The eigenvalues of a Hermitian matrix are entirely real, and if the matrix is also positive definite the eigenvalues are all greater than zero (Noble, 1969). The eigenvectors corresponding to distinct (i.e. different) eigenvalues of a Hermitian matrix are orthogonal, i.e.

$$\mathbf{q}_i^{\mathrm{H}}\mathbf{q}_j = 0 \text{ if } \lambda_i \neq \lambda_j. \tag{A4.3}$$

It is convenient to normalise the eigenvectors so that

$$\mathbf{q}_i^{\mathrm{H}}\mathbf{q}_i = 1, \tag{A4.4}$$

and so the eigenvectors form an orthonormal set and can be expressed as columns of a unitary matrix

$$\mathbf{Q} = (\mathbf{q}_1, \mathbf{q}_2, \ldots, \mathbf{q}_N). \tag{A4.5}$$

It therefore follows that

$$\mathbf{Q}^{\mathrm{H}}\mathbf{Q} = \mathbf{Q}\mathbf{Q}^{\mathrm{H}} = \mathbf{I}. \tag{A4.6}$$

The set of equations defining the eigenvectors and eigenvalues:

$$\mathbf{A}\mathbf{q}_1, \mathbf{A}\mathbf{q}_2, \ldots, \mathbf{A}\mathbf{q}_N = \lambda_1\mathbf{q}_1, \lambda_2\mathbf{q}_2, \ldots \lambda_N\mathbf{q}_N, \tag{A4.7}$$

can now be expressed in the form

$$\mathbf{A}\mathbf{Q} = \mathbf{Q}\mathbf{\Lambda} \tag{A4.8}$$

where $\mathbf{\Lambda}$ is the diagonal matrix of eigenvalues:

$$\mathbf{\Lambda} = \begin{bmatrix} \lambda_1 & 0 & . & . \\ 0 & \lambda_2 & & \vdots \\ \vdots & & \ddots & 0 \\ \vdots & . & . \; 0 & \lambda_N \end{bmatrix}. \tag{A4.9}$$

Post-multiplying both sides of equation (A4.8) by $\mathbf{Q}^{\mathrm{H}}$ gives

$$\mathbf{A} = \mathbf{Q}\,\mathbf{\Lambda}\,\mathbf{Q}^{\mathrm{H}}, \tag{A4.10}$$

and pre-multiplying both sides of equation (A4.8) by $\mathbf{Q}^{\mathrm{H}}$ gives

$$\mathbf{Q}^{\mathrm{H}}\mathbf{A}\mathbf{Q} = \mathbf{\Lambda}. \tag{A4.11}$$

A particularly important property of this form of a Hermitian matrix is that

$$\mathbf{A}^n = \mathbf{Q}\,\mathbf{\Lambda}^n\,\mathbf{Q}^{\mathrm{H}} \tag{A4.12}$$

where

$$\mathbf{\Lambda} = \begin{bmatrix} \lambda_1^n & 0 & . & . \\ 0 & \lambda_2^n & & \vdots \\ \vdots & & \ddots & 0 \\ \vdots & . & . \; 0 & \lambda_N^n \end{bmatrix}. \tag{A4.13}$$

If $n = -1$ in the expression above, the inverse exists provided none of the eigenvalues are zero (the matrix is *full rank*). Even if none of the eigenvalues are zero, $\mathbf{A}^{-1}$ clearly has some large terms if any eigenvalue is close to zero, i.e. it is *ill-conditioned*.

A.5 The Hermitian quadratic form

We define the general *Hermitian quadratic form* to be

$$J = \mathbf{x}^{\mathrm{H}}\mathbf{A}\mathbf{x} + \mathbf{x}^{\mathrm{H}}\mathbf{b} + \mathbf{b}^{\mathrm{H}}\mathbf{x} + c \tag{A5.1}$$

where $\mathbf{A}$ is a Hermitian matrix and $\mathbf{x}$ is a vector of complex variables, so that $\mathbf{x}^{\mathrm{H}}\mathbf{A}\mathbf{x}$ is a real scalar. The matrix $\mathbf{A}$ plays a similar role to the Hessian matrix in quadratic functions with multiple real variables (Adby and Dempster, 1974); we will sometimes refer to it here as the complex Hessian matrix. $\mathbf{b}$ is a vector of complex constants and since $\mathbf{x}^{\mathrm{H}}\mathbf{b} + \mathbf{b}^{\mathrm{H}}\mathbf{x} = \mathbf{x}^{\mathrm{H}}\mathbf{b} + [\mathbf{x}^{\mathrm{H}}\mathbf{b}]^*$, the two terms which are linear in $\mathbf{x}$ also sum to give a real scalar. The term c is a real scalar constant. The quantity J is thus the sum of three real scalar terms and so is itself a real scalar.

The general Hermitian quadratic form may be expressed as a function of entirely real quantities by defining the real and imaginary parts of $\mathbf{x}$, $\mathbf{A}$ and $\mathbf{b}$ to be

$$\mathbf{x} = \mathbf{x}_{\mathrm{R}} + j\mathbf{x}_{\mathrm{I}}; \quad \mathbf{A} = \mathbf{A}_{\mathrm{R}} + j\mathbf{A}_{\mathrm{I}}; \quad \mathbf{b} = \mathbf{b}_{\mathrm{R}} + j\mathbf{b}_{\mathrm{I}} . \tag{A5.2}$$

Since $\mathbf{A}$ is Hermitian, $\mathbf{A}_{\mathrm{R}} = \mathbf{A}_{\mathrm{R}}^{\mathrm{T}}$, $\mathbf{A}_{\mathrm{I}} = -\mathbf{A}_{\mathrm{I}}^{\mathrm{T}}$, and so the general Hermitian quadratic form above can be expressed, after some manipulation, as (Nelson *et al.*, 1987a)

$$J = \mathbf{x}_{\mathrm{R}}^{\mathrm{T}}\mathbf{A}_{\mathrm{R}}\mathbf{x}_{\mathrm{R}} + \mathbf{x}_{\mathrm{I}}^{\mathrm{T}}\mathbf{A}_{\mathrm{R}}\mathbf{x}_{\mathrm{I}} - 2\mathbf{x}_{\mathrm{R}}^{\mathrm{T}}\mathbf{A}_{\mathrm{I}}\mathbf{x}_{\mathrm{I}} + 2\mathbf{b}_{\mathrm{R}}^{\mathrm{T}}\mathbf{x}_{\mathrm{R}} + 2\mathbf{b}_{\mathrm{I}}^{\mathrm{T}}\mathbf{x}_{\mathrm{I}} + c. \tag{A5.3}$$

This real scalar quantity can now be differentiated with respect to each of the real and imaginary components of $\mathbf{x}$. The vectors of derivatives of J with respect to each of these real and imaginary components are defined to be

$$\frac{\partial J}{\partial \mathbf{x}_{\mathrm{R}}} = \left(\frac{\partial J}{\partial x_{1\mathrm{R}}}, \frac{\partial J}{\partial x_{2\mathrm{R}}}, \ldots, \frac{\partial J}{\partial x_{M\mathrm{R}}}\right)^{\mathrm{T}}, \qquad \frac{\partial J}{\partial \mathbf{x}_{I}} = \left(\frac{\partial J}{\partial x_{1I}}, \frac{\partial J}{\partial x_{2\mathrm{I}}}, \ldots, \frac{\partial J}{\partial x_{M\mathrm{I}}}\right)^{\mathrm{T}}. \tag{A5.4a,b}$$

Two properties of such derivatives (Noble, 1969) are, for entirely real $\boldsymbol{\alpha}$, $\boldsymbol{\beta}$, $\mathbf{B}$:

$$\frac{\partial(\boldsymbol{\beta}^{\mathrm{T}}\boldsymbol{\alpha})}{\partial \boldsymbol{\alpha}} = \boldsymbol{\beta}, \qquad \frac{\partial(\boldsymbol{\alpha}^{\mathrm{T}}\mathbf{B}\boldsymbol{\alpha})}{\partial \boldsymbol{\alpha}} = (\mathbf{B} + \mathbf{B}^{\mathrm{T}})\boldsymbol{\alpha}. \tag{A5.5a,b}$$

Since $\boldsymbol{\alpha}^{\mathrm{T}}\boldsymbol{\beta} = \boldsymbol{\beta}^{\mathrm{T}}\boldsymbol{\alpha}$, its derivative must be the same as that of $\boldsymbol{\beta}^{\mathrm{T}}\boldsymbol{\alpha}$. If $\mathbf{B}$ is symmetric, i.e. $\mathbf{B} = \mathbf{B}^{\mathrm{T}}$, the derivative of $\boldsymbol{\alpha}^{\mathrm{T}}\mathbf{B}\boldsymbol{\alpha}$ with respect to $\boldsymbol{\alpha}$ is $2\mathbf{B}\boldsymbol{\alpha}$. Differentiating equation (A5.3) with respect to the real and imaginary parts of $\mathbf{x}$ and using these properties gives

$$\frac{\partial J}{\partial \mathbf{x}_{\mathrm{R}}} = 2\mathbf{A}_{\mathrm{R}}\mathbf{x}_{\mathrm{R}} - 2\mathbf{A}_{\mathrm{I}}\mathbf{x}_{\mathrm{I}} + 2\mathbf{b}_{\mathrm{R}}, \tag{A5.6}$$

$$\frac{\partial J}{\partial \mathbf{x}_{\mathrm{I}}} = 2\mathbf{A}_{\mathrm{R}}\mathbf{x}_{\mathrm{I}} + 2\mathbf{A}_{\mathrm{I}}\mathbf{x}_{\mathrm{R}} + 2\mathbf{b}_{\mathrm{I}}, \tag{A5.7}$$

in which the vectors of derivatives have entirely real elements. A *complex gradient vector* $\mathbf{g}$ is now defined (Haykin, 1986) by the equation

$$\mathbf{g} = \frac{\partial J}{\partial \mathbf{x}_{\mathrm{R}}} + j\frac{\partial J}{\partial \mathbf{x}_{\mathrm{I}}}. \tag{A5.8}$$

Using equations (A5.6) and (A5.7), the complex gradient vector $\mathbf{g}$ can be written

$$\mathbf{g} = 2(\mathbf{A}\mathbf{x} + \mathbf{b}). \tag{A5.9}$$

Note that we are *not* implying that $\mathbf{g} = \partial J/\partial \mathbf{x}$ where $\mathbf{x}$ is a complex variable, and so we avoid the problems discussed by Huang and Chen (1989).

If $\mathbf{A}$ is positive definite, J will have a unique global minimum for some set of

$\mathbf{x}_R + j\mathbf{x}_I = \mathbf{x}_0$. This can be evaluated by setting the derivatives of J with respect to $\mathbf{x}_R$ *and* $\mathbf{x}_I$ (equation (A5.7)) to zero, or equivalently by setting the combined, complex, gradient vector $\mathbf{g}$ to zero. It therefore follows that

$$\mathbf{A}\mathbf{x}_0 + \mathbf{b} = \mathbf{0}, \qquad \mathbf{x}_0 = -\mathbf{A}^{-1}\mathbf{b}, \tag{A5.10a,b}$$

where it is assumed that $\mathbf{A}$ is not singular (which is guaranteed if $\mathbf{A}$ is positive definite). The set of equations defined by (A5.10a) are known as the *normal equations*. Substituting the expression for $\mathbf{x}_0$ into the general expression for J, the minimum value of J is found to be

$$J_0 = c - \mathbf{b}^{\mathrm{H}}\mathbf{A}^{-1}\mathbf{b}. \tag{A5.11}$$

The general Hermitian quadratic form can now be written as

$$J = J_0 + (\mathbf{x} - \mathbf{x}_0)^{\mathrm{H}}\mathbf{A}(\mathbf{x} - \mathbf{x}_0), \tag{A5.12}$$

which emphasises that if $\mathbf{x} = \mathbf{x}_0$, then $J = J_0$ and $J > J_0$ for any $\mathbf{x} \neq \mathbf{x}_0$, if $\mathbf{A}$ is positive definite. Upon defining $J_1 = J - J_0$ and $\mathbf{x}_1 = \mathbf{x} - \mathbf{x}_0$, the expression (A5.12) can be put in the form

$$J_1 = \mathbf{x}_1^{\mathrm{H}}\mathbf{A}\mathbf{x}_1 = (\mathbf{x}_1, \mathbf{A}\mathbf{x}_1), \tag{A5.13}$$

which is the usual inner product representation of a Hermitian quadratic form (Noble, 1969; Bellman, 1960). Also, upon defining a rotated, or transformed, set of variables as

$$\mathbf{v} = \mathbf{Q}^{\mathrm{H}}(\mathbf{x} - \mathbf{x}_0), \tag{A5.14}$$

then

$$J = J_0 + \mathbf{v}^{\mathrm{H}}\,\mathbf{Q}^{\mathrm{H}}\mathbf{A}\mathbf{Q}\mathbf{v} = J_0 + \mathbf{v}^{\mathrm{H}}\mathbf{\Lambda}\mathbf{v}, \tag{A5.15}$$

where, since $\mathbf{\Lambda}$ is diagonal, there is no cross-coupling between the variation of J with each element of $\mathbf{v}$. Equation (A5.15) is known as the *canonical* form of the Hermitian quadratic function. Consider the case when the *overdetermined* least squares problem is put into Hermitian quadratic form, i.e.

$$J = \mathbf{e}^{\mathrm{H}}\mathbf{e} \quad \text{where} \quad \mathbf{e} = \mathbf{d} + \mathbf{C}\mathbf{x}, \tag{A5.16}$$

then in terms of the general Hermitian quadratic form (A5.1) $\mathbf{A} = \mathbf{C}^{\mathrm{H}}\mathbf{C}$, $\mathbf{b} = \mathbf{C}^{\mathrm{H}}\mathbf{d}$, $c = \mathbf{d}^{\mathrm{H}}\mathbf{d}$. If there are more elements of $\mathbf{e}$ than $\mathbf{x}$, and all the individual equations for the elements of $\mathbf{e}$ are independent, the equation $\mathbf{e} = \mathbf{d} + \mathbf{C}\mathbf{x}$ is said to be overdetermined since there are more independent equations (one for each element of $\mathbf{e}$) than there are unknowns (the elements of $\mathbf{x}$) and the matrix $\mathbf{A}$ is positive definite. Physically, in the context of control, the elements of $\mathbf{e}$ correspond to the outputs of a number of error sensors and the elements of $\mathbf{x}$ correspond to the inputs to a smaller number of secondary sources. J is then the sum of the squared moduli of the error sensor outputs, and $\mathbf{x}^{\mathrm{H}}\mathbf{A}\mathbf{x}$ is this sum of squares when the secondary sources only are in operation ($\mathbf{b} = \mathbf{0}$, $c = 0$). In the overdetermined case the sum of squared errors must always be greater than zero for any set of secondary source strengths ($\mathbf{x}$), so $\mathbf{A}$ must be positive definite.

Substituting the expressions for $\mathbf{A}$ and $\mathbf{b}$ above for the least squares problem into the normal equations for the minimum of the Hermitian quadratic form, equation

(A5.10b) gives the optimal vector of secondary source inputs given by

$$\mathbf{x}_0 = -[\mathbf{C}^{\mathrm{H}}\mathbf{C}]^{-1}\mathbf{C}^{\mathrm{H}}\mathbf{d}, \tag{A5.17}$$

as used above. Although (A5.17) is the mathematically correct solution to this least squares minimisation problem, direct computation of this equation is not the best numerical method of calculating $\mathbf{x}_0$ in this case. Stewart (1973) and Noble and Daniel (1977), for example, discuss the *QR decomposition* of the $L \times M$ matrix $\mathbf{C}$. This is given by

$$\mathbf{C} = \tilde{\mathbf{Q}}\,\tilde{\mathbf{R}} \tag{A5.18}$$

where $\tilde{\mathbf{Q}}$ is an $L \times L$ unitary matrix and $\tilde{\mathbf{R}}$ is an $L \times M$ upper triangular matrix. These authors show that the least squares solution can be expressed as

$$\mathbf{x}_0 = \tilde{\mathbf{R}}_0^{-1}\,\tilde{\mathbf{Q}}_0^{\mathrm{H}}\mathbf{d} \tag{A5.19}$$

where $\tilde{\mathbf{Q}}_0$ is the $L \times M$ matrix of the first M columns of $\tilde{\mathbf{Q}}$, and $\tilde{\mathbf{R}}_0$ is the $M \times M$ upper triangular matrix formed from the first M rows of $\tilde{\mathbf{R}}$. Equation (5.19) represents a much better numerical procedure for calculating $\mathbf{x}_0$ than (A5.17); Stewart (1973), for example, shows that the condition number of (A5.19) can be the square root of that of (A5.17). The QR decomposition is the standard method of solving the least squares problem used in such software packages as LINPAC and MATLAB.

If there are a smaller number of independent equations for the elements of $\mathbf{e}$ than variables $\mathbf{x}$ (the equation for $\mathbf{e}$ is *undetermined*), the matrix $\mathbf{A} = \mathbf{C}^{\mathrm{H}}\mathbf{C}$ is still Hermitian but now is positive semi-definite, i.e.

$$\mathbf{x}^{\mathrm{H}}\mathbf{A}\mathbf{x} = 0 \quad \text{for some } \mathbf{x} \neq 0. \tag{A5.20}$$

As explained in Chapter 11, this situation can arise when there are fewer error sensors than secondary sources. Physically, $\mathbf{x}^{\mathrm{H}}\mathbf{A}\mathbf{x}$ is again the sum of the squares of the errors with only the secondary sources in operation, but now this may be driven to zero if some subset of all the secondary sources exactly cancel the effects of the others. If $\mathbf{A}$ is positive semi-definite it is also singular (Noble, 1969) and the optimum defined by equation (A5.10b) is no longer valid. Under these circumstances all the error signals, and hence J, can be driven to zero by an infinite number of combinations of secondary sources, and an additional constraint must be introduced to obtain a unique solution. One physically reasonable constraint discussed in Chapter 11 is to minimise the sum of the squares of the control variables $\mathbf{x}^{\mathrm{H}}\mathbf{x}$ (the control "effort") while setting all the errors to zero. The problem now becomes one of constrained optimisation which can be analysed using Lagrange multipliers (Noble, 1969). If the error is to be set to zero then the constraint imposed is that

$$\mathbf{e} = \mathbf{d} + \mathbf{C}\mathbf{x} = \mathbf{0}. \tag{A5.21}$$

We now seek to minimise the new cost function.

$$J = \mathbf{x}^{\mathrm{H}}\mathbf{x} + (\mathbf{d} + \mathbf{C}\mathbf{x})^{\mathrm{H}}\boldsymbol{\lambda} + \boldsymbol{\lambda}^{\mathrm{H}}(\mathbf{d} + \mathbf{C}\mathbf{x}), \tag{A5.22}$$

where $\boldsymbol{\lambda}$ is a vector of complex Lagrange multipliers. This equation is of Hermitian quadratic form in both $\mathbf{x}$ and $\boldsymbol{\lambda}$, and the complex vectors of derivatives of J with respect to the real and imaginary parts of both of these quantities can be identified as

$$\frac{\partial J}{\partial \mathbf{x}_R} + j\frac{\partial J}{\partial \mathbf{x}_I} = 2\mathbf{x} + 2\mathbf{C}^H \boldsymbol{\lambda}, \tag{A5.23}$$

$$\frac{\partial J}{\partial \lambda_R} + j\frac{\partial J}{\partial \lambda_I} = 2[\mathbf{d} + \mathbf{C}\mathbf{x}]. \tag{A5.24}$$

The minimum value of $\mathbf{x}$, within the constraint, is given by setting both these vectors of derivatives to zero so that

$$\mathbf{x}_0 = -\mathbf{C}^H\boldsymbol{\lambda} \quad \text{and} \quad \mathbf{C}\mathbf{x}_0 = -\mathbf{d}, \tag{A5.25a,b}$$

and therefore

$$\mathbf{C}\mathbf{C}^H\boldsymbol{\lambda} = \mathbf{d}, \quad \boldsymbol{\lambda} = (\mathbf{C}\mathbf{C}^H)^{-1}\mathbf{d}, \tag{A5.26a,b}$$

assuming $\mathbf{C}\mathbf{C}^H$ is not singular. Substituting the equation for $\boldsymbol{\lambda}$ into equation (A5.25a) gives the optimal set of control variables given by

$$\mathbf{x}_0 = -\mathbf{C}^H(\mathbf{C}\mathbf{C}^H)^{-1}\mathbf{d}, \tag{A5.27}$$

which is the result used in Chapters 11 and 12.

References

(References to patents are dated when the specification is published, not filed or granted)

Adby, P.R. and Dempster, M.A.H. (1974). *Introduction to Optimization Methods*. Chapman and Hall, London.

Anthony, D.K. and Elliott, S.J. (1991). A comparison of three methods of measuring the volume velocity of an acoustic source. *Journal of the Audio Engineering Society* **39**, 355–366.

Arzamasov, S.N., Malakhov, A.N. and Mal'tsev, A.A. (1982). Adaptive system for the active suppression of noise fields in a multimode waveguide. *Soviet Physics Acoustics* **28**, 346–348.

Åström, K.J. and Eykhoff, P. (1971). System identification — a survey. *Automatica* **7**, 123–162.

Baker, B.B. and Copson, E.T. (1950). *The Mathematical Theory of Huygens Principle*. Clarendon Press, Oxford.

Balas, M.J. (1978). Active control of flexible systems. *Journal of Optimization Theory and Applications* **25**, 415–436.

Beatty, L.G. (1964). Acoustic impedance in a rigid-walled cylindrical sound channel terminated at both ends with active transducers. *Journal of the Acoustical Society of America* **36**, 1081–1089.

Bellanger, M. (1984). *Digital Processing of Signals: Theory and Practice*. John Wiley and Sons, Chichester.

Bellman, R. (1960). *Introduction to Matrix Analysis*. McGraw-Hill, New York.

Bendat, J.S. and Piersol, A.G. (1986). *Random Data*, 2nd edn. John Wiley and Sons, New York.

Beranek, L.L. (1954). *Acoustics*. McGraw-Hill, New York (reprinted in 1986 by the Acoustical Society of America).

Berengier, M. and Roure, A. (1980). Broad-band active sound absorption in a duct carrying uniformly flowing fluid. *Journal of Sound and Vibration* **68**, 437–449.

Bershad, N. and Machi, O. (1989). Comparison of RLS and LMS algorithms for tracking a chirped signal. *Proceedings of the Institute of Electrical and Electronics Engineers International Conference on Acoustics, Speech and Signal Processing* **2**, pp. 896–899.

Bobber, R.J. (1962). Impedance tube for underwater sound transducer evaluation. *Journal of the Acoustical Society of America* **31**, 832–833.

Boiko, A.I. and Ivanov, V.P. (1976). Suppression of the field generated by a pulsating sphere in a rectangular waveguide. *Soviet Physics Acoustics* **22**, 465–468.

Boucher, C.C. and Elliott, S.J. (1989). The application of adaptive filters to active sound control. *Proceedings of the Institution of Electrical Engineers Colloquium on Adaptive Filters*, paper 1989/46.

Boucher, C.C., Elliott, S.J. and Nelson, P.A. (1991). The effect of errors in the plant model on the performance of algorithms for adaptive feedforward control. *Special Issue of the Proceedings of Institution of Electrical Engineers* **138**, 313–319 *Pt. F on Adaptive Filters*.

Bozic, S.M. (1979). *Digital and Kalman Filtering*. Edward Arnold, London.

Brebbia, C.A., Telles, J.C. and Wrobel, L.C. (1984). *Boundary Element Techniques, Theory and Applications in Engineering*. Springer, Heidelberg.

Bull, M.K. (1968). Boundary layer pressure fluctuations. In: *Noise and Acoustic Fatigue in Aeronautics* (E.J. Richards and D.J. Mead, eds), Ch. 8. John Wiley.

Bullmore, A.J. (1988). The active minimisation of harmonic enclosed sound fields with particular application to propeller induced cabin noise. Ph.D. Thesis, University of Southampton, England.

Bullmore, A.J., Nelson, P.A., Curtis, A.R.D. and Elliott, S.J. (1987). The active minimization of harmonic enclosed sound fields, Part II: A computer simulation. *Journal of Sound and Vibration* **117**, 15–33.

Bullmore, A.J., Nelson, P.A. and Elliott, S.J. (1990). Theoretical studies of the active control of propeller-induced cabin noise. *Journal of Sound and Vibration* **140**, 191–217.

Burgess, J.C. (1981). Active adaptive sound control in a duct: a computer simulation. *Journal of the Acoustical Society of America* **70**, 715–726.

Canévet, G. (1978). Active sound absorption in an air conditioning duct. *Journal of Sound and Vibration* **58**, 333–345.

Carme, C. (1987). Absorption acoustique active dans les cavites. These presentée pour obtenir le titre de Docteur de l'Universite D'Aix-Marseille II, Faculte des Sciences de Luminy.

Chaplin, G.B.B. (1983). Anti-sound — the Essex breakthrough. *Chartered Mechanical Engineer* **30**, 41–47.

Chaplin, G.B.B. and Smith, R.A. (1979). Improvements in and relating to "active" methods for attenuating compression waves. UK Patent 1 555 760.

Chaplin, G.B.B., Smith, R.A. and Bearcroft, R.G. (1980). Active attenuation of recurring vibrations. UK Patent 1 577 322.

Conover, W.B. (1956). Fighting noise with noise. *Noise Control* **2**, 78–82.

Conover, W.B. and Ringlee, R.J. (1955). Recent contributions to transformer audible noise control. *Transactions of the AIEE Part III Power Apparatus and Systems* **74**, 77–90.

Cook, R.K., Waterhouse, R.V., Berendt, R.D., Edelman, S. and Thompson, M.C. (1955). Measurement of correlation coefficients in reverberant sound fields. *Journal of the Acoustical Society of America* **27**, 1072–1077.

Cooley, J.W. and Tukey, J.W. (1965). An algorithm for the machine calculation of complex Fourier series. *Mathematics of Computation* **19**, 297–301. (Reprinted in A.V. Oppenheim (ed.) *Papers on Digital Signal Processing*, MIT Press, Cambridge, Mass., 1969).

Costin, M.H. and Elzinga, D.R. (1989). Active reduction of low-frequency tire impact noise using digital feedback control. *Institute of Electrical and Electronics Engineers Control Systems Magazine*, August 3–6.

Cowan, C.F.N. and Grant, P.M. (1985). *Adaptive Filters*. Prentice Hall, Englewood Cliffs, New Jersey.

Cunefare, K.A. and Koopmann, G.H. (1991a). A boundary element approach to optimization of active noise control sources on three dimensional structures. *American Society of Mechanical Engineers Journal of Acoustics and Vibration* (to appear).

Cunefare, K.A. and Koopmann, G.H. (1991b). Global optimum active noise control: surface and far-field effects. *Journal of the Acoustical Society of America* **90**, 365–373.

Curtis, A.R.D. (1988). The theory and application of quadratic minimisation in the active reduction of sound and vibration. Ph.D. Thesis, University of Southampton, England.

Curtis, A.R.D., Nelson, P.A. and Elliott, S.J. (1985). Active control of one-dimensional enclosed sound fields. *Proceedings of Inter-Noise 85*, Munich, pp. 579–582.

Curtis, A.R.D., Nelson, P.A., Elliott, S.J. and Bullmore, A.J. (1987). Active suppression of acoustic resonance. *Journal of the Acoustical Society of America* **81**, 624–631.

Darlington, P. (1987). Applications of adaptive filters in active noise control. Ph.D. Thesis, University of Southampton, England.

Daughaday, H. (1968). Suppression of transmitted harmonic rotor loads by blade pitch control. *American Helicopter Society Journal* **13**, 1–18.

David, A. (1991). Investigation of the near field zone of quiet. M. Eng. Honours Dissertation, University of Southampton, England.

Davidson, A.R. and Robinson, T.G.F. (1977). Noise cancellation apparatus. U.S. Patent no. 4025724.

Davis, M.W. (1984). Refinement and evaluation of helicopter real time self-adaptive active vibration controller algorithms. NASA CR 3821.

Deffayet, C. and Nelson, P.A. (1988). Active control of low frequency harmonic sound radiated by a finite panel. *Journal of the Acoustical Society of America* **84**, 2192–2199.

Dines, P.J. (1984). Active control of flame noise. Ph.D. Thesis, University of Cambridge, England.

Distefano III, J.J., Stubberud, A.R. and Williams, I.J. (1967). *Schaum's Outline of Theory and Problems of Feedback and Control Systems*. McGraw Hill, New York.

Doak, P.E. (1973a). Fundamentals of aerodynamic sound theory and flow duct acoustics. *Journal of Sound and Vibration* **28**, 527–561.

Doak, P.E. (1973b). Excitation, transmission and radiation of sound from source distributions in hard-walled ducts of finite length (II): The effects of duct length. *Journal of Sound and Vibration* **31**, 137–174.

Dohner, J.L. and Shoureshi, R. (1989). A method for active noise control using a source-point model. *Journal of the Acoustical Society of America* **86**, 1053–1059.

Dorf, R.C. (1967). *Modern Control Systems*. Addison-Wesley, Reading, Mass.

Dorling, C.M., Eatwell, G.P., Hutchins, S.M., Ross, C.F. and Sutcliffe, S.G.C. (1989). A demonstration of active noise reduction in an aircraft cabin. *Journal of Sound and Vibration* **128**, 358–360.

Dowling, A.P. and Ffowcs-Williams, J.E. (1983). *Sound and Sources of Sound*. Ellis Horwood, Chichester.

Durrani, T.S. and Nightingale, J.M. (1972). Data windows for digital spectral analysis. *Proceedings of the Institution of Electrical Engineers* **119**, 343–352.

Eatwell, G.P. and Hughes, I.J. (1988). Active control of helicopter cabin noise. *Proceedings of the Fourteenth European Rotorcraft Forum*, pp. 5.1–5.8.

Ebeling, K.J. (1984). Statistical properties of random wave fields. *Physical Acoustics* **XVII**, 233–310 (Eds W.P. Mason and R.N. Thurston).

Eghtesadi, Kh. and Leventhall, H.G. (1981). Comparison of active attenuators of noise in ducts. *Acoustics Letters* **4**, 204–209.

Eghtesadi, Kh. and Leventhall, H.G. (1982). Active attenuation of noise. The monopole system. *Journal of the Acoustical Society of America* **71**, 608–611.

Eghtesadi, Kh. and Leventhall, H.G. (1983). A study of n-source active attenuator arrays for noise in ducts. *Journal of Sound and Vibration* **9**, 11–19.

Eghtesadi, Kh., Hong, W.K.W. and Leventhall, H.G. (1983). The tight-coupled monopole active attenuator in a duct. *Noise Control Engineering Journal* **20**, 16–20.

Elliott, S.J. (1989). The influence of modal overlap in the active control of sound and vibration. ISVR Memorandum No. 695.

Elliott, S.J. and Darlington, P. (1985). Adaptive cancellation of periodic, synchronously sampled interference. *Institute of Electrical and Electronics Engineers Transactions Acoustics Speech and Signal Processing* **ASSP 33**, 715–717.

Elliott, S.J. and Nelson, P.A. (1984). Models for describing active noise control in ducts. ISVR Technical Report No. 127.

Elliott, S.J. and Nelson, P.A. (1985a). Algorithm for multichannel LMS adaptive filtering. *Electronics Letters* **21**, 979–981.

Elliott, S.J. and Nelson, P.A. (1985b). Error surfaces in active noise control. *Proceedings of the Institute of Acoustics* **7**, 65-72.

Elliott, S.J. and Nelson, P.A. (1985c). The active minimisation of sound fields. *Proceedings of Inter-Noise '85*, pp. 583–586.

Elliott, S.J. and Nelson, P.A. (1988). An adaptive algorithm for IIR filters used in multichannel active sound control systems. ISVR Memorandum No. 681.

Elliott, S.J. and Nelson, P.A. (1989). The behaviour of multichannel LMS algorithms in active control applications. *Proceedings of the Institute of Electrical and Electronics Engineers Acoustics Speech and Signal Processing Workshop on Applications of Signal Processing to Audio and Acoustics*, New York.

Elliott, S.J. and Stothers, I.M. (1986). A multichannel adaptive algorithm for the active control of start-up transients. *Proceedings of Euromech 213*, Marseille.

Elliott, S.J., Stothers, I.M. and Nelson, P.A. (1987a). A multiple error LMS algorithm and its application to the active control of sound and vibration. *Institute of Electrical and Electronics Engineers Transactions on Acoustics, Speech and Signal Processing* **ASSP-35**, 1423–1434.

Elliott, S.J., Curtis, A.R.D., Bullmore, A.J. and Nelson, P.A. (1987b). The active minimisation of harmonic enclosed sound fields. Part III: Experimental verification. *Journal of Sound and Vibration* **117**, 35–58.

Elliott, S.J., Joseph, P., Bullmore, A.J. and Nelson, P.A. (1988a). Active cancellation at a point in a pure tone diffuse sound field. *Journal of Sound and Vibration* **120**, 183–189.

Elliott, S.J., Joseph, P. and Nelson, P.A. (1988b). Active control in diffuse sound fields. *Proceedings of the Institute of Acoustics* **10**, 605–614.

Elliott, S.J., Stothers, I.M., Nelson, P.A., McDonald, A.M., Quinn, D.C. and Saunders, T. (1988c). The active control of engine noise inside cars. *Proceedings of Inter-Noise '88*, Avignon, pp. 987–990.

Elliott, S.J., Boucher, C.C. and Nelson, P.A. (1988d). The effect of transducer spatial distribution on the convergence time of active noise control algorithms. *Proceedings of Inter-Noise '88*, Avignon, pp. 1077–1082.

Elliott, S.J., Nelson, P.A., Stothers, I.M. and Boucher, C.C. (1989a). Preliminary results of in-flight experiments on the active control of propeller-induced cabin noise. *Journal of Sound and Vibration* **128**, 355–357.

Elliott, S.J., Nelson, P.A., Stothers, I.M., Boucher, C.C., Evers, J. and Chidley, B. (1989b). In-flight experiments on the active control of propeller-induced cabin noise. *Proceedings of the American Institute of Aeronautics and Astronautics 12th Aeroacoustics Conference*, San Antonio.

Elliott, S.J., Nelson, P.A., Stothers, I.M. and Boucher, C.C. (1990). In-flight experiments on the active control of propeller-induced cabin noise. *Journal of Sound and Vibration* **140**, 219–238.

Elliott, S.J., Boucher, C.C. and Nelson, P.A. (1991a). The behaviour of a multiple channel active control system. *Institute of Electrical and Electronics Engineers Transactions on Signal Processing* (to appear).

Elliott, S.J., Joseph, P., Nelson, P.A. and Johnson, M.E. (1991b). Power output minimisation and power absorption in the active control of sound. ISVR Technical Report 191. (Also to appear in *Journal of the Acoustical Society of America.*)

Eriksson, L.J. and Allie, M.C. (1987). A digital sound control system for use in turbulent flows. *Proceedings of Noise-Con '87*, pp. 365–370. Pennsylvania State University.

Eriksson, L.J. and Allie, M.C. (1988). System considerations for adaptive modelling applied to active noise control. *Proceedings of Institute of Electrical and Electronics Engineers International Symposium on Circuits and Systems*, Helsinki, pp. 2387–2390.

Eriksson, L.J. and Allie, M.C. (1989). Use of random noise for on-line transducer modelling in an adaptive active attenuation system. *Journal of the Acoustical Society of America* **85**, 797–802.

Eriksson, L.J., Allie, M.C. and Greiner, R.A. (1987). The selection and application of an IIR adaptive filter for use in active sound attenuation. *Institute of Electrical and Electronics Engineers Transactions on Acoustics, Speech and Signal Processing* **ASSP–35**, 433–437.

Eriksson, L.J., Allie, M.C., Bremigan, C.D. and Greiner, R.A. (1988). Active noise control using adaptive digital signal processing. *Institute of Electrical and Electronics Engineers, Proceedings of International Conference on Acoustics, Speech and Signal Processing*, New York, Vol. **5**, pp. 2594–2595.

Eriksson, L.J., Allie, M.C., Hoops, R.H. and Warner, J.V. (1989). Higher order mode cancellation in ducts using active noise control. *Proceedings of Inter-Noise '89*, Newport Beach, Vol. **1**, pp. 495–500.

Fahy, F.J. (1986). Statistical energy analysis. In *Noise and Vibration*, 2nd edn (R.G. White and J.G. Walker, eds). Ellis Horwood, Chichester.

Fedoryuk, M.V. (1975). The suppression of sound in acoustic waveguides. *Soviet Physics Acoustics* **21**, 174–176.

Feintuch, P.L. (1976). An adaptive recursive LMS filter. *Proceedings of the Institute of*

Electrical and Electronics Engineers **64**, 1622–1624.

Ffowcs-Williams, J.E. (1984). Review lecture: Anti-sound. *Proceedings of the Royal Society London* **A395**, 63–88.

Ffowcs-Williams, J.E., Roebuck, I. and Ross, C.F. (1985). Antiphase noise reduction. *Physics in Technology* **6**, 19–24.

Firestone, F.A. (1938). The mobility method of computing the vibrations of linear mechanical and acoustical systems: Mechanical-electrical analogies. *Journal of Applied Physics* **9**, 373–387.

Firestone, F.A. (1954). Twixt earth and sky with rod and tube. *Journal of the Acoustical Society of America* **26**, 140.

Flockton, S.J. (1989a). Adaptive filters in active noise control systems with acoustic feedback. *Proceedings of the Institute of Electrical and Electronics Engineers, Acoustics, Speech and Signal Processing Workshop on Applications of Signal Processing to Audio and Acoustics*, New York.

Flockton, S.J. (1989b). The use of FIR and IIR adaptive filtering algorithms in the active control of acoustic noise. *Proceedings of 32nd Mid West Symposium on Circuits and Systems*, Illinois.

Ford, R.D. (1983). Where does the power go? *Proceedings of 11th International Congress on Acoustics*, Paris, Vol. **8**, 277–280.

Ford, R.D. (1984). Power requirements for active noise control in ducts. *Journal of Sound and Vibration* **92**, 411–417.

Friedrich, J. (1967). Ein quasischallunempfindliches Mikrofon fur Geräuschmessungen in turbulenten Luftströmungen. *Tech. Mitt. RFZ* **11**, 30.

Fuller, C.R. (1988). Analysis of active control of sound radiation from elastic plates by force inputs. *Proceedings of Inter-Noise '88*, Avignon, Vol. **2**, pp. 1061–1064.

Fuller, C.R. (1990). Active control of sound transmission/radiation from elastic plates by vibration inputs: I. Analysis. *Journal of Sound and Vibration* **136**, 1–15.

Fuller, C.R., Hansen, C.H. and Snyder, S.D. (1989). Active control of structurally radiated noise using piezoceramic actuators. *Proceedings of Inter-Noise '89*, Newport Beach, Vol. **1**, pp. 509–512.

Galland, M.A. and Sunyach, M. (1986). Prédiction des performances d'un absorbeur acoustique actif. *Comptes Rendues Acad. Sci. Paris* 302, Série II, 863–866.

Glendinning, A.G., Elliott, S.J. and Nelson, P.A. (1988). A high intensity acoustic source for the active attenuation of exhaust noise. ISVR Technical Report 156.

Goodwin, G.C. and Sin, K.S. (1984). *Adaptive Filtering: Prediction and Control*. Prentice Hall, Englewood Cliffs, New Jersey.

Guicking, D. and Karcher, K. (1984). Active impedance control for one-dimensional sound. *American Society of Mechanical Engineers Journal of Vibration, Acoustics, Stress and Reliability in Design* **106**, 393–396.

Guicking, D., Karcher, K. and Rollwage, M. (1985). Coherent active methods for applications in room acoustics. *Journal of the Acoustical Society of America* **78**, 1426–1434.

Guicking, D., Melcher, J. and Wimmel, R. (1989). Active impedance control in mechanical structures. *Acustica* **69**, 39–52.

Hall, D.E. (1987). *Basic Acoustics*. John Wiley and Sons, New York.

Hamada, H., Miura, T., Takahashi, M. and Oguri, Y. (1988). Adaptive noise control system in air-conditioning ducts. *Proceedings of Inter-Noise '88*, Avignon, Vol. **2**, pp. 1017–1020.

Harris, F.J. (1978). On the use of windows for harmonic analysis with the discrete Fourier transform. *Proceedings of the Electrical and Electronics Engineers* **66**, 51–83.

Haykin, S. (1986). *Adaptive Filter Theory*. Prentice Hall, Englewood Cliffs, New Jersey.

Hesselmann, M. (1978). Investigation of noise reduction on a 100 kVA transformer tank by means of active methods. *Applied Acoustics* **11**, 27–34.

Hong, W.K.W., Eghtesadi, Kh. and Leventhall, H.G. (1982). The tandem tight-coupled active noise attenuator in a duct. *Acoustics Letters* **6**, 19–24.

Hong, W.K.W., Eghtesadi, Kh. and Leventhall, H.G. (1987). The tight-coupled monopole (TCM) and tight-coupled tandem (TCT) attenuators: theoretical aspects and experimental attenuation in an air duct. *Journal of the Acoustical Society of America* **81**, 376–388.

Honig, M.L. and Messerschmitt, D.G. (1984). *Adaptive Filters: Structures, Algorithms and Applications*. Kluwer Academic Publishers, Boston.

Hough, S.P. (1988). Some implications of causality in the active control of sound. Ph.D. Thesis, University of Southampton, England.

Huang, Y-D. and Chen, C-T. (1989). A derivation of the normal equation in FIR Wiener filters. *Institute of Electrical and Electronics Engineers Transactions on Acoustics, Speech and Signal Processing* **ASSP–37**, 759–760.

Jacobsen, F. (1979). The diffuse sound field. Acoustics Laboratory, Technical University of Denmark Report No. 27.

Jessel, M.J.M. (1968). Sur les absorbeurs actifs. *Proceedings 6th International Congress on Acoustics, Tokyo*. Paper F–5–6, 82.

Jessel, M.J.M. and Mangiante, G. (1972). Active sound absorbers in an air duct. *Journal of Sound and Vibration* **23**, 383–390.

Johnson, W. (1982). Self-tuning regulators for multicyclic control of helicopter vibration. NASA Technical Paper 1996.

Joplin, P.M. and Nelson, P.A. (1990). Active control of low frequency random sound in enclosures. *Journal of the Acoustical Society of America* **87**, 2396–2404.

Joseph, P. (1990). Active control of high frequency enclosed sound fields. Ph.D. Thesis, University of Southampton, England.

Joseph, P., Nelson, P.A. and Elliott, S.J. (1989). Near field zones of quiet in the pure tone diffuse sound field. *Proceedings of Inter-Noise '89*, Newport Beach, Vol. **1**, pp. 489–494.

Junger, M.C. and Feit, D. (1986). *Sound, Structures and their Interaction*, 2nd edn. MIT Press, Cambridge, Mass.

Kabal, P. (1983). The stability of adaptive minimum mean square error equalisers using delayed adjustment. *Institute of Electrical and Electronic Engineers Transactions of Communications* **COM-31**, 430–432.

Kazakia, J. (1986). A study of active attenuation of broadband noise. *Journal of Sound and Vibration* **110**, 495–509.

Kempton, A.J. (1976). The ambiguity of acoustic sources — a possibility for active control? *Journal of Sound and Vibration* **48**, 475–483.

Kido, K. (1975). Reduction of noise by use of additional sound sources. *Proceedings of Inter-Noise '75*, Sendai, pp. 647–650.

Kido, K. and Onoda, S. (1972). Automatic control of acoustic noise emitted from power transformer by synthezising directivity. *Science Reports of the Research Institutes, Tohoku University (RITU)*, Sendai, Japan. Series B: Technology. Part I: Reports of the Institute of Electrical Communication (RIEC), Vol. **23**, pp. 97–110.

Kinsler, L.E., Frey, A.R., Coppens, A.B. and Sanders, J.V. (1982). *Fundamentals of Acoustics*, 3rd edn. John Wiley, New York.

Kleshchev, A.A. and Klyukin, I.I. (1974). Theoretical aspects of the synthesis of complex acoustic arrays to compensate scattered fields. *Soviet Physics Acoustics* **20**, 490–491.

Kleshchev, A.A. and Klyukin, I.I. (1977). Compensation of the diffracted field formed by rigid scatterers in the field of a harmonic point sound source. *Soviet Physics Acoustics* **23**, 274–275.

Konyaev, S.I. and Fedoryuk, M.V. (1988). Spherical Huygens surfaces and their discrete approximation. *Soviet Physics Acoustics* **33**, 622–625.

Konyaev, S.I., Lebedev, V.I. and Fedoryuk, M.V. (1977). Discrete approximation of a spherical Huygens surface. *Soviet Physics Acoustics* **23**, 373–374.

Kuo, B.C. (1980). *Digital Control Systems*. Holt, Rinehart and Winston, New York.

Kuo, F.F. (1966). *Network Analysis and Synthesis*, 2nd edn. John Wiley and Sons, New York.

La Fontaine, R.F. and Shepherd, I.C. (1985). The influence of wave guide reflections and system configuration on the performance of an active noise attenuator. *Journal of Sound and Vibration* **100**, 569–579.

Leitch, R.R. and Tokhi, M.O. Active noise control systems. *Proceedings of the Institution of Electrical Engineers* **134** A(6), 525–546.

Levine, H. (1980a). Output of acoustical sources. *Journal of the Acoustical Society of America* **67**, 1935–1946.
Levine, H. (1980b). On source radiation. *Journal of the Acoustical Society of America* **68**, 1199–1205.
Lewers, T.H. (1984). The active control of steady single frequency sound in enclosures. M.Sc. Dissertation, University of Southampton, England.
Lighthill, M.J. (1952). On sound generated aerodynamically: I. General theory. *Proceedings of the Royal Society* **A211**, 564–578.
Lighthill, M.J. (1978). *Waves in Fluids*. Cambridge University Press, Cambridge.
Ljung, L. and Söderström, T. (1983). *Theory and Practice of Recursive Identification*. MIT Press, Boston.
Lueg, P. (1936). Process of silencing sound oscillations. US Patent No. 2,043,416.
Lynn, P.A. (1982). *An Introduction to the Analysis and Processing of Signals*, 2nd edn. Macmillan Press, London.
Lyon, R.H. (1969). Statistical analysis of power injection and response in structures and rooms. *Journal of the Acoustical Society of America* **45**, 545–565.
Lyon, R.H. (1975). *Statistical Energy Analysis of Dynamical Systems: Theory and Application*. MIT Press, Boston.
Maling, G.C. (1967). Calculation of the acoustic power radiated by a monopole in a reverberation chamber. *Journal of the Acoustical Society of America* **42**, 859–865.
Malyuzhinets, G.D. (1969). An inverse problem in the theory of diffraction. *Soviet Physics Doklady* **14**, 118–119.
Mangiante, G. (1977). Active sound absorption. *Journal of the Acoustical Society of America* **61**, 1516–1523.
Mangiante, G. and Vian, J.P. (1977). Application du principe de Huygens aux absorbeurs acoustiques actifs. II: Approximations du principe de Huygens. *Acustica* **37**, 175–182.
Markel, J.D. and Gray Jr, A.H. (1976). *Linear Prediction of Speech*. Springer, Berlin.
Mazanikov, A.A. and Tyutekin, V.V. (1976). Autonomous active systems for the suppression of sound fields in single-mode waveguides. *Soviet Physics Acoustics* **22**, 409–412.
Mazanikov, A.A., Tyutekin, V.V. and Ukolov, A.T. (1977). An active system for the suppression of sound fields in a multimode waveguide. *Soviet Physics Acoustics* **23**, 276–277.
Mazzanti, S. and Piraux, J. (1983). An experiment of active noise attenuation in three-dimensional space. *Proceedings of Inter-Noise '83*, Edinburgh, Vol. **1**, pp. 427–430.
McDonald, A.M., Quinn, D.C., Saunders, T.J., Elliot, S.J., Nelson, P.A. and Stothers, I.M. (1988). Adaptive noise control of automobile interior noise. *Proceedings of the Institution of Mechanical Engineers Conference on Vehicle Noise, C353*.
Meirovitch, L. (1987). Control of distributed structures. *Proceedings Noise-Con '87*, Pennsylvania State University, pp. 3–14.
Meirovitch, L. (1990). *Dynamics and Control of Structures*. John Wiley, New York.
Metzger, F.B. (1981). Strategies for reducing propeller aircraft cabin noise. *Automotive Engineering* **89**, 107–113.
Miyoshi, M. and Kaneda, Y. (1988a). Inverse filtering of room acoustics. *Institute of Electrical and Electronics Engineers Transactions on Acoustics, Speech and Signal Processing* **ASSP-36**, 145–152.
Miyoshi, M. and Kaneda, Y. (1988b). Active noise control in a reverberant three-dimensional sound-field. *Proceedings of Inter-Noise '88*, Avignon, Vol. **2**, pp. 983–986.
Molo, C.G. and Bernhard, R.J. (1987). Generalised method of predicting optimal performance of active noise controllers. *American Institute of Aeronautics and Astronautics Journal* **27**, 1473–1478.
Mood, A.M., Graybill, F.A. and Boes, D.C. (1976). *Introduction to the Theory of Statistics*, 3rd edn. McGraw-Hill, New York.
Morgan, D.R. (1980). An analysis of multiple correlation cancellation loops with a filter in the auxiliary path. *Institute of Electrical and Electronics Engineers Transactions on Acoustics, Speech and Signal Processing* **ASSP-28**, 454–467.
Morrow, C.T. (1971). Point-to-point correlation of sound pressures in reverberation chambers. *Journal of Sound and Vibration* **16**, 29–42.

Morse, P.M. (1948). *Vibration and Sound*, 2nd edn. McGraw Hill, New York (reprinted in 1981 by the Acoustical Society of America).

Morse, P.M. and Bolt, R.H. (1944). Sound waves in rooms. *Reviews of Modern Physics* **16**, 69–150.

Morse, P.M. and Feshbach, H. (1953). *Methods of Theoretical Physics*, Part I (international student edition). McGraw-Hill, New York.

Morse, P.M. and Ingard, K.U. (1961). Linear acoustic theory. In: *Handbuch der Physik* (S. Flügge, ed.), Vol. **11**, Part 1. Springer, Berlin.

Morse, P.M. and Ingard, K.U. (1968). *Theoretical Acoustics*. McGraw-Hill, New York.

Munjal, M.L. and Eriksson, L.J. (1988). An analytical, one-dimensional, standing-wave model of a linear active noise control system in a duct. *Journal of the Acoustical Society of America* **84**, 1086–1093.

Munjal, M.L. and Eriksson, L.J. (1989). Analysis of a linear one-dimensional active noise control system by means of block diagrams and transfer functions. *Journal of Sound and Vibration* **129**, 443–455.

Nadim, M. and Smith, R.A. (1983). Synchronous adaptive cancellation in vehicle cabins. *Proceedings of Inter-Noise '83*, Edinburgh, Vol. **1**, pp. 461–464.

Neise, W. and Stahl, B. (1979). The flow noise level at microphones in flow ducts. *Journal of Sound and Vibration* **63**, 561–579.

Nelson, P.A. and Elliott, S.J. (1986). The minimum power output of a pair of free field monopole sources. *Journal of Sound and Vibration* **105**, 173–178.

Nelson, P.A. and Elliott, S.J. (1987). Active minimisation of acoustic fields. *Journal of Theoretical and Applied Mechanics* **6** (suppl.), 39–89.

Nelson, P.A. and Elliott, S.J. (1989a). Active control of sound and vibration. *ISVR Short Course Notes*.

Nelson, P.A. and Elliott, S.J. (1989b). Multichannel adaptive filters for the reproduction and active control of sound fields. *Proceedings of the Institute of Electrical and Electronics Engineers, Acoustics Speech and Signal Processing Workshop on Applications of Signal Processing to Audio and Acoustics*, New York.

Nelson, P.A. and Elliott, S.J. (1991). Active noise control; a tutorial review. *Proceedings of the Acoustical Society of Japan International Conference on the Active Control of Sound and Vibration*, Tokyo, pp. 45–74.

Nelson, P.A., Curtis, A.R.D. and Elliott, S.J. (1985). Quadratic optimisation problems in the active control of free and enclosed sound fields. *Proceedings of the Institute of Acoustics* **7**, 45–53.

Nelson, P.A., Curtis, A.R.D. and Elliott, S.J. (1986a). Optimal multipole source distributions for the active suppression and absorption of acoustic radiation. *Proceedings of Euromech Colloquium 213*, Marseille.

Nelson, P.A., Curtis, A.R.D. and Elliott, S.J. (1986b). On the active absorption of sound. *Proceedings of Inter-Noise '86*, Boston, Vol. **1**, pp. 601–606.

Nelson, P.A., Curtis, A.R.D., Elliott, S.J. and Bullmore, A.J. (1987a). The minimum power output of free field point sources and the active control of sound. *Journal of Sound and Vibration* **116**, 397–414.

Nelson, P.A., Curtis, A.R.D., Elliott, S.J. and Bullmore, A.J. (1987b). The active minimization of harmonic enclosed sound fields, Part I: Theory. *Journal of Sound and Vibration* **117**, 1–13.

Nelson, P.A., Hammond, J.K., Joseph, P. and Elliott, S.J. (1988a). The calculation of causally constrained optima in the active control of sound. ISVR Technical Report 147.

Nelson, P.A., Hammond, J.K. and Elliott, S.J. (1988b). Linear least squares estimation problems in the active control of stationary random sound fields. *Proceedings of Institute of Acoustics* **10**, 589–604.

Nelson, P.A., Hammond, J.K. and Elliott, S.J. (1988c). Analytical approaches to the active control of stationary random enclosed sound fields. *Proceedings of Inter-Noise '88*, Avignon, Vol. **2**, pp. 959–962.

Nelson, P.A., Hammond, J.K., Joseph, P. and Elliott, S.J. (1990a). Active control of stationary random sound fields. *Journal of the Acoustical Society of America* **87**, 963–975.

Nelson, P.A., Sutton, T.J. and Elliott, S.J. (1990b). Performance limits for the active control of random sound fields from multiple primary sources. *Proceedings of the Institute of Acoustics* **12**, 677–687.
Nelson, P.A., Hamada, H. and Elliott, S.J. (1991a). Inverse filters for multi-channel sound reproduction. Paper presented to the *Japanese Institute of Electronics, Information and Communication Engineers*, April 1991, Tokyo Denki University.
Nelson, P.A., Hamada, H. and Elliott, S.J. (1991b). Adaptive inverse filters for stereophonic sound reproduction. *Institute of Electrical and Electronics Engineers Transactions on Signal Processing* (to appear).
Newland, D.E. (1984). *An Introduction to Random Vibrations and Spectral Analysis*, 2nd edn. Longman Scientific and Technical, Harlow.
Noble, B. (1969). *Applied Linear Algebra*. Prentice Hall, Englewood Cliffs, New Jersey.
Noble, B. and Daniel, J.W. (1977). *Applied Linear Algebra*. Prentice Hall, Englewood Cliffs, New Jersey.
Olson, H.F. (1943). *Dynamical Analogies*. Van Nostrand, New York.
Olson, H.F. (1956). Electronic control of noise vibration and reverberation. *Journal of the Acoustical Society of America* **28**, 966–972.
Olson, H.F. (1957). *Acoustical Engineering*. Van Nostrand, New York.
Olson, H.F. and May, E.G. (1953). Electronic sound absorber. *Journal of the Acoustical Society of America* **25**, 1130–1136.
Oppenheim, A.V. and Schafer, R.W. (1975). *Digital Signal Processing*. Prentice Hall, Englewood Cliffs, New Jersey.
Orduna-Bustamante, F. and Nelson, P.A. (1991). An adaptive controller for the active absorption of sound. *Journal of the Acoustical Society of America* (to appear).
Orfanides, S.J. (1985). *Optimum Signal Processing*. Macmillan, New York.
Oswald, L.J. (1984). Reduction of diesel engine noise inside passenger compartments using active, adaptive noise control. *Proceedings of Inter-Noise '84*, Honolulu, pp. 483–488.
Owens, D.H. (1981). *Multivariable and Optimal Systems*. Academic Press, London.
Papoulis, A. (1981). *Signal Analysis*. McGraw Hill, New York.
Perry, D.C., Elliott, S.J., Stothers, I.M. and Oxley, S.J. (1989). Adaptive noise cancellation for road vehicles. *Proceedings of the Institution of Mechanical Engineers Conference on Automotive Electronics*, pp. 150–163.
Pierce, A.D. (1981). *Acoustics: An Introduction to its Physical Properties and Applications*. McGraw-Hill, New York.
Pierce, R.D. (1985). An algorithm for active adaptive control of periodic interference. David W. Taylor Naval Ship Research and Development Center Report No. 85/047.
Piraux, J. and Nayroles, B. (1980). A theoretical model for active noise attenuation in three-dimensional space. *Proceedings of Inter-Noise '80*, Miami, pp. 703–706.
Poole, L.A., Warnaka, G.E. and Cutter, R.C. (1984). The implementation of digital filters using a modified Widrow–Hoff algorithm for the adaptive cancellation of acoustic noise. *Proceedings of International Conference on Acoustics Speech and Signal Processing*, Vol. **2**, pp. 21.7.1–21.7.4.
Qureshi, S.K.H. and Newhall, E.E. (1973). An adaptive receiver for data transmission of time dispersive channels. *Institute of Electrical and Electronics Engineers, Transactions on Information Theory* **IT-19**, 448–459.
Rabiner, L.R. and Gold, B. (1975). *Theory and Application of Digital Signal Processing*. Prentice Hall, Englewood Cliffs, New Jersey.
Randall, R.B. (1987). *Frequency Analysis*, 3rd edn. Brüel and Kjaer.
Richards, R.J. (1979). *An Introduction to Dynamics and Control*. Longman, New York.
Robinson, E.A. (1978). *Multichannel Time Series Analysis with Digital Computer Programs* (revised edition). Holden Day, San Francisco.
Ross, C.F. (1980). Active control of sound. Ph.D. Thesis, University of Cambridge, England.
Ross, C.F. (1981). A demonstration of active control of broadband sound. *Journal of Sound and Vibration* **74**, 411–417.
Ross, C.F. (1982a). An algorithm for designing a broadband active sound control system. *Journal of Sound and Vibration* **80**, 373–380.

Ross, C.F. (1982b). An adaptive digital filter for broadband active sound control. *Journal of Sound and Vibration* **80**, 381–388.

Roure, A. (1984). Systeme d'absorption acoustique active large bande auto-adaptive. Internal Report No. LMA 2040, CNRS Marseille.

Roure, A. (1985). Self adaptive broadband active sound control system. *Journal of Sound and Vibration* **101**, 429–441.

Sabine, W.C. (1898). Architectural acoustics. *Eng. Rec.* **38**, 520–522.

Salloway, A.J. and Twiney, R.C. (1985). Earphone active noise reduction systems. *Proceedings of the Institute of Acoustics* **7**, 95.

Saxon, M.J. (1986). Improvements in electrical transfer filters. U.K. Patent Application GB 2172762A.

Schenck, H.A. (1968). Improved integral formulation for acoustic radiation problems. *Journal of the Acoustical Society of America* **44**, 41–58.

Schroeder, M.R. (1954). Die statistischen Parameter der Frequenzkurven von grossen Räumen. *Acustica* **41**, 594–600. For an English translation see Schroeder, M.R. (1987). Statistical parameters of the frequency response curves of large rooms. *Journal of the Audio Engineering Society* **35**, 299–306.

Schroeder, M.R. and Kuttruff, K.H. (1962). On frequency response curves in rooms. Comparison of experimental, theoretical and Monte Carlo results for the average frequency spacing between maxima. *Journal of the Acoustical Society of America* **34**, 76–80.

Shepherd, I.C., Cabelli, A. and LaFontaine, R.F. (1986). Characteristics of loudspeakers operating in an active noise attenuator. *Journal of Sound and Vibration* **110**, 471–481.

Shepherd, I.C., LaFontaine, R.F. and Cabelli, A. (1989). The influence of turbulent pressure fluctuations on an active attenuator in a flow duct. *Journal of Sound and Vibration* **130**, 125–135.

Short, W.R. (1980). Global low frequency active noise attenuation. *Proceedings of Inter-Noise '80*, Miami, pp. 695–698.

Silcox, R.J. (1990). Active control of multi-modal random sound in ducts. Paper presented at the Institute of Acoustics Spring Meeting, Southampton.

Silcox, R.J. and Elliott, S.J. (1985). Applicability of superposition and source impedance models of active noise control systems. *Proceedings of Inter-Noise '85*, Munich, pp. 587–590.

Skudrzyk, E. (1971). *The Foundations of Acoustics*. Springer, New York.

Smith, R.A. and Chaplin, G.B.B. (1983). A comparison of some Essex algorithms for major industrial applications. *Proceedings of Inter-Noise '83*, Edinburgh, Vol. **1**, pp. 407–410.

Sondhi, M.M. and Berkley, D.A. (1980). Silencing echoes on the telephone network. *Proceedings of the Institute of Electrical and Electronics Engineers* **68**, 948–963.

Spencer, A.J.M., Parker, D.F., Berry, D.S., England, A.H., Faulkner, T.R., Green, W.A., Holden, J.T., Middleton, D. and Rogers, T.G. (1977). *Engineering Mathematics*. Van Nostrand Reinhold, New York.

Stewart, G.W. (1973). *Introduction to Matrix Computations*. Academic Press, London.

Sutton, T.J., Elliott, S.J., Nelson, P.A. and Moore, I. (1990). Active control of multiple-source random sound in enclosures. *Proceedings of the Institute of Acoustics* **12**, 689–693.

Swinbanks, M.A. (1973). The active control of sound propagation in long ducts. *Journal of Sound and Vibration* **27**, 411–436.

Swinbanks, M.A. (1982). The active control of low frequency sound in a gas turbine compressor installation. *Proceedings of Inter-Noise '82*, San Francisco, Vol. **2**, pp. 423–426.

Swinbanks, M.A. (1985). Active noise and vibration control. *Proceedings of DAGA 85*.

Tamm, K. and Kurtze, G. (1954). Ein neuartiges Mikrophon grosser Richtungsselektivität. *Acustica* **4**, 469–470.

Texas Instruments (1987). *DSP Talk*.

Thomas, D.R., Nelson, P.A. and Elliott, S.J. (1990). Letter to the Editor: Experiments on the active control of sound transmission. *Journal of Sound and Vibration* **135**, 351–355.

Thornton, H.M. (1988). Active power minimisation in the free field using a finite number of error sensors. B. Eng. Honours Dissertation, University of Southampton, England.

Tichy, J., Warnaka, G.E. and Poole, L.A. (1984). A study of active control of noise in ducts. *American Society of Mechanical Engineers Journal of Vibration, Acoustics, Stress and Reliability in Design* **106**, 399–404.

Tobey, G.E., Graeme, J.G. and Huelsman, L.P. (1971). *Operational Amplifiers — Design and Applications*, p. 429. McGraw Hill, New York.

Tohyama, M. and Suzuki, A. (1987). Letter to the Editor: Active power minimization of a sound source in a closed space. *Journal of Sound and Vibration* **119**, 562–564.

Treichler, J.R., Richard Johnson Jr., C. and Larimore, M.G. (1987). *Theory and Design of Adaptive Filters*. Wiley Interscience, New York.

Trinder, M.C.J. and Nelson, P.A. (1983). Active noise control in finite length ducts. *Journal of Sound and Vibration* **89**, 95–105.

Tsujino, M. and Elliott, S.J. (1991). A globally optimal formulation for feedforward active sound control. *Mechanical Systems and Signal Processing* **5**, 167–181.

Urusovskii, I.A. (1977a). Active noise suppression in a waveguide. *Soviet Physics Acoustics* **23**, 170–174.

Urusovskii, I.A. (1977b). Self-excitation of an active noise-suppression system in a waveguide. *Soviet Physics Acoustics* **23**, 243–246.

Urusovskii, I.A. (1980). Active noise suppression in a waveguide with monopole radiators and dipole receivers. *Soviet Physics Acoustics* **26**, 153–157.

Urusovskii, I.A. (1986). Active synthesis of a scattered sound field. *Soviet Physics Acoustics* **32**, 354–355.

Van Trees, H.L. (1965). *Detection and Estimation Theory*. John Wiley, New York.

Van Trees, H.L. (1968). *Detection, Estimation and Modulation Theory*. Part I. John Wiley, New York.

Vyalyshev, A.I., Dubinin, A.I. and Tartakovskii, B.D. (1986). Active acoustic reduction of a plate. *Soviet Physics Acoustics* **32**, 96–98.

Walker, L.A. and Yaneske, P.P. (1976a). Characteristics of an active feedback system for the control of plate vibrations. *Journal of Sound and Vibration* **46**, 157–176.

Walker, L.A. and Yaneske, P.P. (1976b). The damping of plate vibrations by means of multiple active control systems. *Journal of Sound and Vibration* **46**, 177–193.

Wanke, R.L. (1976). Acoustic abatement method and apparatus. US Patent no. 3936606.

Warnaka, G.E., Zalas, J.M., Tichy, J. and Poole, L.A. (1983). Active control of noise in interior spaces. *Proceedings of Inter-Noise '83*, Edinburgh, Vol. **1**, pp. 415–418.

Warnaka, G.E., Poole, L.A. and Tichy, J. (1984). Active acoustic attenuator. US Patent no. 4473906.

Wernicke, R.K. and Drees, J.M. (1963). Second harmonic control. *Proceedings of the Nineteenth National Forum of the American Helicopter Society*, pp. 1–7.

Wheeler, P.D. (1986). Voice communications in the cockpit noise environment — the role of active noise reduction. Ph.D. Thesis, University of Southampton, England.

White, A.D. and Cooper, D.G. (1984). An adaptive controller for multivariable active noise control. *Applied Acoustics* **17**, 99–109.

Widrow, B. (1971). Adaptive filters. in: *Aspects of Network and System Theory* (R.E. Kalman and M. DeClaris, eds). Holt, Rinehart and Winston, New York.

Widrow, B. and Hoff, M. (1960). Adaptive switching circuits. *Proceedings IRE WESCON Convention Record*, Part 4, Session 16, pp. 96–104.

Widrow, B. and Stearns, S.D. (1985). *Adaptive Signal Processing*. Prentice Hall, Englewood Cliffs, New Jersey.

Widrow, B., Glover, J.R., McCool, J.M., Kaunitz, J., Williams, C.S., Hern, R.H., Zeidler, J.R., Dong, E. and Goodlin, R.C. (1975). Adaptive noise cancelling: principles and applications. *Proceedings of the Institute of Electrical and Electronics Engineers* **63**, 1692–1716.

Widrow, B., McCool, J.M., Larimore, M.G. and Johnson, C.R. (1976). Stationary and nonstationary learning characteristics of the LMS adaptive filter. *Proceedings of the Institute of Electrical and Electronics Engineers* **64**, 1151–1162.

Widrow, B., Shur, D. and Shaffer, S. (1981). On adaptive inverse control. *Proceedings of the 15th ASILOMAR Conference on Circuits, Systems and Computers*, pp. 185–195.

Wiener, N. (1949). *Extrapolation, Interpolation and Smoothing of Stationary Time Series*. John Wiley, New York.

Wilby, J.F., Rennison, D.C., Wilby, E.G. and Marsh, A.H. (1980). Noise control prediction for high control speed propeller-driven aircraft. *Proceedings of the American Institute of Aeronautics and Astronautics 6th Aeroacoustics Conference*, Connecticut. Paper No. AIAA-80-0999.

Zavadskaya, M.P., Popov, A.V. and Egelskii, B.L. (1975). An approximate solution of the problem of active suppression of sound fields by the Malyuzhinets method. *Soviet Physics Acoustics* **21**, 541–544.

Zavadskaya, M.P., Popov, A.V. and Egelskii, B.L. (1976). Approximation of wave potentials in the active suppression of sound fields by the Malyuzhinets method. *Soviet Physics Acoustics* **21**, 451–454.

Index

Absorption coefficient, 35
Acoustic admittance, 313
Acoustic compliance, 74
Acoustic energy density, 17
Acoustic impedance, characteristic, 8
Acoustic impedance, specific, 16
Acoustic intensity, 18
Acoustic mobility of dipole source, 295
Acoustic power output, 27
Acoustic propagation, 1
Acoustic resistance, 74
Acoustic transfer impedance, 26
Adaptive algorithms, 195
Adaptive digital filters, 113
Adjoint of a matrix, 414
Aliasing, 61
Analogue-to-digital conversion, 59
Anti-aliasing filter, 59
Autocorrelation function, 54
Axial modes, 317

Bilinear transform, 107
Block Toeplitz matrix, 406
Broad band random process, 56

Canonical form of quadratic function, 418
Causality, 85, 247
Characteristic equation, 415
Chi-squared distribution, 367
Chi-squared statistical distribution, 70
Coefficients of digital filters, 99
Cofactor of a matrix, 414
Coherence functions for performance prediction, 177, 402
Coherence function (ordinary), 89
Coherence function (multiple), 403
Coherent output power, 89
Compensation filter, 219
Complex gradient vector, 390, 417
Complex notation, 11
Complex pressure, 12
Compliant mode, 317
Condition number, 387
Conditioning of control algorithms, 386
Continuous source layers, 275, 284
Control of aircraft propeller noise, 351
Control of car engine noise, 354
Control spillover, 329
Controller design in the frequency domain, 179
Controller, electronic, 116, 161
Convergence coefficient, 114
Convolution integral, 84
Correlation coefficient, 52
Cost function, 381
Cross-correlation function, 54
Cross-spectral density, 58

Damped least squares algorithm, 393
Damping of modes in enclosures, 316
Decibel scale, 4
Delayed LMS algorithm, 196
Delays in the controller, 194
Desired signal, 108
Detection sensor, 161
Determinant of a matrix, 414
Deterministic signals, 37
Difference equation, 99
Diffuse field acoustic intensity, 35
Diffuse field acoustic transfer impedance, 360
Diffuse field energy balance, 35
Diffuse field energy density, 33
Diffuse field zone of quiet, 362
Diffuse sound field, 33
Digital filter, 92
Dirac delta function, 45
Direct field impedance, 369
Directivity of sources and sensors, 184
Discrete convolution, 98
Discrete Fourier transform, 65

Discrete source layers, 302
Dominant mode, 324
Dummy controller, 173, 398

Effort in control algorithm, 383
Effort term in cost function, 395
Eigenfunctions of the Helmholtz equation, 30
Eigenfunctions of enclosures, 311
Eigenvalue spread limitations, 391
Eigenvalue/eigenvector decomposition, 415
Eigenvalue/eigenvector structure of Hermitian matrices, 416
Eigenvalues of a positive definite matrix, 415
Enclosed sound fields in one dimension, 152, 322
Enclosed sound fields, 29
Energy balance equation, 293
Energy in a closed duct, 153
Energy minimisation in a closed duct, 156
Equivalent absorbing area, 295
Equivalent block diagram, 161
Error criterion, 381
Error path, 163
Error sensor, 162
Error surface, 110, 381
Estimates of frequency response, 88
Estimation of power spectra, 69
Evanescent waves, 178
Expected (mean) value, 49

F(2,2) distribution, 367
Far field, 233
Fast Fourier transform (FFT), 66
Feedback cancellation, 193
Feedback control, 204
Feedback control in a cavity, 207
Feedback path, 172
Feedforward control, 171
Filtered reference signal, 170
Filtered-x LMS algorithm, 196
FIR filter, 102
Force distribution, 276
Fourier integral, 42
Fourier series, 38
Free field control, 231
Free space Green function, 281
Frequency response, 78
Frequency response of digital systems, 99
Fully determined systems, 382

Gain margin, 217
Gauss-Newton algorithm, 393
Gaussian distribution of diffuse field pressure, 358
Global control in an enclosure at high frequencies, 348
Global control in duct, 322
Global control in enclosures, 310
Global control in two dimensions, 330
Global control of random excitation in enclosures, 334
Global control with multiple secondary sources, 327
Global control with one secondary source, 319
Global system model, 395
Globally optimal control, 410
Gradient descent algorithms, 387
Green function, 277
Green function for an enclosure, 314
Green's theorem, 279

Hard limiting, 368
Harmonic synthesis, 165
Helmholtz equation, 23
Helmholtz equation, one-dimensional, 13
Hermitian matrix, 413
Hermitian quadratic form, 416
Hermitian transpose of a matrix, 413
Hermitian transpose of a vector, 411
Higher harmonic control, 394
Huygen's Principle, 275, 284

IIR filter, 103
Ill-conditioning, 386
Impulse response, of analogue systems, 79
Impulse response, of digital systems, 98
Inner product, 411
Interference between two monopoles, 232
Internal acoustic impedance of a source, 190
Inverse of a matrix, 414
Iterative algorithms, 195

Kempton's suggestion, 262
Kirchhoff–Helmholtz integral equation, 282
Kronecker delta function, 95

Lag network, 222
Lagrange multipliers, 419
Laplace transform, 75
Lateral quadrupole, 258
Leak in filter coefficients, 171
Least squares solutions, 381
Linearity, 13
LMS algorithm, 114
Loading, 231

Loading of primary source, 342
Local active control, 356
Local system model, 395
Locally reacting surface, 313
Longitudinal quadrupole, 257
Loop gain, 208
Lumped acoustical elements, 73

Manual control, 403
Mass conservation equation, 8, 22
Matrix, 412
Matrix product, 412
Measurement noise, 176
Method of least squares, 52
Minimisation of signals from microphones in an enclosure, 336
Minimisation of signals from multiple microphones, 244
Minimisation of signals from two microphones in a diffuse field 365
Minimum phase system, 82
Minimum power output of a source in a duct, 133, 159
Minimum power output of source arrays, 271
Minimum power output of two monopoles, 240
Minor of a matrix, 414
Modal control, 339
Modal damping ratio, 319
Modal density, 344
Modal expansion of pressure, 315
Modal overlap, 344
Mode shape, 21, 30
Mode shapes in one dimension, 323
Model shapes in two dimensions, 333
Modes of a control system, 390
Momentum conservation equation, 10, 23, 276
Monopole source, 26
Monopole/dipole source layer, 283
Motional impedance, 141
Multichannel adaptive filters, 407
Multichannel control of periodic sound, 379
Multichannel control of random sound, 396
Multichannel control with digital filters, 405
Multiple coherence function, 402
Multiple Error LMS algorithm, 408
Multipole analysis, 260

Narrow band random process, 56
Natural frequency, 21, 30
Near field components of a loudspeaker, 228
Near field zone of quiet, 369
Negative frequency, 40
Non-causal action of a secondary source, 133, 243
Non-linearity , 73
Normal equations, 111
Normalised frequency, 94
Null matrix, 412
Nyquist frequency, 64
Nyquist stability criterion, 213

Oblique modes, 317
Observable modes, 338
Octopole source, 273
Off-line adjustment, 195
Optimal analogue filtering, 89
Optimal digital filtering, 108
Orthogonal vectors, 411
Orthogonality of eigenfunctions, 313
Orthogonality principle, 90
Overdetermined quadratic form, 418
Overdetermined systems, 382

Panel radiation control, 289
Particle velocity, 3
Performance index, 381
Persistent excitation, 109
Phase margin, 217
Physical interpretation of optimal controller, 183
Plane dipole source, 143
Plane monopole source, 118
Plane wave, 3
Point dipole source, 253
Point quadrupole sources, 257
Pole-zero diagram, of analogue system, 81
Pole-zero diagram, of digital system, 99
Positive definite matrix, 414
Positive semi-definite matrix, 414
Power absorption by a source in a duct, 128, 157
Power absorption by loudspeakers, 140
Power absorption by monopole arrays, 307
Power absorption by monopoles and dipoles, 294
Power absorption by quadrupoles, 304
Power absorption in a diffuse field, 376
Power absorption in enclosures, 342
Power output for local control, 374
Power output in enclosures, 340
Power output of source arrays, 269
Power output of sources in a duct, 127
Power output of two monopoles, 235
Power sensing algorithm, 388
Power spectral density, 55
Prediction, 91
Predictor, 249

Pressure release boundary condition, 125
Primary path, 172
Principle coordinates, 387
Probability density function, 47

QR decomposition, 419
Quantisation noise, 59

Radiation from two monopoles, 236
Radius of reverberation, 357
Random incidence absorption coefficient, 318
Random process, 46
Random signals, 37
Rank of a matrix, 416
Rate term in a cost function, 395
Reciprocity, 277
Recursive controllers, 199
Reference signal, 108, 162
Reflections from a primary source in a duct, 148, 189
Relative stability, 216
Residual modes, 324
Response of a linear system to random inputs, 85
Reverberant field impedance, 369
Reverberation time, 36
RLMS algorithm, 201

Sampled signals, 93
Schroeder frequency, 348, 359
Sensor location, 338
Singular matrix, 414
Soft limiting, 368
Sommerfeld radiation condition, 279
Sound power level, 29
Sound pressure level, 5
Source array in a duct, 135
Source strength, 26
Source strength distribution, 276
Spatial average, 32
Spatial cross-correlation function, 359
Spatial orthogonality, 362
Speed of sound, 1, 8
Spherical wave, 24
Stability, 82
Standing waves, 19
State equation, 8
Stationarity, 50
Statistical behaviour of source strength, 366
Statistical properties of diffuse fields, 358
Steepest descent algorithm, 389
Stochastic gradient algorithm, 114, 409
Stochastic Newton algorithm, 409
Superposition, 13, 72
Suppression of radiation, 287
Suppression of scattered radiation, 290

Tandem tight coupled monopole, 230
Tangential modes, 317
Tight coupled monopole, 204
Time invariance, 73
Toeplitz matrix, 415
Total acoustic potential energy, 321
Transfer function, 78
Transfer impedance matrix, 264
Transpose of a matrix, 413
Transpose of a vector, 411
Travelling wave, 11
Trial and error algorithm, 388
Turbulence as a source of measurement noise, 178

Underdetermined systems, 383
Underdetermined quadratic form, 419
Unidirectional radiation in a duct, 146
Unique global minimum, 388
Unit circle, 101
Unit matrix, 412
Unitary matrix, 413

Variance (mean square) value, 50
Vector, 411
Vector dipole strength, 256
Virtual earth, 204

Wave equation, inhomogeneous, 276
Wave equation, one-dimensional, 9
Wave equation, three-dimensional, 22
Waveform synthesis, 166
Wavelength, 3
Wavenumber, 11
White noise, 56
Widrow-Hoff algorithm, 114
Wiener filter, 111
Wiener-Hopf integral equation, 90
Window method of filter design, 105
Windowing of data, 67

z-transform, 96
Zone of quiet, 358
Zone of quiet experiments, 364, 372
Zone of quiet near a monopole source, 371

Printed and bound by CPI Group (UK) Ltd, Croydon, CR0 4YY
11/06/2025
01899189-0011